Peroxyl Radicals

THE CHEMISTRY OF FREE RADICALS

Peroxyl Radicals

Edited by
Zeev B. ALFASSI
Department of Nuclear Engineering, Ben Gurion University of the Negev, Israel

JOHN WILEY & SONS
Chichester · New York · Weinheim · Brisbane · Singapore · Toronto

Baffins Lane, Chichester,
West Sussex PO19 1UD, England

National 01243 779777
International (+44) 1243 779777
e-mail (for orders and customer service enquiries): cs-books@wiley.co.uk
Visit our Home Page on http://www.wiley.co.uk
or http://www.wiley.com

Other Wiley Editorial Offices

John Wiley & Sons, Inc., 605 Third Avenue,
New York, NY 10158-0012, USA

VCH Verlagsgesellschaft mbH, Pappelallee 3,
D-69469 Weinheim, Germany

Jacaranda Wiley Ltd, 33 Park Road, Milton,
Queensland 4064, Australia

John Wiley & Sons (Asia) Pte Ltd, Clementi Loop #02-01,
Jin Xing Distripark, Singapore 129809

John Wiley & Sons (Canada) Ltd, 22 Worcester Road,
Rexdale, Ontario M9W 1L1, Canada

Library of Congress Cataloging-in-Publication Data

Peroxyl radicals / edited by Zeev B. Alfassi
p. cm.
Includes bibliographical references (p. –) and index.
ISBN 0-471-97065-4 (cloth : alk. paper)
1. Peroxides. 2. Radicals (Chemistry) I. Alfassi, Zeev B.
QD305.E7P395 1997 96-34040
547'.63—dc20

British Library Cataloguing in Publication Data

A catalogue record for this book is available from the British Library

ISBN 0 471 97065 4

Typeset in 10 on 12pt Times by Mackreth Media Services, Hemel Hempstead
Printed and bound in Great Britain by Biddles Ltd, Guildford, Surrey
This book is printed on acid-free paper responsibly manufactured from sustainable forestation, for which at least two trees are planted for each one used for paper production

Dedication

This book is dedicated to
Dr P. Neta
of the Chemical Kinetics division of NIST, for his friendship, guidance and hospitality during numerous visits in NIST

CONTENTS

CONTRIBUTORS

Zeev B. Alfassi
Department of Nuclear Engineering, Ben Gurion University of the Negev, Beer Sheva 84102, Israel

L.R.C. Barclay
Department of Chemistry, Mount Allison University, Sackville, NB E0A 3C0, Canada

S.W. Benson
Loker Hydrocarbon Research Institute, University of Southern California, University Park, Los Angeles, CA 90089-1661, USA

Diane E. Cabelli
Department of Chemistry, Brookhaven National Laboratory, Upton, NY 11973-5000, USA

N. Cohen
Thermochemical Kinetics Research, 2604 SE 32nd Avenue, Portland, OR 97202-1412, USA

Nikola Getoff
Institute für Theoretische Chemie und Strahlenchemie, Universität Wien, Althanstrasse 14, A-1090 Wien, Austria

Yasuro Hori
Department of Computing and Electrical Engineering, Nagoya Institute of Technology, Showa, Nagoya 466, Japan

J.A. Howard
Steacie Institute for Molecular Sciences, National Research Council, Ottawa, PQ K1A OR6, Canada

R.E. Huie
National Institute of Standards and Technology, Gaithersburg, MD 20899, USA

Robert Lesclaux
Laboratoire de Photophysique et Photochimie Moléculaire, Université Bordeaux I—URA 348 CNRS, 33405 Talence Cedex, France

P. Neta
National Institute of Standards and Technology, Gaithersburg, MD 20899, USA

Ole J. Nielsen
Ford Research Center Aachen, Ford Motor Company, Dennewartstrasse 25, D-52068 Aachen, Germany

Christopher J. Rhodes
School of Pharmacy and Chemistry, John Moores University, Byrom Street, Liverpool L3 3AF, UK

Heinz-Peter Schuchmann
Max-Planck-Institut für Strahlenchemie, Stiftstrasse 34–36, P.O. Box 101365, D-45413 Mülheim an der Ruhr, Germany

J. Sehested
Section for Chemical Reactivity, Environmental Science and Technology Department, Risø National Laboratory, DK-4000 Roskilde, Denmark

Clemens von Sonntag
Max-Planck-Institut für Strahlenchemie, Stiftstrasse 34–36, P.O. Box 101365, D-45413 Mülheim an der Ruhr, Germany

Timothy J. Wallington
Ford Motor Company, Mail Drop SRL-3083, P.O. Box 2053, Dearborn, MI 48121-2053, USA

1 Formation of Peroxyl Radicals in Solution

ZEEV B. ALFASSI
Department of Nuclear Engineering, Ben Gurion University of the Negev, Beer Sheva, Israel

1 THE RATE CONSTANTS OF FORMATION

Most carbon-centred radicals react with molecular oxygen with rate constants which are close to the diffusion-controlled limit, equal to $k = 2\text{–}4 \times 10^9\ M^{-1}\,s^{-1}$, to form peroxyl radicals—the peroxylation reaction:

$$R + O_2 \longrightarrow RO_2 \quad (1)$$

However, close observation of all rate constants shows that there are some exceptions, both for the higher and lower values. Some of the rate constants are above $10^{10}\ M^{-1}\,s^{-1}$, as was found for some OH-adducts of purines. Thus, Vieira and Steenken [1] found that the 8-OH adduct of N^6, N^6-dimethyl adenosine reacts with a rate constant which is greater than $1.6 \times 10^{10}\ M^{-1}\,s^{-1}$, while the 5-OH adduct of the same molecule reacts with oxygen molecules with a rate constant greater than $2.2 \times 10^{10}\ M^{-1}\,s^{-1}$. Another OH-adduct of the same molecule, i.e. the 4-OH adduct, reacts with O_2 with a rate constant which is lower than that usually observed, namely $k \leqslant 4 \times 10^8\ M^{-1}\,s^{-1}$. An even lower rate constant was found for the reaction of the thymidine 6-OH adduct with O_2, i.e. $k = 2.5 \times 10^7\ M^{-1}\,s^{-1}$ [2]. The low rate constants were explained as due to these radicals not being carbon-centred radicals, since the single electron is located mainly on the nitrogen atom. No explanation was given for the very high rate constants. Another example of a very high rate constant is that of the methyl acetate radical ($CH_3CO_2CH_2^{\bullet}$) in aqueous solution, i.e. $k = 1.4 \times 10^{10}\ M^{-1}\,s^{-1}$ [3].

Other low rate constants were found for the reactions of O_2 with radicals obtained by OH reacting with anions of polyunsaturated fatty acids (linoleate, linolenate and arachidonate); in these cases the rate constants are $2\text{–}3 \times 10^8\ M^{-1}\,s^{-1}$ [4,5]. For the singly unsaturated oleate anion the rate constant is higher, i.e. $1 \times 10^9\ M^{-1}\,s^{-1}$. These low rate constants are suggested to result from the high stability of the doubly allylic radical. A low value for the reaction of O_2 with the doubly allylic radical was also found for the

Peroxyl Radicals. Edited by Z. B. Alfassi

reaction of the hydroxy-cyclohexadienyl radical (obtained by the addition of an OH radical to benzene). Three different laboratories have studied this reaction in aqueous solution, with the following results being obtained: $5 \times 10^8 M^{-1} s^{-1}$ [6], $3.9 \times 10^8 M^{-1} s^{-1}$ [7], and $3.1 \times 10^8 M^{-1} s^{-1}$ [8]. Similarly close values were also obtained for the OH-adducts of other aromatic hydrocarbons [9], such as toluene ($4.5 \times 10^8 M^{-1} s^{-1}$), naphthalene ($5 \times 10^8 M^{-1} s^{-1}$) and β-methylnaphthalene ($4.2 \times 10^8 M^{-1} s^{-1}$). A lower rate constant for the cyclohexadienyl type was found also for the OH adduct of di-tert-butyl-4-methyl phenol, i.e. $k = 9 \times 10^7 M^{-1} s^{-1}$ [10]. However, not all of the cyclohexadienyl types of radical were found to react slowly. Maillard, Ingold and Scaiano [11] found that the cyclohexadienyl radical (the H-adduct of benzene) reacts with O_2 in benzene solution with $k = 1.64 \times 10^9 M^{-1} s^{-1}$. Some older data gave a low rate constant for the resonance-stabilized benzyl radical, but later studies, however, showed this value to be wrong. Zimina, Kovacs and Putirskaja [12] found $k = 2.5 \times 10^7 M^{-1} s^{-1}$ for the benzyl radical, whereas Maillard and coworkers [11] found $k = 1\text{–}3 \times 10^9 M^{-1} s^{-1}$, with the different values being for different solvents. The normal high value for the benzyl radical was confirmed in other studies, i.e. $2.0 \times 10^9 M^{-1} s^{-1}$ in water [13], $2.6 \times 10^9 M^{-1} s^{-1}$ in n-hexane [14] and $2.8 \times 10^9 M^{-1} s^{-1}$ in water [15]. We can conclude that it might not be just the resonance stabilization of the radical alone which decelerates the reaction with O_2, but rather a combination of the resonance stabilization with a retardation of the O_2 by the OH group in the OH adducts. Tokumara, Nosaka and Ozaki [14] studied the rate constant for the peroxylation reaction of various *p*-substituted benzyl radicals. They found that for substituents with a negative Fisher–Meierhoefer free-radical substituent constant, σ (e.g. F, OCH_3, and CH_3) the rate constant is independent of the substituent within experimental error, having a value of $3 \times 10^9 M^{-1} s^{-1}$. However, for positive substituent constants the rate constant decreases with increasing σ, with log k being a decreasing linear function of σ. This decrease in the rate constant is explained as being due to the stabilization of the benzyl radical by the *p* substituent, which attracts electrons (e.g. CN, NO_2, and C_6H_5).

Older results also show low rate constants for cyclopentyl- and cyclohexyl-type radicals [16]; however, later studies have usually found high rate constants, i.e. 4.9×10^9 [17] and $3.5 \times 10^9 M^{-1} s^{-1}$ [15] for the cyclopentyl radical and $2 \times 10^9 M^{-1} s^{-1}$ for the cyclohexyl radical [18].

Another low value, which was found for the radical from lactate ($3.5 \times 10^8 M^{-1} s^{-1}$) [19], does not agree with the usual high value measured earlier, i.e. $2.6 \times 10^9 M^{-1} s^{-1}$ [20]. It can be concluded that every low value reported should be treated very carefully and checked in more than one laboratory, preferably by more than one method.

A very low rate constant was found by Asmus *et al.* [21] for the peroxylation of the OH-adduct to nitrobenzene, namely $2.5 \times 10^6 M^{-1} s^{-1}$. The more than two orders of magnitude difference from other OH-adducts was explained as being due to the high electronegativity of the reactant. This decrease is considerably higher than that found for the benzyl radical [14].

Nitrogen centred radicals react with oxygen more slowly than carbon-centred radicals, as can be seen from a comparison of the reaction with oxygen of CH_3 compared to NH_2 (k values are 3×10^8 [22], 1×10^7 [23], 3.4×10^7 [24], and $1.2 \times 10^8\ M^{-1}s^{-1}$ [25]). However, for N_2H_3 a rate constant $\geqslant 3 \times 10^9\ M^{-1}s^{-1}$ [26] was found. Oxygen-centred radicals react with O_2 very slowly, if at all. For the tyrosyl radical it was found that the rate constant is less than $1 \times 10^3\ M^{-1}s^{-1}$ [27,28].

2 SOLVENT DEPENDENCE

Since the rate constants measured for the reactions of C-centred radicals with O_2 are essentially diffusion controlled, the values obtained can be expected to be inversely proportional to the viscosity of the solvent which is employed. However, no experiments have been carried out with high-viscosity solvents, and the few cases of solvent-dependence studies reported do not show any correspondence with the viscosity of the solvent, as can be seen in Table 1. For example, the viscosity of 2-propanol is seven times higher than that of n-hexane, but the rate constant for the reaction of $C_6H_5CH_2$ with O_2 differs only by 10%, when measured in these solvents.

Table 1 Rate constants for peroxyl formation measured in various solvents

Radical	Solvent	k ($M^{-1}s^{-1}$)	Viscosity[a] (10^{-3} Pa s)	Ref.
$C_6H_5CH_2O_2$	n-Hexane	2.8×10^9	0.299	11
	Cyclo-hexane	1.0×10^9	0.898	11
	n-Hexadecane	1.0×10^9	3.24	11
	Benzene	2.9×10^9, 2.6×10^9	0.603	11,14
	Acetonitrile	3.4×10^9	0.341	11
	2-Propanol	2.5×10^9	2.073	11
	Water	2.8×10^9, 2.0×10^9	0.890	15,13
CH_2C1O_2	Water	1.9×10^9	0.898	15
	Dichromethane	6.3×10^9	0.411	31
$4\text{-}NO_2C_6H_4CH_2O_2$	Water/2-propanol	9.0×10^8	1.1	13
	n-Hexane	8.9×10^8	0.299	14
$(CH_3)_2C(OH)O_2$	Water	4.2×10^9, 3.5×10^9, 4.5×10^9	0.898	20,32,33
	2-Propanol	3.9×10^9	2.073	11

[a] Viscosities were measured at 25°C [29], except for n-hexadecane which was measured at 20°C [30].

3 METHODS FOR MEASURING THE RATE CONSTANTS FOR PEROXYL FORMATION

Most measurements of peroxyl radical formation are direct measurements in which the radical R is formed in a short time by either a pulse of energetic electrons from an accelerator (pulse radiolysis) or with a pulse of light from a flash lamp or a laser (flash photolysis or laser flash photolysis). In most studies the concentrations were determined by optical absorption. Very few studies have used electron spin resonance (ESR) detection, conductivity measurements or gas chromatography (GC) measurements of the final products. The rate constant can be measured if the concentration of R or RO_2 can be determined as a function of the time. These data are analyzed as a pseudo-first-order-process, as long as the O_2 concentration is considerably larger than the concentration of the radical R. The radical R can react via the following reactions:

$$R + O_2 \longrightarrow RO_2\ (k_1) \tag{2}$$

$$R + R \longrightarrow R_2\text{, or disproportionation products } (k_c) \tag{3}$$

$$R + A \longrightarrow \text{products } (k_2) \tag{4}$$

where A is any compound, a competitor or an impurity.

$$RO_2 \longrightarrow R + O_2\ (k_{-1}) \tag{5}$$

Analyzing the build-up of RO_2 concentration or the decay of R concentration as a pseudo-first-order process with an observed rate constant k_{obs}, gives, according to the above mechanism, an approximate value of k_{obs} [34] as follows:

$$k_{obs} = k_1[O_2] + k_c[R]_{ave} + k_2[A] + k_{-1} \tag{6}$$

where $[R]_{ave}$ is the average concentration of the radicals. Keeping [A] constant thus leads to:

$$k_{obs} = k_0 + k_1[O_2] \tag{7}$$

where

$$k_0 = k_c[R]_{ave} + k_2[A] + k_{-1} \tag{8}$$

From equation (7) it can be seen that measuring k_{obs} as a function of O_2 will allow the measurement of k_1. One should be warned that measuring k_{obs} at only one concentration, neglecting k_0, will give only an approximate value of k_1. In this case it is better to use as high a concentration of O_2 as possible, with the limits on the O_2 concentration being due to solubility factors; care should also be taken that the reaction is not too fast to be measured accurately.

Rabani, Pick and Simic [17] measured k_{obs} as a function of the O_2 concentration for cyclopentyl radicals in aqueous solution. Instead of plotting k_{obs} vs. $[O_2]$ they assumed k_0 to be due only to radical–radical combination and corrected k_{obs} only for this contribution from the measured k_c and $[R]_{ave}$. However, their results for k_1 show clearly higher values at low oxygen concentrations. A more preferable method would be to plot the k_{obs} values (obtained by multiplying k_1 by the appropriate $[O_2]$ concentration) vs. $[O_2]$, as shown in Figure 1, rather than by calculating k_1 for each O_2 concentration (assuming $k_0 = 0$) and averaging. This procedure gives $k_1 = (3.9 \pm 0.4)$ whereas Rabani *et al.* obtained $(4.9 \pm 0.6) \times 10^9\,M^{-1}\,s^{-1}$ by averaging. Other studies using measurements taken at a single O_2 concentration were usually made at high O_2 concentrations, with lower relative contributions of k_0.

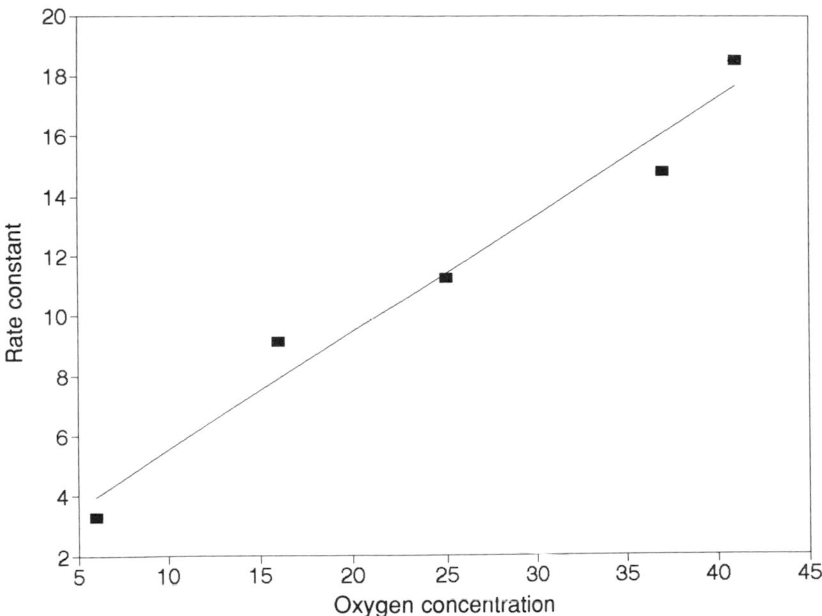

Figure 1 Plotting of data from Rabani, Pick and Simic [17] for values of the rate constant ($10^5\,M^{-1}\,s^{-1}$) for the reaction of cyclopentyl radicals with O_2 vs. O_2 concentration ($10^{-5}\,M$).

Von Sonntag and coworkers [8,35] took k_0 to represent a measure of the reversibility of the reaction with O_2 (the peroxylation reaction), i.e. by equating k_0 with k_{-1}. This derivation should be considered with caution, as equation (8) shows that there are two additional contributing factors to k_0. The contribution of $k_c[R]_{ave}$ to k_0 can be calculated by using different values of $[R]_{ave}$, by applying pulses with varying doses. However, the contribution of any reactions with contaminants is something which is difficult to estimate.

Even assuming the absence of any reacting impurity, k_0 could be due to N_2O (present in solution in order to convert e^-_{aq} to OH radicals) reacting with the radicals. A rate constant of $5 \times 10^5\,M^{-1}\,s^{-1}$ for R + N_2O will explain a k_0 value of $1.2 \times 10^4\,M^{-1}\,s^{-1}$, and a rate constant of 2×10^6 will lead to a k_0 value of $5 \times 10^4\,M^{-1}\,s^{-1}$. This can be compared to $2.1 \times 10^6\,M^{-1}\,s^{-1}$ measured for $H^\bullet + N_2O$ [36], although it is usually accepted that R + N_2O is a considerably slower reaction. The best way to equate k_{obs} with k_{-1} is if the decay of $R^\bullet$ in the absence of O_2 can be measured and shown to be much smaller than k_0. The remaining absorption of (supposedly) RO_2, which is inversely proportional to $[O_2]$ [35], can be explained as being due to the product of the reaction of R with the contaminants. It should be emphasized that it is not stated here that k_0 is not due to k_{-1}, but only that this statement should be considered with care. Benson [37] estimated the thermodynamic parameters of peroxyl radicals and concluded that the latter are quite stable at room temperature, and their reverse decomposition is insignificant at this temperature. According to his estimates, the resonance-stabilized allyl radical forms the least stable radical, but even the allylperoxyl radical decomposes only insignificantly near room temperature. Ruiz et al. [38] and Morgan *et al.* [39] studied the reaction of allyl radicals with oxygen at various temperatures. At 296 K, Ruiz and coworkers [38] observed the allyl radicals' signal to decay to completion with a single exponent. Morgan and coworkers [39] obtained the following for allyl peroxyl radical: log k_{-1} $(M^{-1}\,s^{-1}) = 10.20 \pm 2755/T(K)$. Thus, at room temperature $k^{-1} = 10\,M^{-1}\,s^{-1}$, which is negligibly small. Janzen, Johnston and Ayers [40] studied the reaction of the highly resonance-stabilized triphenylmethyl radical, $(C_6H_5)_3C^\bullet$, with oxygen in a solid lattice permeable to oxygen. They found that at room temperature and atmospheric pressure of O_2 the equilibrium, $(C_6H_5)_3C^\bullet + O_2 \longrightarrow (C_6H_5)_3CO_2^\bullet$, is almost completely to the right and that only a very small concentration of triphenylmethyl radicals exists. At higher temperatures the reverse decomposition of the triphenylmethylperoxyl radical is the predominant reaction. The same is true for the allylperoxyl radical at higher temperatures. Porter and coworkers [41,42] assumed a reversibility of the reaction of oxygen with radicals of polyunsaturated fatty acids and phospholipids in order to explain the distribution of the products of autoxidation.

For many systems, neither R nor RO_2 can be detected by optical absorption. Since this is the main detection method in pulse radiolysis or flash photolysis, the use of additional reactions besides the peroxylation is required.

3.1 DETECTING A CONSECUTIVE REACTION

Peroxyl radicals are, in most cases, stronger oxidants than the parent radicals themselves. Polyhaloperoxyl radicals are quite strong oxidants, and can oxidize many compounds in which their oxidized form absorbs optically over a convenient range. Commonly used compounds are various antioxidants such as ascorbate (measured at 360 nm), chlorpromazine, metiazinic acid, and promethazine (measured at 520–530 nm), 2,2′-azinobis(3-ethylbenzthiazoline-6-sulfonate), known usually as ABTS (measured at 415 nm), propyl gallate, and various porphyrins:

$$R + O_2 \longrightarrow RO_2\ (k_1);\ RO_2 + AO \longrightarrow RO_2^- + AO^+\ (k_3) \qquad (9)$$

where AO = antioxidant.

The build-up of the optically active (i.e. absorbing optically at a convenient wavelength) AO^+ (or after its rapid hydrolysis) is measured and analyzed as a pseudo-first-order reaction. Plotting k_{obs} as a function of the concentration of AO gives a linear dependence for low AO concentrations, from which k_3 can be calculated. Increasing the concentration of AO increases the rate of the (RO_2 + AO) step, such that the latter is no longer the rate determining step. For high concentrations of AO, k_{obs} reaches a saturation value which cannot be increased further by increasing the concentration of AO. This saturation value is due to the first step, i.e. $R + O_2 \longrightarrow RO_2$, being the rate determining step. The saturation value can be used to calculate the value of k_1. It should be remembered that the saturation value when AO^+ was measured is equal to k_{obs} and not to $k_1[O_2]$. However, in most studies only one oxygen concentration was used and k_0 was assumed to be negligible. An important warning for these kind of studies in pulse radiolysis is that at high concentrations of antioxidants, reaction might also occur with e^-_{aq}, and this point should be taken into consideration.

Instead of using the reaction of RO_2 with a substrate to form a species with observable physical properties, the decomposition of RO_2 (usually catalyzed by OH), when forming observable species, can be used. If this decomposition is sufficiently rapid to render the first step ($R + O_2 \longrightarrow RO_2$) as the rate determining step, the k_{obs} for the build-up of a property is equal to that measured if RO_2 could be detected. This method has been used mainly for cases of decomposition of RO_2 to form O_2^- (catalyzed by high pH), where O_2^- is measured either by optical absorption [41] or by conductivity measurements [42].

3.2 THE 'PROBE' METHOD—USE OF A COMPETITOR

When both R and RO_2 are not optically active over a convenient wavelength range, and when RO_2 is not a strong enough oxidizant to rapidly oxidize an antioxidant, the rate constant for ($R + O_2$) can be measured by competition with a reaction (R + C) if the product of this reaction can be detected optically. This method is sometimes referred to as the 'probe' method. An important point to note in this case is to ascertain that the reaction of RO_2 with C is either considerably slower or does not lead to a product which absorbs optically in a close wavelength range:

$$R + O_2 \longrightarrow RO_2\ (k_1) \tag{10}$$

$$R + C \longrightarrow \text{product}\ (k_3) \tag{11}$$

where C = competitor.
The build-up of the product is a pseudo-first-order reaction, and the observed rate constant is given by:

$$k_{obs} = k_0 + k_1[O_2] \tag{12}$$

Plotting k_{obs} vs. $[O_2]$ gives k_1 from the slope. In this case of a competing reaction, k_1 can also be deduced from the final absorption due to the competing reaction:

$$\frac{(OD)_{O_2}}{OD_O} = \frac{k_3}{k_3 + k_1[O_2]} \Longrightarrow \frac{OD_O}{(OD)_{O_2}} = 1 + \frac{k_1}{k_3}[O_2] \tag{13}$$

$(OD)_{O_2}$ is the final absorption for a system with O_2, and OD_O is the final absorption in the absence of oxygen; k_3 is the pseudo-first-order rate constant of formation of the absorption in the absence of oxygen.

However, there are only a few optically active substrates which react faster with R than with RO_2. Species having optical bands in the visible range of the spectrum usually have a labile electron, i.e. they are oxidants and are the product of a reaction with reductants. This means that the competitor C should be a reductant (antioxidant), but RO_2 is usually a much stronger oxidant than R and thus a reducing C is not compatible. In some cases, C can be a molecule from which R abstracts hydrogen to form an optically active radical. However, this method should also be checked carefully, since in many cases RO_2 abstracts hydrogen faster than R [43].

In some cases of competition studies, a competitor ('probe') does not need to be added as the competition is achieved by the monomolecular decay of R to produce optically absorbing species:

$$R + O_2 \longrightarrow RO_2\ (k_1);\ R \longrightarrow P_1 + P_2\ (k_3) \tag{14}$$

where at least one of P_1 and P_2 can be detected. Equations (12) and (13) remain the same for this case.

Another version of the competition method is to use an optically active competitor C and then to measure its decay. The equation for the observed pseudo-first-order decay rate constant is the same as the build up-equation (12). The final absorption equation, equation (13), is slightly changed, however; instead of writing OD, we now use $\Delta\ (OD)$, i.e. the decrease in the absorption. This version of the 'probe' method is advantageous from the point of view that C can be an oxidant, and consequently will not react with

RO_2, but will react with R. This method can be used for many radicals, whereas the case where C is non-detectable, while the product is detectable, is limited to only a few radicals. A disadvantage of this technique is the lower accuracy associated with the measurement of the decrease of *OD* when compared to its increase; $\Delta(OD)$ cannot be higher than 10–20% of the original *OD*, since otherwise the data cannot be treated as a pseudo-first-order reaction. The original *OD* cannot be too high, however, as the transmitted light will then be too low to be measured accurately. Thus $\Delta(OD)$ in the measurement of the decay is smaller than the *OD* in the measurement of the build-up, and in addition the accuracy of the measurement is lower, due to less light reaching the detector.

4 MEASUREMENTS OF THE PEROXYLATION RATE CONSTANT

4.1 FORMATION OF RO_2

Most peroxyl radicals absorb at wavelengths lower than 300 nm; however, peroxyl radicals which contain C=C double bonds (either alkenic or aromatic) absorb in the visible range [44,45].

Alfassi, Marguet and Neta [44] and Mertens and von Sonntag [45] used this visible absorption of RO_2 to detect the rate constants for the reactions with oxygen of phenyl [44,45] and various vinyl radicals [45]. Both studies used varying O_2 concentrations, plotting k_{obs} vs. $[O_2]$ to obtain k_1. For the only similar radicals measured in both studies the agreement is not particularly good. For the 4-carboxyphenyl radical Alfassi and coworkers [44] obtained $1.6 \times 10^9\,M^{-1}\,s^{-1}$, whereas for phenyl radicals Mertens and von Sonntag [45] obtained $4.6 \times 10^9\,M^{-1}\,s^{-1}$ both obtained in aqueous solution. Mertens and von Sonntag [45] measured $k = 3.8\text{–}4.6 \times 10^9\,M^{-1}\,s^{-1}$ for all of the vinylic radicals studied in their work.

Rabani, Pick and Simic [17] measured the peroxylation reaction of the cyclopentyl radical, detecting the build-up of the absorption at 270 nm due to the peroxyl radical.

Schuchmann, Zegota and von Sonntag [46] found that while the acetate radical absorbs with a peak at 350 nm, the acetate peroxyl radical absorbs at 280 nm. Pulse radiolysis of an aqueous solution of acetate saturated with N_2O/O_2 (4:1 vol/vol; the role of N_2O is to convert e^-_{aq} to OH) yields a very fast formation (within the pulse) of an absorption at around 350 nm. This absorption decays rapidly to give rise to a new transient absorption with a maximum at around 280–290 nm. These authors measured the rate constant from the build-up at 275 nm. If possible, it is preferable to measure both the decay of R and the build-up of RO_2, and to show that they are equal; in this case, the build-up was measured only at one O_2 concentration.

Zegota *et al.* [47] found absorption of the acetonylperoxyl radical over the range from 270–360 nm, although they did not find the peak. Both the acetonyl and the acetonyl peroxyl radical absorb in this range. Since the molar absorptivity ϵ of the acetonylperoxyl radical is higher than that of the acetonyl radical over the whole wavelength range, each wavelength in this range will show a build-up due to RO_2 formation. These authors chose to make measurements at 350 nm, although the difference in ϵ is larger at 270 nm, probably due to less absorption by the acetone itself and to the higher intensity of the light at this wavelength. They used an O_2 concentration range of $(2.6–9.1) \times 10^{-5}$ M. A similar situation exists for long-chain alkyl radicals, where both the alkyl radical and the peroxyl radical absorb over the range from 230 to 300 nm (but none have peaks in this range). Since the peroxyl radical has a higher absorption over all of this range, each wavelength will show a build-up. Brede, Herman and Mehnert [10] used the build-up at 270 nm to measure k_1, since the difference in ϵ is the largest at this wavelength. Mieden, Schuchmann and von Sonntag [48] observed that the radicals obtained by H-atom abstraction from cyclic dipeptides (either glycine anhydride or alanine anhydride) absorb strongly at 265 nm and weakly at 360 nm. The corresponding peroxyl radicals absorb strongly over all of the range from 265–360 nm and consequently these authors used the build-up at 360 nm to measure the formation of the peroxyl radicals.

4.2 DECAY OF R

Abramovitch and Rabani [49] found that while both the acetate radical $^{\bullet}CH_2CO_2^-$ and its corresponding peroxy radical $^{\bullet}O_2CH_2CO_2^-$ absorb at around 370 nm, the absorption of the acetate radical is higher in this region. Thus, the decay of the absorption at 366 nm was used for measuring k_1. These authors used different concentrations of O_2 and then corrected for radical combination and O_2 depletion; however, they did not report whether the results given are obtained by averaging of the results from different O_2 concentrations, or by plotting as a function of the O_2 concentration.

The OH-adducts of aromatic hydrocarbons absorb at 310–320 nm (with a peak at around 313 nm), with the absorptions of corresponding peroxyl radicals being only about one third of the former. It was found that in the presence of O_2 the absorptions at 313 nm decay by first-order processes both for benzene [6,7,11] and other aromatic hydrocarbons [9]. For benzene, Dorfman, Taub and Bühler [6] and Ramanan [7] used only one O_2 concentration (1.04 mM), whereas Maillard, Ingold and Scaiano [11] used various O_2 concentrations and plotted k_{obs} vs. O_2. Their results show k_0 to be 5% of the rate measured at $[O_2] = 1.04$ mM. Roder, Wojnarovits and Földiak [9] used one O_2 concentration (1.25 mM). Pan, Schuchmann and von Sonntag [50] obtained the cyclohexadienyl radical by hydrogen-atom abstraction from 1,4-cyclohexanediene with H and OH radicals from pulse radiolysis of aqueous solutions, and measured their reaction with O_2 by the rate of the decay of the absorption at 310 nm. They used relatively low concentrations of O_2 of (3×10^{-5}–3×10^{-4} M) and a plot of k_{obs} vs. O_2. For their highest O_2 concentration the contribution of k_0 is 7–8%. A similar study from the same group was carried out for the $(CH_3)_2NCH_2^{\bullet}$ radical. They measured the first-order decay of the absorption at 260 nm. The observed rate constants were found to be linearly dependent on the O_2 concentration (7.5×10^{-5}–1.4×10^{-4} M).

Measurements of the decay of the absorption due to R as a result of its reaction with O_2 is the most common method reported in the literature. The following are a few further examples of this approach. Fel, Zaozerskaya and Dolin [51] studied the peroxylation reactions of the H- and OH-adducts of uracil. These adducts are absorbing in the range from 300 to 400 nm, whereas the peroxyl radicals absorb at lower wavelengths with a maximum at 300 nm, and are practically non-absorbing at 380–400 nm. These authors measured k_{obs} in the range 1–4×10^{-4} M of O_2, and by plotting k_{obs} vs. $[O_2]$ obtained $k = 1.6 \times 10^9\,M^{-1}\,s^{-1}$ with $k_0 \sim 0$. Josimovic, Draganic and Markovic [52] studied the rate of peroxylation of the acetate radical by measuring the kinetics of the decay of the absorption at 370 nm at two oxygen concentrations. They obtained $k = 3 \times 10^9\,M^{-1}\,s^{-1}$, compared to a value of $2.1 \times 10^9\,M^{-1}\,s^{-1}$ obtained by Abramovitch and Rabani [49]. Hasegawa and Patterson [4] found that the peroxyl derivatives of the radicals of polyunsaturated fatty acids are less absorbing at 280 nm than the radicals themselves, and hence used the decay at 280 nm to measure the rate of peroxylation. For a singly unsaturated acid (oleic acid) the peroxyl radical absorbs more strongly at 280 nm than the parent radical itself, and the build-up of the absorption at 280 nm was therefore used to measure the rate constant.

4.3 MEASUREMENT OF BOTH R AND RO_2

The best method for measuring the rate of peroxylation is by following both RO_2 and R. However, this has been carried out in only a very few cases. Kartasheva *et al.* [53] measured the rate of peroxylation of the OH-adduct of *o*-iodohippurate. Immediately after the pulse the absorption has a peak at 340 nm, which then decays with the subsequent formation of an absorption at 410 nm. These authors measured both the decay at 340 nm and the build-up at 400 nm and obtained values for the disappearance of $k_1 = 6 \times 10^6\,M^{-1}\,s^{-1}$ and for the build-up of $k_1 \leqslant 1 \times 10^7\,M^{-1}\,s^{-1}$. Smaller, Remko and Avery [16] measured the rate of peroxylation of cyclopentyl and cyclohexyl radicals by following the ESR signals of both the R and the RO_2 radicals. The rate of decay of R was equal within experimental error to the rate of formation of RO_2. However, their measured values ($4.3 \times 10^7\,M^{-1}\,s^{-1}$ for cyclohexyl at 25°C, and $3.9 \times 10^6\,M^{-1}\,s^{-1}$ for cyclopentyl at −40°C), are considerably slower than values measured later.

4.4 MEASUREMENT OF A CONSECUTIVE REACTION

This method consists of measuring a consecutive reaction, which must be very much faster in order that the previous one, the peroxylation reaction, being the rate determining step. Asmus and coworkers [54], measured the rate of peroxylation of various haloalkyl radicals by using this method. The peroxyl derivatives of haloalkyl radicals are strong oxidants and can oxidize very rapidly even at low concentrations of antioxidants. Mönig *et al.* [55] used ABTS for studying the reaction of $CF_3CHCl^{\cdot}$ with O_2 by following the first-order rate constant of the formation of $ABTS^{+\cdot}$, which absorbs at 415 nm:

$$R^{\cdot} + O_2 \longrightarrow RO_2^{\cdot}; RO_2^{\cdot} + ABTS \longrightarrow RO_2^- + ABTS^{+\cdot} \quad (15)$$

For low concentrations k_{obs} is proportional to [ABTS]; however, for $[ABTS] > 3.5 \times 10^{-4}$ M, a plot of k_{obs} vs. [ABTS] is curved downwards and above [ABTS] = 3 mM it reaches a plateau, due to $(R + O_2)$ being the rate determining step. Mönig, Bahnemann and Asmus [56] studied the reaction of CCl_3 and O_2 by using methiazinic acid an an antioxidant. Asmus and coworkers [54] measured the peroxylation reaction of various haloalkyl and haloethers radicals, using five different antioxidants. The only low value was found for $^{\cdot}CCl_2CN$, where $k_1 = 4 \times 10^8\, M^{-1}\, s^{-1}$.

The consecutive reaction can also be a decomposition of the peroxyl radical to give a detectable species. This decomposition is usually too slow to allow the peroxylation reaction to be the rate determining step, but it can sometimes be catalyzed by OH^- or H^+. Bothe and Schulte-Frohlinde [57] measured the peroxylation of $^{\cdot}CH_2(OH)$ at pH = 3.5–6.5 by measuring the increase in the conductivity. The observed first-order rate constant is proportional to $[O_2]$, thus proving that the peroxylation reaction is the rate determining step. The conductivity build-up is due to the following decay reaction:

$$H_2C(OH)O_2 \longrightarrow H_2CO + H^+ + O_2^- \quad (16)$$

Rabani, Klug-Roth and Henglein [58] measured the peroxylation rate constant of $^{\cdot}CH_2OH$ at pH = 10.7 due to the decomposition of the ionized form of the peroxyl radical:

$$\begin{aligned} ^{\cdot}O_2CH_2OH &\longrightarrow O_2CH_2O^- + H^+ \ (pK > 8.8) \\ O_2CH_2O^- &\longrightarrow CH_2O + O_2^- \end{aligned} \quad (17)$$

The O_2^- formed in the decomposition was detected by its absorption at 248 nm. At this pH $^{\cdot}CH_2OH$ is half-ionized to $^{\cdot}CH_2^-$ and the measured rate constant ($4.2 \times 10^9\, M^{-1}\, s^{-1}$) is really a composite of two constants.

4.5 THE 'PROBE' METHOD—MEASUREMENT OF A COMPETITOR

Ingold and coworkers [59] studied the peroxylation of phenyl radicals in both the gaseous phase and in aqueous solution. In aqueous solution they used phenol as the competitor, i.e. the 'probe':

$$\begin{aligned} C_6H_5^{\bullet} + O_2 &\longrightarrow C_6H_5O_2^{\bullet}\ (k_1) \\ C_6H_5^{\bullet} + C_6H_5OH &\longrightarrow C_6H_6 + C_6H_5O^{\bullet}\ (k_2) \end{aligned} \qquad (18)$$

These authors measured the build-up at 380 nm due to the formation of the phenoxyl radical. This particular study can be used as a warning that careful checking is necessary of all of the radicals present in a system. As a source of phenyl radicals (4-carboxyphenyl radicals being used as a representative for phenyl radicals in this case) they used 4-iodobenzoate and ruptured the C—I bond by laser flash photolysis (at 308 nm). The rupturing of the bond produces iodine atoms in addition to phenyl radicals. Ingold and coworkers completely ignored the possibility that reaction of the iodine atoms will also lead to formation of phenoxyl radicals [60], or to other species absorbing at the same wavelength. If the iodine atoms also react with O_2 the measured dependence on the O_2 concentration is therefore not due only to peroxyl formation of the phenyl radical. It should be mentioned that in the 'probe' method both k_{obs} and OD can be used to abstract k_1. If both give the same k_1 this is supportive evidence for the suggested mechanism.

The use of a competitor sometimes allows the measurement of k_1 from stable products, rather than from short-lived intermediates. Russel and Bridger [61] produced phenyl radicals in cyclohexane/CCl_4 solutions by thermal decomposition (at 333 K) of phenylazotriphenylmethane, in both the presence and absence of oxygen, and studied the effect of oxygen on the yields of benzene and/or chlorobenzene. From the yields obtained they were able to calculate the ratio of the rate constants for the following three reactions:

$$\begin{aligned} {}^{\bullet}C_6H_5 + O_2 &\longrightarrow C_6H_5O_2^{\bullet}\ (k_1) \\ {}^{\bullet}C_6H_5 + CCl_4 &\longrightarrow C_6H_5Cl + {}^{\bullet}CCl_3\ (k_2) \\ C_6H_5 + \text{cyclo-}C_6H_{12} &\longrightarrow C_6H_6 + {}^{\bullet}\text{cyclo-}C_6H_{11}\ (k_3) \end{aligned} \qquad (19)$$

These workers found $k_1 : k_2 : k_3 = 1200 : 1 : 1$.

Kryger *et al.* [62] estimated k_3 from competitive kinetics, and assuming the reactions ($C_6H_5 + Br_2$) or ($C_6H_5 + I_2$) to be diffusion controlled, found a value of $4 \times 10^6\,M^{-1}\,s^{-1}$. From this value and the above mentioned ratios, $k_1 = 5 \times 10^9\,M^{-1}\,s^{-1}$. Scaiano and Stewart [63] measured k_2 by laser flash photolysis and obtained $k_2 = 8 \times 10^6\,M^{-1}\,s^{-1}$, which together with the above ratios gives $k_1 = 9 \times 10^9\,M^{-1}\,s^{-1}$.

Adams, McNaughton and Michael [64] measured the rate constant for ($^{\bullet}CH_2OH + O_2$) by the competition method, using cysteamine as the competitor:

$$^{\bullet}CH_2OH + RSH \longrightarrow CH_3OH + RS^{\bullet}\ (k_2);\ RS^{\bullet} + RS^- \longrightarrow RSSR^- \quad (20)$$

By measuring the absorption due to $RSSR^-$ and then plotting OD_0/OD_{O_2} vs. $[O_2]$, where OD_0 and OD_{O_2} are the absorptions in the absence and presence of oxygen, respectively, they obtained $k_1 = 2.3 \times 10^9\,M^{-1}\,s^{-1}$. Adams and Wilson [20] used ferricyanide as a competitor in measuring the bleaching of the absorption due to ferricyanide in the following reaction:

$$^{\bullet}CH_2OH + [Fe(CN)_6]^{3-} \longrightarrow CH_2O + [Fe(CN)_6]^{-4} + H^+\ (k_2) \quad (21)$$

The addition of O_2 reduced the bleaching according to $OD_0/\Delta(OD_{O_2}) = 1 + k_1[O_2]/(k_2[\text{ferricyanide}])$, where $\Delta(OD_{O_2})$ stands for the bleached absorption.

4.6 SPECIAL METHODS

Meyerstein and coworkers [65] developed a special method which used the equilibrium of a nickel cyclam complex with radicals to measure the rate constant for the (methyl + O_2) reaction. Espenson and coworkers [15] have used this method for the measurement of the rate constants of several (R + O_2) reactions for various radicals. Meyerstein and coworkers found that when methyl radicals were formed by pulse radiolysis of an aqueous solution containing a cyclam complex of nickel, LNi^{2+} (L represents a ligand, in this case cyclam), there is build-up of absorption with a peak at 550 nm. This absorption is due to a relatively long-lived species which decays in the 100 s region. If the solution contains oxygen at low concentration, the decay is replaced by an increase which is due to a first-order process. These observations have been explained as being due to a reversible reaction of the methyl radicals with the nickel cyclam, and an irreversible reaction of the peroxyl radical with the same species. This explanation assumed that the molar absorption of the $RO_2Ni(cyclam)^{2+}$ species is larger than that of the complex with the parent radical $RNi(cyclam)^{2+}$.

$$R + Ni(cyclam)^{2+} \longleftrightarrow RNi(cyclam)^{2+}\ (k_2, k_{-2}) \quad (22)$$

$$R + O_2 \longrightarrow RO_2\ (k_1) \quad (23)$$

$$RO_2 + Ni(cyclam)^{2+} \longrightarrow RO_2Ni(cyclam)^{2+}\ (k_3) \quad (24)$$

The concentration of oxygen should be low enough so that at the start of the reaction most of the radicals will react with the nickel cyclam complex and not with the oxygen ($k_2[Ni(cyclam)] > k_1[O_2]$). However, as reaction (22) is reversible, while reactions (23) and (24) are not, all of the radicals will eventually form $RO_2Ni(cyclam)^{2+}$. The rate of the second absorption step (the first step being due to $RNi(cyclam)^{2+}$) is due to formation of $RO_2Ni(cyclam)^{2+}$ and the first-order observed rate constant obeys the equation:

$$k_{obs} = \frac{k_{-2}}{1 + k_2 [Ni(cyclam)^{2+}]/k_1[O_2]}$$

$$\frac{1}{k_{obs}} = \frac{1}{k_{-2}} + \frac{k_2}{k_1 k_{-2}} \frac{[Ni(cyclam)^{2+}]}{[O_2]} \tag{25}$$

$K_2(= k_2/k_{-2})$ was measured by Meyerstein and coworkers [65] and thus a plot of $1/k_{obs}$ vs. $[O_2]$ yielded k_1; they obtained a rate constant for CH_3 radicals which agreed with other measurements [66,67]. Espenson and coworkers [15] used this method for several radicals produced by the photolysis of various organocobalt (III) complexes, and obtained rate constants similar to those obtained by other methods.

Ladygin and Revina [68] studied the rate of the peroxylation of the radicals produced by radiolysis of cycloalkanols by breaking the α-(C—H) bond. They found that the molar absorptions of the alkyl radicals and their corresponding peroxyl radicals were very close, and neither the decay nor the build-up could be measured. However, they observed that the decay of the absorption, due to radical combination, is slowed down in the prsence of oxygen. They explained this as being due to lower rate constants for combination of the peroxyl radicals than for combination of the alkyl radicals:

$$R + R \longrightarrow R_2 \text{ or disproportionation products } (k_2) \tag{26}$$

$$R + O_2 \longrightarrow RO_2 \tag{27}$$

$$RO_2 + RO_2 \longrightarrow \text{products } (k_3) \tag{28}$$

By assuming $k_2 > k_3$, this leads to the following kinetic expression:

$$[R] = [R]_0 \exp [-(k_2[R]_{ave} + k_1[O_2])] \tag{29}$$

which gives, for first-order analysis:

$$k_{obs} = k_1[O_2] + k_2[R]_{ave} \tag{30}$$

These authors measured k_{obs} at $[O_2] = 0$ (absence of air) and in an oxygen atmosphere, and from the results calculated k_1. For the α-hydroxycyclododecyl radical they found $k_1 = 2 \times 10^6\, M^{-1} s^{-1}$ at 95°C, while for the α-hydroxycyclohexyl radical they obtained $k_1 = (4 \times 10^6–2 \times 10^7)\, M^{-1} s^{-1}$ for the temperature range of 26–114°C.

REFERENCES

1. A.J.S.C. Vieira and S. Steenken, *J. Am. Chem. Soc.* **109**, 7441 (1987).
2. P. O'Neil and J.E. Davies, *Int. J. Radiat. Biol. Relat. Stud. Phys. Chem. Med.* **52**, 577 (1987).
3. M.T. Nenadovic and O.I. Micic, *Radiat. Phys. Chem.* **12**, 85 (1978).
4. K. Hasegawa and L.K. Patterson, *Photochem. Photobiol.* **28**, 817 (1978).
5. M. Erben-Russ, W. Bors and M. Saran, *Int. J. Radiat. Biol. Relat. Stud. Phys. Chem. Med.* **52**, 393 (1987).
6. L.M. Dorfman, I.A. Taub and R.E. Bühler, *J. Chem. Phys.* **36**, 3056 (1962).
7. G. Ramanan, *J. Indian Chem. Soc.* **53**, 957 (1976).
8. X.M. Pam and C. von Sonntag, *Z. Naturforsch.* **45B**, 1337 (1990).
9. M. Roder, L. Wojnarovits and G. Földiak, *Radiat. Phys. Chem.* **36**, 175 (1990).
10. O. Brede, R. Herman and R. Mehnert, *J. Chem. Soc. Faraday Trans. 1* **83**, 2365 (1987).
11. B. Maillard, K.U. Ingold and J.C. Scaiano, *J. Am. Chem. Soc.* **105**, 5095 (1983).
12. G.M. Zimina, L. Kovacs and G.V. Putirskaja, *Magy. Kem. Foly.* **87**, 395 and 569 (1981); G.M. Zimina, L. Kovacs and G.V. Putirskaja, *Radiochem. Radioanal. Lett.* **44**, 413 (1980).
13. P. Neta, R.E. Huie, S. Mosseri, L.V. Shastri, J.P. Mittal, P. Maruthamuthu and S. Steenken, *J. Phys. Chem.* **93**, 4099 (1989).
14. K. Tokumara, H. Nosaka and T. Ozaki, *Chem. Phys. Lett.* **169**, 321 (1990).
15. A. Marchaj, D.G. Kelley, A. Bakac and J.H. Espenson, *J. Phys. Chem.* **95**, 4440 (1991).
16. B. Smaller, J.R. Remko and E.C. Avery, *J. Phys. Chem.* **48**, 5174 (1968).
17. J. Rabani, M. Pick and M. Simic, *J. Phys. Chem.* **78**, 1049 (1974).
18. O. Brede and L. Wojnarovits, *Radiat. Phys. Chem.* **37**, 537 (1991).
19. E. Hayon and M. Simic, *J. Am. Chem. Soc.* **95**, 6681 (1973).
20. G.E. Adams and R.L. Wilson, *Trans. Faraday Soc.* **65**, 2981 (1969).
21. K.D. Asmus, B. Cercek, M. Ebert, A. Henglein and A. Wigger, *Trans. Faraday Soc.* **63**, 2435 (1967).
22. P.B. Pagsberg, *Report Risø-250*, 209 (1972).
23. P. Neta, P. Maruthamuthu, P.M. Carton and R.W. Fessenden, *J. Phys. Chem.* **82**, 1875 (1978).
24. B.G. Ershov, T.L. Mikhailova, A.V. Gordeev and V.I. Spitsyn, *Dokl. Phys. Chem.* **300**, 506 (1988).
25. V.B. Men'kin, I.E. Makarov and A.K. Pikaev, *High Energy Chem.* **25**, 48 (1991).
26. B.G. Ershov and T.L. Mikhailova, *Bull Acad. Sci. USSR Div. Chem. Sci.* **40**, 288 (1991).
27. E.P.L. Hunter, M.F. Desrosiers and M.G. Simic, *Free Radicals Biol. Med.* **6**, 581 (1989).
28. F. Jin, J. Leitich and C. von Sonntag, *J. Chem. Soc. Perkin Trans.* **2**, 1583 (1993).
29. Y. Marcus, *Ion Solvation*, John Wiley & Sons, Chichester (1985), p. 136.
30. *Handbook of Chemistry and Physics*, 69th Edn, CRC Press, Boca Raton (1988), p. F-41.
31. S.S. Emmi, G. Beggiato, G. Casalbore and P.G. Fuochi, in *Proceedings of the Fifth Tihany Symposium on Radiation Chemistry*, edited by J. Dobo, P. Hedvig and R. Schiller, Vol. 1, Akad. Kiado, Budapest (1983), p. 677.
32. R.L. Wilson, *Trans. Faraday Soc.* **67**, 3008 (1971).
33. J. Butler, G.G. Jayson and A.J. Swallow, *J. Chem. Soc. Faraday Trans. 1* **70**, 1394 (1974).
34. X. Zhang, N. Zhang, H.P. Schuchmann and C. von Sonntag, *J. Phys. Chem.* **98**, 6541 (1994).

35. D. Wang, H.P. Schuchmann and C. von Sonntag, *Z. Naturforsch* **48B**, 761 (1993).
36. G. Czapski and E. Peled, *Isr. J. Chem.* **6**, 421 (1968).
37. S.W. Benson, *J. Am. Chem. Soc.* **87**, 972 (1965).
38. R.P. Ruiz, K.D. Bayes, M.T. Macpherson and M.J. Pillig, *J. Phys. Chem.* **85**, 1622 (1981).
39. C.A. Morgan, M.J. Pilling, J.M. Tulloch, R.P. Ruiz and K.D. Bayes, *J. Chem. Soc. Faraday Trans. 2* **78**, 1323 (1982).
40. E.G. Janzen, F.J. Johnston and C.L. Ayers, *J. Am. Chem. Soc.* **89**, 1176 (1967).
41. N.A. Porter, B.A. Wever, H. Weenen and J.A. Khan, *J. Am. Chem. Soc.* **102**, 5597 (1980).
42. N.A. Porter, L.S. Lehman, B.A. Weber and K.J. Smith, *J. Am. Chem. Soc.* **103**, 6447 (1981).
43. S. Mosseri, Z.B. Alfassi and P. Neta, *Int. J. Chem. Kinet.* **19**, 309 (1987).
44. Z.B. Alfassi, S. Marguet and P. Neta, *J. Phys. Chem.* **98**, 8019 (1994).
45. R. Mertens and C. von Sonntag, *Angew. Chem. Int. Ed. Engl.* **33**, 1262 (1994).
46. M.N. Schuchmann, H. Zegota and C. von Sonntag, *Z. Naturforsch.* **40B**, 215 (1985).
47. H. Zegota, M.N. Schuchmann, D. Schulz and C. von Sonntag, *Z. Naturforsch.* **41B**, 1015 (1986).
48. O.J. Mieden, M.N. Schuchmann and C. von Sonntag, *J. Phys. Chem.* **97**, 3783 (1993).
49. S. Abramovitch and J. Rabani, *J. Phys. Chem.* **80**, 1562 (1976).
50. X.M. Pan, M.N. Schuchmann and C. von Sonntag, *J. Chem. Soc. Perkin Trans. 2* 1021 (1993).
51. N.S. Fel, L.A. Zaozerskaya and P.I. Dolin, *Radiat. Effects* **9** 145 (1971).
52. L.R. Josimovic, I.G. Draganic and V.M. Markovic, *Bull. Soc. Chim. Beograd.* **41**, 75 (1976).
53. L.I. Kartasheva, Z.M. Potapova, V.T. Kharlamov and A.K. Pikaev, *High Energy Chem.* **11**, 446 (1977).
54. M. Lahl, C. Schöneich, J. Mönig and K.D. Asmus, *Int. J. Radiat. Biol.* **54**, 773 (1988).
55. J. Mönig, K.D. Asmus, M. Schaeffer, T.F. Slater and R.L. Wilson, *J. Chem. Soc. Perkin Trans. 2* 1133 (1983).
56. J. Mönig, D. Bahnemann and K.D. Asmus, *Chem. Biol. Interact.* **47**, 15 (1983).
57. E. Bothe and D. Schulte-Frohlinde, *Z. Naturforsch.* **35B**, 1035 (1980).
58. J. Rabani, D. Klug-Roth and A. Henglein, *J. Phys. Chem.* **78**, 2089 (1974).
59. P.M. Sommeling, P. Mulder, R. Louw, D.V. Avila, J. Lisztyk and K.U. Ingold, *J. Phys. Chem.* **97**, 8361 (1993).
60. Z.B. Alfassi, R.E. Iluie, S. Marguet, E. Natarajan and P. Neta, *Int. J. Chem. Kinet.* **27**, 181 (1995).
61. G.A. Russel and R.F. Bridger, *J. Am. Chem. Soc.* **85**, 3765 (1963).
62. R.G. Kryger, J.P. Lorand, N.R. Stevens and N.R. Herron, *J. Am. Chem. Soc.* **99**, 7589 (1977).
63. J.C. Sciano and L.C. Stewart, *J. Am. Chem. Soc.* **105**, 3609 (1983).
64. G.E. Adams, G.S. McNaughton and B.D. Michael, *Trans. Faraday Soc.* **64**, 902 (1968).
65. A. Sauer, H. Cohen and D. Meyerstein, *Inorg. Chem.* **27**, 4578 (1988).
66. J.K. Thomas, *J. Phys. Chem.* **71**, 1919 (1967).
67. D. Meyerstein and H.A. Schwarz, *J. Chem. Soc. Faraday Trans. 1* **84**, 2933 (1988).
68. B.Ya. Ladygin and A.A. Revina, *Bull. Acad. Sci. USSR Div. Chem. Sci.* **34**, 2013 (1985).

2 Methods of Preparing Organic Peroxy Radicals for Laboratory Studies

OLE J. NIELSEN
Research Center, Ford Motor Company, Aachen, Germany

and

TIMOTHY J. WALLINGTON
Research Laboratory, Ford Motor Company, Dearborn, USA

1 INTRODUCTION

Two different methods can be used to prepare organic peroxy radicals, RO_2, for laboratory studies. The first and most widely used approach is to generate an organic radical, R, in the presence of O_2. The rapid reaction of R with O_2 then forms the RO_2 radicals:

$$R + O_2 + M \longrightarrow RO_2 + M \tag{1}$$

where M respresents a third body capable of removing the excess energy associated with the formation of the $R–O_2$ bond.

The second method involves releasing RO_2 radicals from species which possess the RO_2 moiety, such as by the decomposition of a peroxy nitrate, RO_2NO_2:

$$RO_2NO_2 + M \longrightarrow RO_2 + NO_2 + M \tag{2}$$

or by H-atom abstraction from a hydroperoxide, ROOH:

$$ROOH + X \longrightarrow RO_2 + HX \tag{3}$$

As discussed below, there are a great variety of methods (e.g. photolysis, abstraction, addition) that provide convenient sources of R radicals. In all cases attention should be paid to possible unwanted side reactions which may influence the kinetics of the RO_2 radicals. While the rate constant for reaction (1) is dependent on the total pressure, for all but the smallest R radicals, reaction (1) approaches the high-pressure limit for pressures above 20 torr. Values of the high-pressure limiting rate constant, k_1, are typically in the range $(1–10) \times 10^{-12}\,cm^3\,molecule^{-1}\,s^{-1}$. In most experimental systems it is

Peroxyl Radicals. Edited by Z.B. Alfassi

practicable to have at least 10 torr of O_2, and under such circumstances the R radicals react to give RO_2 radicals in about 1 μs. For experiments conducted at a low total pressure (<10 torr) there is a competing channel for reaction (1) which needs consideration, namely reaction to give an alkene and HO_2 radical. The importance of this channel is well established for C_2H_5 radicals [1].

At room temperature, the C_2H_4 yield from the reaction of a C_2H_5 radical with O_2 decreases with increasing total pressure, following a $P^{-0.8\pm0.1}$ dependence. At a total pressure of air of 10 torr, 2% of the C_2H_5 radicals react to give C_2H_4, while at 1 torr the C_2H_4 yield is 12% [1]. Recent experimental and computational studies have shown that C_2H_4 is not formed by a direct hydrogen abstraction mechanism but rather via an excited ethylperoxy radical intermediate [1–3]:

$$C_2H_5 + O_2 \longleftrightarrow C_2H_5O_2^* \tag{4}$$

$$C_2H_5O_2^* \longrightarrow C_2H_4 + HO_2 \tag{5}$$

$$C_2H_5O_2^* + M \longrightarrow C_2H_5O_2 + M \tag{6}$$

M represents a third body which is capable of removing excess energy from the excited $C_2H_5O_2$ radical. R radicals larger than C_2H_5 possess a greater number of internal degrees of freedom, and for a given total pressure, the relative importance of the alkene-producing channel is expected to be less than that observed for the C_2H_5 radicals.

2 METHODS OF PRODUCING R RADICALS IN THE PRESENCE OF O_2

2.1 PHOTOLYSIS

Photolysis has been used for many years for generating organic fragments. Azoalkanes can be readily photolyzed in their n-π* bands at 300–400 nm to produce R radicals [4]:

$$RN{=}NR + h\nu \longrightarrow 2R + N_2 \tag{7}$$

The photolysis of azoalkanes was used as a source of peroxy radicals in the first kinetic and mechanistic studies of CH_3O_2 [5–10], $C_2H_5O_2$ [11,12], $C_3H_7O_2$ [13,14], and t-butylperoxy radicals [15].

This method provides an important source of radicals that are difficult to produce by abstraction or addition reactions, e.g. i-C_3H_7 radicals, and has been used recently to produce CF_3O_2 radicals from $CF_3N_2CF_3$ [16]. Photolysis of azoalkanes at wavelengths shorter than 300 nm, e.g. 193 nm, should be used with great care as excited states and fragments other than R radicals may be formed [17]. The limited commercial availability of azo compounds restricts their use for generating RO_2 radicals.

Photolysis of the more easily available organic halides:

$$RX + h\nu \longrightarrow R + X \quad (8)$$

in the presence of O_2 provides a more convenient source of R, and hence RO_2 radicals. Iodides are readily photolyzed. Photolysis of CH_3I and $HOCH_2CH_2I$ at 254 nm in the presence of O_2 leads to the formation of CH_3O_2 and $HOCH_2CH_2O_2$, respectively [18]. Photolysis of chlorides and bromides at 193 nm has also been used to form RO_2 radicals. Thus, CF_2ClBr, CCl_4, and $CFCl_3$ have been used to generate CF_2ClO_2, CCl_3O_2, and $CFCl_2O_2$, respectively [19].

Photolysis of CF_3NO by visible light has been used to produce CF_3 radicals and CF_3O_2 in the presence of O_2 [20].

Carbonyl compounds, e.g. acetone and halogenated acetones, can be photolyzed to give R radicals and CO and hence, peroxy radicals in the presence of O_2 [21–23].

$$CH_3COCH_3 + h\nu \longrightarrow 2CH_3 + CO \quad (9)$$

Photolysis at 193 nm of 1,5-hexadiene is a source of the allyl radical, $CH_2{=}CHCH_2$ [24]:

$$1,5\text{-}C_6H_{10} + h\nu \longrightarrow 2C_3H_5 \quad (10)$$

The thermochemistry of the corresponding RO_2 radical has also been studied using this method [25].

Finally, photolysis of aldehydes produces R and HCO radicals:

$$RCHO + h\nu \longrightarrow R + HCO \quad (11)$$

Since HCO radicals react with O_2 to produce HO_2, aldehyde photolysis is a convenient method of producing RO_2 and HO_2 simultaneously for the investigation of (RO_2 + HO_2) reactions [26].

2.2 ABSTRACTION

The most frequently used method of generating R radicals is by hydrogen-atom abstraction:

$$RH + X \longrightarrow R + HX \quad (12)$$

To generate only one isomer of R, RH needs to have either chemically identical H atoms, e.g. CH_4, C_2H_6, $C(CH_3)_4$, or one significantly more reactive H atom, e.g. CH_3CHO. If neither of these conditions are fulfilled, a mixture of radical isomers will be obtained. There are several candidates for X, e.g. F,

Cl, Br, OH, NO_3, O(^{1}D), O(^{3}P) and H atoms. F and Cl atoms are used frequently to generate RO_2 radicals by abstraction of H atom from RH in the presence of O_2. F and Cl atoms may be produced in a variety of photolysis and discharge flow systems [27–30]:

$$F_2 \text{ (or } Cl_2) + h\nu \longrightarrow 2F \text{ (or 2Cl)} \quad (13)$$

F atoms can also be produced by pulse radiolysis of SF_6 [31,32]:

$$SF_6 + 2\text{ MeV } e^- \longrightarrow F + \text{products} \quad (14)$$

As a result of their ease of generation and rapid reaction with most organic compounds, F and Cl atoms are by far the most widely used precursors for RO_2 radicals.

Bromine atoms can be produced by photolysis, microwave discharge or thermal decomposition of Br_2 and used to abstract aldehydic hydrogen atoms. This method has been used to generate $CH_3C(O)O_2$ [33] and $C_2H_5C(O)O_2$ radicals [34].

The use of OH radicals for H-atom abstraction is somewhat problematic. The abstraction reactions involving OH radicals are normally slower than the corresponding F- and Cl-atom reactions, and clean sources of OH radicals are scarce. Photolysis of H_2O_2 in the presence of C_2H_6 has been used to generate $C_2H_5O_2$ radicals [35].

Photolysis of HNO_3 can be used to simultaneously produce RO_2 and NO_3 radicals, and hence study their reaction [36]:

$$HNO_3 + h\nu \longrightarrow OH + NO_2 \quad (15)$$

$$OH + HNO_3 \longrightarrow H_2O + NO_3 \quad (16)$$

There are no reports in the literature on the use of O(^{1}D) for the production of organic peroxy radicals. O(^{1}D) atoms were used in one study of the reaction of HO_2 with IO [37].

Abstraction reactions using H and O(^{3}P) atoms are normally too slow at room temperature to be used as a source of RO_2 radicals. There is one study reported in the literature using O(^{3}P) atoms generated from the photolysis of O_2 to generate CH_3O_2 and HO_2 [38].

2.3 ADDITION

Addition of an atom or a radical to an unsaturated carbon–carbon bond can be used to create a new radical:

$$RC{=}CR' + X + M \longrightarrow RXC{-}C^{\cdot}R' + M \quad (17)$$

This is not only a useful method in the laboratory, but a very important pathway in the atmosphere, where OH and NO_3 radicals add to unsaturated compounds and the subsequent reaction with atmospheric O_2 leads to RO_2 formation. As in the case of abstraction, there are serious limitations to the use of this method. In general, cyclic compounds and compounds with more than one unsaturated carbon–carbon bond will again lead to the formation of a mixture of isomeric RO_2 radicals. In addition, care must be taken to ensure that the reaction of X with the unsaturated compound does not have a substantial abstraction channel. However, careful choice of RC=CR′ and X can lead to useful methods of preparing RO_2 radicals.

Addition of OH to C_2H_4 has been used to study $HOCH_2CH_2O_2$ radicals [33,39]. Addition of H atoms, from pulse radiolysis of H_2, to C_2H_4 and propene was used to study $C_2H_5O_2$ [40] and i-$C_3H_7O_2$ radicals [41], respectively. $ClCH_2CH_2O_2$ radicals have been prepared by adding Cl atoms to C_2H_4 in the presence of O_2 [42].

3 METHODS OF FORMING RO_2 RADICALS DIRECTLY

3.1 RO_2NO_2 DECOMPOSITION

The equilibrium between RO_2 radicals, NO_2 and peroxynitrates:

$$RO_2NO_2 + M \longleftrightarrow RO_2 + NO_2 + M \tag{18}$$

provides a source of RO_2 radicals which does not require the presence of O_2. For this method to be useful the peroxy nitrate must be synthesized conveniently. This method has been used to generate $CH_3C(O)O_2$ [43] and $C_6H_5C(O)O_2$ [44] radicals have been generated using this method. More often RO_2 is generated in the presence of NO_2 to study the RO_2NO_2 compounds [45].

3.2 ROOH + X $\longrightarrow$ RO_2 + XH

Reaction (19) is a potential source of RO_2 radicals. For example, H_2O_2 can be used as a source of HO_2 radicals by using X=F, Cl or OH:

$$ROOH + X \longrightarrow RO_2 + XH \tag{19}$$

For higher hydroperoxides, several pathways for abstraction exist. In the case of methylhydroperoxide:

$$CH_3OOH + X \longrightarrow CH_3O_2 + HX \tag{19a}$$

$$CH_3OOH + X \longrightarrow HCHO + OH + HX \tag{19b}$$

For X=OH, both channels (19a) and (19b) are significant, with channel (19a) accounting for 70% of the reaction [46,47]. Chlorine has been reacted with t-C_4H_9OOH [48]:

$$Cl + t\text{-}C_4H_9OOH \longrightarrow t\text{-}C_4H_9OO + HCl \tag{20a}$$

$$Cl + t\text{-}C_4H_9OOH \longrightarrow CH_2(CH_3)_2COOH + HCl \tag{20b}$$

In this case, 35% of the reaction proceeds through channel (20a). However, in the absence of O_2 the radical produced in channel (20b) regenerates Cl atoms by reaction with the molecular chlorine used in the photolysis system:

$$CH_2(CH_3)_2COOH + Cl_2 \longrightarrow ClCH_2(CH_3)_2COOH + Cl \tag{21}$$

A major disadvantage of using hydroperoxides is, that even though they can be synthesized readily they tend to be thermally unstable and explosive.

REFERENCES

1. E.W. Kaiser, I.M. Lorkovic and T.J. Wallington, *J. Phys. Chem.* **94**, 3352.
2. I.R. Slagle, Q. Feng and D. Gutman, *J. Phys. Chem.* **88**, 3648 (1984).
3. A.F. Wagner, I.R. Slagle, D. Sarzynski and D. Gutman, *J. Phys. Chem.* **94**, 1853 (1990).
4. J.G. Calvert and J.N. Pitts, *Photochemistry*, John Wiley & Sons, New York, (1966).
5. P.L. Hanst and J.G. Calvert, *J. Phys. Chem.* **63**, 71 (1959).
6. D.A. Parkes, *Proceedings of the 15th International Symposium on Combustion, Tokyo, 1974*, The Combustion Institute, Pittsburgh, PA (1974) p. 795.
7. W.G. Alcock and B. Mills, *Combust. Flame* **24**, 125 (1975).
8. J. Weaver, J. Meagher, R. Shortridge and J. Heicklen, *J. Photochem.* **4**, 341 (1975).
9. C.S. Kan, J.G. Calvert and J.H. Shaw, *J. Phys. Chem.* **84**, 3411 (1980).
10. H. Niki, P.D. Maker, C.M. Savage and L.P. Breitenbach, *J. Phys. Chem.* **85**, 877 (1981).
11. H. Niki, P.D. Maker, C.M. Savage and L.P. Breitenbach, *J. Phys. Chem.* **86**, 3825 (1982).
12. C. Anastasi, D.J. Waddington and A. Woolley, *J. Chem. Soc. Faraday Trans 1.* **79**, 505 (1983).
13. L.J. Kirsch, D.A. Parkes, D.J. Waddington and A. Woolley, *J. Chem. Soc. Faraday Trans. 1* **75**, 2678 (1979).
14. L.T. Crowley, D.J. Waddington and A. Woolley, *J. Chem. Soc. Faraday Trans. 1* **78**, 2535 (1982).
15. L.J. Kirsch and D.A. Parkes, *J. Chem. Soc. Faraday Trans. 1* **77**, 293 (1981).
16. T.J. Wallington and J.C. Ball, *Chem. Phys. Lett.* **234**, 187 (1995).
17. S. Yamashita and T. Hayakawa, *Bull. Chem. Soc. Jpn.* **46**, 2290 (1973).
18. M.E. Jenkin and R.A. Cox, *J. Phys. Chem.* **95**, 3229 (1991).
19. A.M. Dognon, F. Caralap and R. Lesclaux, *J. Chim. Phys.* **82**, 349 (1985).
20. J. Chen, T. Zhu, H. Niki and G.J. Mains, *Geophys. Res. Lett.* **19**, 2215 (1992).
21. M. Keiffer, M.J. Pilling and M. Smith, *J. Phys. Chem.* **91**, 6028 (1987).
22. R. Lesclaux and F. Caralp, *Int. J. Chem. Kinet.* **16**, 1117 (1984).

23. R.W. Carr, D.G. Peterson and F.K. Smith, *J. Phys. Chem.* **90**, 607 (1986).
24. J.M. Tulloch, M.T. Macpherson, C.A. Moergan and M.J. Pilling, *J. Phys. Chem.* **86**, 3812 (1982).
25. I.R. Slagle, E. Ratajczak and D. Gutman, *J. Phys. Chem.* **90**, 407 (1986).
26. G.K. Moortgat, R.A. Cox, G. Schuster, J.P. Burrows and G.S. Tyndall, *J. Chem. Soc. Faraday Trans. 2* **85**, 809 (1989).
27. C.J. Howard, *J. Phys. Chem.* **83**, 3 (1979).
28. L. Elmaimouni, R. Minetti, J.P. Sawerysyn and P. Devolder, *Int. J. Chem. Kinet.* **25**, 399 (1993).
29. T.J. Wallington and S. Japar, *Chem. Phys. Lett.* **167**, 513 (1990).
30. T.J. Wallington, L.M. Skewes, W.O. Siegl, C.H. Wu and S.M. Japar, *Int. J. Chem. Kinet.* **20**, 867 (1988).
31. O.J. Nielsen, J. Munk, G. Locke and T.J. Wallington, *J. Phys. Chem.* **95**, 8714 (1991).
32. P. Pagsberg, O.J. Nielsen and C. Anastasi, in *Spectroscopy in Environmental Science*, edited by R.J.H. Clark and R.E. Hester, John Wiley & Sons, Chichester (1995), p. 263.
33. H. Niki, P.D. Maker, C.M. Savage and L.P. Breitenbach, *Int. J. Chem. Kinet.* **17**, 525 (1985).
34. I. Bridier, F. Caralp, H. Loirat, R. Lesclaux, B. Veyret, K.H. Becker, A. Reimer and F. Zabel, *J. Phys. Chem.* **95**, 3594 (1991).
35. T.P. Murrells, M.E. Jenkin, S.J. Shalliker and G.D. Hayman, *J. Chem. Soc. Faraday Trans.* **87**, 2351 (1991).
36. J.N. Crowley, J.P. Burrows, G.K. Moortgat, G. Poulet and G. Le Bras, *Int. J. Chem. Kinet.* **22**, 673 (1990).
37. M.E. Jenkin, R.A. Cox and G.D. Hayman, *Chem. Phys. Lett.* **177**, 272 (1991).
38. P.D. Lightfoot, P. Roussel, F. Caralp and R. Lesclaux, *J. Chem. Soc. Faraday Trans.* **87**, 3213 (1991).
39. C. Anastasi, V. Simpson, J. Munk and P. Pagsberg, *J. Phys. Chem.* **94**, 6327 (1990).
40. J. Munk, P. Pagsberg, E. Ratajczak and A. Sillesen, *J. Phys. Chem.* **90**, 2752 (1986).
41. J. Munk, P. Pagsberg, E. Ratajczak and A. Sillesen, *Chem. Phys. Lett.* **132**, 417 (1986).
42. P. Dagaut, T.J. Wallington and M.J. Kurylo, *Chem. Phys. Lett.* **146**, 589 (1988).
43. E.C. Tuazon, W.P.L. Carter and R. Atkinson, *J. Phys. Chem.* **95**, 2434 (1991).
44. R.A. Kenley and D.G. Hendry, *J. Am. Chem. Soc.* **104**, 220 (1982).
45. T.J. Wallinton, J. Sehested, and O.J. Nielsen, *Chem. Phys. Lett.* **226**, 563 (1994).
46. H. Niki, P.D. Maker, C.M. Savage and L.P. Breitenbach, *J. Phys. Chem.* **87**, 2190 (1983).
47. G.L. Vaghjiani and A.R. Ravishankara, *J. Phys. Chem.* **93**, 1948 (1989).
48. P.D. Lightfoot, P. Roussel, B. Veyret and R. Lesclaux, *J. Chem. Soc. Faraday Trans.* **84**, 2927 (1990).

3 Structure of Organic Peroxyl Radicals

L. R. C. BARCLAY
Mount Allison University, Sackville, Canada

1 CARBON-BASED PEROXYL RADICALS

Interest in the structure and properties of peroxyl radicals has continued unabated since an earlier review appeared more than 25 years ago [1]. The significance of oxygen-centred radicals in biological systems has had a remarkable effect on the activity and interest in this field. In this present chapter, structural information on organic peroxyl radicals provided by (1) ground state examination, e.g. dipole moment measurements, (2) electron spin resonance (ESR) and ultraviolet–visible spectra, and (3) quantum chemical calculations, will be reviewed selectively where insight is provided on the ground state structure. Detailed spectroscopic properties are discussed elsewhere in this book. The intrinsic structure of the R—O—O$^{\bullet}$ group and its bond properties result in some unique dynamic behaviour. Rearrangements of allyl peroxyls and within cumyl peroxyl will be used to illustrate this dynamic property for some of these radicals. The structures and dynamics of *lipid* peroxyl radicals are discussed briefly in the context of their stereochemistry, diffusion dynamics, and mechanism of peroxidation and antioxidant action in lipid membranes.

1.1 THE ELECTRONIC STRUCTURE

Peroxyl radicals possess a π-type electronic structure which can be represented by the simple p-orbital formulae shown in Figure 1. Such a structure is not unexpected since peroxyls are usually formed by the combination of a σ-carbon radical and ground state O_2 which has two unpaired electrons in its π-orbitals. The structures of various peroxyl radicals have been examined by experimental (e.g. spectroscopic) and extensive theoretical methods. Information bearing on the electronic structure of peroxyl radicals will be reviewed briefly in this section. Some details of the structures and significant related properties will be given for typical examples of known organic peroxyl radicals in the following sections on saturated and unsaturated hydrocarbon peroxyl radicals.

Peroxyl Radicals. Edited by Z. B. Alfassi

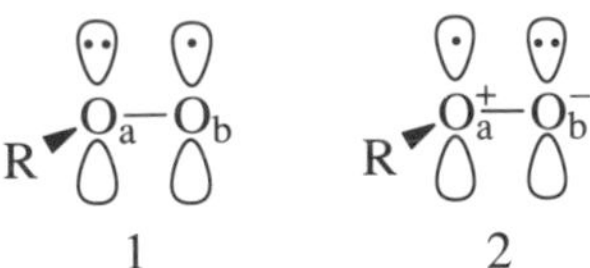

Figure 1 The p-orbital structures of a peroxyl radical, where the relative contributions of **1** and **2** and resulting properties; e.g. spin density distribution, and dipole moment, depend on the nature of R (from Ref. 2).

1.1.1 Spin density

The valence bond structures **1** and **2** of Figure 1 imply a partitioning of the unpaired electron between the p orbitals of the inner oxygen 'a', and the outer oxygen, 'b'. The hyperfine coupling constants from ESR spectra of ^{17}O-enriched peroxyl radicals with various R groups indicate that the spin density on the terminal oxygen averages approximately twice that on the inner oxygen [2–4], although theoretical studies predict that a higher proportion of spin density resides on the terminal oxygen [5–7].

1.1.2 Polarity

Peroxyl radicals are highly polarized. Calculations using semi-empirical methods gave values for the dipole moments of around 2 D for simple alkylperoxyl radicals [8] and a value of 2.6 D from spin density considerations [9]. The first experimental measurement, carried out by Fessenden, Hitachi and Nagarajan by microwave dielectric absorption on the benzylperoxyl radical, gave a dipole moment of 2.4 ± 0.2 D [10], confirming the earlier predictions. Subsequent higher-order calculations of the dipole moments of alkyl-, cyclopropyl-, and alkenylperoxyl radicals gave dipole moments which were all in the range from 2.0–2.6 D for various conformations of these radicals, whereas the result obtained for the alkynylperoxyl radical was 1.41 D [7]. Calculations of charge distribution between the two peroxyl oxygens of these radicals showed that there is greater negative charge on the *inner* oxygen (O_a), with none of the radicals having any significant negative charge on the *terminal* oxygen (O_b) [7]. This result appears surprising in view of the shift in π-electron density and of spin density on to the outer oxygen which is implied in Figure 1. However, the apparent ambiguity was obviated by indicating that there is a corresponding shift in sigma (σ) electron density in the direction from O_a towards O_b because adjacent R groups are electron donating relative to oxygen. Thus it is the total electron density which determines the magnitude and orientation of the polarity of peroxyl radicals. The result of these considerations gave the calculations for orientations of the dipole moments for typical classes of peroxyl radicals shown schematically in Figure 2 [7].

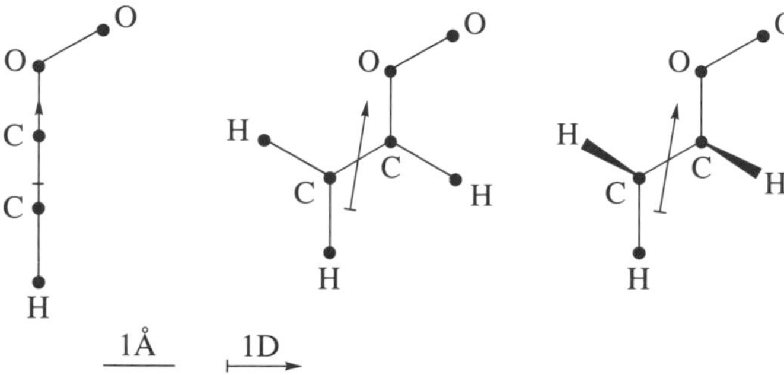

Figure 2 A schematic representation of the magnitude and orientation of the dipole moment vectors of the ethylperoxyl, ethenylperoxyl, and ethynylperoxyl radicals. In each case the length of the vector, denoted by the bold arrow, is proportional to the magnitude of the dipole moment (from Ref. 7).

1.1.3 Effect of substituents

Electron-withdrawing substituents are known to increase the chemical reactivity of alkylperoxyl- [2,11] and aralkylperoxyl radicals [12]. Sevilla, Becker and Yan rationalized this effect in terms of the effect of electronegative groups on 'hyperfine coupling and radical structure' [2]. They observed an increase in hyperfine coupling on the terminal oxygen while the coupling decreased on the inner oxygen as substituents became more electronegative. A linear relationship was obtained between the Taft σ^* parameter and ^{17}O hyperfine couplings on the terminal oxygen for a series of substituted alkylperoxyl radicals. The effects on spin density and reactivity were interpreted in terms of the effect of the electronegative substituents at R on the contributions of the p-orbital structures **1** and **2** (Figure 1). As R becomes more electron-attracting it was postulated that structure **2** becomes less important due to a build up of positive charge. Consequently, structure **1** becomes more important, resulting in a higher spin density at the terminal oxygen and higher reactivity at this site [2].

1.2 ALKYLPEROXYL RADICALS (CONFORMATIONAL GEOMETRY)

The simplest alkylperoxyl radical, CH_3O_2, is a bent radical, similar to HO_2 which has been studied extensively [6, and references therein]. The smallest organic peroxyl, methylperoxyl, has been studied in detail by *ab initio* calculations [5–7]. Two minima conformations, referred to as *cis*- and *trans*- [5], are given in Figure 3. The calculated geometry for this radical (Table 1), carried out by two independent groups, shows excellent agreement. The small energy difference between the two conformers, ca. 5 kJ mol^{-1}, indicates that it would behave as a free rotor.

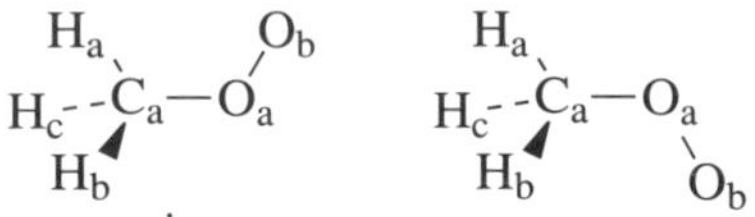

Figure 3 *Cis*- and *trans*- forms of the methylperoxyl radical; calculated properties are given in Table 1.

Table 1 Structures and relative energies of some alkylperoxyl radicals[a]

Radical	Bond length (Å)	Bond angle (deg)	Dipole moment (D)	Relative energy (kJ/mol^{-1})
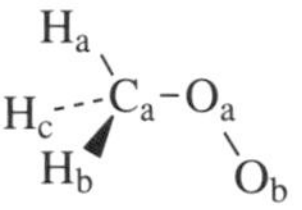 H_a, H_c--C_a-O_a, H_b, O_b *trans*	H_aC_a 1.080 (1.0797) H_bC_a 1.081 (1.081) C_aO_a 1.416 (1.4166) O_aO_b 1.301 (1.3011)	$H_aC_aO_a$ 105.6 (105.6) $H_bC_aO_a$ 110.0 (110.0) $H_aC_aH_b$ 110.4 $H_bC_aH_c$ 110.3 $C_aO_aO_b$ 110.9 (110.9)	2.45	0
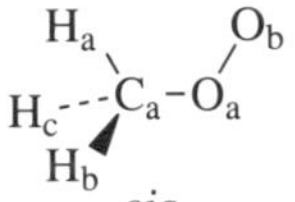 H_a, O_b, H_c--C_a-O_a, H_b *cis*	H_aC_a 1.079 (1.0789) H_bC_a 1.081 (1.0811) C_aO_a 1.421 (1.4213) O_aO_b 1.301 (1.2979)	$H_aC_aO_a$ 109.3 (109.1) $H_bC_aO_a$ 108.2 (108.1) $H_aC_aH_b$ 109.0 $H_bC_aH_c$ 107.4 $C_aO_aO_b$ 112.5 (112.4)	2.49	5.6 (5.4)
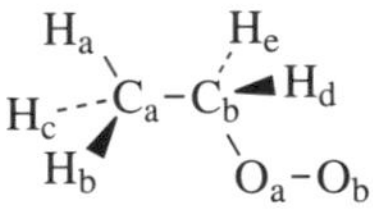 H_a, H_e, H_c--C_a-C_b-H_d, H_b, O_a-O_b	H_aC_a 1.084 H_bC_a 1.084 C_aC_b 1.514 C_bH_d 1.083 C_bO_a 1.401 O_aO_b 1.301	$H_aC_aC_b$ 109.8 $H_bC_aC_b$ 110.8 $H_aC_aH_b$ 108.7 $H_bC_aH_c$ 108.1 $C_aC_bH_d$ 111.6 $C_aC_bO_a$ 407.8 $H_dC_bH_a$ 107.3 $H_dC_bO_a$ 109.3 $C_bO_aO_b$ 111.3	2.53	0
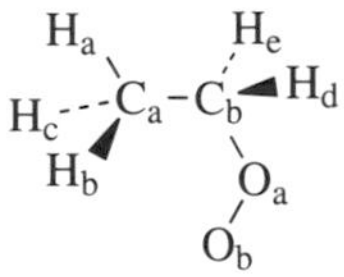 H_a, H_e, H_c--C_a-C_b-H_d, H_b, O_a, O_b	H_aC_a 1.085 H_bC_a 1.083 C_aC_b 1.516 C_bH_d 1.083 C_bO_a 1.435 O_aO_b 1.294	$H_aC_aC_b$ 108.4 $H_bC_aC_b$ 111.5 $H_cC_aC_b$ 109.0 $H_bC_aH_c$ 107.4 $C_aC_bH_d$ 111.0 $C_aC_bO_a$ 115.2 $H_dC_bH_e$ 107.9 $H_dC_bO_a$ 105.7 $C_bO_aO_b$ 114.7	2.57	14

[a] Structures and data are taken from Ref. 7; values in parentheses are from Ref. 6, for comparison.

The calculated structural parameters for two important conformers of the ethylperoxyl radical are also shown in Table 1. In this case the barrier to rotation about the C—O—O group is higher (14 kJ mol^{-1}), compared to that in methylperoxyl, apparently due to additional hindrance to rotation provided by the adjacent methyl group. The ESR hyperfine coupling constants of secondary peroxyl radicals are reported to vary with the nature of the R-substituent due to restricted rotation about the C—O bond [13].

Some recent calculations on the effect of fluorine substituent(s) on the properties of the methylperoxyl radical are given in Table 2. The effect of the

Table 2 Structures and calculated properties of some fluoromethylperoxyl radicals[a]

Radical	Spin density	Dipole moment (D)	Relative energy ($kJ\ mol^{-1}$)
H H--C—O_a H O_b	O_a + 0.105 O_b + 0.901	2.66	0
H O_b H--C—O_a H	O_a + 0.120 O_b + 0.891	2.68	3.0
F O H--C—O_a H	O_a + 0.067 O_b + 0.939	2.56	0
F H--C—O_a H O_b	O_a + 0.095 O_b + 0.910	2.36	8.6
F O_b H--C—O_a H	O_a + 0.100 O_b + 0.911	2.66	25.5
H O F--C—O_a F	O_a + 0.067 O_b + 0.939	1.94	0
H F--C—O_a F O	O_a + 0.046 O_b + 0.957	2.03	3
H O_b F--C—O_a F	O_a + 0.081 O_b + 0.928	1.93	1.5
F F--C—O_a F O_b	O_a + 0.041 O_b + 0.962	0.132	0
F O_b F--C—O_a F	O_a + 0.060 O_b + 0.947	0.301	12.0

[a]Susan L. Boyd, unpublished results. Department of Chemistry, Mount St. Vincent University, Halifax, NS (1995).

increasing substitution of fluorine for hydrogen results in a higher relative spin density being transferred to the terminal oxygen (O_b) so that the relative spin density at this site is approximately 95% for the trifluoromethylperoxyl radical. This calculated result supports the conclusion of Sevilla, Becker and Yan [2] regarding the effect of electron-attracting groups on spin density on the peroxyl oxygens. Furtherfore, one would expect the trifluoromethylperoxyl radicals to be very reactive in H-atom abstraction reactions. The fluorine substituents cause very significant reductions in the dipole moments, whereby most of the dipole is cancelled in the trifluoro derivative (Table 2). This is a reflection of the effect of the electronegative fluorine in reducing the contribution of the polar structure **2** (Figure 1) to the radical.

Larger hydrocarbon peroxyl radicals are important as motional probes and models for the behaviour of polyperoxyl radicals in polymers such as polyethylene. Of particular interest is the main rotational motion of the C—O—O group in polymer chains [14–20]. This motion has been interpreted as either (1) rotation about the C—O bond axis, as noted in Figure 3 for the methylperoxyl radical or (2) motion of the whole C—O—O group about the axis of the chain in larger molecules [18,19]. Some calculations of hindered rotation in the isopropylperoxyl and 3-pentylperoxyl radicals have estimated the barrier to rotation about the C—O bond in longer chains. For example, by using the structure for the 3-pentylperoxyl radical shown in Figure 4, Sevilla and coworkers have estimated the barrier to internal rotation of the O—O group about the C—O axis to be 39 kcal mol^{-1} [6]. This relatively high rotational barrier was attributed to a close approach of the terminal oxygen of the O—O group (1.5 Å) to a methylene hydrogen two sites along the chain. If applicable to the more complex situation in polymers, such a barrier would be important in determining the main rotational motion of peroxyl radicals in polymers, either C—O bond rotation or chain axis rotation [17].

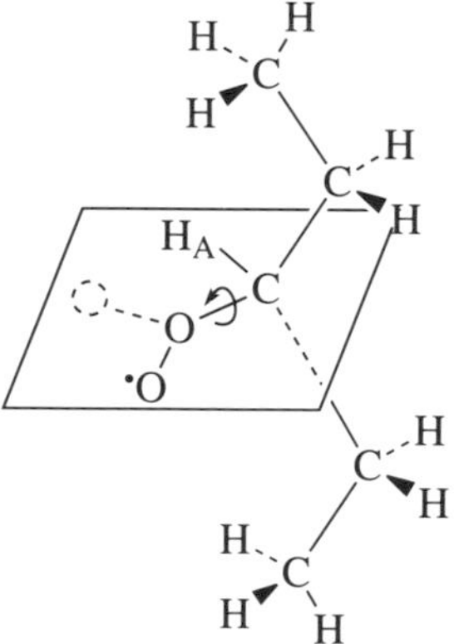

Figure 4 Structure of the 3-pentylperoxyl radical showing the *cis*- and *trans*- forms; the plane shown contains the H_ACOO group and is perpendicular to the plane that contains the five carbon atoms (from Ref. 6).

1.3 UNSATURATED HYDROCARBON PEROXYL RADICALS

1.3.1 Ethenylperoxyl and ethynylperoxyl radicals

Some of the structural parameters that have been calculated [6] for the ethenylperoxyl and the ethynylperoxyl radicals are given in Table 3. The C—O bond length for oxygen bonded to a saturated carbon (1.42–1.44 Å, Table 1) decreases for the ethenylperoxyl radical to 1.38 Å, and to 1.32 Å for the ethynylperoxyl radical, consistent with an expected trend of decreasing bond length with increasing s-character of the bonding orbital. The O—O bond length is essentially constant (1.30 Å) for the structures studied, with the exception being the ethynylperoxyl radical where the value deviated slightly (1.32 Å).

Table 3 Structures and relative energies of ethenyl- and ethynylperoxyl radicals

Radical	Bond length (Å)	Bond angle (deg)	Dipolemoment (D)	Relative energy (kJ mol⁻¹)
$H_a(H_b)C_a{=}C_b(H_c)O_a{-}O_b$	H_aC_a 1.072 H_bC_a 1.074 C_aC_b 1.316 C_bH_c 1.071 C_bO_a 1.375 O_aO_b 1.298	$H_aC_aC_b$ 119.8 $H_aC_aH_b$ 118.5 $H_bC_aC_b$ 121.7 $C_aC_bH_c$ 126.3 $C_aC_bO_a$ 120.2 $H_cC_bO_a$ 113.5 $C_bO_aO_b$ 112.6	2.06	0
$H_a(H_b)C_a{=}C_b(H_c)O_a{-}O_b$	H_aC_a 1.073 H_bC_a 1.071 C_aC_b 1.317 C_bH_c 1.071 C_bO_a 1.376 O_aO_b 1.300	$H_aC_aC_b$ 118.8 $H_aC_aH_b$ 118.5 $H_bC_aC_b$ 122.7 $C_aC_bH_c$ 123.9 $C_aC_bO_a$ 126.8 $H_cC_bO_a$ 108.3 $C_bO_aO_b$ 114.6	2.01	3
$H_a{-}C_a{\equiv}C_b{-}O_a{-}O_b$	H_aC_a 1.056 C_aC_b 1.180 C_bO_a 1.318 O_aO_b 1.322	$H_aC_aC_b$ 180.0 $C_aC_bO_a$ 180.0 $C_bO_aO_b$ 112.1	1.41	-

These unsaturated peroxyl radicals display high chemical reactivity and remarkable spectroscopic properties which are the subject of considerable interest and discussion [21–23]. Vinylperoxyl radicals and various chloro derivatives exhibit a characteristic broad absorption in the *visible* region (500–600 nm) [21]. The transition causing this absorption is made possible by conjugative interaction of the unpaired π-electron of the peroxyl group with the π-system of the double bond, which provides a low-energy excited state [23]. Additional conjugation, for example in the 2-phenylvinylperoxyl radical, extends the absorption to near 700 nm [22]. Corresponding spectra for the alkynylperoxyls have not yet been reported.

Vinylperoxyls are reported to be more reactive than similarly substituted alkylperoxyls and electron-attracting groups increase the reactivity [22]. These effects are probably similar in origin to those found for substituted alkylperoxyls [2].

The allylperoxyl radical

Unsaturated peroxyl radicals containing an adjacent double bond, e.g. the allylperoxyl radical and its derivatives, are of particular significance. Allylperoxyls are among the intermediates formed during the peroxidation of unsaturated lipid membranes, both from polyunsaturated fatty acid (PUFA)

Table 4 Conformers and relative energies of the allylperoxyl radical

Conformer	Dipole moment (D)	Relative energy ($kJ\,mol^{-1}$)
1	2.51	0
2	2.34	3
3	2.42	36
4	2.39	8
5	2.43	27

chains and cholesterol. In addition the structure and reactions of allylperoxyl radicals have received a great deal of attention over several decades.

A detailed theoretical study was made of the simplest member, the allylperoxyl radical [7,24]. The important minima conformations and relative energies are shown in Table 4. These result from rotation about the C_c—O_a bond (**2** ⟶ **3**), the C_b—C_c bond (**3** ⟶ **4**) and about the C_c—O_a bond (**4** ⟶ **5**). Steric interactions between the terminal oxygen of the C—O—O group and an alkene hydrogen, H_b or H_c, raise the relative energy of conformers **3** and **5**, respectively.

The structural rearrangement of allylic hydroperoxides occurs through intermediate peroxyl radicals. This acyclic (3, 2) peroxyl radical rearrangement has received a lot of attention since its discovery almost four decades ago by Schenck and coworkers [25] in the rearrangement of the C-5 α-hydroperoxide of cholesterol to the C-7 α-hydroperoxide isomer. The mechanism of this remarkable rearrangement has been studied by at least three independent groups and is still open to debate. Mechanisms proposed for the rearrangement are outlined briefly in Scheme 1. Three mechanisms have been proposed by various research groups, including a cyclic intermediate, **1**, a transition state, **2**, or β-scission to form an allylic radical, **3**, followed by recombination with oxygen.

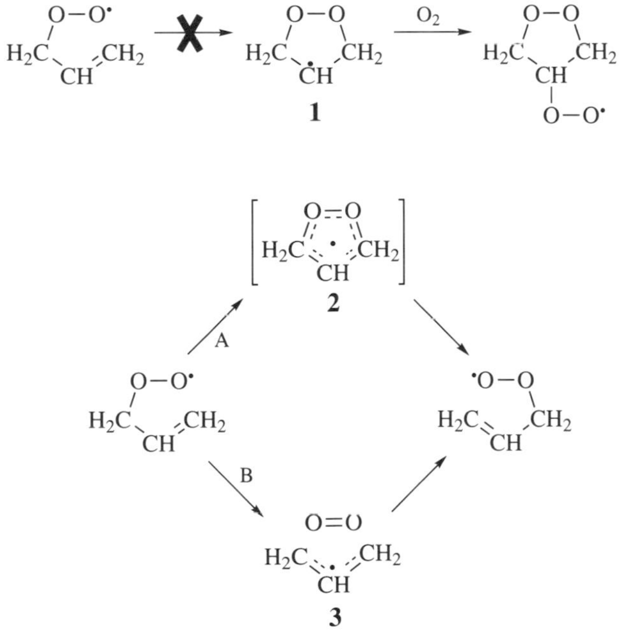

Scheme 1
(adapted from Ref. 24, from original drawings provided by N.A. Porter)

The cyclic carbon-centred radical, proposed initially by Brill [26], was found *not* to be an intermediate for this rearrangement, since neither ordinary molecular oxygen nor ^{18}O-enriched oxygen could be trapped during the course of the rearrangement [27,28]. An intermediate of type **1**, generated by a different method, is known to be trapped by molecular oxygen [29]. Beckwith *et al.* [30] also showed that the allylic rearrangement of the C-5 α-hydroperoxide of cholesterol [25] proceeds *without* incorporation of ^{18}O in the presence of $^{18}O_2$. In addition, ESR spectra showed that separate peroxyl radicals exist in the system, rather than a common intermediate of type **1** [13,30]. In 1987, Porter and Wujek discussed the results published up to that date in terms of a concerted rearrangement of allylic peroxyl radicals passing through a five-membered transition state (e.g. **2** in Scheme 1) [28], and this work was later supported by Beckwith *et al.* [30]. This mechanism received strong support by the results obtained by Porter, Kaplan and Dussault on the rearrangement of chiral allylic hydroperoxides, which were shown to undergo rearrangement through the allylic peroxyl radicals with a high degree of stereoselectivity, with the product having the opposite configuration to the starting material [31], as summarized in Scheme 2.

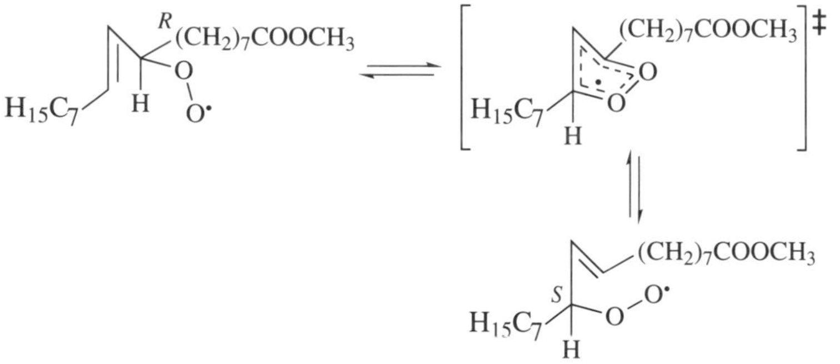

Scheme 2

(adapted from Ref. 31, from original drawings provided by N.A. Porter)

Nevertheless, the mechanism was opened for further examination and debate when theoretical investigations of allylperoxyl radicals failed to find a concerted transition state for the rearrangement, but rather proposed a dissociative process involving an allyl radical intermediate–oxygen pair in a solvent-cage reaction [24], as represented in Scheme 3. Recently, Porter and coworkers provided convincing experimental evidence for such an intermediate allyl radical–dioxygen caged pair in this allylperoxyl rearrangement by employing a combination of solvent viscosity, isotopically labelled oxygen and stereochemistry [32].

The allylperoxyl radical structure is an important component formed during lipid peroxidation and it is of interest to examine the behaviour of this

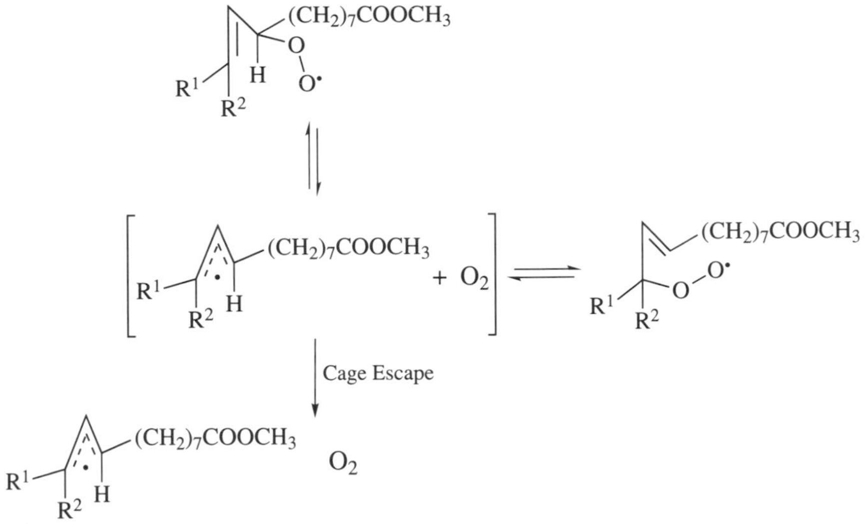

Scheme 3
(adapted from Ref. 32a, from original drawings provided by N.A. Porter)

intermediate when generated on a lipid chain. This has recently been carried out for the peroxidation of oleate where the mechanism has recently been re-examined in the light of the behaviour of intermediate peroxyl radicals [33]. Dienylperoxyl radicals are also important intermediates in lipid peroxidation [34], and these radicals are known to rearrange with incorporation of atmospheric oxygen [34,35]. The difference in behaviour of the conjugated dienylperoxyl radical compared to the allylperoxyl species can now be understood in terms of the increased stability of the conjugated pentadienyl carbon radical so that oxygen readily diffuses from the initial solvent cage.

1.4 ARYLPEROXYL RADICALS

There has been considerable interest recently in the phenylperoxyl radical, mainly concerning the kinetics of formation from the phenyl radical [36–39], and some interesting results are emerging pertaining to the structures of arylperoxyls. Calculations indicate a strong polarization of the phenylperoxyl radical, with a dipole moment of 3.21 D [23]. Such an enhanced dipole moment might be attributed to delocalization of the unpaired electron into the π-aromatic system from the inner oxygen (e.g. via structure **2**, Figure 1), thus resulting in increased charge separation. However, the vinylperoxyl radical does not exhibit such an enhanced dipole moment (see Tables 1–3), although the visible absorption spectra of vinylperoxyls suggest conjugative interaction between the —O—O group and the double bond [21–23].

The phenylperoxyl radical exhibits broad visible absorption in the 450–500 nm region and this has been the subject of some discussion [36–39].

This absorption is also observed in other conjugated peroxyl radicals, such as vinylperoxyl [21–23]. The precise interpretation of this band is yet to be determined. It probably arises from a low-energy transition between the π-type open-shell orbital on the oxygen atoms in the ground state which then couples to a π-orbital or conjugated aromatic π-orbital in the excited state, as has been suggested [23]. This general interpretation is consistent with the bathochromic shift caused by electron-donating *para*-groups [39] and the shift caused by extended aromatic systems; 1-naphthylperoxyl absorbs at around 650 nm and 9-phenanthrylperoxyl at around 700 nm [38].

Arylperoxyl radicals are surprisingly reactive, for example in one-electron oxidations of ionized phenols, ascorbate and chlorpromazine, arylperoxyls are reported to be orders of magnitude more reactive than unsubstituted alkylperoxyls [37,38]. Electron-withdrawing *para*-substituents increase the rate constants, whereas electron-donating substituents decrease the rate constants, providing a linear Hammett correlation for substituent effects in one-electron oxidations [37]. Arylperoxyls with extended π-systems are more reactive than phenylperoxyl in these reactions [38]. These effects on reactivity (as well as the visible spectra) indicate a strong interaction between the aromatic π-system and the unpaired peroxyl π-electron. Thus the effect of an extended π-system on reactivity can be attributed to a compression of the energy levels with a consequent reduction of the energy difference between the singly occupied molecular orbital (SOMO) from the peroxyl group and the penultimate occupied molecular orbital. In addition, a reduction in the energy between the SOMO and the continuum level (ionization potential) is expected, and as the ionization potential decreases, the electron affinity increases [38]. Polycyclic arylperoxyl radicals might act as reactive H-atom abstractors, and in this way initiate damaging peroxidation reactions in lipid membranes.

The interaction of the unpaired electron of the peroxyl group with the aromatic π-system could result in an actual structural change. Quantum chemical calculations indicate a possible internal reaction of the unpaired electron on the terminal oxygen through formation of a spirodioxiranyl radical [40].

1.5 ARYLALKYLPEROXYL RADICALS

Arylalkylperoxyl radicals of the benzylperoxyl and cymylperoxyl classes exhibit some significant properties which distinguish them from the simple saturated peroxyl radicals. For example, the benzylperoxyl radical exhibits ultraviolet absorption, compared to that of alkylperoxyls, indicative of 'coupling' between the O—O chromophore and the π-system of the aromatic ring [41]. Actually this is *not* surprising in view of the interesting observation that the benzyloxyl and cumyloxyl radicals have strong absorptions in the *visible* region [42]. Indeed, if the arylcarbinyloxyl radicals exhibit such absorption due to internal charge transfer of the type

$$\overset{\delta^+}{[C_6H_5}\overset{\delta^-}{CH_2O]},$$

operating through the saturated carbon centre as suggested [42], it would be expected that the corresponding peroxyl radicals might also absorb in the visible region; this, however, has not been observed to date.

Cumylperoxyl radicals exist in equilibrium with cumyl radicals and oxygen at room temperature according to isotopic oxygen exchange studies [43]. Recently, cumylhydroperoxide was synthesized specifically labelled with ^{18}O at the terminal oxygen, and some significant new observations were reported on the derived labelled peroxyl radicals [44]. In particular the 'internal scrambling' (rotation) of the ^{18}O label from the outer position to the inner position was measured as well as external exchange with atmospheric O_2. These reactions were found to depend on solvent viscosity so that a significant cage effect was discovered for the fragmentation reaction forming the cumyl radical and oxygen [44].

1.6 LIPID PEROXYL RADICALS

Peroxyl radicals formed on unsaturated lipid chains during lipid peroxidation are chain-propagating species in the well known autoxidation cycle. Such free-radical processes in biomembranes are implicated in various degenerative diseases, such as cancer, arthritis, heart disease and the ageing process. This more biologically relevant subject is discussed elsewhere in this book (e.g. see Chapter 14). The detailed *structures* and *dynamics* of peroxyl radicals formed in biomembranes are likely important factors in these pathological events. Fundamental studies of lipid peroxyl radicals of a physical-organic nature are receiving more prominence in recent years.

1.6.1 Stereochemistry of lipid peroxyls and mechanism

In 1977 two groups independently reported four major isomers from the free-radical autoxidation of methyl linoleate (18:2) [45,46], and Chan and Levett [45] were able to separate the four *regio*-geometric isomers as the corresponding conjugated hydroxy esters, i.e. the 9- and 13-hydroxy *cis*-, *trans*- and *trans*-, *trans*-isomers shown in Figure 5. In contrast, singlet oxygen oxidation of methyl linoleate is known to yield both non-conjugated as well as conjugated isomers in approximately equal amounts [48,49], presumably by the concerted ene reactions rather than through peroxyl radical intermediates.

In a pioneering detailed study, Porter and coworkers related the structure of intermediate lipid peroxyl radicals to the detailed mechanism of free-radical lipid peroxidation. A review of this mechanism is given in Ref. 34.

1

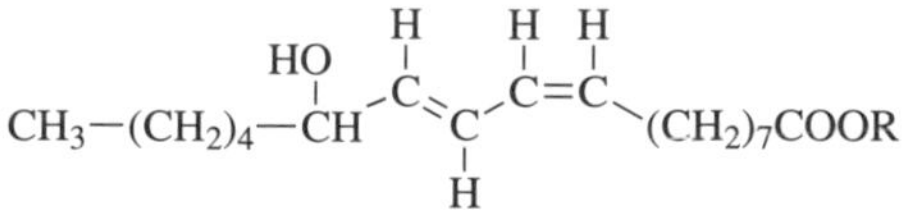

$CH_3-(CH_2)_4-CH(OH)-CH{=}CH-CH{=}CH-(CH_2)_7COOR$

Methyl 13-hydroxy-*cis*-9, *trans*-11-octadecadienoate

2

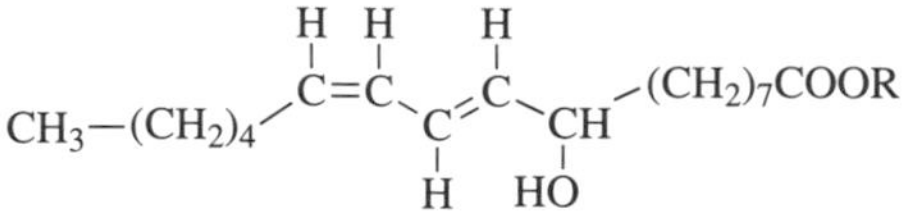

$CH_3-(CH_2)_4-CH{=}CH-CH{=}CH-CH(OH)-(CH_2)_7COOR$

Methyl 9-hydroxy-*trans*-10, *cis*-12-octadecadienoate

3

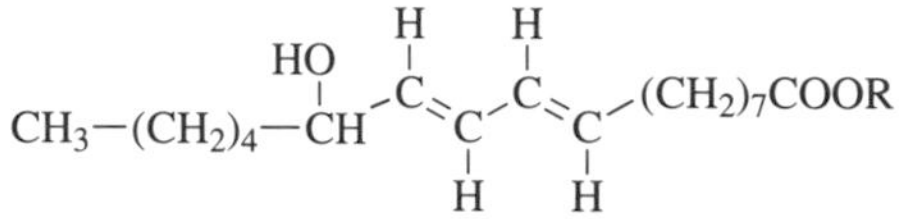

$CH_3-(CH_2)_4-CH(OH)-CH{=}CH-CH{=}CH-(CH_2)_7COOR$

Methyl 13-hydroxy-*trans*-9, *trans*-11-octadecadienoate

4

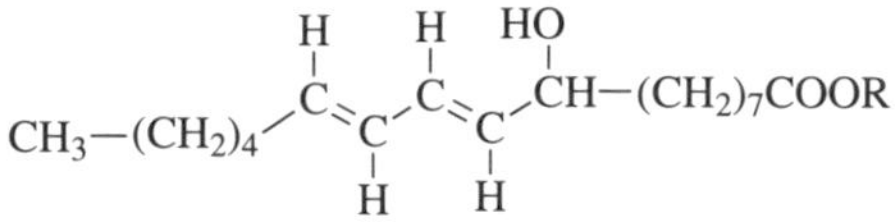

$CH_3-(CH_2)_4-CH{=}CH-CH{=}CH-CH(OH)-(CH_2)_7COOR$

Methyl 9-hydroxy-*trans*-10, *trans*-12-octadecadienoate

A Hydroxy esters for analysis

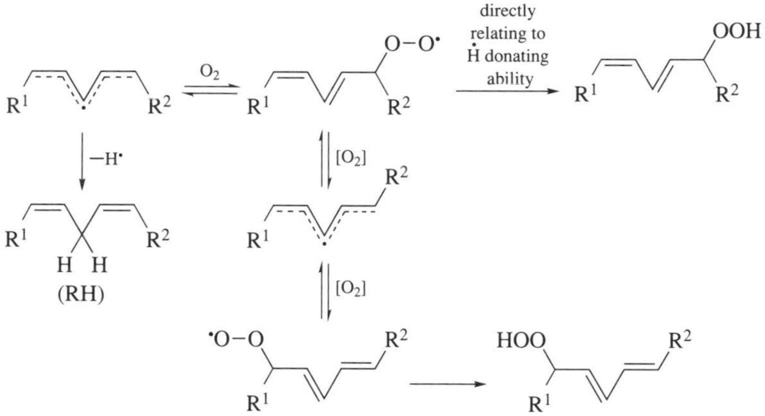

R^1 R^2 O_2 R^1 O–O• R^2 directly relating to H donating ability R^1 OOH R^2

–H•

R^1 R^2 H H (RH)

$[O_2]$ R^2 R^1 $[O_2]$

•O–O R^2 R^1 HOO R^2 R^1

$t, c\ /\ t, t\ =\ a + b \times [RH]$

B

Figure 5 (A) Hydroxy esters: *Cis*-, *trans*- and *trans*-, *trans*- isomers **1–4** prepared by reduction of hydroperoxides derived from linoleate. (B) Mechanistic scheme for peroxidation of linoleate (from Ref. 47).

The 'Porter Mechanism', derived from structure studies, is outlined in part in the scheme shown in Figure 5. The mechanism is based on the discovery of a linear relationship between the ratios of *cis*-, *trans*- to *trans*-, *trans*-geometrical isomers (or kinetic/thermodynamic product ratios) formed and the hydrogen-atom-donating ability of the medium. For example, the ratio of *c*, *t*//*t*, *t* depends directly on the concentration of the bisallylic system used (e.g. linoleate) or on the concentration of an added hydrogen-donor such as a phenolic antioxidant [50]. The competition that accounts for the relative amounts of *cis*-, *trans*- versus *trans*-, *trans*- products is the partitioning of the peroxyl radical initially formed between the corresponding *cis*-, *trans*-product (kinetic) and the β-scission of oxygen which can then lead, after re-oxidation, to the *trans*-, *trans*- product (thermodynamic) (Figure 5). The mechanism was expanded [51] by autoxidation studies of the four isomeric methyl 9,12-octadecadienoates in the presence of a hydrogen donor (cyclohexadiene). These results showed that β-fragmentations of oxygen leading to *transoid* centres occur much faster (16 times) than such fragmentations leading to *cisoid* centres. In addition, oxygen addition occurs more readily at the *transoid* end of a pentadienyl radical than at the *cisoid* end [51]. The mechanistic scheme was shown to be applicable to tri- and tetraene lipids, such as linolenic and arachidonic acids [34,52]. In the latter case, a more complex array of products was found, including six *trans*-, *cis*-conjugated diene hydroperoxides, and cyclic peroxides derived from homoallylic intermediate peroxyl radicals.

Convincing support for the Porter Mechanism is provided by ESR studies which showed that the pentadienyl carbon radical formed either from 1,4-pentadiene [53] in the (*E*)-,(*E*)-conformation or similarly the *cis*-, *cis*-carbon radical formed by hydrogen abstraction from methyl linoleate [54] to *not* convert to (*E*)-,(*Z*)- (*cis*-, *trans*-) or *trans*-, *trans*-conformers. Thus the isomeric hydroperoxides must be formed from the intermediate peroxyl radicals (Figure 5) and not by direct isomerization of the carbon radicals. Nevertheless, an intriguing question remains—*Why is there no evidence for the formation of an intermediate bis-allylic lipid peroxyl radical (Figure 6) expected from direct oxygen addition at the original bis-allylic carbon radical?*

The reaction of various carbon radicals with oxygen is known to be rapid $k_{ox} \geqslant 1 \times 10^9 \, M^{-1} s^{-1}$) and insensitive to structure [55]; thus some of this 'original' peroxyl radical, and derived hydroperoxide, is expected to be present. A possible explanation is provided by the actual formation of

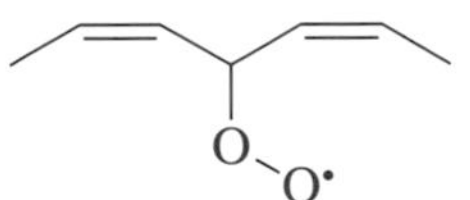

Figure 6 The postulated non-conjugated bis-allylic peroxyl radical.

products apparently derived from such a bis-allylic peroxyl radical formed by autoxidation of a constrained cyclic system, i.e. 1-butylcyclohexa-2,5-diene-1-carboxylate [56]. Formation of a bis-allylic peroxyl radical, at least in this cyclic system, is possible. Then it was suggested that the lack of isolation of the bis-allylic hydroperoxide from acyclic lipids is due to 'sufficient energy difference' (4 kcal mol^{-1}) in favour of the conjugated peroxyl radicals [56].

Weenen and Porter used the *c, t*//*t,t* distribution of hydroperoxides as a probe for the mechanism of autoxidation of phosphatidylcholines aggregated as model membranes in water [57]. Others applied the Porter procedure of peroxidation product studies from linoleate when the lipid is contained in the hydrophobic phases of micelles [58,59] or membranes. We found the same linear trend of *cis*-, *trans*- to *trans*-, *trans*-product ratios for peroxidation of linoleate in sodium dodecyl sulfate (SDS) micelles, independent of the method of initiation of the reaction, including initiation by lipid-soluble or water-soluble azo initiators or by lipid-soluble or water-soluble photoinitiators, such as benzophenone and its derivatives [59]. It was especially interesting to find a similar linear trend in the *c,t*//*t,t* ratios of products from peroxidation of mixed bilayers containing varying amounts of dilinoleoylphosphatidylcholine (DLPC) mixed with dipalmitoylphosphatidylcholine (as non-oxidizing medium), even when the oxidation was initiated by a water-soluble azo initiator [60]. Thus these product studies are remarkable useful probes of lipid peroxidation occurring in a variety of environments under various conditions [47].

1.6.2 Dynamics of lipid peroxyl radicals: effects in membranes

Peroxyl radicals on lipid chains may undergo several significant kinds of molecular motion, including (1) rotation around the C—O bond of the peroxyl group, (2) rotation about the long axis of the molecule (chain axis rotation), which may take the form of segmental rotation or 'crankshaft' rotation about several bonds of the molecule, and (3) diffusion of the whole molecule. The latter motion may be very important when polar peroxyl radicals are generated in the hydrophobic phases of micelles or phospholipid membranes.

Sevilla and coworkers studied the intramolecular rotational motions of lipid peroxyl radicals by using ESR spectroscopy of the peroxyls generated in a low-temperature matrix [61] and in urea clathrates [62]. Results to date on the linoleic acid peroxyl radical could not distinguish between a 180° chain axis rotation or a 90° rotation around the C—O bond. Their results on a triglyceride, i.e. triarachidin, indicate that significant motion involves a segmental rotation which requires a barrier of 18 kJ mol^{-1}. Rotations about the C—O bond appear to involve relatively larger barriers [62].

Diffusional motion of peroxyl radicals in lipid membranes is an important issue because it is expected to be an important cause of the disruption of membrane structure and function which accompanies extensive peroxidation. Overall physical transformations are well known as a result of peroxidation of liposomes [60]. However, *specific information* on the dynamics of individual lipid peroxyls in lipid membranes is lacking. Barclay and Ingold proposed that peroxyl radicals formed in the non-polar hydrocarbon-like environment in a lipid bilayer would diffuse rapidly away from this region into the polar surface region [9]. This could have very significant effects on the bilayer structure and various attempts have been made to check this 'Floating Peroxyl Radical Hypothesis' in biphasic systems of micelles or membranes. In this regard, a very significant decrease in the termination rate constant of peroxyl radicals as observed in aqueous micelles and liposomes was interpreted in terms of the diffusion of polar radicals to the aqueous surface where the radicals are expected to be stabilized by hydrogen bonding by water, as suggested schematically in Figure 7 [47].

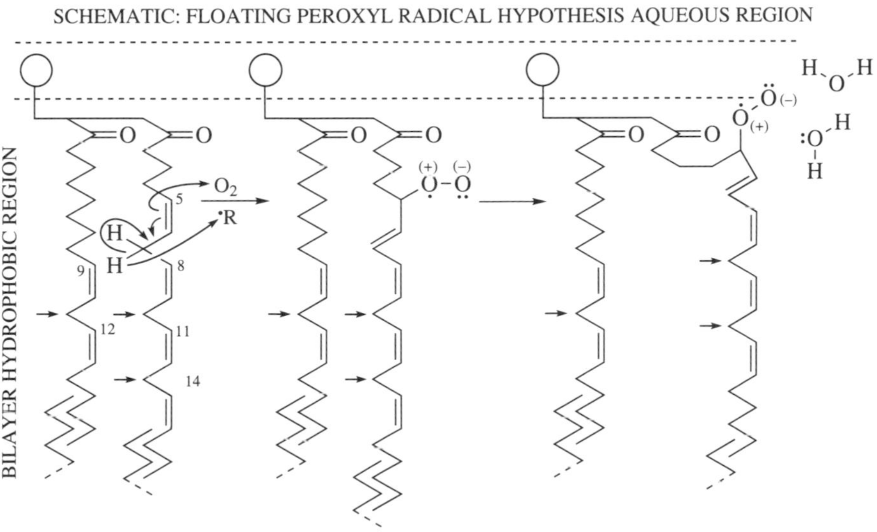

Figure 7 Schematic representation of the diffusion of polar peroxyl radicals from the hydrophobic region to a polar interface during peroxidation (from Ref. 47).

Peroxyl-trapping antioxidants, such as α-tocopherol, which are effective H-atom donors, cause formation predominately of the *cis*-, *trans*-hydroperoxides when added during peroxidation of DLPC liposomes. However an 'expected' difference in trapping radicals at the 9-position compared to the 13-position was not found in this system during inhibited peroxidation [57]. Such a difference might be expected if peroxyl radicals diffuse towards the surface because this is the region where the radical trapping head group of α-tocopherol resides [63]. On the other hand, effective peroxyl radical trapping by a water-soluble antioxidant such as Trolox is in agreement with the trapping of lipid peroxyls at the surface (e.g. see Figure 7), and diffusion phenomena are probably faster than the radical-trapping step so that differential trapping at the 9- and 13-sites is not observed [64]. Peroxyl radicals at the aqueous surface of membranes are expected to be *spin trapped* by a water-soluble spin-trapping agent. This has in fact been observed during oxidation of a lipidic azo initiator, 6, 6′-azo-6-cyanododecanoic acid, in saturated liposomes. After oxidative thermolysis, radicals were trapped in the aqueous phase by a water-soluble derivative of phenyl-tert-butylnitrone [47]. While such results are supportive of the diffusion hypothesis, other more compelling results are yet to be reported in support of this concept.

2 THIYLPEROXYL RADICALS

Peroxyl radicals with oxygen bonded to sulfur form rapidly from the reaction of thiyl radicals and oxygen [65,66]. Sulfur peroxyl radicals are of special interest because they are involved as reactive intermediates during the autoxidation of biologically important sulfur-containing compounds such as amino acids and glutathione. Heteroatom-peroxyl radicals are reviewed more completely elsewhere in this book. It is useful to compare some recent results on the structure of thiylperoxyl radicals with carbon-based peroxyl radicals (*vide supra*).

Because sulfur can expand its valence shell, the product from the combination of oxygen with a thiyl radical can be represented by three different structures:

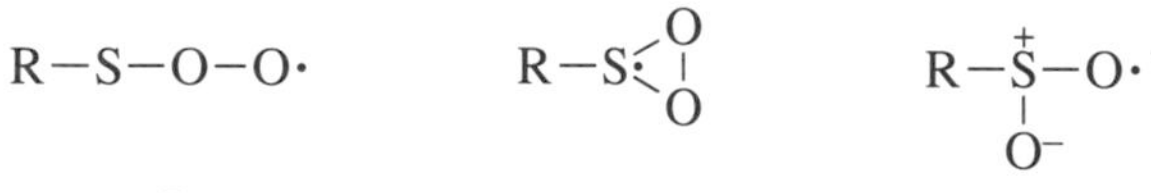

R–S–O–O· R–S·<(O–O) R–S⁺(–O⁻)–O·

I II III

Structure I, the thiylperoxyl radical, is known to rearrange either photochemically [67] or thermally [66] to the known sulfonyl radical, III, which has been calculated to be more stable thermodynamically than I by 150 kJ mol^{-1} [67]. There is no evidence for II, although it might be a transition state between I and III.

The calculated geometry for the simplest alkylthiylperoxyl radical, $CH_3SOO^{\cdot}$ in given in Table 5. This radical has a bent structure with a geometry which is predicted [69] to be very similar to that of the *trans*-methylperoxyl radical; both structures were calculated at the same level

Table 5 Geometry of the methylthiylperoxyl radical

	Bond length (Å)	Bond angle[a] (deg)
	H_aC_c 1.084	$H_aC_cS_d$ 111
	C_cS_d 1.80	$H_bC_cS_d$ 106
	S_dO_a 1.71	$C_cS_dO_a$ 94
trans	O_aO_b 1.29	$S_dO_aO_b$ 111

[a]Data taken from Ref. 68.

(6–31 G*) (compare Table 1). Significant differences appear in the longer bond lengths found for the sulfur peroxyl group, i.e. C_c—S_d(1.80 Å), and S_d—O_a(1.71 Å), compared to the methylperoxyl radical where the calculated C—O bond length is 1.42 Å.

A calculation of the *spin density* showed that only a small part (1%) of the spin is located on the sulfur, with the terminal oxygen being calculated to have the higher proportion of the spin (84%) [69]. Such calculations overestimate the spin on the terminal oxygen, as was also the case for calculations carried out for the methylperoxyl radical. ESR spectroscopic studies using ^{17}O-labelled oxygen indicated that the spin density in $RSOO^\bullet$ is almost equally shared between the two oxygens, in contrast to the alkylperoxyl radicals where the higher proportion appears on the terminal oxygen. However, it was found that the solvent environment had a considerable influence on the ^{17}O hyperfine couplings and on the spin densities on the two oxygens. Unlike the usual case for alkylperoxyls, polar solvents appear to shift the spin densities towards nearly equal values on the oxygens of $RSOO^\bullet$. This was interpreted in terms of additional structures C and D, of a charge-transfer type, as contributors to the structure of the radical. Polar solvents may solvate the charge-separated structures C and D by donating electron pairs [70]:

$$\underset{A}{R-\ddot{\underset{..}{S}}-\ddot{O}-\dot{O}} \qquad \underset{B}{R-\ddot{\underset{..}{S}}-\overset{+}{\dot{O}}-\overset{-}{\ddot{O}}} \qquad \underset{C}{R-\overset{+}{\dot{\underset{..}{S}}}-\ddot{O}-\overset{-}{\ddot{O}}} \qquad \underset{D}{R-\overset{+}{\dot{\underset{..}{S}}}\ (\ddot{O}-\overset{-}{\dot{O}})}$$

The thiylperoxyl radicals are expected to be strongly polarized. The *dipole moments* of these species do not yet appear to have been measured or calculated. The S—O *bond energy* is relatively weak, with measurements, giving a value of only 46 kJ mol^{-1} in CH_3—$SOO^\bullet$ [68], compared to the reported C—O bond energy in $CH_3OO^\bullet$ of 135 kJ mol^{-1} [6]. The weak S—O bond would account for the equilibrium observed in the following reaction [66,70]:

$$R-\dot{S} + O_2 \rightleftharpoons R-S-O-\dot{O}$$

Alkylthiylperoxyl radicals exhibit a broad-band absorption in the *visible* spectrum at around 550 nm [66,70], unlike alkylperoxyl radicals. This interesting visible absorption is probably due to a new electronic transition resulting from the presence of the sulfur atom, and probably originates from a transition from a non-bonding electron on the sulfur to a SOMO π-orbital on the peroxyl oxygens. It appears to involve intramolecular charge separation [70]. Therefore the sulfur atom in these thiylperoxyls has a similar function in the visible spectra as the multiple bond in the spectra of unsaturated carbon-based peroxyl radicals.

REFERENCES

1. K.U. Ingold, *Acc. Chem. Res.* **2**, 1 (1969).
2. M.D. Sevilla, D. Becker and M. Yan, *J. Chem. Soc. Faraday Trans.* **86**, 3279 (1990).
3. K. Adamic, K.U. Ingold and J.R. Morton, *J. Am. Chem. Soc.* **92**, 922 (1970).
4. E. Melamud and B.L. Silver, *J. Phys. Chem.* **15**, 1896 (1973).
5. S. Biskupic and L. Valko, *J. Mol. Struct.* **27**, 97 (1975).
6. B.H. Besler, P. MacNeille and M.D. Sevilla, *J. Phys. Chem.* **90**, 6446 (1986).
7. S.L. Boyd, R.J. Boyd and L.R.C. Barclay, *J. Am. Chem. Soc.* **112**, 5724 (1990).
8. A. Ohkubo and F. Kitagawa, *Bull. Chem. Soc. Jpn.* **48**, 703 (1975).
9. (a) L.R.C. Barclay and K.U. Ingold, *J. Am. Chem. Soc.* **103**, 6478 (1981); (b) L.R.C. Barclay and K.U. Ingold, *J. Am. Chem. Soc.* **102**, 7792 (1980).
10. R.W. Fessenden, A. Hitachi and V. Nagarajan, *J. Phys. Chem.* **88**, 107 (1984).
11. P. Neta, R.E. Huie, S. Mosseri, L.V. Shastri, J.P. Mittal, P. Marathamuthu and S. Steenken, *J. Phys. Chem.* **93**, 4099 (1989).
12. I.A. Opeid, A.F. Dmitruk and R.V. Kucher, *Teor. Eksp. Khim.* **8**, 385 (1972); *Chem. Abstr.* **77**, 163870 (1972).
13. J.E. Bennett and R. Summers, *J. Chem. Soc. Faraday Trans. 2,* **69**, 1043 (1973).
14. H. Kashiwabara, S. Shimada and Y. Hori, *Radiat. Phys. Chem.* **37**, 511 (1991).
15. S. Shimada, Y. Hori and H. Kashiwabara, *Macromolecules* **18**, 170 (1985).
16. S. Shimada, A. Kotake, Y. Hori and H. Kashiwabara, *Macromolecules* **17**, 1104 (1984).
17. L. Kevan and S. Schlich, *J. Phys. Chem.* **90**, 1998 (1986).
18. M. Iwasaki and Y. Sakai, *J. Polym. Sci. (Part A-2)* **6**, 265 (1968).
19. Y. Hori, S. Aoyama and H. Kashiwabara, *J. Chem. Phys.* **75**, 1582 (1981).
20. S. Schlick and L. Kevan, *J. Am. Chem. Soc.* **102**, 4622 (1980).
21. R.F. Mertens and C. von Sonntag, *Angew. Chem. Int. Ed. Engl.* **33**, 1262 (1994).
22. G.I. Khaikin and P. Neta, *J. Phys. Chem.* **99**, 4549 (1995).
23. M. Krauss and R. Osman, *J. Phys. Chem.* **99**, 11387 (1995).
24. S.L. Boyd, R.J. Boyd, Z. Shi, L.R.C. Barclay and N.A. Porter, *J. Am. Chem. Soc.* **115**, 687 (1993).

25. G.O. Schenck, O.A. Neumüller and W. Eisfeld, *Angew. Chem.* **70**, 595 (1958).
26. W.F. Brill, *J. Am. Chem. Soc.* **87**, 3286 (1965).
27. W.F. Brill, *J. Chem. Soc. Perkin Trans. 2*, 621 (1984).
28. N.A. Porter and J.S. Wujek, *J. Org. Chem.* **52**, 5058 (1987).
29. N.A. Porter and P. Zuram, *J. Chem. Soc. Chem. Commun.*, 1472 (1985).
30. A.L.J. Beckwith, A.G. Davies, I.G.E. Davison, A. Maccoll and M.H. Mruzek, *J. Chem. Soc. Perkin Trans. 2*, 815 (1989).
31. N.A. Porter, J.K. Kaplan and P.H. Dussault, *J. Am. Chem. Soc.* **112**, 1266 (1990).
32. (a) K.A. Mills, S.E. Caldwell, G.R. Dubay and N.A. Porter, *J. Am. Chem. Soc.* **114**, 9689 (1992); (b) K.A. Mills, S.E. Caldwell, G.R. Dubay and N.A. Porter, *J. Am. Chem. Soc.* **116**, 6697 (1994).
33. N.A. Porter, K.A. Mills and R.L. Carter, *J. Am. Chem. Soc.* **116**, 6690 (1994).
34. N.A. Porter, *Acc. Chem. Res.* **19**, 262 (1986).
35. H.W.W. Chan, G. Levett and J.A. Matthey, *Chem. Phys. Lipids* **24**, 245 (1979).
36. P.M. Sommeling, P. Mulder, R. Louw, D.V. Avila, J. Lusztyk and K.U. Ingold, *J. Phys. Chem.* **97**, 8361 (1993).
37. Z.B. Alfassi, S. Marguet and P. Neta, *J. Phys. Chem.* **98**, 8019 (1994).
38. Z.B. Alfassi, G.I. Khaikin and P. Neta, *J. Phys. Chem.* **99**, 265 (1995).
39. X. Fang, R. Mertens and C. von Sonntag, *J. Chem. Soc. Perkin Trans. 2.*, 1033 (1995).
40. B.K. Carpenter, *J. Am. Chem. Soc.* **115**, 9806 (1993).
41. B. Noziere, R. Lesclaux, M.D. Hurley, M.A. Dearth and T.J. Wallington, *J. Phys. Chem.* **98**, 2864 (1994).
42. D.V. Avila, J. Lusztyk and K.U. Ingold, *J. Am. Chem. Soc.* **114**, 6576 (1992).
43. J.A. Howard, J.E. Bennett and G. Brunton, *Can. J. Chem.* **59**, 2253 (1981).
44. S.E. Caldwell and N.A. Porter, *J. Am. Chem. Soc.*, **117**, 8676 (1995).
45. H.W.S. Chan and G. Levett, *Lipids* **12**, 99 (1977).
46. E.N. Frankel, W.E. Neff, W.K. Rohwedder, B.P.S. Khambay, R.F. Garwood and B.C.L. Weedon, *Lipids* **12**, 908 (1977).
47. L.R.C. Barclay, *Can. J. Chem.* **71**, 1 (1993).
48. J. Terao and S. Matsushita, *J. Am. Oil Chem. Soc.* **54**, 234 (1977).
49. M.J. Thomas and W.A. Pryor, *Lipids* **15**, 544 (1980).
50. N.A. Porter, B.A. Weber, H. Weener and J.A. Khan, *J. Am. Chem. Soc.* **102**, 5597 (1980).
51. N.A. Porter and D.W. Wujek, *J. Am. Chem. Soc.* **106**, 2626 (1984).
52. N.A. Porter, L.S. Lehman, B.A. Weber and K.J. Smith, *J. Am. Chem. Soc.* **103**, 6447 (1981).
53. D. Griller, K.U. Ingold and J.C. Walton, *J. Am. Chem. Soc.* **101**, 758 (1979).
54. E. Bascetta, F.D. Gunstone and J.C. Walton, *J. Chem. Soc. Perkin Trans. 2*, 603 (1983).
55. B. Maillard, K.U. Ingold and J.C. Scaiano, *J. Am. Chem. Soc.* **105**, 5095 (1983).
56. A.L.J. Backwith, D.M. O'Shea and D.H. Doberts, *J. Am. Chem. Soc.* **108**, 6408 (1986).
57. H. Weenen and N.A. Porter, *J. Am. Chem. Soc.* **104**, 5216 (1982).
58. Y. Yamamoto, S. Haga, E. Niki and Y. Kamiya, *Bull. Chem. Soc. Jpn* **57**, 1260 (1986).
59. L.R.C. Barclay, K.A. Baskin, S.J. Locke and T.D. Schaefer, *Can. J. Chem.* **65**, 2529 (1987).
60. L.R.C. Barclay, K.A. Baskin, D. Kong and S.J. Locke, *Can. J. Chem.* **65**, 2541 (1987).
61. D. Becker, J. Yanez, M.D. Sevilla, M.G. Alonso-Amigo and S. Schlick, *J. Phys. Chem.* **91**, 492 (1987).

62. M.D. Sevilla, M. Champagne and D. Becker, *J. Phys. Chem.* **93**, 2653 (1989).
63. B. Perly, I.C.P. Smith, L. Hughes, G.W. Burton and K.U. Ingold, *Biochim. Biophys. Acta* **819**, 131 (1985).
64. L.R.C. Barclay, J.D. Artz and J.J. Mowat, *Biochem. Biophys. Acta* **1237**, 77 (1995).
65. J. Monig, K.-D. Asmus, L.G. Forni and R.L. Wilson, *Int. J. Radiat. Biol.* **52**, 589 (1978).
66. X. Zhang, N. Zhang, H.P. Schuchmann and C. von Sonntag, *J. Phys. Chem.* **98**, 6541 (1994).
67. M.D. Sevilla, D. Becker and M. Yan, *Int. J. Radiat. Biol.* **57**, 65 (1990).
68. A.A. Turnispeed, S.B. Barone and A.R. Rabishankara, *J. Phys. Chem.* **96**, 7502 (1992).
69. S.G. Swarts, D. Becker, S. DeBolt and M.D. Sevilla, *J. Phys. Chem.* **93**, 155 (1989).
70. Y. Razskazovskii, A.-O. Colson and M.D. Sevilla, *J. Phys. Chem.* **99**, 7993 (1995).

4 The Thermochemistry of Peroxides and Polyoxides, and their Free Radicals

S.W. BENSON
University of Southern California, Los Angeles, USA

and

N. COHEN
Thermochemical Kinetics Research, Portland, USA

1 THE GROUP ADDITIVITY APPROACH

Twenty five years ago the thermochemistry of organic peroxides and polyoxides was reviewed in detail by Benson and Shaw [1], who deduced the necessary parameters required to estimate the gaseous enthalpies of formation by the use of group additivities, a semi-empirical computational procedure that is conceptually simple and easy to apply [2]. Over the following years, the group additivity method has proved highly reliable, considering its relative simplicity; frequently it has been shown that in cases where group additivity predictions disagree with measurements, the latter, rather than the former, are at fault, particularly if the calculated values for other, homologous compounds agree with the experimental data. The original review [1], at least, with respect to peroxides, was updated by Baldwin, but no substantive changes were required [3]. This is still largely true today; in fact, it will be seen that most of the important pieces of data that were originally missing or unreliable then are no better known at the present time. However, recent reviews of polyoxides [4] show that considerable progress has been made in recent years in the understanding of these compounds, whose existence was even doubted a few decades ago.

Each family of peroxides or polyoxides has one or more unique group components (a 'group' is defined as a polyvalent atom in a molecule or radical together with its ligands) that are not found outside of components with the O—O bond. These groups are listed in Table 1. A convenient shorthand notation for such groups is R—(X)(Y)(Z), where R is the polyvalent central atom (or functional group) and X, Y, and Z are ligands. Generally, R, X, Y, and Z are atoms, but exceptions include >C=O, >C=C=O, NO_2, and NO, with only the first of these being important in this

Peroxyl Radicals. Edited by Z. B. Alfassi

Table 1 Groups unique to peroxides and polyoxides

Family	Group(s)
Alkyl hydroperoxides	O—(H)(O) and O—(C)(O)
Dialkylperoxides	O—(C)(O)
Diacylperoxides	O—(O)(CO)
Aryl hydroperoxides	O—(C_B)(O) and O—(H)(O)
Peroxoic acids	O—(O)(CO) and O—(H)(O)
Peroxoic esters	O—(O)(CO) and O—(C)(O)
Polyoxides (RO_nR, $n > 2$)	O—$(O)_2$
Alkylperoxy radicals ($RO_2^{\cdot}$)	O—(C)($O^{\cdot}$)
Acylperoxy radicals	O—(CO)($O^{\cdot}$)
Arylperoxy radicals	O—(C_B)($O^{\cdot}$)
Hydroperoxy radicals ($HO_2^{\cdot}$)	O—(H)($O^{\cdot}$)
Polyoxy radicals ($RO_{n-1}O^{\cdot}$)	O—(O)($O^{\cdot}$)

present discussion. If R is a carbon atom, the notation C, C_d, C_t, or C_B is used to distinguish singly bonded, double bonded, triple bonded, and aromatic C atoms, respectively. For example, the groups comprising tert-butyl hydroperoxide are: C—$(H)_3$(C) (×3), C—$(C)_3$(O), O—(C)(O), and O—(H)(O). The first two groups occur in a great many other compounds besides peroxides and polyoxides, and their evaluations are thus derived from a substantial database. The third and fourth groups are characteristic of peroxides and polyoxides, and their relatives, and do not occur elsewhere.

The characteristic groups of interest here are listed in Table 1. A central concern of this present section will be the re-evaluation of these groups, which can then be used to predict the enthalpies of formation (at 298 K) of compounds and radicals that have not yet been measured.

1.1 PEROXIDES AND HYDROPEROXIDES

Experimental data obtained for the enthalpies of formation ($\Delta_f H^0_{298}$) are listed in Table 2. The first group of compounds are hydroperoxides. As the table shows, there are only four hydroperoxides (including H_2O_2) for which gas-phase data are available. This paucity of results is a result of the instability of the smaller peroxides and hydroperoxides and the consequent experimental difficulties of preparing pure samples and carrying out accurate thermochemical measurements. For a few other compounds, gas-phase values can be estimated by using the liquid-phase enthalpies of formation, corrected by approximate values for $\Delta_{vap}H_{298}$; the latter are estimated from the enthalpies of vaporization of the corresponding alcohols, which are very close to those of the hydroperoxides for the few cases (e.g. ethyl and tert-butyl) for which data for both are available. There are thus direct experimental data available for only four compounds for the determination of the enthalpy contribution, or 'group additivity value' (GAV), from the

Table 2 Values for the enthalpies of formation (at 298 K) of hydroperoxides and peroxides

Compound	Formula	Gas			Liquid			Solid		
		Expt.	Calc.	Δ	Expt.	Calc.	Δ	Expt.	Calc.	Δ
Hydrogen peroxide	H_2O_2	−32.5	−32.5	[0]	−44.9	−45.6	0.7	−47.9	−47.9	[0]
Ethyl hydroperoxide	$C_2H_6O_2$	**−47.5**	−39.4	−8.2	−57.8	−49.7	−8.2		−50.1	
Propyl hydroperoxide	$C_3H_8O_2$	**−64.7**[b]	−44.4	−20.3	−76.0	−55.8	−20.2		−57.1	
tert-Butyl hydroperoxide	$C_4H_{10}O_2$	−58.8	−57.9	−0.9	−70.2	−69.3	−0.9		−71.0	
Cyclohexyl hydroperoxide	$C_6H_{12}O_2$	**−50.5**[b]	−52.8	2.3	−65.3	−67.3	2.0		−71.0	
Hexyl hydroperoxide	$C_6H_{14}O_2$	**−56.9**[b]	−59.4	2.5	−71.6	−74.1	2.5		−78.1	
1-Methylpentyl hydroperoxide	$C_6H_{14}O_2$	**−60.1**[b]	−62.7	2.5	−74.1	−76.9	2.8		−76.8	
1-Ethylbutyl hydroperoxide	$C_6H_{14}O_2$		−61.9		−72.9	−76.4	3.4		−76.4	
1-Methylcyclohexyl hydroperoxide	$C_6H_{14}O_2$		−60.2		−79.0	−74.9	−4.1		−78.1	
1-Heptyl hydroperoxide	$C_7H_{16}O_2$	**−66.0**[b]	−64.4	−1.7	−82.0	−80.2	−1.8		−85.1	
2-Heptyl hydroperoxide	$C_7H_{16}O_2$		−68.5		−82.8	−83.6	0.8		−84.2	
3-Heptyl hydroperoxide	$C_7H_{16}O_2$		−68.5		−82.9	−83.6	0.7		−84.2	
4-Heptyl hydroperoxide	$C_7H_{16}O_2$		−68.5		−79.8	−83.6	3.8		−84.2	
1-Methyl-1-phenylethyl hydroperoxide	$C_9H_{12}O_2$	−18.7	−23.5	4.8	−35.4	−35.8	[0]	**−38.7**	−26.9	−11.7
1,2,3,4-Tetrahydro-5-hydroperoxynaphthalene	$C_{10}H_{12}O_2$		NA			NA		−44.4	−44.4	[0]
(*E*)-4a-Hydroperoxydecahydronaphthalene	$C_{10}H_{18}O_2$		−68.1			−86.5		**−83.5**	−100.2	16.7
1,2,3,4-Tetrahydro-1-methylhydroperoxynaphthalene	$C_{11}H_{14}O_2$		−20.0			−39.6		**−37.6**	−47.9	10.3
4b,9-Dihydro-9-hydroperoxy-4b,9,10-triphenyl-indeno-[1,2,3-fg]naphthacene-9-yl	$C_{42}H_{28}O_2$		133.7			NA		**145.2**	74.6	70.6
Dimethylperoxide	$C_2H_6O_2$	−30.0	−30.0	−0.0	**−34.7**[c]	−35.5	0.9		−42.3	
Diethylperoxide	$C_4H_{10}O_2$	−46.1	−46.2	0.1	−55.8	−53.7	−2.0		−52.3	
Diisopropylperoxide	$C_6H_{14}O_2$	−65.0[d]	−64.4	−0.6	**−72.7**[c]	−72.6	−0.1		−64.6	
Di-tert-butylperoxide	$C_8H_{18}O_2$	−83.4	−83.2	[0]	**−91.0**	−93.0	1.9		−94.1	
Dibenzoylperoxide	$C_{14}H_{10}O_4$	−64.9	−65.0	[0]		−87.4		−88.3	−88.3	[0]
Bis(2-methylbenzoyl)peroxide	$C_{16}H_{14}O_4$		−74.5			−104.9		**−119.6**	−97.9	−21.7
Bis(4-methylbenzoyl)peroxide	$C_{16}H_{14}O_4$		−80.5			−104.9		−107.9	−105.9	−2.0
Bis(1-oxo-3-phenyl-2-propenyl)peroxide (Dicinnamoyl peroxide)	$C_{18}H_{14}O_4$		−32.4			−64.4		**−85.1**	−64.1	−21.0

(continued overleaf)

Table 2 – *continued*

Compound	Formula	Gas			Liquid			Solid		
		Expt.	Calc.	Δ	Expt.	Calc.	Δ	Expt.	Calc.	Δ
(*E*)-4-tert-Butoxy-4-oxo-2-buteneperoxoic acid methyl ester	$C_9H_{14}O_5$		−166.9		−178.5	−184.7	6.		−182.6	
4-tert-Butoxy-4-oxobutaneperoxoic acid methyl ester	$C_9H_{16}O_5$		−193.7		−209.9	−209.9	0.0		−216.8	
Benzoyl phenoxycarbonyl peroxide	$C_{14}H_{10}O_5$	**−92.5**	−101.8	9.3		−123.1		−115.9	−121.0	5.1
Benzoyl(cyclohexyloxy)carbonylperoxide	$C_{14}H_{16}O_5$	−154.2	−155.2	1.0		−177.3	.	−177.2	−172.6	−4.7
Benzenecarboperoxoic acid (Perbenzoic acid)	$C_7H_6O_3$		−48.7			−66.5		**−87.7**	−68.1	−19.6
Dodecaneperoxoic acid	$C_{12}H_{24}O_3$		−129.9			−153.5		−162.6	−162.6	0.0
Tetradecaneperoxoic acid (Pertetradecanoic acid)	$C_{14}H_{28}O_3$		−134.7			−160.5		**−179.2**	−162.0	−17.2
Decaneperoxoic acid tert-butyl ester	$C_{14}H_{28}O_3$		−145.2		−164.5	−164.9	0.4		−171.7	
Benzenecarboperoxoic acid 1-methyl-1-phenylethyl ester	$C_{16}H_{16}O_3$	−45.0	−39.8	−5.2		−56.7		−55.3	−47.2	−8.2
Hexadecaneperoxoic acid (Perhexadecanoic acid)	$C_{16}H_{32}O_3$		−144.7			−172.7		**−191.7**	−176.0	
−15.6										
Dodecaneperoxoic acid tert-butyl ester	$C_{16}H_{32}O_3$		−155.2		−176.5	−177.1	0.7		−185.7	
Octadecaneperoxoic acid	$C_{18}H_{36}O_3$		−154.7			−184.9		**−204.9**	−190.0	−14.9
Tetradecaneperoxoic acid tert-butyl ester	$C_{18}H_{36}O_3$		−165.2		−190.2	−189.4	−0.9		−199.7	
Diacetylperoxide	$C_4H_6O_4$	**−115.5**[e]	−126.8	11.3	**−127.9**	−140.9	13.0		−122.1	
Bis(hydroxymethyl)peroxide (Dioxybismethanol)	$C_2H_6O_4$	−136.6	−118.8	−17.8		−135.6		−159.1	−159.0	[0]
Bis(1-oxypropyl)peroxide (Dipropionylper oxide)	$C_6H_{10}O_4$	**−135.6**[e]	−137.2	1.6	−148.2	−151.3	3.1		−151.3	
Bis(1-oxobutyl)peroxide (Dibutyrylperoxide)	$C_8H_{14}O_4$		−147.2		−160.9	−163.5	2.6		−165.3	

[a] All experimental data are taken from Ref. 5 unless noted otherwise. All calculated values are obtained by group additivity; all data in units of kcal mol^{-1}, with Bold values being estimates or of questionable reliability. Bracketed [0] values in 'Difference' column indicate that a GAV was chosen on the basis of this experimental value, i.e. agreement is forced. NA: indicates one or more group additivity values are not available to evaluate the compound in this phase.

[b] Estimated by assuming $\Delta_{vap}H$ for hydroperoxide is the same as measured $\Delta_{vap}H$ for corresponding alcohol.

[c] Estimated by assuming $\Delta_{vap}H$ for peroxide is the same as measured $\Delta_{vap}H$ for corresponding ether.

[d] Taken from Ref. 6.

[e] Estimted by assuming $\Delta_{vap}H$ for peroxide is the same as measured $\Delta_{vap}H$ for corresponding acid anhydride.

O—(H)(O) group, namely H_2O_2, and ethyl, tert-butyl, and 1-methyl-1-phenylethyl hydroperoxides. The enthalpy of hydrogen peroxide is the most accurately known of the four; in addition, the others involve other groups that are difficult to establish. Therefore, we choose the GAV so as to force agreement with H_2O_2. After selection of other groups, this results in a discrepancy between the calculated and experimental values for tert-butyl of $-0.8\,\text{kcal mol}^{-1}$ (discrepancy, $\Delta = H_{\text{expt}} - H_{\text{calc}}$); however, the discrepancies for ethyl and 1-methyl-1-phenylethyl are -8.1 and $4.9\,\text{kcal mol}^{-1}$, respectively, and these probably reflect experimental errors.

In the liquid phase, data are more plentiful. We choose the GAV so as to match the experimental value for $\Delta_{\text{vap}}H_{298}$ for ethyl, propyl, and tert-butyl hydroperoxides. This leaves H_2O_2 with a discrepancy of approximately 0.7 kcal mol^{-1}; however, like H_2O, H_2O_2 has an anomalous boiling point and, consequently, enthalpy of vaporization. This still leaves discrepancies in the cases of the liquid-phase enthalpies, in particular for ethyl and propyl ($-20.4\,\text{kcal mol}^{-1}$); the experimental values are almost certainly erroneous. In the solid phase, there are five compounds, H_2O_2 again, and four others with between 9 and 11 carbon atoms. If we again use hydrogen peroxide to fix the GAV (even though it may result in an error of less than $1\,\text{kcal mol}^{-1}$, as in the liquid phase), then all four of the other compounds show large errors (9 to $18\,\text{kcal mol}^{-1}$), which again we attribute mainly to experimental errors. The resulting GAVs for the O—(H)(O) group in the liquid and solid phases are -22.8 and $-23.9\,\text{kcal mol}^{-1}$, respectively. In general, the large discrepancies in the case of many hydroperoxides may be a result of the difficulty in preparing pure samples of the compounds; the principal likely impurity is the corresponding alcohol, which is always more stable than the hydroperoxide, resulting in an experimental $\Delta_f H^\circ_{298}$ value that is too low.

The O—(C)(O) group occurs in a handful of hydroperoxides and dialkylperoxides; the value chosen here for the gas phase ($-5.0\,\text{kcal mol}^{-1}$) gives good agreement (within $0.6\,\text{kcal mol}^{-1}$) for dimethyl-, diethyl-, diisopropyl-, and di-tert-butylperoxides, and thus seems well established. (The agreement in the case of di-tert-butylperoxide may be fortuitous; the experimental value for the liquid phase listed in Table 2 has an uncertainty assigned by the original workers of $5\,\text{kcal mol}^{-1}$, and there are two other measurements that differ by 5 and $14\,\text{kcal mol}^{-1}$ from the one cited. Furthermore, the gas-phase value depends on a measurement of $\Delta_{\text{vap}}H$, combined with the liquid value, and the heat of vaporization reported is $7.6\,\text{kcal mol}^{-1}$, i.e. approximately $1.4\,\text{kcal mol}^{-1}$ smaller than that expected on the basis of reliable estimation techniques.) In the liquid phase the same four compounds provide data (if we assume, in two cases, that $\Delta_{\text{vap}}H$ is the same as for the corresponding ethers) that are in accord within $\pm 2\,\text{kcal mol}^{-1}$. We are not so well off in the solid phase, as the five compounds for which data are available are wildly discordant. We base the GAV in this case on dioxybismethanol, although the chance of a large error is considerable.

With the GAVs established for these groups, we can predict, within ca. 1 kcal mol^{-1}, $\Delta_f H^\circ_{298}$(g) for CH_3OOH, which has not been measured:

$$\begin{aligned}\Delta_f H^\circ_{298}(\mathrm{CH_3OOH, g}) &= \Delta_f H^\circ_{298}[\mathrm{C{-}(H)_3(O)}] + \Delta_f H^\circ_{298}[\mathrm{O{-}(C)(O)}] + \Delta_f H^\circ_{298}[\mathrm{O{-}(H)(O)}] \\ &= -10.0 + (-5.1) + (-16.3) \\ &= -31.4\ \mathrm{kcal\ mol^{-1}}\end{aligned}$$

The O—(O)(CO) group occurs in four molecules in the gas phase, namely dibenzoylperoxide, benzoylcarboxyperoxide phenyl ester, benzoyl(cyclohexyloxy)carbonylperoxide, and benzenecarboperoxoic acid 1-methyl-1-phenylether ester. The GAVs chosen give good agreement (within 1 kcal mol^{-1}) for the first and third of these species, while the other two are in considerable error. There are several compounds containing this group in the solid phase, but for only one (dibenzoylperoxide) are there two independent measurements that agree within less than 0.1 kcal mol^{-1}; hence we use this compound to determine the GAV. With this GAV, the other compounds exhibit an unacceptably large scatter. An alternative is to increase the GAV by ca. 5 kcal mol^{-1}, giving a somewhat lower average absolute error, but worsening considerably the discrepancy for dibenzoylperoxide (and a few others). We favour the former choice at this time, although clearly there are considerable experimental errors and much room for improvement with replicated measurements.

The O—(C_B)(O) group occurs only in one compound whose enthalpy has been measured, namely 1,2,3,4-tetrahydro-5-hydroperoxynaphthalene, and then only in the solid phase. Since $\Delta_f H^\circ_{298}$ is known only for the solid phase, the GAVs for this group are not yet established in the gas and liquid phases. However, if we compare the gas-phase GAVs for O—(C)(C) (−23.8 kcal mol^{-1}) with O—(C)(C_B) (−21.6 kcal mol^{-1}), and O—(C)(H) (−37.9 kcal mol^{-1}) with O—(C_B)(H) (−38.5 kcal mol^{-1}) we see that replacing the C ligand with a C_B ligand changes the GAV by only 1 to 2 kcal mol^{-1}; hence we can assume that the GAV for O—(C_B)(O) in the gas phase is -5.1 ± 2 kcal mol^{-1}.

The derived (or estimated, shown in parentheses) values for the relevant groups are listed in Table 3. Other groups that are required for the calculations discussed in this paper are included as well.

1.2 POLYOXIDES

The existence of polyoxides (RO_nR) and hydropolyoxides (HO_nH and RO_nH) has been established conclusively in recent years. All of these compounds require but one additional group in order to estimate enthalpies by group additivity, i.e. the O—$(O)_2$ group. Establishing the GAV for this group is a considerably more roundabout procedure than in the case of the

Table 3 Group additivity values[a]

Group	GAV (gas)	GAV (liquid)	GAV (solid)
$C-(C)(H)_3$	−10.0[b]	−11.6[b]	−10.1[b]
$C-(C)_2(H)_2$	−5.0[b]	−6.1[b]	−7.0[b]
$C-(C)_3(H)$	−2.4[b]	−2.2[b]	−4.0[b]
$C-(C)_4$	−0.1[b]	0.75[b]	−1.5[b]
$C_d-(H)_2$	6.3[b]	4.9[b]	6.0[b]
$C_d-(C)(H)$	8.5[b]	7.6[b]	6.1[b]
$C_d-(C)_2$	10.2[b]	9.7[b]	NA
$C-(C_d)(H)_3$	−10.0[b]	−11.6[b]	−10.1[b]
O–(H)(O)	−16.3[c]	−22.8[c]	−23.9[c]
O–(C)(O)	−5.1[c]	−6.2[c]	−11.0[c]
O–(O)(CO)	−18.2[c]	−20.8[c]	−15.4[c]
$O-(C_B)(O)$	(−5 ± 2)[c]	NA	−4.5[c]
$O-(O)_2$	14.7 ± 1.5[c]	(13.4 ± 2.5)[c]	NA
$O-(O)(O^\cdot)$	34.5[c]	NA	NA
$O-(C)(O^\cdot)$	14.7 ± 1[c]	NA	NA
$O-(CO)(O^\cdot)$	1.6[c]	NA	NA
$O-(C_B)(O^\cdot)$	(14.8 ± 2)[c]	NA	NA

[a] NA indicates value not available.
[b] From Ref. 7.
[c] This work.

groups already discussed. We will follow basically the trail blazed by Nangia and Benson [8], but with several modifications and revisions. It will be seen that there are several segments along this path that can be traversed via different routes, leading to differing numerical results, each of which therefore contributes some measure of uncertainty to the final result. Briefly, our task will be to deduce $\Delta_f H[O-(O)_2]$ from $\Delta_f H$(t-BuO$_4$t-Bu), $\Delta_f H$(t-BuO$_2^\cdot$), and $\Delta_f H$(t-BuO$^\cdot$).

Consider first the following set of equilibria:

$$\text{t-BuO}_4\text{t-Bu} \rightleftharpoons 2\text{t-BuO}_2^\cdot \quad (1)$$

$$\text{t-BuO}_3\text{t-Bu} \rightleftharpoons \text{t-BuO}_2^\cdot + \text{t-BuO}^\cdot \quad (2)$$

$$\text{t-BuO}_2\text{t-Bu} \rightleftharpoons 2\text{t-BuO}^\cdot \quad (3)$$

The enthalpies of reaction are given by:

$$\Delta_r H(1) = 2\Delta_f H(\text{t-BuO}_2^\cdot) - \Delta_f H(\text{t-BuO}_4\text{t-Bu}) \quad (4)$$

$$\Delta_r H(2) = \Delta_f H(\text{t-BuO}_2^\cdot) + \Delta_f H(\text{t-BuO}^\cdot) - \Delta_f H(\text{t-BuO}_3\text{t-Bu}) \quad (5)$$

$$\Delta_r H(3) = 2\Delta_f H(\text{t-BuO}^\cdot) - \Delta_f H(\text{t-BuO}_2\text{t-Bu}) \quad (6)$$

It can be shown, using group additivity, that:

$$\Delta_f H(\text{t-BuO}_4\text{t-Bu}) + \Delta_f H(\text{t-BuO}_2\text{t-Bu}) = 2\Delta_f H(\text{t-BuO}_3\text{t-Bu}) \quad (7)$$

from which:

$$\Delta_r H(1) + \Delta_r H(3) = 2\Delta_r H(2) \tag{8}$$

It also follows from the preceding relationships that:

$$\Delta_r H(3) - \Delta_r H(2) = \Delta_f H(\text{t-BuO}^\bullet) - \Delta_f H(\text{t-BuO}_2^\bullet) + \Delta_f H(\text{t-BuO}_3\text{t-Bu}) - \Delta H_f(\text{t-BuO}_2\text{t-Bu}) \tag{9}$$

and since, from group additivity:

$$\Delta_f H(\text{t-BuO}_3\text{t-Bu}) - \Delta_f H(\text{t-BuO}_2\text{t-Bu}) = \Delta_f H[\text{O—(O)}_2] \tag{10}$$

we obtain:

$$\Delta_r H(3) - \Delta_r H(2) = \Delta_f H(\text{t-BuO}^\bullet) - \Delta_f H(\text{t-BuO}_2^\bullet) + \Delta_f H[\text{O—(O)}_2] \tag{11}$$

Hence, from equations (8) and (11):

$$\Delta_f H[\text{O—(O)}_2] = {}^1\!/\!_2\Delta_r H(3) - {}^1\!/\!_2\Delta_r H(1) - \Delta_f H(\text{t-BuO}^\bullet) + \Delta_f H(\text{t-BuO}_2^\bullet) \tag{12}$$

We thus need to evaluate the four terms on the right-hand side of equation (12) to reach our goal.

For $\Delta_r H(3)$ we accept the value of 38.0 ± 1 kcal mol^{-1} determined by Batt *et al.* [9]. $\Delta_f H(\text{t-BuO}_2^\bullet)$ is derived by considering the dissociation of t-BuO$_2$H:

$$DH^\circ(\text{t-BuO}_2\text{—H}) = \Delta_f H(\text{t-BuO}_2^\bullet) + \Delta_f H(\text{H}) - \Delta_f H(\text{t-BuO}_2\text{H}) \tag{13}$$

where DH° is the bond dissociation enthalpy (BDE). $DH^\circ(\text{t-BuO}_2\text{–H})$ was determined by Heneghan and Benson [10] to be 89.4 ± 0.2 kcal mol^{-1}. Nevertheless, we believe it may be more reliable to give more weight to the value of $DH^\circ(\text{t-BuO}_2\text{—H}) = DH^\circ(\text{HOO—H})$, which is well established as 88.2 kcal mol^{-1}. Taking both into account, we shall use a universal BDE for ROO—H of 88.6 ± 0.5 kcal mol^{-1}. This assumption is defended further in Section 2. Measurement of $\Delta_f H(\text{t-BuO}_2\text{H})$ gives a value of -58.8 kcal mol^{-1} [5]; from group additivity [11] we calculate -58.0 kcal mol^{-1}, which we expect is more reliable than the measured value. Hence:

$$\begin{aligned}\Delta_f H(\text{t-BuO}_2^\bullet) &= DH^\circ(\text{t-BuO}_2\text{—H}) + \Delta_f H(\text{t-BuO}_2\text{H}) - \Delta_f H(\text{H}) \\ &= (88.6 \pm 0.5) - (58.0 \pm 0.8) - 52.1 \\ &= -21.5 \pm 0.9\,\text{kcal mol}^{-1}\end{aligned} \tag{14}$$

$\Delta_f H(\text{t-BuO}^\bullet)$ can be deduced by various methods. It can be evaluated if we assume that $DH^\circ(\text{t-BuO—H}) = DH^\circ(\text{CH}_3\text{O—H})$. If this relationship is accepted, then it can be shown that:

$$\Delta_f H(\text{t-BuO}_2\text{H}) - \Delta_f H(\text{t-BuO}^\bullet) = \Delta_f H(\text{CH}_3\text{OOH}) - \Delta_f H(\text{CH}_3\text{O}^\bullet) \tag{15}$$

$\Delta_f H(CH_3O^{\cdot})$ was determined by Batt and coworkers [12] to be 3.9 kcal mol^{-1}. $\Delta_f H(CH_3OOH)$ has not been measured, but can be calculated by group additivity (see above) to be −31.4 kcal mol^{-1}. Hence:

$$\begin{aligned} \Delta_f H(\text{t-BuO}^{\cdot}) &= -58.0 + 31.4 + 3.9 \\ &= -22.3 \text{ kcal mol}^{-1} \end{aligned} \tag{16}$$

Alternatively, $\Delta_f H(\text{t-BuO}^{\cdot})$ can be calculated from the experimental data of Batt and Milne [13] on the equilibrium t-BuONO $\rightleftharpoons$ t-BuO$^{\cdot}$ + NO. Using GAVs to estimate $\Delta_f H$(t-BuONO), we find $\Delta_f H(\text{t-BuO}^{\cdot}) = -23.0$ kcal mol^{-1}; we prefer the former calculation. We can now substitute three of the terms on the right-hand side of equation (12) and obtain:

$$\Delta_f H[\text{O—(O)}_2] = (19.0 \pm 0.5) - {}^1\!/_2\Delta_r H(1) + (22.3 \pm 1) - (21.5 \pm 0.9) \tag{17}$$

It remains only to evaluate $\Delta_r H(1)$.

For this determination we start with several experimental measurements in liquid solution on the following equilibrium (18):

$$\text{ROOOOR} \rightleftharpoons 2\text{RO}_2^{\cdot} \tag{18}$$

This was originally carried out for R = t-Bu at a mean temperature of 140 K in solution by Bartlett and Guaraldi [14], and later confirmed by Adamic, Howard and Ingold [15]. Since then, other systems have been studied and all showed very similar thermodynamic quantities, as would be expected if the reaction is a simple bond cleavage as shown above, with the R group unchanged in the reaction. The various systems studied, with the resulting values reported for ΔH and ΔS, are listed in Table 4. All the entropy changes are between 29 and 44 gibbs mol^{-1}, with reported uncertainties being between

Table 4 Experimental data obtained for the ROOOOR $\rightleftharpoons$ $2RO_2$ equilibria

R	ΔH (kcal mol^{-1})	ΔS	Ref.
2-Methyl-2-butyl	−8.9 ± 1	−36 ± 8	16
2-Methyl-2-pentyl	−8.9 ± 0.7	−39 ± 6	16
3-Methyl-2-pentyl	−9.7 ± 0.5	−44 ± 4	16
2,3-Dimethyl-2-butyl	−8.2 ± 0.5	−34 ± 4	16
2,3,3-Trimethyl-2-butyl	−8.7 ± 1	−38 ± 7	16
i-Propyl	−8.0	−25	17
t-Butyl	−8.2	−27	14
	−8.8 ± 0.4	−34 ± 1	15
	−8.0 ± 0.2	−31 ± 1	15
	−8.4	−30	15
2-Ethyl-2-propyl	−7.5 ± 0.3	−29 ± 1	15
2-i-Propyl-2-propyl	−8.6 ± 0.4	33 ± 1	15

1 and 8 gibbs mol^{-1}. In most cases, the entropies are derived from van't Hoff extrapolations of equilibrium data and entail considerable imprecision in the resulting values. The expected entropy change for the reaction can be estimated by calculating the entropies of suitable model compounds for the reactant and products. The estimation procedure is itself subject to various uncertainties, but it seems fair to conclude that the expected ΔS should be ca. 29 ± 3 gibb mol^{-1}. The deduced enthalpies should be adjusted accordingly, but the corrections will generally be smaller than the reported uncertainties. Hence, it seems reasonable to conclude that $\Delta_r H_{140}$(18, 1M soln) is 8.5 ± 0.5 kcal mol^{-1}.

This value of $\Delta_r H_{140}$(18, soln) needs to be corrected by a cycle of processes to give $\Delta_r H^\circ_{298}$(18, gas). This requires three steps: (soln, 140 K) $\longrightarrow$ (liq., 140 K) (i) $\longrightarrow$ (liq., 298 K) (ii) $\longrightarrow$ (gas, 298 K) (iii).

(i) Nangia and Benson [8] argued that the heat of mixing in a relatively non-polar solvent such as was used in the experiments should be small, i.e. of the order of a few tenths of a kcal mol^{-1}. Thus, taking into account one reactant and two product species, this contributes 0.5 kcal mol^{-1} or less to $\Delta_r H^0_{298}$(18, gas), and consequently can be neglected.

(ii) The temperature correction again requires C_p(liq.) for the products and reactant over the temperature range 140–298 K. Approximate values can be obtained by using hydrocarbon model compounds for ROOOOR and for $RO_2^{\cdot}$. Values of C_p(liq.) for both of these can be evaluated by using polynomial expansions for the group additivity values for liquid hydrocarbons calculated by Luria and Benson [18] (we take R = t-Bu and use $R(CH_2)_4R$ and RCH_2CH_3 as models). We thus estimate ΔC_p(liq.) = 3.6 gibbs mol^{-1}, so the contribution to $\Delta_r H(18)$ is $\langle C_p \rangle \Delta T =$ 0.6 kcal mol^{-1}.

(iii) The remaining correction term involves $\Delta_{vap}H$ for the reactant and products and is not small. For this step, we use t-Bu$(CH_2)_4$t-Bu to model the reactant. We estimate T_b for this alkane to be 448 K by comparison with other alkanes. Then, Hildebrand's rule [19] ($\Delta_{vap}H^\circ_{298} = -2950 + 23.7T_b + 0.02(T_b)^2$) gives $\Delta_{vap}H(298) = 9.4$ kcal mol^{-1}. If we then replace the —CH_2—CH_2— groups by —O—O—, we can calculate, by group additivity, that this should increase $\Delta_{vap}H$ by 1.4 kcal mol^{-1}. We cannot calculate accurately the effect of changing two more CH_2 groups into O groups, but the change is probably the same, within ± 1 kcal mol^{-1}. This would give $\Delta_{vap}H$(t-BuO$_4$t-Bu) $= 12.5 \pm 1$ kcal mol^{-1}.

Evaluating $\Delta_{vap}H$(t-$BuO_2^{\cdot}$) is somewhat more problematic. We could model it with neopentyl fluoride, as we did in estimating C_p in order to correct the entropy, but $\Delta_{vap}H$ for this fluoride is not known. We can estimate $\Delta_{vap}H$ from Trouton's rule if we know the boiling point; unfortunately, this is also unknown. Borrowing from the closest compound for which T_b is known, namely 2-fluoro-2-methylbutane, we assume T_b to be 315 K. From Trouton's rule, $\Delta_{vap}H = 10.4RT_b = 6.5$ kcal mol^{-1}. If instead of neopentyl fluoride we used neohexane, since replacing a CH_3 group by F has little effect on T_b, we would calculate $\Delta_{vap}H = 10.4R(323) = 6.7\,kcal\,mol^{-1}$. Alternatively, we could model the peroxyl radical with pivaldehyde, t-Bu-CHO. By using group additivities, we can calculate $\Delta_{vap}H$ for this compound to be $7.5\,kcal\,mol^{-1}$ [20]. By considering the three methods, then, we estimate $\Delta_{vap}H^{\circ}_{298}(\text{t-BuO}_2^{\bullet}) = 7 \pm 0.5\,kcal\,mol^{-1}$. The effect on $\Delta_r H^{\circ}_{298}(18,\text{ gas})$ is $(-12.5 \pm 1) + 2(6.8 \pm 0.5) = (1.1 \pm 1.1)\,kcal\,mol^{-1}$. Hence, by taking all corrections into account, $\Delta_r H^{\circ}_{298}(18,\text{ gas}) = (8.5 \pm 0.5) + 0.6 + (1.1 \pm 1.1) = 10.2 \pm 1.3\,kcal\,mol^{-1}$.

We can now substitute this value into equation (17):

$$\begin{aligned} \Delta_f H[\text{O—(O)}_2](\text{g}) &= (19.8 \pm 1.4) - {}^1\!/_2\Delta_r H(1) \\ &= 14.7 \pm 1.5\,\text{kcal mol}^{-1} \end{aligned} \tag{19}$$

We can estimate the corresponding GAV for the liquid phase by estimating $\Delta_{vap}H(CH_3OOOCH_3)$ at 298 K using Hildebrand's rule. We estimate T_b by comparing the boiling points of n-C_5H_8 (309 K), $CH_3OC_3H_7$ (312), and $CH_3OCH_2OCH_3$ (318), and assume $T_b = 325$ K. From other (known) GAVs we find that $\Delta_f H[\text{O—(O)}_2](\text{liq.})$ must be $13.8\,kcal\,mol^{-1}$. With the value of $\Delta_f H[\text{O—(O)}_2]$ now established, we can calculate the enthalpies for RO_nR and RO_nH for any n. The values for several molecules thus calculated are shown in Table 5.

Table 5 Estimated values of $\Delta_f H^0_{298}(g)$ (kcal mol^{-1})[a]

Group	H	CH_3	t-Bu
R	52.1[b]	35.1 ± 0.2[c]	9.4 ± 1.5[d]
RO	9.3[b]	3.9[e]	−22.3[f]
RO_2	3.5 ± 0.5[g]	5.1 ± 1.1[h]	−21.9[f]
RO_3	18.0	19.4	−7.2
RO_4	32.8	34.1	7.5
RO_5	47.5	48.8	22.2
RO_2H	−32.6[b]	−31.2[f]	−58.0
RO_3H	−17.8	−16.5	−43.3
RO_4H	−3.1	−1.8	−28.6
RO_5H	11.6	12.8	−13.9
RO_2R		−30.0	−83.2
RO_3R		−15.3	−68.5
RO_4R		−0.6	−53.8
RO_5R		14.1	−39.1

[a] All values estimated by group additivity unless noted otherwise.
[b] See Ref. 21.
[c] See Ref. 22.
[d] There is still considerable disagreement regarding this quantity; see Refs 11 and 22 for discussion.
[e] See Ref. 9.
[f] See text.
[g] See Refs 23 and 24, and also text.
[h] See Ref. 25 (revised), and text.

1.3 FREE RADICALS

The enthalpies for alkylperoxy radicals ROO$^{\cdot}$ can be calculated by using the group additivity value for one additional group, i.e. O—(C)(O$^{\cdot}$). Acylperoxy and arylperoxy radicals have the O—(CO)(O$^{\cdot}$) and the O—(C_B)(O$^{\cdot}$) groups, respectively. (The hydroperoxyl radical is identical to the O—(H)(O$^{\cdot}$) group.) To evaluate each of these free-radical groups, we need one experimental radical enthalpy in each case. From equation (14) we find that $\Delta_f H^\circ_{298}$(t-$BuO_2^{\cdot}$) = -21.9 ± 1 kcal mol^{-1}. Replacing the radical by its groups, we have:

$$3\Delta_f H[\mathrm{C—(H)_3}] + \Delta_f H[\mathrm{C—(C)_3(O)}] + \Delta_f H[\mathrm{O—(C)(O^{\cdot})}] = -22\ \mathrm{kcal\,mol^{-1}} \quad (20)$$

Using the established GAVs for the first two groups (-10 and -6.6 kcal mol^{-1}, respectively), we obtain:

$$\Delta_f H[\mathrm{O—(C)(O^{\cdot})}] = 30 + 6.6 - 21.9 = 14.7 \pm 1\ \mathrm{kcal\,mol^{-1}} \quad (21)$$

In the absence of experimental data on acylperoxy and arylperoxy radicals, we can derive the GAVs by again assuming the constancy of the ROO—H bond strength, i.e. DH°(RC(O)OO—H) = DH°(PhOO—H) = DH°(ROO—H). These assumptions give the following relationships:

$$\Delta_f H(\mathrm{RC(O)OO^{\cdot}}) - \Delta_f H(\mathrm{RC(O)OOH}) = \Delta_f H(\mathrm{HOO^{\cdot}}) - \Delta_f H(\mathrm{HOOH}) \quad (22)$$

and

$$\Delta_f H(\mathrm{PhOO^{\cdot}}) - \Delta_f H(\mathrm{PhOOH}) = \Delta_f H(\mathrm{HOO^{\cdot}}) - \Delta_f H(\mathrm{HOOH}) \quad (23)$$

From equation (22) we obtain:

$$\begin{aligned}\Delta_f H[\mathrm{O—(CO)(O^{\cdot})}] &= -\Delta_f H(\mathrm{HOOH}) + \Delta_f H(\mathrm{HO_2^{\cdot}}) + \\ &\qquad \Delta_f H[\mathrm{O—(O)(CO)}] + \Delta_f H[\mathrm{O—(O)(H)}] \\ &= 32.6 + 3.5 - 18.2 - 16.3 = 1.6\ \mathrm{kcal\,mol^{-1}} \quad (24)\end{aligned}$$

while from equation (23) we obtain:

$$\begin{aligned}\Delta_f H[\mathrm{O—(C_B)(O^{\cdot})}] - \Delta_f H[\mathrm{O—(C_B)(O)}] - \Delta_f H[\mathrm{O—(O)(H)}] &= \Delta_f H(\mathrm{HOO^{\cdot}}) - \Delta_f H(\mathrm{HOOH}) \\ \Delta_f H[\mathrm{O–(C_B)(O^{\cdot})}] = \Delta_f H[\mathrm{O–(C_B)(O)}] - \Delta_f H[\mathrm{O–(O)(H)}] &+ \Delta_f H(\mathrm{HOO^{\cdot}}) \\ = -5 + 16.3 + 3.5 = 14.8\ \mathrm{kcal\,mol^{-1}} & \quad (25)\end{aligned}$$

Polyoxide radicals require the O–(O)(O$^{\cdot}$) group, in addition to molecular groups. This group can be evaluated if we assume that DH°(t-BuOOO–H) = DH°(t-BuOO–H). From this assumption:

$$\begin{aligned}&\Delta_f H(\mathrm{t\text{-}BuOOO^{\cdot}}) - \Delta_f H(\mathrm{t\text{-}BuOOOH}) \\ &= \Delta_f H(\mathrm{t\text{-}BuOO^{\cdot}}) - \Delta_f H(\mathrm{t\text{-}BuOOH}) \quad (26)\end{aligned}$$

from which:

$$\Delta_f H[\text{O—(O)(O}^\cdot)] = \Delta_f H[\text{O—(O)}_2] + \Delta_f H[\text{O—(O)(H)}] + \Delta_f H(\text{t-BuOO}^\cdot) - \Delta_f H(\text{t-BuOOH})$$
$$= 14.7 - 16.3 - 21.9 + 58.0 = 34.5\,\text{kcal mol}^{-1} \quad (27)$$

2 EXPERIMENTAL STUDIES OF THE THERMOCHEMISTRY OF PEROXY RADICALS

Because of the usually transient existence of free radicals in chemical reactions, their thermochemistry has been among the most elusive properties to establish. Such stable radicals as O_2, NO and NO_2 are notable exceptions. Among the best established values have been those for atoms, usually arrived at theoretically from precise spectroscopic measurements. Peroxy radicals have been particularly elusive, in part because of their manifold ways of reacting. Currently accepted values, which we shall discuss here, have been established from the observation of equilibrium or near-equilibrium reactions. Two equilibria which have been employed are:

$$RO_2H + X \rightleftharpoons RO_2^\cdot + HX \quad (28)$$

$$R + O_2 \rightleftharpoons RO_2^\cdot \quad (29)$$

where X is usually an atom or radical, generally chosen to make reaction (28) as close to thermoneutral as possible. Br is the ideal candidate in this respect.

If $\Delta_r H_T^\circ$ can be measured or established for reaction (28), $\Delta_f H_T^\circ(RO_2^\cdot)$ can be deduced only if the values of $\Delta_f H_T^\circ$ are known for the other species, RO_2H, HX and X. Of these $\Delta_f H_T^\circ(RO_2H)$ is the least well known. Similarly, equilibrium (29) requires the knowledge of $\Delta_f H_T^\circ(R^\cdot)$ in order to deduce $\Delta_f H_T^\circ(RO_2^\cdot)$.

Since the last review of peroxy radical thermochemistry [26] there has been much theoretical work carried out on the structures of the peroxy radicals of both hydroperoxides and peroxides. Very extensive *ab initio* calculations of the structures, dipole moments and energies of different conformations have been made by Boyd, Boyd and Barclay [27]. Such calculations are generally quite reliable for structural properties (bond angles, bond lengths and dipole moments) but less accurate for total energies.

2.1 THE $HO_2^\cdot$ RADICAL

The value of $\Delta_f H^\circ(HO_2^\cdot)$ is probably the best established for all of the peroxy radicals. The most recent review [23] gives a best value for $\Delta_f H_{298}^\circ(HO_2^\cdot)$ of 3.5 (+1.0, −0.5) kcal mol^{-1}. Over eight different types of measurement and equilibria have been used to obtain values and they all agree to within their experimental uncertainties with this value. These include the following chemical equilibria:

$$Br + H_2O_2 \rightleftharpoons HBr + HO_2\ (-1\,\text{kcal mol}^{-1}\ (+4.5\,\text{eu})) \quad (30)$$

$$NO + HO_2 \rightleftharpoons NO_2 + OH\ (+7.8\,\text{kcal mol}^{-1}\ (-3.9\,\text{eu})) \quad (31)$$

$$Cl + HO_2 \rightleftharpoons HO + ClO\ (-1.1\,\text{kcal mol}^{-1}\ (+3.7\,\text{eu})) \quad (32)$$

The heats of reaction at 298 K are shown, together with standard entropy changes; 1 cal mol^{-1} K^{-1} = 1 eu at 298 K (shown in parentheses).

For both the Br/H_2O_2 and the Cl/HO_2 reactions the standard free energy changes, $\Delta G^\circ = \Delta H^\circ - T\Delta S^\circ$, are close to zero at 298 K, so that the forward and reverse rate constants should be nearly equal at 298 K. For the NO/HO_2 reaction, $\Delta_r G^\circ_{298} = 6.6$ kcal mol^{-1} so that the latter must be studied at much higher temperatures, i.e. above 600 K.

None of the systems are free from competing side reactions, so that their study requires demonstrations that these side reactions do not interfere with the equilibrium measurement. In the Br/H_2O_2 system there is a very rapid reaction between Br and HO_2 to produce HBr + O_2, thus preventing equilibrium from being attained. In the Cl/HO_2 system, the fastest reaction is to produce HCl + O_2 [28]. We shall adopt the value of 3.5 ± 0.5 kcal mol^{-1} for $\Delta_f H^\circ_{298}(HO_2^{\bullet}$, gas).

2.2 THE $CH_3O_2^{\bullet}$ RADICAL

Khachatryan *et al.* [29] have measured the equilibrium constant for the following equilibrium:

$$CH_3 + O_2 \rightleftharpoons CH_3O_2^{\bullet} \quad (33)$$

from 706–786 K by quenching the mixture in liquid N_2 and analysing for radicals. They reported a value of ΔH in this range and used a mean value for $\langle \Delta C_p \rangle$ of 1.8 cal mol^{-1} K^{-1}. This gives $\Delta_r H^\circ_{298} = -32.2 \pm 1.5$ kcal mol^{-1} for the above equilibrium. Together with the value of $\Delta_f H^\circ_{298}(CH_3)$ of 35.1 kcal mol^{-1}, this gives $\Delta_r H^\circ_{298}(CH_3O_2^{\bullet}) = 2.9 \pm 1.5$ kcal mol^{-1}.

Kondo and Benson [25] measured the reaction of Br atoms with CH_3O_2H at low pressures. They observed a relatively slow rate constant with an activation energy of 3.2 ± 0.4 kcal mol^{-1}:

$$CH_3O_2H + Br \rightleftharpoons HBr + CH_3O_2^{\bullet} \quad (34)$$

Attempts to observe the reverse reaction by adding HBr led only to a lower limit of $K(34) \geq 0.46$ at 333 K. Estimating $\Delta_r S^\circ_{298}(34) = 3.6 \pm 1.1$ cal mol^{-1} K^{-1}, we calculate $\Delta_r H^\circ_{298}(34) \leq 0.7$ kcal mol^{-1}, which fixed $DH^\circ_{298}(CH_3O_2\text{–}H) \geq 88.2$ kcal mol^{-1}. The numbers that Kondo and Benson reported are slightly different because these authors used the upper limit of 4.7 cal mol^{-1} K^{-1} for $\Delta_r S^\circ_{298}(34)$. From these values we can calculate $\Delta_f H^\circ_{298}(CH_3O_2^\bullet) \leqslant 4.8$ kcal mol^{-1}.

Slagle and Gutman [30] have reported on a study of equilibrium (33) from 694 to 811 K in flash-photolysis-type experiments. They measured the relative CH_3 radical concentrations by mass spectrometric sampling using threshold photoionization at an ion mass of 15 amu. Since they were unable to determine the absolute concentrations of CH_3, they assumed that in the presence of O_2 some CH_3O_2 is formed, reducing CH_3 to a new stationary level, and that all of the CH_3 used up is in the form of CH_3O_2, which, however, they did not analyse for independently. They obtained $\Delta_r H^\circ_{298}(33) = -30.9 \pm 2.5$ kcal mol^{-1} from a van't Hoff plot of $\ln K(33)$ against $1/T$; $\Delta_r S^\circ_{298}(33) = -29.1 \pm 3.5$ cal mol^{-1} K^{-1}. By using calculated values of $\Delta C^\circ_p(33)$ and $\Delta S^\circ_{298}(33)$ over the range 300–800 K they made third-law estimates of $\Delta_r H^\circ_{298}(33)$ of -32.4 ± 0.7 kcal mol^{-1}. These authors used $\Delta_r S^\circ_{298}(33) = -31.0$ eu and obtained $\Delta_f H^\circ_{298}(CH_3O_2^\bullet) = 2.7 \pm 0.8$ kcal mol^{-1}, in very good agreement with Khachatryan *et al.* [29].

We can compare all of these values with a value obtained from an assigned bond strength for (ROO—H) of 88.6 ± 0.5 kcal mol^{-1} for all hydroperoxides. This gives $\Delta_f H^\circ_{298}(CH_3O_2^\bullet) = 5.2 \pm 1.1$ kcal mol^{-1} and $DH^\circ_{298}(CH_3O_2) = 30.0 \pm 1.1$ kcal mol^{-1}.

2.3 THE $C_2H_5O_2^\bullet$ RADICAL

Studies of the equilibrium:

$$C_2H_5^\bullet + O_2 \rightleftharpoons C_2H_5O_2^\bullet \qquad (35)$$

have been reported by Slagle, Ratajazak and Gutman [31] using the same technique described by this group for $CH_3 + O_2$ [30]; the temperature range was 609–654 K. They were unable to measure $C_2H_5O_2^\bullet$ radicals directly, and so had to rely solely on the temporal behavior of the C_2H_5 signal.

By fitting the observed ethyl signal to a double exponential decay they deduced the ratio of forward and reverse rate constants from which they obtain $K_{eq}(T)$ with an estimated uncertainty of not more than a factor of 2. Using third-law estimates of $\Delta_r S_{298}(35)$ and $\Delta C_p(35)$ over the temperature range 300–800 K they deduced $\Delta_r H^\circ_{298}(35) = -35.2 \pm 1.5$ kcal mol^{-1}. In a later study [26] they revised their estimate of $\Delta_r S^\circ_{298}(35)$ by 1.5 eu, leading to a new value of $\Delta_r H^\circ_{298}(35)$ of -34.1 ± 0.5 kcal mol^{-1}. This can be compared to a value estimated by using group additivity to obtain $\Delta_f H^\circ_{298}(EtO_2(EtO_2H) = -39.5$ kcal mol^{-1} and a bond strength of 88.6 ± 0.5 kcal mol^{-1} for alkyl hydroxperoxides. From these values, we find $\Delta_f H^\circ_{298}(EtO_2^\bullet) = -2.9$ kcal mol^{-1} and $\Delta_r H^\circ_{298}(35) = DH^\circ_{298}(Et—O_2^\bullet) = 31.3 \pm 1.1$ kcal mol^{-1}.

2.4 THE i-PrO$_2^{\bullet}$ RADICAL

Studies have been made by Slagle *et al.* [32] of the following equilibrium (at 592–692 K):

$$\text{i-Pr}^{\bullet} + O_2 \rightleftharpoons \text{i-PrO}_2^{\bullet} \qquad (36)$$

by using the same techniques described in their earlier work on CH_3 and C_2H_5 radicals with O_2 [30, 31].

Using a third-law treatment of their data with an estimated $\Delta_r S^{\circ}_{298}(36) = -39.9\ \text{cal mol}^{-1}\text{K}^{-1}$ and an average $\Delta C_p(36) = 1.3\ \text{cal mol}^{-1}\text{K}^{-1}$ they estimated $\Delta_r H^{\circ}_{298}(36)$ to be $-39.7\ \text{kcal mol}^{-1}$.

We estimate $\Delta_r S^{\circ}_{298}(36) = -37.0\ \text{cal mol}^{-1}\text{K}^{-1}$, while from group additivity estimates of $\Delta_f H^{\circ}_{298}(\text{i-PrO}_2\text{H}) = -48.5 \pm 0.6\ \text{kcal mol}^{-1}$ and our universal $DH^{\circ}_{298}(\text{RO}_2\text{–H}) = 88.6 \pm 0.5$ we obtain $\Delta_f H^{\circ}_{298}(\text{i-PrO}_2^{\bullet}) = 12.2 \pm 0.8\ \text{kcal mol}^{-1}$, and with $\Delta_f H^{\circ}_{298}(\text{i-Pr}) = 20.0 \pm 1.0\ \text{kcal mol}^{-1}$ we obtain $\Delta_r H^{\circ}_{298}(36) = 32.2 \pm 1.3\ \text{kcal mol}^{-1}$, in very marked disagreement with the above studies.

2.5 THE t-BuO$_2^{\bullet}$ RADICAL

The following equilibrium:

$$\text{t-Bu} + O^{\bullet}_2 \rightleftharpoons \text{t-BuO}_2^{\bullet} \qquad (37)$$

has been studied and reported at 550–580 K using the techniques already described by Slagle, Ratajazak and Gutman [31]. Using an estimated $\Delta_r S^{\circ}_{298}(37) = -41 \pm 2\ \text{cal mol}^{-1}\text{K}^{-1}$ they calculated from their measurements of $K(37)$ that $\Delta_r H^{\circ}_{298}(37) = -36.7 \pm 1.9\ \text{kcal mol}^{-1}$. If we use our value of $\Delta_f H^{\circ}_{298}(\text{t-Bu}^{\bullet}) = 9.5 \pm 0.6\ \text{kcal mol}^{-1}$ [33], this gives $\Delta_f H^{\circ}_{298}(\text{t-BuO}_2^{\bullet}) = -27.2\ \text{kcal mol}^{-1}$. Using the value of $\Delta_f H^{\circ}_{298}(\text{t-BuO}_2\text{H}) = -57.6 \pm 0.6\ \text{kcal mol}^{-1}$ from group additivity we find $DH^{\circ}_{298}(\text{t-BuO}_2\text{—H}) = 84.8\ \text{kcal mol}^{-1}$, i.e. $3.8\ \text{kcal mol}^{-1}$ different from our universal value of $88.6 \pm 0.5\ \text{kcal mol}^{-1}$. On the other hand, by using their own suggested value of $\Delta_f H^{\circ}_{298}(\text{t-Bu}^{\bullet}) = 12.3\ \text{kcal mol}^{-1}$, Slagle and coworkers [31] obtain $\Delta_f H^{\circ}_{298}(\text{t-BuO}_2^{\bullet}) = -24.4\ \text{kcal mol}^{-1}$ and $DH^{\circ}_{298}(\text{t-BuO}_2\text{—H}) = 82.0\ \text{kcal mol}^{-1}$, which are in even worse agreement with the group additivity values.

2.6 ALKYLPEROXY RADICALS: GENERAL COMMENTS

It is clear that there are systematic discrepancies between all of the values reported from the flash photolysis studies [30–32] and the group additivity estimates. It is our feeling that the flash photolysis studies, which follow a single component, are susceptible to major errors. One of these is the occurrence of alternate reactions at the temperatures employed to measure the equilibrium. Perhaps the most important is the isomerization reaction.

Taking the ethyl radical as an example:

$$\mathrm{Et^0 + O_2 \underset{}{\overset{(35)}{\rightleftharpoons}} EtO_2^{\bullet} \overset{(i)}{\rightleftharpoons} {}^{\bullet}CH_2CH_2O_2H \rightleftharpoons C_2H_4 + HO_2^{\bullet}} \quad (38)$$

the rate constant for reaction (i) has been estimated as $k_i = 10^{12.5-27/\theta}\,\mathrm{s^{-1}}$, where $\theta = 2.303\,RT$ in kcal mol^{-1}. At 630 K, which is about the middle of the range used to study equilibrium (35), $k_i = 10^{3.13}\,\mathrm{s^{-1}}$. The competing reaction k_{35} has a high-pressure rate constant at 630 K, estimated as $10^{14.6-38.8/\theta}\,\mathrm{s^{-1}}$. At 630 K this becomes $k_{35} = 10^{4.9}\,\mathrm{s^{-1}}$, so that it is larger than k_i by a factor of approximately 60. However, k_{35} is very much below its high-pressure limit at the pressures used [34] so that k_i can introduce a significant error. This can also happen for the i-PrO_2 and t-BuO_2 species, where the isomerization rate constants are expected to be two- and threefold faster, respectively, than that of the ethyl species on statistical grounds (i.e. number of H atoms). Note that if we use the higher reported bond dissociation energy, k_{35} becomes $10^{14.6-34.6/\theta}\,\mathrm{s^{-1}} \approx 10^4\,\mathrm{s^{-1}}$ at 630 K, i.e. only a factor of seven faster than k_i.

However, perhaps the greatest difficulty with the flash photolysis data comes with the data themselves. The authors [30–32] used a double exponential decay relationship to describe the decrease in the ethyl radical decay:

$$[C_2H_5^{\bullet}]/[C_2H_5^{\bullet}]_0 = A\exp(-m_1t) + B\exp(-m_2t) \quad (39)$$

where at $t = 0$, $[Et] = [Et]_0$ so that $A + B = 1$, with both A and $B > 0$. It can be shown that $(m_1 + m_2) = (k_{35} + k_{-35} + k_w)$ and $m_1m_2 = k_{35}k_w$, where k_w is the supposed rate of destruction of ethyl radicals on the walls of the vessel. Since they report values of m_1, m_2 and k_w, it is a very simple exercise to use the above relationships to estimate k_{35} and k_{-35}. After carrying this out, negative values are found for k_{35}, and also in some cases for k_{-35}. There is thus an inconsistancy between the reported data and the equations used to describe them, which throws the techniques used and/or the interpretation into question.

2.7 THE ALLYLPEROXY RADICAL

In contrast to the high-temperature flash photolysis studies of the alkylperoxy radicals in the presence of O_2, is the much simpler study of the equilibrium of allylperoxy radicals at 382–453 K [35] with O_2 and allyl radicals.

Using flash photolysis of biallyl at 193 nm to generate allyl radicals in the presence of O_2, Morgan and coworkers [35] could follow the allyl radical decay by its optical absorption at 223 nm. Because of the short times (1–3 ms) and high pressures used, any wall reactions were unimportant and the allyl radical concentrations were found to decay to a constant level. No competing reactions were seen and the system showed the typical decay of a reversible reaction:

$$\text{Allyl} + O_2 \rightleftharpoons \text{Allyl } O_2^{\bullet} \quad (40)$$

From the data it was possible to obtain values of $K(40)$ from 382–453 K. A van't Hoff plot of $\ln K(40)$ against $1/T$, using the appropriate correction for $\langle \Delta C_p(40) \rangle$ gave $\Delta_r H^\circ_{298}(40) = -18.2 \pm 0.5\,\text{kcal mol}^{-1}$ and $\Delta_r S^\circ_{298}(40) = 29.2 \pm 1.2\,\text{cal mol}^{-1}\,\text{K}^{-1}$. Using a value for $\Delta_f H^\circ_{298}(\text{allyl}) = 41 \pm 0.6\,\text{kcal mol}^{-1}$ this gives a value of $\Delta_f H^\circ_{298}(\text{allyl } O_2^{\bullet}) = 22.8 \pm 0.8\,\text{kcal mol}^{-1}$. Using $DH^\circ_{298}(\text{allyl } O_2\text{—H}) = 88.6 \pm 0.5\,\text{kcal mol}^{-1}$ this gives $\Delta_f H^\circ_{298}(\text{allyl } O_2H) = -13.7 \pm 1.0\,\text{kcal mol}^{-1}$. From group additivity we obtain a value of $-14.1 \pm 0.6\,\text{kcal mol}^{-1}$, which is in excellent agreement. The value of $\Delta_r S^\circ_{298}(40)$ is also in excellent agreement with that obtained from group additivity. This simple study shows an extraordinary consistency between group additivity and the thermochemistry of allyl- and allylperoxy radicals. It constitutes a strong confirmation of group additivity and the assumption of a universal bond dissociation energy for hydroperoxides.

2.8 THE HYDROXYMETHYLPEROXY RADICAL

This rather unusual peroxy radical is of importance in the chemistry of air pollution. It is formed in a complex chemical system by the addition of the HO_2 radical to CH_2O [36]:

$$HO_2^{\bullet} + CH_2O \overset{a}{\rightleftharpoons} HOOCH_2O^{\bullet} \overset{b}{\rightleftharpoons} {}^{\bullet}OOCH_2OH \quad (41)$$

The radical undergoes a reversible isomerization back to the oxy-radical intermediate, and over the temperature range from 300–33 K reaches equilibrium in a few milliseconds with ($HO_2^{\bullet} + CH_2O$). It has been possible to study this equilibrium by following the optical absorption of $RO_2^{\bullet}$ at 200–280 nm. The HO_2 absorption has a maximum at 210 nm, while the $RO_2^{\bullet}$ maximum is at 230 nm, so the two can be measured simultaneously. From group additivity it is possible to estimate $\Delta_f H^\circ_{298}(RO_2^{\bullet}) = -38.2\,\text{kcal mol}^{-1}$. Using $\Delta_f H^\circ_{298}(HO_2^{\bullet}) = 3.5 \pm 0.5$ and $\Delta_f H^0_{298}(CH_2O) = -26.0\,\text{kcal mol}^{-1}$, this gives for the overall equilibrium $(a + b)$ a value for $\Delta H^\circ_{298}(a, b)$ of $-15.7\,\text{kcal mol}^{-1}$. The experimental value is $-16.3 \pm 0.3\,\text{kcal mol}^{-1}$, which is in excellent agreement, and once again provides strong support for group additivity and the universal RO_2—H bond dissociation energy.

REFERENCES

1. S.W. Benson and R. Shaw, in *Organic Peroxides,* edited by D. Swern, Vol. 1, John Wiley & Sons, New York (1970), p. 105.

2. See, for example: (a) S.W. Benson and J.H. Buss, *J. Chem. Phys.* **29,** 546 (1958); (b) S.W. Benson, *Chem. Rev.* **69,** 279 (1969); (c) S.W. Benson, *Thermochemical Kinetics*, 2nd Ed., John Wiley & Sons, New York (1976).
3. A.C. Baldwin, in *The Chemistry of Functional Groups, Peroxides*, edited by S. Patai, John Wiley & Sons, New York (1983), p. 97.
4. B. Plesnicar, in *Organic Peroxides*, edited by W. Ando, John Wiley & Sons, New York (1992), p. 479; B. Plesnicar, in *The Chemistry of Functional Groups, Peroxides*, edited by S. Patai, John Wiley & Sons, New York (1983), p. 483; J.S. Francisco and I.H. Williams, *Int. J. Chem. Kinet.* **20,** 455 (1988).
5. J.B. Pedley, R.D. Naylor and S.P. Kirby, *Thermochemical Data of Organic Compounds*, Chapman and Hall, London (1986).
6. S. Stein, *NIST Structures and Properties,* NIST-SRD DB 25, version 2.0, Washington, DC (1994).
7. N. Cohen, *J. Phys. Chem. Ref. Data* **25** (1996), in press.
8. P.S. Nangia and S.W. Benson, *J. Phys. Chem.* **83,** 1138 (1979).
9. L. Batt, K. Christie, R.T. Milne and A.J. Summers, *Int. J. Chem. Kinet.* **6,** 877 (1974).
10. S.P. Heneghan and S.W. Benson, *Int. J. Chem. Kinet.* **15,** 815 (1983).
11. N. Cohen and S.W. Benson, *Chem. Rev.* **93,** 2419 (1993); some of this work has been updated by N. Cohen, Ref. 7.
12. See Ref. 9. The later report of 4.1 kcal mol^{-1} in L. Batt, J.P. Burrows and G.N. Robinson, *Chem. Phys. Lett.* **78,** 467 (1981), citing unpublished results, seems not to be accepted in Batt's later reviews; see, for example, L. Batt, Reactions of alkoxy and alkyl peroxy radicals. *Int. Rev. Phys. Chem.* **6,** 53 (1987).
13. L. Batt and R.T. Milne, *Int. J. Chem. Kinet.* **6,** 945 (1974).
14. P.D. Bartlett and G. Guaraldi, *J. Am. Chem. Soc.* **89,** 4799 (1967).
15. K. Adamic, J.A. Howard and K.U. Ingold, *Can. J. Chem.* **47,** 3803 (1969).
16. J.E. Bennett, D.M. Brown and B. Mile, *Trans. Faraday Soc.* **66,** 397 (1970).
17. E. Furimsky, J.A. Howard and J. Selwyn, *Can. J. Chem.* **58,** 677 (1980).
18. M. Luria and S.W. Benson, *J. Chem. Eng. Data* **22,** 90 (1977).
19. See J.H. Hildebrand and R.L. Scott, *Regular Solutions*, Prentice-Hall, Englewood Cliffs, NJ (1962), p. 168.
20. N. Cohen, unpublished results.
21. See M.W. Chase, Jr., *JANAF Thermochemical Tables*, 3rd Edn, *J. Phys. Chem. Ref. Data* **14,** Supplement 1 (1985).
22. N. Cohen, *J. Phys. Chem.* **96,** 9052 (1992).
23. L.G. Shum and S.W. Benson, *J. Phys. Chem.* **87,** 3479 (1983).
24. L.G. Shum and S.W. Benson, *Int. J. Chem. Kinet.* **15,** 341 (1983).
25. O. Kondo and S.W. Benson, *J. Phys. Chem.* **88,** 6675 (1984).
26. A.F. Wagner, I.R. Slagle, D. Sarzyuski and D. Gutman, *J. Phys. Chem.* **94,** 1853 (1990).
27. S.L. Boyd, R.J. Boyd and L.R.C. Barclay, *J. Am. Chem. Soc.* **112,** 5724 (1980).
28. O. Dobis and S.W. Benson, *J. Am. Chem. Soc.* **115,** 8798 (1993).
29. L.A. Kahachatryan, O.M. Niazyan, A.A. Mantashyan, V.I. Vedenew and M.S. Teitel'Boim, *Int. J. Chem. Kinet.* **14,** 1231 (1982).
30. I.R. Slagle and D. Gutman, *J. Am. Chem. Soc.* **107,** 5354 (1985).
31. I.R. Slagle, E. Ratajazak and D. Gutman, *J. Phys. Chem.* **90,** 402 (1986).
32. I.R. Slagle, E. Ratajazak, M.C. Heaven, D. Gutman and A.F. Wagner, *J. Am. Chem. Soc.* **107,** 1838 (1985).
33. T.J. Mitchell and S.W. Benson, *Int. J. Chem. Kinet.* **25,** 931 (1993).
34. I.C. Plumb and K.R. Ryan, *Int. J. Chem. Kinet.* **13,** 1011 (1981).

35. C.A. Morgan, M.J. Pilling and J.M. Tulloch, *J. Chem. Soc. Faraday Trans. 2*, **78,** 1323 (1982).
36. F. Su, F.G. Calvert and F.H. Shaw, *J. Phys. Chem.* **83,** 3185 (1979); F. Su, F.G. Calvert, F.H. Shaw, H. Niki, P.D. Maker, C.M. Savage and L.D. Breitenbach, *Chem. Phys. Lett.* **65,** 221 (1979).

5 Ultraviolet Absorption Spectra of Peroxy Radicals in the Gas Phase

OLE J. NIELSEN
Research Center, Ford Motor Company, Aachen, Germany

and

TIMOTHY J. WALLINGTON
Research Laboratory, Ford Motor Company, Dearborn, USA

1 INTRODUCTION

Peroxy radicals have been detected and monitored by using several different optical techniques, e.g. ultraviolet (UV) absorption, infrared (IR) and electron spin resonance (ESR) spectroscopy. In the vast majority of cases, the monitoring of RO_2 in kinetic experiments has been performed by using time-resolved UV absorption spectroscopy. Conventional UV lamps are conveniently used as the light source. Because of the importance of the UV absorption spectroscopic technique, only UV absorption spectra of alkyl and substituted alkylperoxy radicals are reviewed below.

The near-UV absorption spectra of peroxy radicals consist of one or, in the case of radicals with one additional chromophore such as a carbonyl group, e.g. $CH_3C(O)O_2$, two broad featureless absorptions in the 200–350 nm wavelength region. The lack of any observable structure suggests that the peroxy radical undergoes rapid dissociation following transition to the excited electronic state. Consistent with this suggestion, *ab initio* calculations performed for HO_2 [1, 2] and CH_3O_2 [3] radicals show that the excited electronic surfaces are repulsive along the O—O coordinate and are thus continuous in nature. Peroxy radicals may be described as π-radicals, formed by the bonding of a σ-alkyl radical with the two unpaired π-electrons of the O_2 di-radical, leaving the unpaired electron of the RO_2 radical in a π-molecular orbital. The transition responsible for the peroxy radical absorption in the 200–300 nm region is a $\pi \longrightarrow \pi^*$ transition from the X^2A'' ground state to the $2^2A''$ excited state. As discussed by Herzberg [4], the shape of a continuous spectrum resulting from a transition from a bound to a continuous electronic state is similar to the probability density distribution of the lower bound state. For transitions from the lowest vibrational level of the

Peroxyl Radicals. Edited by Z. B. Alfassi

ground electronic state ($v'' = 0$) the shape of the UV absorption is then Gaussian with one maximum. Transitions from higher vibrational levels ($v'' > 0$) give rise to absorption spectra with $v'' + 1$ maxima. For ambient temperatures, the population of levels of $v'' > 0$ for most peroxy radicals is negligible. Accordingly, the UV spectra of peroxy radicals at ambient temperature are expected to consist of one Gaussian-shaped band.

As discussed elsewhere [5, 6] the experimentally determined UV spectra of peroxy radicals under ambient conditions can, in most cases, be well described by a Gaussian distribution function of the form:

$$\sigma = \sigma_{max} \exp(-a[\ln(\lambda_{max}/\lambda)]^2) \quad (1)$$

The spectral data in the figures have been fitted by using this expression. The resulting σ_{max}, λ_{max} and 'a' values are given in Table 1. Models for the temperature dependence of peroxy radical spectra have been developed by several workers [6–8].

The UV absorption spectra of peroxy radicals was the subject of two

Table 1 Summary of resulting parameters from a Gaussian fitting of peroxy radical spectra

Radical	$\sigma_{max} \times 10^{20}$ (cm^2)	λ_{max} (nm)	a
CH_3O_2	439	237	41
$C_2H_5O_2$	423	240	53
n-$C_3H_7O_2$	417	242	53
i-$C_3H_7O_2$	497	240	61
t-$C_4H_9O_2$	449	240	62
neo-$C_5H_{11}O_2$	548	244	48
cyclo-$C_5H_9O_2$	532	247	51
cyclo-$C_6H_{11}O_2$	477	249	48
CH_2=$CHCH_2O_2$	618	233	38
$HOCH_2O_2$	376	227	32
$HOCH_2CH_2O_2$	265	239	54
$(CH_3)_2C(OH)CH_2O_2$	412	241	32
$(CH_3)_2C(OH)C(O_2)(CH_3)_2$	403	247	29
$HOC(CH_3)_2CH_2O_2$	345	239	29
$CH_3OCH_2O_2$	410	226	33
$(CH_3)_3COC(CH_3)_2CH_2O_2$	440	242	30
$CH_3SCH_2O_2$	463	240	55
CH_2FO_2	529	221	40
CHF_2O_2	453	218	58
CF_3O_2	349	208	50
$CF_3CF_2O_2$	381	211	51
CF_3CFHO_2	569	212	40
$CHF_2CF_2O_2$	390	218	58

independent reviews in 1992 [6, 9]. We present here a brief review of the available data including selected work published through 1995. Despite a large number of studies, there are still significant uncertainties associated with the absolute values of the absorption cross-sections of several important peroxy radicals.

2 ALKYLPEROXY RADICALS

The absorption spectrum of CH_3O_2 radicals has been measured in 16 different studies [10–25], as shown in Figure 1. While all studies are in good agreement concerning the overall shape of the spectrum there are significant differences in the reported absolute values of $\sigma(CH_3O_2)$. These differences reflect the fundamental experimental difficulty associated with the absolute calibration of low concentrations of these reactive radicals.

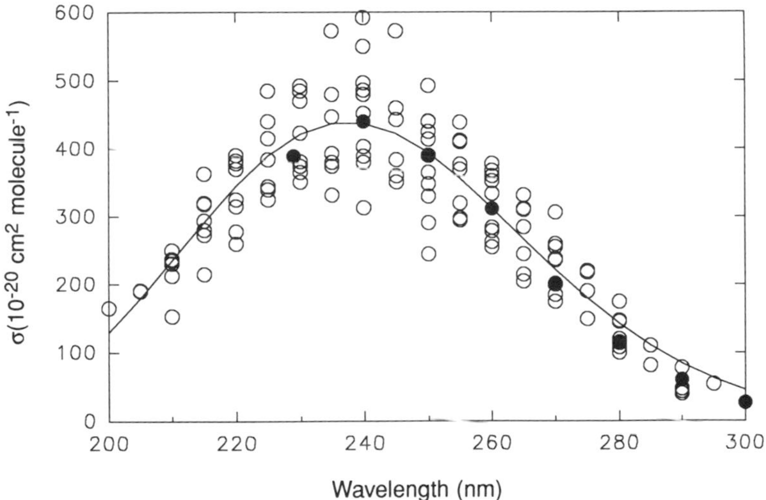

Figure 1 Summary of cross-section data for CH_3O_2 from Ref. 9; continuous line is a Gaussian fit to all of the open points, while filled circles represent data from Ref. 26.

The conventional approach to this problem has been to conduct two experiments. First, the absorption of the peroxy radical in question is measured. Then, in a separate experiment, the yield of radicals in the system is determined by using actinometry, either by following the loss of the photolytic precursor or by monitoring the appearance of a product.

One new approach to this problem in which the peroxy radical is converted stoichiometrically into an alkyl nitrite by reaction with NO has been reported

[26, 27]. For example, for methyl peroxy radicals:

$$CH_3O_2 + NO \longrightarrow CH_3O + NO_2 \quad (2)$$

$$CH_3O + NO + M \longrightarrow CH_3ONO + M \quad (3)$$

Alkyl nitrites absorb strongly in the UV. Hence, as peroxy radicals are converted into the nitrite, the initial UV absorption attributable to the peroxy radical is replaced by absorption by the nitrite. From a single experiment it is therefore possible to relate the absorption cross-section of these two species. By varying the monitoring wavelength, the isobestic point, i.e. where the absorption cross-section of the proxy radical is equal to the sum of those of the nitrite and NO_2, is readily determined.

Once the isobestic point has been established, the problem of measuring the absorption cross-section of the peroxy radical is simplified to measuring that of the nitrite and NO_2 at that wavelength. Alkyl nitrites are, in general, stable compounds under ambient conditions, and are readily synthesized in high purity. Hence, measurement of the UV spectra of these compounds can be performed with great accuracy by using conventional UV spectrometers.

Providing that conditions can be chosen to ensure stoichiometric conversion of the peroxy radicals into the nitrite, then this approach offers a significant advantage over conventional methods, as the nitrite acts as an internal standard. In Figure 1 the results of such an approach are plotted together with all the reported cross-section values for CH_3O_2.

Figure 2 shows the recommended spectra [9] of $C_2H_5O_2$, i-$C_3H_7O_2$,

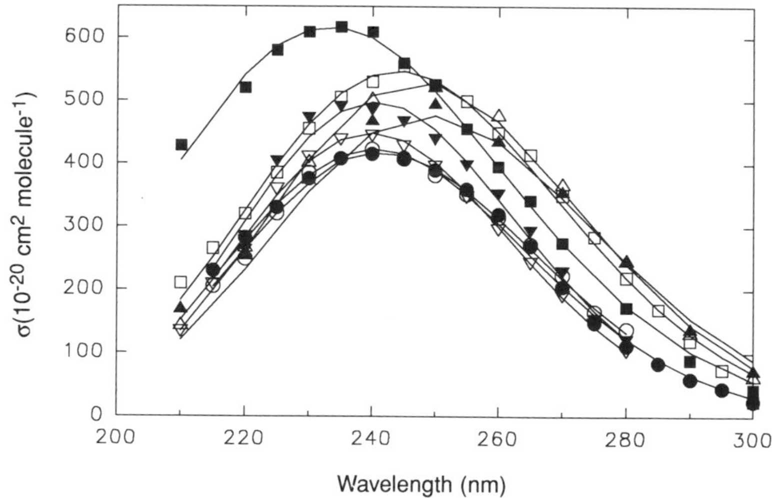

Figure 2 Spectra of alkylperoxy radicals: (●) $C_2H_5O_2$; (▼) i-$C_3H_7O_2$; (○) n-$C_3H_7O_2$; (▽) t-$C_4H_9O_2$; (□) neo-$C_5H_{11}O_2$; (△) cyclo-$C_5H_9O_2$; (▲) cyclo-$C_6H_{11}O_2$; (■) CH_2=$CHCH_2O_2$.

n-$C_3H_7O_2$, t-$C_4H_9O_2$, and neo-$C_5H_{11}O_2$, together with the recently reported spectra of cyclo-$C_5H_9O_2$ [28], cyclo-$C_6H_{11}O_2$ [29] and $CH_2{=}CHCH_2O_2$ [30]. As can be seen in Figure 2 and Table 1, with the exception of the spectrum of $CH_2{=}CHCH_2O_2$, the spectra of all unsubstituted alkylperoxy radicals are very similar. Maximum absorption cross-sections lie in the 237–242 nm region, with $\sigma_{max} = (4.9 \pm 0.6) \times 10^{-18}$ cm^2 molecule^{-1}. Compared to the average data for other alkylperoxy radicals, the absorption spectrum of $CH_2CHCH_2O_2$ is 25% more intense and blue-shifted by 7 nm. The cause of this is unknown. An additional investigation is needed to confirm the $CH_2CHCH_2O_2$ spectrum.

3 ACYLPEROXY RADICALS

Four studies of the spectrum of the acetylperoxy radical, $CH_3C(O)O_2$, have been reported in the literature [31–33]. As seen in Figure 3, there are significant discrepancies between the reported spectra, and possible reasons for these differences have been discussed [33]. These spectra all have two absorption maxima. The longer-wavelength maxima is typical of a peroxy radical and the second is attributed to a $\pi \longrightarrow \pi^*$ transition associated with the C=O group [9]. Further work is needed to better define this spectrum.

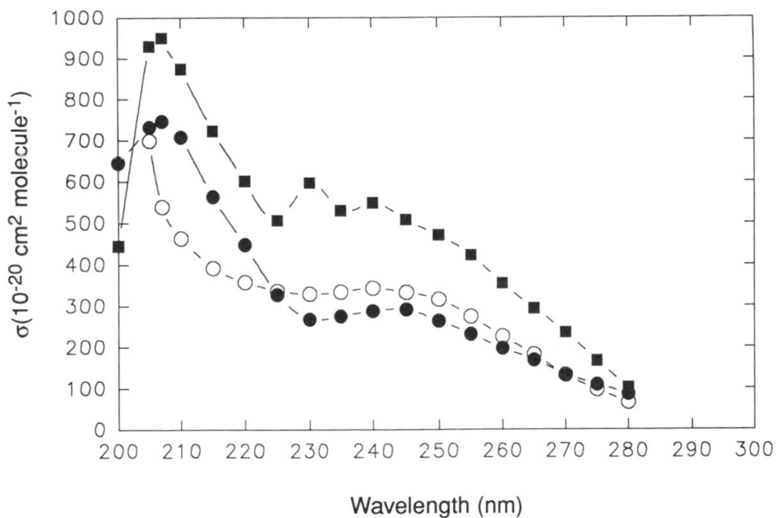

Figure 3 Spectra of the acetylperoxy radical: (○) Ref. 31; (■) Ref. 32; (●) Ref. 33.

4 OXYGENATED PEROXY RADICALS

Spectra of oxygenated peroxy radicals containing OH groups, ether linkages and carbonyl groups have been published. In Figure 4 the spectra of $HOCH_2O_2$ [34, 35], $HOCH_2CH_2O_2$ [36], $(CH_3)_2C(OH)CH_2O_2$ [37], $(CH_3)_2C(OH)C(O_2)(CH_3)_2$ [37] and $(CH_3)_2C(OH)CH_2O_2$ [38] radicals are shown. It is seen that the presence of an OH group broadens the spectrum and lowers the value of the maximum absorption.

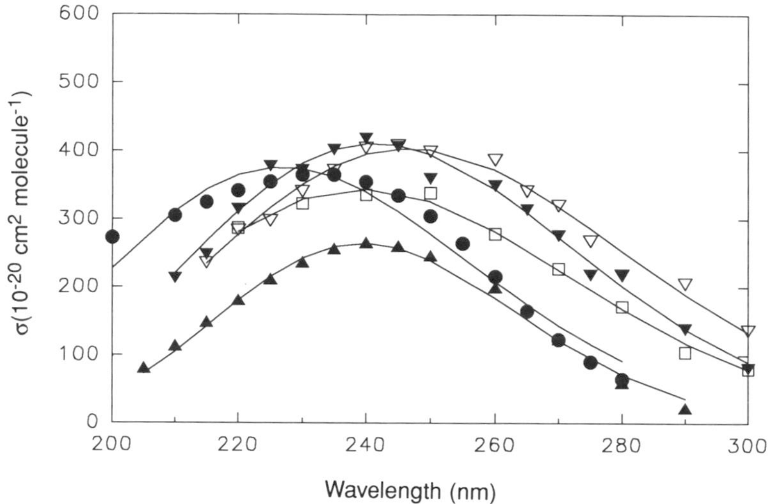

Figure 4 Spectra of hydroxyalkylperoxy radicals: (●) $HOCH_2O_2$; (▲) $HOCH_2CH_2O_2$; (▼) $(CH_3)_2C(OH)CH_2O_2$; (△) $(CH_3)_2C(OH)C(O_2)(CH_3)_2$; (□) $(CH_3)_2C(OH)CH_2O_2$.

The spectra of the peroxy radical derived from dimethylether [39] and di-*t*-butyl ether (DTBE) [40] shown in Figure 5 have the characteristic, one-broad-band peroxy radical type of spectrum, whereas the peroxy radical of acetone [41] has two broad features, also seen for the acylperoxy radical in Figure 3.

5 SULFUR-SUBSTITUTED PEROXY RADICALS

There is one spectrum of a sulfur-containing peroxy radical in the literature. A pulse radiolysis technique was used to measure the UV absorption spectrum of $CH_3SCH_2O_2$ [42]. As seen from Figure 6, the spectrum of $CH_3SCH_2O_2$ is indistinguishable (within the experimental uncertainties) from that of CH_3O_2, showing that the CH_3S group has little effect on the spectrum.

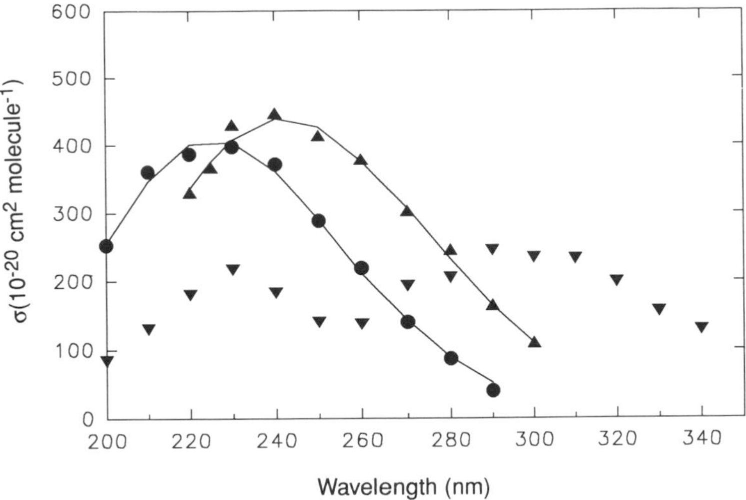

Figure 5 Spectra of (●) $CH_3OCH_2O_2$; (▲) $(CH_3)_3COC(CH_3)_2CH_2O_2$ and (▼) $CH_3COCH_2O_2$ radicals.

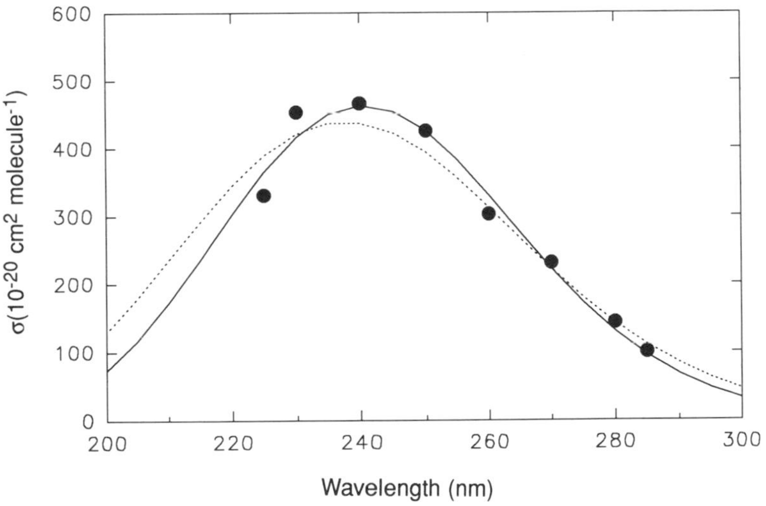

Figure 6 The spectrum of $CH_3SCH_2O_2$ compared to that of CH_3O_2 (dashed line).

6 HALOGENATED PEROXY RADICALS

Recognition of the adverse effect of chlorofluorocarbon (CFC) release into the atmosphere [43, 44] has led to an international effort to replace CFCs with environmentally acceptable alternatives. Hydrofluorocarbons (HFCs) and hydrochlorofluorocarbons (HCFCs) are two classes of CFC substitutes.

As part of studies to determine the atmospheric chemistry, and hence environmental impact of HFCs and HCFCs, there have been many investigations of the UV spectra derived from these compounds.

Here we have chosen to give a little more detailed description of work performed at Risø National Laboratory concerning some peroxy radicals originating from HFCs. The UV spectra of many peroxy radicals derived from HFCs have been measured by using the pulse radiolysis technique. Pulse radiolysis of SF_6 is a convenient source of F atoms. In the presence of HFC and O_2 the F atoms convert HFCs into RO_2 radicals. After the pulse radiolysis of HCF/O_2/SF_6 mixtures a rapid increase in absorption in the ultraviolet was observed, followed by a slower decay. No absorption was observed in the absence of SF_6. The UV absorption following radiolysis of HCF/O_2/SF_6 mixtures was ascribed to the formation of fluorinated peroxy radicals and their subsequent loss by self-reaction.

To illustrate some typical data we present here the detailed results for CH_2FO_2 [45]. Results for CHF_2O_2 [46], CF_3O_2 [47], $CF_3CF_2O_2$ [48], $CF_2HCF_2O_2$ [49], CF_3CFHO_2 [50], $CF_3CH_2O_2$ [51], $CF_3CF_2CFHO_2$ [52], $CH_3CF_2CH_2O_2$ [53] and $CF_3CHO_2CF_3$ [54] radicals are available elsewhere.

Measurement of absolute absorption spectra requires calibration of the initial F-atom concentration. Additionally, experimental conditions have to be chosen such that there is stoichiometric conversion of F atoms to the RO_2 radical. The yield of F atoms was established by two techniques. First, by monitoring the transient absorption at 216.4 nm due to methyl radicals produced by radiolysis of SF_6/CH_4 mixtures and using a value of $4.12 \times 10^{-17}\,\mathrm{cm^2\,molecule^{-1}}$ for $\sigma(CH_3)$ at 216.4 nm [55, 56], and secondly, by monitoring the transient absorption at 250 nm due to CH_3O_2 radicals following radiolysis of SF_6/CH_4/O_2 mixtures and using $\sigma(CH_3O_2) = 3.92 \times 10^{-18}\,\mathrm{cm^2\,molecule^{-1}}$ at 250 nm [9]. Values of $[F]_0$ derived from both methods differed by less than 10%, with $[F]_0 = 2.7 \times 10^{15}\,\mathrm{cm^{-3}}$ at 1000 mbar SF_6, using full dose.

In order to work under conditions where the F atoms are converted stoichiometrically into CH_2FO_2 radicals, it is necessary to consider the potential interfering secondary chemistry. Potential complications include: (i) competition for the available F atoms by reaction with molecular oxygen:

$$F + O_2 + M \longrightarrow FO_2 + M \tag{4}$$

(ii) reaction of CH_2F radicals with CH_2FO_2 radicals:

$$CH_2F + CH_2FO_2 \longrightarrow CH_2FO + CH_2FO \tag{5}$$

and (iii) reaction of F atoms with CH_2F and/or CH_2FO_2 radicals:

$$F + CH_2F \longrightarrow \text{products} \tag{6}$$

$$F + CH_2FO_2 \longrightarrow CH_2FO + FO \tag{7}$$

Under 600 mbar of SF_6 diluent the rate constant for the reaction of F atoms with CH_3F is 264 times greater than that for reaction with O_2 [57, 58]. There are no literature data concerning the kinetics of reactions (5–7) and their importance cannot be calculated. In order to check for the presence of complications in these experiments caused by such reactions, the transient absorption at 240 nm was monitored in experiments using mixture of 10 mbar of CH_3F, 40 mbar of O_2, and 500 mbar of SF_6, with the radiolysis dose varied by over an order of magnitude. The initial F-atom concentration is linearly proportional to the radiolysis dose. It was observed that, with the exception of the data obtained by using full dose, the initial absorption was linear with the radiolysis dose. This linearity shows that the secondary reactions (5–7) are not important at low radiolysis doses. For experiments employing the full radiolysis dose, initial absorptions were 10–20% lower than expected based upon extrapolation of the data obtained at lower doses. This behaviour is ascribed to secondary chemistry at high F-atom concentrations resulting in incomplete conversion of F into CH_2FO_2.

Having established the necessary experimental conditions that ensure stoichiometric conversion of F atoms into the desired peroxy radical (CH_2FO_2 in this example) the observed absorption can be related to the initial F-atom concentration to give the absorption cross-section of the peroxy radical. Before the introduction of diode array devices the acquisition of spectra was achieved by the laborious sequential recording of many experimental transients at different wavelengths. Examples of such spectra are given in Figure 7. Fast inexpensive diode array systems are now commercially available and most spectra are now acquired by using such systems.

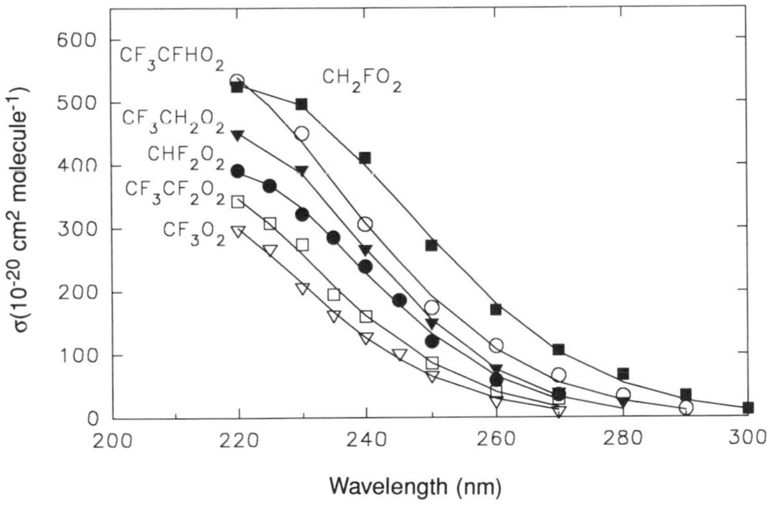

Figure 7 Spectra of CH_2FO_2, CHF_2O_2, $CF_3CF_2O_2$, $CHF_2CF_2O_2$, and CF_3CFHO_2; solid lines are Gaussian fits to the data.

To map out the absorption spectrum of CH_2FO_2 radicals, experiments were performed to measure the initial absorption between 220 and 300 nm following the pulsed irradiation of $SF_6/CH_3F/O_2$ mixtures with $SF_6 = 550$ mbar. Initial absorptions were then scaled to that at 240 nm and converted into absolute absorption cross-sections. The results obtained are shown in Figure 7.

The UV spectra of CH_2FO_2, CHF_2O_2, CF_3O_2, $CF_3CF_2O_2$, $CHF_2CF_2O_2$, CF_2CFHO_2 are presented in Figure 7. As expected, the spectra of these closely related peroxy radicals are very similar in shape. There is some evidence for a slight shift to the red on progressing along the series, CF_3O_2, $CF_3CF_2O_2$, CH_2FO_2, CHF_2O_2, CF_3CFHO_2 to $CF_3CH_2O_2$. This shift is consistent with the literature spectra [9] concerning halogenated alkylperoxy absorption spectra where substitution of increasingly electron-withdrawing groups on the carbon bearing the —O—O* group results in a shift of the spectrum to the blue.

In general, there is reasonable agreement between the various studies. For CF_3CFHO_2 the spectra of Ellermann *et al.* [58] and Maricq and Szente [59] are in agreement. However, the data from Jemi-Alade, Lightfoot and Lesclaux [60] are approximately 40% lower. Ellermann *et al.* [58] and Maricq and Szente [59] both used the reaction of F atoms with HFC-134a as a source of CF_3CFHO_2 radicals, while Jemi-Alade, Lightfoot and Lesclaux [60] employed the reaction of Cl atoms with HFC-134a. Chlorine atoms react slowly with HFC-134a, with a rate constant of 1.4×10^{-15} cm^3 molecule^{-1} s^{-1} [61]. Fluorine-atom attack is three orders of magnitude more rapid [62]. It seems likely that the anomalously low absorption cross-sections reported by Jemi-Alade and coworkers are a result of loss of Cl atoms via reaction with a reactive species in the HFC-134a sample used by this group [60].

The spectra of fluorinated peroxy radicals are all well described by the Gaussian distribution function given above. The intensity of the spectra are all very similar to those of the unsubstituted alkylperoxy radicals. There is a progressive blue shift with increasing fluorination. As given in Table 1, when compared to CH_3O_2, the spectra of CH_2FO_2, CHF_2O_2, and CF_3O_2 are blue shifted by 16, 19, and 29 nm, respectively. Similar trends are evident in the data obtained for the C_2-fluorinated radical.

7 CONCLUSIONS

The gas-phase UV spectra presented here allow for some general conclusions to be drawn. All of the spectra are continuous, indicating loosely bound radicals with dissociative excited electronic states. They can be well fitted by a Gaussian distribution function. The shape and position of the 'peroxy' absorption band depends on the substituents (OH, —O—, or halogen atoms). If the peroxy radical contains other chromophores, several absorption bands may be present. In the case of acetyl- and acetonylperoxy radicals, two bands are observed. In general, substitution has a larger impact on λ_{max} than on σ_{max} and the actual shape of the spectra.

REFERENCES

1. S.R. Langhoff and R.L. Jaffe, *J. Chem. Phys.* **71,** 1475 (1979).
2. G.J. Vazquez, S.D. Peyerimhoff and R.J. Buenker, *J. Chem. Phys.* **99,** 239 (1985).
3. J.A. Jafri and D.H. Philips, *J. Am. Chem. Soc.* **112,** 2586 (1990).
4. G. Herzberg, *Molecular Spectra and Molecular Structure*, Vol 1, 2nd Ed., Van Nostrand Reinhold, New York (1950), p. 388.
5. J.S. Francisco and M.M. Maricq, *Adv. Photochem.* **20,** 79 (1995).
6. P.D. Lightfoot, R.A. Cox, J.N. Crowley, M. Destriau, G.D. Hayman, M. Jenkin, G. Moortgat and F. Zabel, *Atmos. Environ.* **26A,** 1805 (1992).
7. D.C. Astholz, L. Brouwer and J. Troe, *Ber. Bunsenges. Phys. Chem.* **85,** 559 (1981).
8. J.A. Joens, *J. Phys. Chem.* **98,** 1394 (1994).
9. T.J. Wallington, P. Dagaut and M.J. Kurylo, *Chem. Rev.* **92,** 667 (1992).
10. D.A. Parkes, D.M. Paul, C.P. Quinn and R.C. Robinson, *Chem. Phys. Lett.* **23,** 425 (1973).
11. D.A. Parkes, *Int. J. Chem. Kinet.* **9,** 451 (1977).
12. C.S. Kan, R.D. McQuigg, M.R. Whitbeck and J.C. Calvert, *Int. J. Chem. Kinet.* **11,** 921 (1979).
13. H. Adachi, N. Basco and D.G.L. James, *Int. J. Chem. Kinet.* **12,** 949 (1980).
14. M.J. Pilling and M.J.C. Smith, *J. Phys. Chem.* **89,** 4713 (1985).
15. C.J. Hochanadel, J.A. Ghormley, J.W. Boyle and P.J. Ogren, *J. Phys. Chem.* **81,** 3 (1977).
16. R.A. Cox and G.S. Tyndall, *J. Chem. Soc. Faraday Trans. 2* **76,** 153 (1980).
17. S.P. Sander and R.T. Watson, *J. Phys. Chem.* **85,** 2960 (1981).
18. M.E. Jenkin, R.A. Cox, G.D. Hayman and L.J. Whyte, *J. Chem. Soc. Faraday Trans. 2* **84,** 913 (1988).
19. P. Dagaut and M.J. Kurylo, *J. Photochem. Photobiol. Chem.* **51A,** 133 (1990).
20. F. Simon, W. Schneider and G.K. Moortgat, *Int. J. Chem. Kinet.* **22,** 791 (1990).
21. M.E. Jenkin and R.A. Cox, *J. Phys. Chem.* **95,** 3229 (1991).
22. G.K. Moortgat, B. Veyret and R. Lesclaux, *J. Phys. Chem.* **93,** 2362 (1989).
23. K. McAdam, B. Veyret and R. Lesclaux, *Chem. Phys. Lett.* **133,** 39 (1987).
24. M.J. Kurylo, T.J. Wallington and P.A. Ouellette, *J. Photochem.* **39,** 201 (1987).
25. T.J. Wallington, P. Dagaut and M.J. Kurylo, *J. Photochem. Photobiol. Chem.* **42A,** 173 (1988).
26. T.J. Wallington, M.M. Maricq, T. Ellermann and O.J. Nielsen, *J. Chem. Phys.* **96,** 982 (1992).
27. M.M. Maricq and T.J. Wallington, *J. Phys. Chem.* **96,** 987 (1992).
28. D.M. Rowley, P.D. Lightfoot, R. Lesclaux and T.J. Wallington, *J. Chem. Soc. Faraday Trans.* **88,** 1369 (1992).
29. D.M. Rowley, P.D. Lightfoot, R. Lesclaux and T.J. Wallington, *J. Chem. Soc. Faraday Trans.* **87,** 3221 (1991).
30. M.E. Jenkin, T.P. Murrells, S.J. Shalliker and G.D. Hayman, *J. Chem. Soc. Faraday Trans.* **89,** 433 (1993).
31. M.C. Addison, J.P. Burrows, R.A. Cox and R. Patrick, *Chem. Phys. Lett.* **73,** 283 (1980).

32. N. Basco and S.S. Parmar, *Int. J. Chem. Kinet.* **17,** 891 (1985).
33. G.K. Moortgat, B. Veyret and R. Lesclaux, *J. Phys. Chem.* **93,** 2362 (1989).
34. B. Veyret, R. Lesclaux, M.-T. Rayez, J.-C. Rayez and R.A. Cox, *J. Phys. Chem.* **93,** 2368 (1989).
35. J.P. Burrows, G.K. Moortgat, G.S. Tyndall, R.A. Cox, M.E. Jenkin, G.D. Hayman and B. Veyret, *J. Phys. Chem.* **93,** 2375 (1989).
36. P. Dagaut, T.J. Wallington and M.J. Kurylo, *Chem. Phys. Lett.* **146,** 589 (1988).
37. M.E. Jenkin and G. Hayman, *J. Chem. Soc. Faraday Trans.* **91,** 1911 (1995).
38. S. Langer, E. Ljungström, J. Sehested and O.J. Nielsen, *Chem. Phys. Lett.* **226,** 165 (1994).
39. P. Dagaut, T.J. Wallington and M.J. Kurylo, *J. Photochem. Photobiol. Chem.* **48A,** 187 (1989).
40. O.J. Nielsen, J. Sehested, S. Langer, E. Ljungström and I. Wängberg, *Chem. Phys. Lett.* **238,** 359 (1995).
41. I. Bridier, B. Veyret, R. Lesclaux and M.E. Jenkin, *J. Chem. Soc. Faraday Trans.* **89,** 2993 (1993).
42. T.J. Wallington, T. Ellermann and O.J. Nielsen, *J. Phys. Chem.* **97,** 8442 (1993).
43. J.D. Farman, B.G. Gardiner and J.D. Shanklin, *Nature* (*London*) **315,** 207 (1985).
44. S. Solomon, *Nature* (*London*) **347,** 6291 (1990), and references therein.
45. T.J. Wallington, J.C. Ball, O.J. Nielsen and E. Bartkiewicz, *J. Phys. Chem.* **96,** 1241 (1992).
46. O.J. Nielsen, T. Ellermann, E. Bartkiewicz, T.J. Wallington and M. Hurley, *Chem. Phys. Lett.* **192,** 82 (1992).
47. O.J. Nielsen, T. Ellermann, J. Sehested, E. Bartkiewicz, T.J. Wallington and M. Hurley, *Int. J. Chem. Kinet.* **24,** 1009 (1992).
48. J. Sehested, T. Ellermann, O.J. Nielsen, T.J. Wallington and M.D. Hurley, *Int. J. Chem. Kinet.* **25,** 701 (1993).
49. O.J. Nielsen, T. Ellermann, J. Sehested and T.J. Wallington, *J. Phys. Chem.* **96,** 10875 (1992).
50. T.J. Wallington and O.J. Nielsen, *Chem. Phys. Lett.* **187,** 33 (1991).
51. O.J. Nielsen, E. Gamborg, J. Sehested, T.J. Wallington and M.D. Hurley, *J. Phys. Chem.* **98,** 9518 (1994).
52. T.E. Møgelberg, A. Feilberg, A.M.B. Giessing, J. Sehested, M. Bilde, T.J. Wallington and O.J. Nielsen, *J. Phys. Chem.* **99,** 17386 (1995).
53. T.E. Møgelberg, O.J. Nielsen, J. Sehested, T.J. Wallington and M.D. Hurley, *J. Phys. Chem.* **99,** 1995 (1995).
54. T.E. Møgelberg, J. Platz, O.J. Nielsen, J. Sehested and T.J. Wallington, *J. Phys. Chem.* **99,** 5373 (1995).
55. D.A. Parkes, *Int. J. Chem. Kinet.* **9,** 451 (1977).
56. T. Macpherson, M.J. Pilling and M.J.C. Smith, *J. Chem. Phys.* **89,** 2268 (1985).
57. T.J. Wallington, M.D. Hurley, J. Shi, M.M. Maricq, J. Sehested, T. Ellermann and O.J. Nielsen, *Int. J. Chem. Kinet.* **25,** 651 (1993).
58. T. Ellermann, J. Sehested, O.J. Nielsen, P. Pagsberg and T.J. Wallington, *Chem. Phys. Lett.* **218,** 287 (1994).
59. M.M. Maricq and J.J. Szente, *J. Phys. Chem.* **96,** 10862 (1992).
60. A.A. Jemi-Alade, P.D. Lightfoot and R. Lesclaux, *Chem. Phys. Lett.* **179,** 119 (1991).
61. T.J. Wallington and M.D. Hurley, *Chem. Phys. Lett.* **189,** 437 (1992).
62. T.J. Wallington, M.D. Hurley, J. Shi, M.M. Maricq, J. Sehested, O.J. Nielsen and T. Ellermann, *Int. J. Chem. Kinet.* **25,** 651 (1993).

6 Combination of Peroxyl Radicals in the Gas Phase

ROBERT LESCLAUX
Université Bordeaux I, Talence, France

1 INTRODUCTION

Peroxyl radicals are key intermediate species in oxidation processes of hydrocarbons and other organic compounds containing oxygen, nitrogen or halogens. Their importance has been known for a long time but very little information has been available until recently on the kinetics and mechanisms of their reactions. The increasing need to model the chemical processes occurring in combustion or in the atmosphere has motivated a large amount of work during the last ten years, resulting in significant improvement of our knowledge of the reactivity of this class of radicals. As a result, three important reviews dealing with peroxyl radical reactions have been published during the last few years [1, 2, 3]. However, a number of new results have since appeared in the literature and, consequently, these reviews need to be updated.

In general, peroxyl radicals exhibit fairly low reactivity, particularly towards molecular (closed-shell) species and as a result, combination reactions with themselves or with other radical (open-shell) species are their most common reactions. The kinetics and mechanisms of such combination reactions will be examined in this chapter, excluding those reactions with nitrogen oxides which are dealt with separately. Similarly, since liquid-phase reactions are also the subject of a separate chapter, only gas-phase reactions, relevant to the atmospheric or combustion chemistry, will be examined here.

Most oxidation processes in the gas phase involve hydrocarbons and thus emphasis will be given to alkylperoxyl radicals. However, the reactions of peroxyl radicals which result from partially oxidized hydrocarbons (alcohols, ethers, aldehydes, ketones, etc.), bearing the various corresponding functional groups, will also be discussed, including unsaturated and aromatic radicals. With the recent ban of chlorofluorocarbons (CFCs), which are potentially harmful to the stratospheric ozone layer, the oxidation mechanisms of a number of halogenated molecules, which have been considered as possible substitutes to CFCs, have been investigated. As a result, studies of the reaction kinetics of various halogenated peroxyl radicals will also be included in this review.

Peroxyl Radicals. Edited by Z. B. Alfassi

A very large number of peroxyl radicals are involved in oxidation processes of hydrocarbons and other compounds mentioned above, whereas the reactivity of only a few model radicals can be investigated. Consequently, available kinetic data must be extrapolated to radicals having similar structures or bearing the same functional groups. The main purpose of this chapter will not be to discuss each reaction in detail, but rather to discuss the general reactivity of peroxyl radicals and when possible, to propose structure–reactivity relationships which can be used in modelling the oxidation of volatile organic compounds in various gas-phase systems.

2 EXPERIMENTAL METHODS USED IN LABORATORY STUDIES OF PEROXYL RADICAL COMBINATION REACTIONS

The investigation of the kinetics of a particular peroxyl radical reaction requires the reaction mechanism to be known and hence both kinetic and mechanistic studies, involving end-product analyses, are generally necessary for a comprehensive description of the reaction.

Radical–radical combination reactions of peroxyl radicals are often difficult to study as they generally involve kinetics related to complex reaction mechanisms. Complications arise mainly from the production of alkoxyl radicals, which can react with oxygen to produce HO_2, or decompose into a carbonyl compound and a new alkyl radical, thus forming a new peroxyl radical in the presence of oxygen. For example, even in the study of the simplest self-reaction, i.e. $CH_3O_2 + CH_3O_2$, three radicals, five reactions and four products have to be taken into account [4, 5]:

$$CH_3O_2 + CH_3O_2 \longrightarrow 2CH_3O + O_2 \quad (1a)$$

$$\longrightarrow CH_3OH + CH_2O + O_2 \quad (1b)$$

$$CH_3O + O_2 \longrightarrow CH_2O + HO_2 \quad (2)$$

$$CH_3O_2 + HO_2 \longrightarrow CH_3OOH + O_2 \quad (3)$$

$$HO_2 + HO_2 \longrightarrow H_2O_2 + O_2 \quad (4)$$

The reaction mechanism can be even more complex when successive alkoxyl radical decompositions occur, as shown in Table 1 for the particular case of the self-reaction of the acetonylperoxyl radical [6, 7]. This shows that an elementary combination reaction of a particular peroxyl radical cannot generally be separated from a set of subsequent secondary reactions.

Table 1 The full reaction mechanism occurring in laboratory studies of the acetonylperoxyl radical self-reaction

Reactions[a]		
$2CH_3C(O)CH_2O_2$	$\longrightarrow$	$2CH_3C(O)CH_2O + O_2$
	$\longrightarrow$	$CH_3C(O)CH_2OH + CH_3C(O)CHO + O_2$
$CH_3C(O)CH_2O + M\ (+O_2)$	$\longrightarrow$	$CH_3C(O)O_2 + CH_2O + M$
$CH_3C(O)CH_2O_2 + CH_3C(O)O_2$	$\longrightarrow$	$CH_3C(O)CH_2O + CH_3C(O)O + O_2$
	$\longrightarrow$	$CH_3C(O)CHO + CH_3C(O)OH + O_2$
$2CH_3C(O)O_2$	$\longrightarrow$	$2CH_3C(O)O + O_2$
$CH_3C(O)O + M\ (+O_2)$	$\longrightarrow$	$CH_3O_2 + CO_2 + M$
$CH_3C(O)CH_2O_2 + CH_3O_2$	$\longrightarrow$	$CH_3C(O)CH_2O + CH_3O + O_2$
	$\longrightarrow$	$CH_3C(O)CH_2OH + CH_2O + O_2$
	$\longrightarrow$	$CH_3C(O)CHO + CH_3OH + O_2$
$CH_3C(O)O_2 + CH_3O_2$	$\longrightarrow$	$CH_3C(O)O + CH_3O + O_2$
	$\longrightarrow$	$CH_3C(O)OH + CH_2O + O_2$
$2CH_3O_2$	$\longrightarrow$	$2CH_3O + O_2$
	$\longrightarrow$	$CH_3OH + CH_2O + O_2$
$CH_3O + O_2$	$\longrightarrow$	$CH_2O + HO_2$
$CH_3C(O)CH_2O_2 + HO_2$	$\longrightarrow$	$CH_3C(O)CH_2OOH + O_2$
$CH_3C(O)O_2 + HO_2$	$\longrightarrow$	$CH_3C(O)OOH + O_2$
	$\longrightarrow$	$CH_3C(O)OH + O_3$
$CH_3O_2 + HO_2$	$\longrightarrow$	$CH_3OOH + O_2$
$HO_2 + HO_2$	$\longrightarrow$	$H_2O_2 + O_2$

[a] See Tables 2–5 for the rate coefficients of peroxyl radical reactions.

2.1 KINETIC STUDIES

Most of the common experimental methods used in chemical kinetics have been applied to peroxyl radical reactions, e.g. modulation spectroscopy, flash and laser flash photolysis, pulse radiolysis, fast flow reactors, shock tubes, relative-rate methods, etc., coupled to various detection methods, namely ultraviolet (UV) and infrared (IR) absorption, electron paramagnetic resonance (EPR), laser magnetic resonance (LMR), mass spectrometry, fluorescence, etc. The most common methods have been reviewed recently [1] and will not be detailed again here.

The most appropriate experimental conditions for kinetic studies of radical combination reactions generally involve the building-up of high enough radical concentrations to favour the radical–radical reactions of interest. Pulsed techniques, such as flash photolysis, laser flash photolysis or pulse radiolysis, have been widely used to generate the appropriate high radical concentrations. The chemical systems used to produce radicals are detailed in a separate chapter concerning the generation of peroxyl radicals.

2.2 MECHANISTIC STUDIES

As emphasized above, combination reactions of peroxyl radicals often initiate complex reaction mechanisms and consequently kinetic studies must generally be complemented by mechanistic studies. The latter generally involve the analyses of stable reaction products which allow the determination of branching ratios in multichannel reactions and of the way in which the alkoxyl radicals react. The most widely used method for end-product analyses has been the photochemical reaction chamber, coupled to Fourier-transform infrared (FTIR) analysis. Such a method allows *in situ* analysis of the accumulated products during the progression of the reaction. This is particularly advantageous since the usual products of peroxyl radical reactions, such as hydroperoxides, peracids, and carbonyl halides, are generally unstable or are efficiently degraded on the walls of the reaction vessel.

Other methods, such as UV absorption or mass spectrometry, have also provided valuable mechanistic information in some particular cases, e.g. determination of branching ratios of self-reactions [8], and characterization of products in peroxyl radical reactions with chlorine atoms, ClO or NO_3 radicals [9–14].

2.3 ROLE OF ALKOXYL RADICALS IN COMBINATION REACTION STUDIES OF PEROXYL RADICALS

As shown above, the alkoxyl radicals produced in peroxyl radical reactions can induce complex reaction mechanisms which must be taken into account in the kinetic studies of elementary combination reactions of peroxyl radicals. In polluted atmospheres, alkoxyl radicals, RO, are mainly produced by $RO_2 + NO$ reactions. In clean atmospheres, in combustion, or in laboratory studies, they are mainly produced by $RO_2 + RO_2$, $RO_2 + R$, and $RO_2 +$ atom reactions. Alkoxyl radicals generally undergo three types of reactions, namely reaction with oxygen, decomposition and isomerization:

$$\mathrm{RO + O_2 \longrightarrow carbonyl\ compound + HO_2} \tag{5}$$

$$\mathrm{RO + M \longrightarrow carbonyl\ compound + R'\ or\ atom} \tag{6}$$

$$\mathrm{RO + M \longrightarrow R_{(-H)}OH\ (internal\ H\text{-}transfer)} \tag{7}$$

with R' and $R_{(-H)}OH$ being new radicals which, in the presence of oxygen, form the new peroxyl radicals $R'O_2$ and $R_{(-H)}(OH)O_2$, respectively. Such radicals then also undergo self-reactions or cross-reactions with the original RO_2 radical. When the decomposition reaction yields an atom, which may occur in the case of chlorine- or bromine-substituted radicals, chain reactions may result. For example, the following fast chain reaction occurs in the chlorine-atom-initiated oxidation of chloroform [15, 16]:

$$Cl + CHCl_3 \longrightarrow HCl + CCl_3 \tag{8}$$

$$CCl_3 + O_2 + M \longrightarrow CCl_3O_2 + M \tag{9}$$

$$CCl_3O_2 + CCl_3O_2 \longrightarrow 2CCl_3O + O_2 \tag{10}$$

$$CCl_3O + M \longrightarrow CCl_2O + Cl + M \tag{11}$$

to produce phosgene, CCl_2O. In this case, the chain is long enough to derive kinetic information on the CCl_3O_2 self-reaction from the formation rate of phosgene [16]. In other cases, when termination reactions are more efficient, the occurrence of chain processes may be hidden by concurrent reactions, thus resulting in erroneous kinetic data. This is why the determination of accurate kinetic data requires the reaction mechanism to be well known.

3 SELF REACTIONS OF PEROXYL RADICALS

As emphasized above, peroxyl radicals typically exhibit low reactivity towards molecular species and therefore combination reactions, and in particular self-reactions, are important reaction pathways for this class of radicals. Consequently, self-reactions must be taken into account in certain reaction systems, such as:

(a) in the combustion of hydrocarbons and flame ignition, where high radical concentrations can build-up;
(b) in atmospheric oxidation of hydrocarbons and halocarbons, under conditions of low-nitrogen-oxide concentrations;
(c) in laboratory studies of reaction kinetics and mechanisms, where self-reactions cannot generally be separated from other reactions.

In addition, self-reactions of peroxyl radicals have been largely studied for their academic interest, as their rate constants can vary by several orders of magnitude from one radical to another and can exhibit large variations when containing particular functional groups. At present, such large changes in reactivity are not understood in terms of the potential energy profile of the reaction path and can only be described by establishing detailed structure–reactivity relationships. The major part of this section will be devoted to discussing such relationships.

The HO_2 self-reaction has already been described in detail in the literature [2, 17] and thus it will not be discussed further here.

3.1 MECHANISM OF SELF-REACTIONS

Three reaction pathways are generally considered for peroxyl radical self-reactions:

$$RO_2 + RO_2 \longrightarrow RO + RO + O_2 \qquad (12a)$$

$$\longrightarrow ROH + RO_{-H} + O_2 \qquad (12b)$$

$$\longrightarrow ROOR + O_2 \qquad (12c)$$

with RO being an oxyl radical, ROH the corresponding alcohol and RO_{-H} the corresponding carbonyl compound (aldehyde, ketone or acid). Reaction path (12a) always occurs and its importance generally increases with temperature. It produces the oxyl radical RO and is therefore a non-terminating channel. Note that the branching ratio of this channel is nearly one for most halogenated peroxyl radicals. Reaction path (12b) involves a hydrogen-atom transfer from one RO_2 radical to the other (see reaction mechanism below), and consequently cannot exist for those radicals not bearing a hydrogen atom at the α-position (such as tertiary, acyl, or perhalogenated peroxyl radicals). Pathway (12b) is an important reaction pathway for most alkylperoxyl radicals and its importance decreases with increasing temperature. It produces molecular species and is therefore a terminating channel. Reaction pathway (12c), corresponding to the equivalent reaction of HO_2 to form HOOH, has never been clearly characterized. It is likely that this pathway is always negligible due to the excess energy of the reaction which prevents formation of the O—O bond in the peroxide ROOR (e.g., $\Delta H_{12c} \approx -180\,\mathrm{kJ\,mol^{-1}}$ for the CH_3O_2 reaction, whereas the O—O bond dissociation energy in CH_3OOCH_3 is ca. $150\,\mathrm{kJ\,mol^{-1}}$). The reverse situation prevails in the reaction of HO_2, i.e. $D(HO\text{–}OH) \approx 215\,\mathrm{kJ\,mol^{-1}}$ for a similar ΔH of reaction as for CH_3O_2.

The branching ratio of the non-terminating channel (12a) is generally defined as $\alpha = k_{12a}/k_{12}$. The room temperature values of α have been determined for a number of radicals and are reported in Table 2. The temperature dependence of the branching ratios is generally expressed as $k_{12}/k_{12b} = \alpha/(1-\alpha) = A\exp(-B/T)$. Such temperature dependencies have already been discussed in detail in a previous review [1], and will not be discussed further here as no new results have since appeared. Let us only recall that the expression for CH_3O_2 is: $\alpha/(1-\alpha) = 25\exp(-1165K/T)$ and that similar expressions have been obtained for some other alkylperoxyl radicals; note that α always approaches the value of 1 above 500 K.

It is now recognized that self-reactions of peroxyl radicals proceed through an intermediate tetroxide ROOOOR*, which can rearrange and dissociate through the product channel, or go back to the reactants [1]:

$$RO_2 + RO_2 \longleftrightarrow ROOOOR^* \longrightarrow RO + RO + O_2 \qquad (12a)$$

$$\longrightarrow ROH + R_{-H}O + O_2 \qquad (12b)$$

Table 2 Recommended rate coefficients for RO_2 self-reactions

Radical	$k = A \exp(-E/RT)$					Ref./notes
	$k_{298\ K}$ ($cm^3\ molecule^{-1}\ s^{-1}$)	$\alpha_{298\ K}$[a]	A ($cm^3\ molecule^{-1}\ s^{-1}$)	(E/R) (K)	T range (K)	
Primary alkylperoxyl radicals						
CH_3O_2	3.7×10^{-13}	0.33	9.5×10^{-14}	−410	248–650	1–3
$C_2H_5O_2$	7.0×10^{-14}	0.63	7.0×10^{-14}	0	228–460	1–3, 18, 19[b]
n-$C_3H_7O_2$[c]	3×10^{-13}					1–3
neo-$C_5H_{11}O_2$	1.2×10^{-12}	0.39–0.43	1.7×10^{-15}	−1960	248–373	1, 20[d]
Primary peroxyl radicals bearing functional groups						
$HOCH_2O_2$	6.2×10^{-12}	0.89	5.7×10^{-14}	−745	275–323	1, 2
$HOCH_2CH_2O_2$	2.3×10^{-12}	0.50	8.0×10^{-14}	−1000	298–470	21–23[e]
$HOC(CH_3)_2CH_2O_2$	4.8×10^{-12}	0.59	1.4×10^{-14}	−1740	306–398	24, 25[f]
$CH_3OCH_2O_2$	2.1×10^{-12}					26[g]
$(CH_3)_3COC(CH_3)_2CH_2O_2$[c]	2.7×10^{-12}					28
$CH_3SCH_3O_2$[c]	7.9×10^{-12}					29
$CH_3C(O)CH_2O_2$	8.0×10^{-12}	0.75				6, 7, 30[h]
$CH_2{=}CHCH_2O_2$	7.0×10^{-13}	0.61	5.4×10^{-14}	−760	286–394	31, 32
$C_6H_5CH_2O_2$	7.7×10^{-12}	0.40	2.75×10^{-14}	−1680	273–450	33
Primary halogenated peroxyl radicals						
CH_2FO_2	4.0×10^{-12}	>0.77	3.8×10^{-13}	−700	228–380	1, 2
CH_2ClO_2	3.7×10^{-12}	>0.8	1.95×10^{-13}	−875	251–588	34, 35[i]
CH_2BrO_2	1.0×10^{-12}	≈1				36[j]
CH_2IO_2[c]	9×10^{-11}					39[k]
$CHCl_2O_2$	3.9×10^{-12}	>0.95	2.65×10^{-13}	−800	286–440	16[l]
CCl_3O_2	4.0×10^{-12}	>0.98	3.3×10^{-13}	−745	273–460	16[m]
CHF_2O_2[c]	5.0×10^{-12}	≈1				41[n]
CF_3O_2	1.8×10^{-12}					42, 43[o]
$CH_2ClCH_2O_2$	2.7×10^{-12}	0.69	3.5×10^{-14}	−1300	228–380	44, 45[p]
$CH_2ClCHClO_2$		>0.9				46

(continued overleaf)

Table 2 — *continued*

Radical	$k = A \exp(-E/RT)$					Ref./notes
	$k_{298\ K}$ (cm^3 molecule^{-1} s^{-1})	$\alpha_{298\ K}$[a]	A (cm^3 molecule^{-1} s^{-1})	(E/R) (K)	T range (K)	
$CH_2BrCH_2O_2$	5.8×10^{-12}	0.57	5.5×10^{-14}	−1390	275–370	47, 48–50[q]
$CHF_2CF_2O_2$[c]	2.7×10^{-12}					51
$CCl_3CH_2O_2$[c]	4.7×10^{-12}					52
$CF_3CH_2O_2$[c]	8.4×10^{-12}					53, 54
CF_3CClHO_2[c]	4.4×10^{-12}					55
CF_3CFHO_2	5.9×10^{-12}	0.84	7.8×10^{-13}	−605	211–372	56, 57
$CF_3CCl_2O_2$[c]	3.3×10^{-12}					58[r]
CF_3CFClO_2[c]	2.6×10^{-12}					59[r]
$CF_3CF_2O_2$[c]	2.1×10^{-12}					53
$CH_3CF_2CH_2O_2$[c]	8.6×10^{-12}					60
Acylperoxyl radicals						
$CH_3C(O)O_2$	1.4×10^{-11}	1	2.3×10^{-12}	−530	253–368	See note[s]
$C_2H_5C(O)O_2$	1.2×10^{-11}	1				63
$C_6H_5C(O)O_2$	1.4×10^{-11}	1				64
$FC(O)O_2$	6.5×10^{-12}	>0.80	2.5×10^{-12}	−285	213–258	See note[t]
Secondary alkylperoxyl radicals						
$CH_3CH(O_2)CH_3$	1.1×10^{-15}	0.56	1.7×10^{-12}	2200	298–373	1, 2
cyclo-$C_5H_9O_2$[c]	4.5×10^{-14}		2.9×10^{-13}	555	243–373	1
cyclo-$C_6H_{11}O_2$	4.2×10^{-14}	0.30	7.7×10^{-14}	184	253–373	1
Substituted secondary peroxyl radicals						
$CH_3CH(OH)CH(O_2)CH_3$	6.6×10^{-13}	<0.30	8.4×10^{-15}	−1300	298–470	21, 22[u]
$CH_3CHBrCH(O_2)CH_3$	7.3×10^{-13}	≈0.5	1.3×10^{-14}	−1200	275–370	47, 50[v]
$CF_3CH(O_2)CF_3$[c]	5.6×10^{-12}					68
Tertiary alkylperoxyl radicals						
$(CH_3)_3CO_2$	3.0×10^{-17}	1	4.1×10^{-11}	4200	293–418	1–3

Substituted tertiary peroxyl radicals					
$(CH_3)_2C(OH)C(O_2)(CH_3)_2$	5.0×10^{-15}	1	3.0×10^{-13}	1220	21, 22
$(CH_3)_2CBrC(O_2)(CH_3)_2$	$\approx 2 \times 10^{-14}$				50

[a] Branching ratio for the non-terminating channel at 298 K (see text).
[b] All determinations are in good agreement; the measured temperature dependencies are not significant and thus, we recommend $E/R = 0$.
[c] $k_{observed}$ (rate constant for the apparent second-order kinetics); the values of $k_{observed}$ should be taken only as a rough approximation of the real rate constants, as complex reaction mechanisms may be initiated by the products of the self-reaction. They are generally upper limits for the real rate constants, with $k_{observed} \geqslant k_{real} \geqslant k_{observed}/2$; however, this may not be the case when chain reactions occur (in particular when elimination of halogen atoms from oxyl radicals takes place). The expression $k_{observed} = k_{real}(1 + \alpha)$, which has often been used, is only valid when the self-reaction rate controls the global kinetics of the whole reaction mechanism, which is not generally the case for fast reactions ($k > 2$–3×10^{-12} cm^3 molecule^{-1} s^{-1}).
[d] Rate parameters proposed in Ref. 1 and confirmed in a subsequent study [20].
[e] Based on the results of Jenkin and Hayman [21] with the temperature dependence taken from Ref. 22 and the α value from Refs 22 and 23.
[f] Good agreement obtained with the earlier results from Ref. 25, ($k_{observed}(298) = 7.8 \times 10^{-12}$ cm^3 molecule^{-1} s^{-1}), as large amounts of HO_2 are formed in the reaction.
[g] The preceding value reported in Refs 1 and 2 (from Ref. 27) must be ignored as the decomposition of the corresponding oxyl radical, CH_3OCH_2O, into $CH_3OCHO + H$, was not taken into account in the reaction mechanism.
[h] The rate constants determined in Refs 7 and 30 are in good agreement; the α value is taken from Ref. 7, in agreement with product studies [6].
[i] Recommended rate coefficients from Ref. 34, in good agreement with $k_{observed}(298) = 4.0 \times 10^{-12}$ [35].
[j] This value is preferred to that reported in Ref. 37, in which the reported UV spectrum of CH_2BrO_2 was very different from the usual spectra of peroxyl radicals. Recent results [38] confirm the UV spectrum reported in Ref. 36.
[k] The rate constant seems too large and the UV spectrum of CH_2IO_2 is very different from the usual spectra of peroxyl radicals.
[l] Assumed expression, by comparison with those of CH_2ClO_2 and CCl_3O_2.
[m] The value reported previously of $k(298) = 1.6 \times 10^{-12}$ cm^3 molecule^{-1} s^{-1} [1, 40] was too low as problems with the purity of the precursor have been identified [16].
[n] The only product of the self-reaction is FCOF, consistent with $\alpha = 1$; no conclusion could be derived about the reaction of CHF_2O_2 with HO_2 and thus no modelling could be performed to derive the real rate constant.
[o] Two studies in good agreement; only CF_3OOOCF_3 is observed as a stable product, indicating that the only product of the self-reaction is CF_3O, which combines quantitatively with CF_3O_2 to give CF_3OOOCF_3; consequently, $k \approx \frac{1}{2}k_{observed}$.
[p] $k_{observed} = 1.1 \times 10^{-13} \exp(1020K/T)$ was reported in Ref. 44 and α is from Ref. 45; the recommended rate coefficients have been estimated by using rate constant values halfway between $k_{observed}$ and $k_{observed}/(1 + \alpha)$, with $\alpha = 0.69$ at 298 K and $\alpha \approx 1$ at 380 K.
[q] α is derived from product studies [48, 49]; two kinetic studies, at room temperature [47] and as a function of temperature [50], are in very good agreement.
[r] Derived from a complex mechanism due to the elimination of Cl atoms from CF_3CFClO, so that the rate constant includes the reactions $RO_2 + RO_2$, $RO_2 + Cl$, and $RO_2 + ClO$, and is probably overestimated.
[s] Previous recommendations [1, 2] were based on the rate expression reported by Moortgat and coworkers [6]; the present review recommends a lower value for the rate constant in view of the last study by Roehl and coworkers [62] and the new UV absorption cross-sections reported for the radical, which are lower than those used in Ref. 61; the temperature dependence reported in Ref. 61 has been retained.
[t] The recommendation is based on the rate expression reported by Maricq *et al.* [65]; it is in good agreement with $k_{observed}$ reported by Wallington *et al.* [66]; the determination of α is by Wallington, Hurley and Mericq [67]; all products are accounted for by the formation of the FC(O)O radical which reacts with itself to give FC(O)OO(O)CF or decomposes into $F + CO_2$.
[u] The two studies [21, 22] are in good agreement for $k(298)$; the temperature dependence and the α value are taken from Ref. 22.
[v] α was assumed to be the same as for the reaction of $CH_2BrCH_2O_2$; two kinetic studies [47, 50] are in good agreement.
[w] Due to the complex reaction mechanism, only an estimate of the rate constant is given.

As already discussed previously [1], only indirect evidence exists in favour of such a mechanism, in addition to the fact that it is difficult to imagine another reaction pathway leading to the observed products. Most peroxyl radical self-reactions exhibit a fairly strong negative temperature dependence, indicating the formation of an intermediate adduct. This behaviour is similar to that of the $HO_2 + HO_2$ reaction, for which experimental evidence has been obtained, in particular the effect of pressure on the rate constant, showing that the first step of the reaction is an association reaction [2, 17]. A pseudo-steady-state radical concentration has been observed in the liquid phase in the case of the slow self-reaction of the isopropylperoxyl radical and tertiary peroxyl radicals, and was assigned to the presence of an equilibrium between the peroxyl radicals and the tetroxide [69–71]. The temperature dependence of the equilibrium constant has led to $\Delta H = -35 \pm 2\,\mathrm{kJ\,mol^{-1}}$ for the liquid-phase association reaction forming the tetroxide, for the radicals investigated.

It should be noted that, in contrast to the HO_2 self-reaction, peroxyl radical self-reactions have not been observed to exhibit any pressure dependence, even in the case of the simplest radical, CH_3O_2. It would therefore appear that, due to the large number of degrees of freedom in the tetroxide, the rate constant is at the high-pressure limit in the pressure ranges that have been explored ($P > 10$ torr) [1, 2].

The tetroxide intermediate can directly dissociate to RO radicals and O_2 in the case of the non-terminating channel (12a) but, energetic considerations show that the dissociation process must be concerted [72]. However, it must dissociate through a cyclic transition state to ensure an internal hydrogen-atom transfer and form the disproportionation products of the terminating channel (12b), according to Russell's mechanism [73]. For example, the transition state structure for the CH_3O_2 reaction could be:

```
        .H.     CH3
 H2C´´     `O´
  |         :
  O.       .O
    `O´
```

As emphasized above, such a structure of the transition state is not possible for tertiary, acyl and perhalogenated radicals and the linear tetroxide must dissociate directly to form the RO radicals. For example:

$$\begin{aligned} 2(CH_3)_3COO &\longrightarrow (CH_3)_3C{-}O{-}O{-}O{-}O{-}C(CH_3)_3^* \\ &\longrightarrow 2(CH_3)_3CO + O_2 \end{aligned} \tag{13}$$

and

$$\begin{aligned} 2CH_3C(O)OO &\longrightarrow CH_3C(O){-}O{-}O{-}O{-}O{-}(O)CCH_3^* \\ &\longrightarrow 2CH_3C(O)O + O_2 \end{aligned} \tag{14}$$

It is not understood why, in these two similar cases, the first reaction involves a significant activation energy, which makes the reaction quite slow at around room temperature, whereas the second reaction is very fast and proceeds with a negative temperature dependence.

The large differences in reactivity observed in peroxyl radical self-reactions (Table 2) can only be explained qualitatively at the present time. It is thought that the height of the exit barrier from the tetroxide to the products, relative to the height of the entrance barrier, is very sensitive to the structure of the radical and to substituent effects [1, and references therein]. As a result, large variations of the ratio of the forward and backward rates are expected, particularly if both barriers have similar heights.

3.2 KINETICS OF PEROXYL RADICAL SELF-REACTIONS

The kinetic parameters that can be recommended to date for self-reactions are presented in Table 2. The branching ratio α at room temperature for the non-terminating channel is also included. The criteria used for recommendation will not be discussed here, and only the principal arguments are presented in the notes of the table. Recommendations given in preceding reviews [1, 2] have been retained when no further results have since been reported.

The most striking feature is the large variation in the rate constants observed for radicals of different structures, or bearing different substituents or functional groups. In view of the very large number of possible peroxyl radicals related to hydrocarbons and derivatives, it is clear that extrapolation of data to radicals which have not been investigated is very difficult. This is why most of the recent work has been aimed at establishing relationships between structure and reactivity for the most common classes of radicals. It must be emphasized that building such relationships, which are discussed below, is a difficult task for self-reactions of peroxyl radicals and that recommendations about such relationships still require further improvement.

The most important variation of reactivity with structure is that observed between primary, secondary and tertiary alkylperoxyl radicals, with about two orders of magnitude between the rate constants of each type of radical at room temperature. This is mainly as a result of the temperature dependence of the rate constants, which are small and generally negative for primary radicals, and usually positive for others. As discussed above, such temperature dependences are not presently understood. Due to this large variation in reactivity, primary, secondary and tertiary radicals are presented separately in Table 2 and further relationships between structure and reactivity are discussed below for each of these classes.

3.2.1 Primary peroxyl radicals

Alkylperoxyl

Only four primary unsubstituted alkylperoxyl radicals have been investigated, namely CH_3O_2, $C_2H_5O_2$, n-$C_3H_7O_2$ and neo-$C_5H_{11}O_2$. It is not easy to make a general recommendation for this class of radicals as the rate constants vary by a factor of about 15 from $C_2H_5O_2$ to neo-$C_5H_{11}O_2$. More data would be necessary to clarify the trends and to confirm, in particular, a possible enhancement of reactivity, either with the number of carbon atoms in the radical or, more likely, with the degree of branching in the alkyl group in the case of large radicals (>C_4). A tentative recommendation might be $k_{298} \approx 3 \times 10^{-13}$ molecule^{-1} s^{-1} for linear radicals (as for CH_3O_2 and n-$C_3H_7O_2$), with a slight negative temperature coefficient ($E/R \approx -500$ K) and $k_{298} \approx 1.2 \times 10^{-12}$ cm^3 molecule^{-1} s^{-1} for branched radicals (as for neo-$C_5H_{11}O_2$), with a stronger negative temperature dependence ($E/R \approx -1500$ K). Both the terminating and non-terminating channels are efficient near room temperature ($\alpha_{298} \approx 0.5 \pm 0.15$) and the branching ratios should exhibit a temperature dependence similar to that of CH_3O_2 (see above).

Substituted alkylperoxyl

Substitution of a hydrogen atom in alkylperoxyl radicals, either in the α- or in the β-position, always results in a significant increase in the self-reaction rate constant, whatever the nature of the substituent, as seen in Table 2.

A number of substituents and functional groups have now been investigated: OH, F, Cl, Br, I, OCH_3, SCH_3, $CH_3C(O)$, vinyl and phenyl groups in the α-position, and OH, F, Cl, and Br in the β-position. Concerning the α-position, most rate constants are increased by a factor of 10 to 20 compared to CH_3O_2. The principal exception is for the allylperoxyl radical whose rate constant is hardly increased by a factor of 2 compared to both CH_3O_2 and n-$C_3H_7O_2$. Thus, the effect of a simple double bond appears quite limited, while the effect of the aromatic ring in the benzyl radical is about tenfold higher. Another interesting feature is the much smaller effect of subsequent substitutions of H atoms, compared to that of the first one. This is particularly clear in the series of chlorine-substituted methylperoxyl radicals, CH_3O_2, CH_2ClO, $CHCl_2O_2$ and CCl_3O_2, where the rate constants are 0.37, 3.7, 3.9 and 4.0×10^{-12} cm^3 molecule^{-1} s^{-1}, respectively. The same trend is observed for fluorine-substituted radicals, with even a slight decrease of the rate constant for multiple fluorine-atom substitutions (in CF_3CFClO_2 and $CF_3CF_2O_2$, for example).

The most interesting fact that must be pointed out about substituted primary peroxyl radicals is that the rate constants of self-reactions vary over a much narrower range than is the case of unsubstituted radicals. Indeed, most rate constants are within a factor of 2 from the average value of 4×10^{-12} cm^3 molecule^{-1} s^{-1}, and no clear trend can be seen for the rate constants with the nature of the substituent. As a result, this latter value constitutes a reasonable recommendation for any substituted primary peroxyl radical whose self-reaction rate constant has not yet been measured. The exponential factors (E/R) of the Arrhenius expressions vary over a fairly large range, approximately from -700 to -1700 K, again without any clear trend, and hence a reliable recommendation is not possible. An average value of ca. -1200 K can be used.

Acylperoxyl radicals

Acylperoxyl radicals, formed from the oxidation of aldehydes, possess a different structure compared to other peroxyl radicals in that the same carbon atom bears both the carbonyl function and the peroxyl function. This structure confers specific properties to these radicals: UV absorption spectra exhibit an additional absorption band at around 200 nm, the corresponding peroxynitrates (PANs) are considerably more stable than others, and the reactivity is significantly enhanced in all of their reaction types. This can be seen in the chapter devoted to the reactions with nitrogen oxides and in this chapter for other combination reactions.

The self-reactions of only four acylperoxyl radicals have been studied to date, including the acetylperoxyl radical, one of the most important radical in various reaction systems, namely $CH_3C(O)O_2$, $C_2H_5C(O)O_2$, $C_6H_5C(O)O_2$, and $FC(O)O_2$. With the exception of the reaction of $FC(O)O_2$, which is slower than the others, all reactions of radicals derived from organic precursors are fast, with rate constants close to 1.5×10^{-11} cm^3 molecule^{-1} s^{-1}. Even though this value should be confirmed by some additional studies, it can be recommended for all radicals of this class. A consequence is that self-reactions of acylperoxyl radicals can play an important role in reaction systems, such as the oxidation of hydrocarbons in clean atmospheres (low NO_x concentrations).

The temperature dependence has only been measured for the acetylperoxyl radical, with $E/R = -530$ K, a value which can be reasonably recommended for all other radicals.

3.2.2 Secondary peroxyl radicals

Secondary alkylperoxyl radicals

The self-reactions of only three alkylperoxyl radicals have been investigated to date, namely the isopropyl-, cyclopentyl- and cyclohexylperoxyl radicals. As already pointed out, the rate constant for the isopropylperoxyl radical is more than two orders of magnitude lower than that of the n-propylperoxyl radical. It exhibits, in addition, a fairly high positive activation energy. To date, the same rate constant has been assumed for all non-cyclic secondary alkylperoxyl radicals, but this should be verified for larger radicals.

Both the cyclic peroxyl radicals investigated have very similar rate constants and temperature dependences. However, these rate constants are about 40 times larger than that measured for the isopropylperoxyl radical, and have only a small positive activation energy. It is not clear whether this large increase in the rate constant is due to the cyclic structure of the radicals or to their larger size, but very recent results suggest that the latter effect may be more important (unpublished results of author).

Substituted secondary alkylperoxyl radicals

As in the case of primary radicals, the self-reaction rate constants are substantially increased by the presence of a substituent, e.g. by more than 700 times the rate constant of the isopropylperoxyl radical for the β-substituted secondary radicals obtained by the addition of OH or Br to 2-butene in the presence of O_2, with a large decrease in the activation energy, which is now negative (Table 2). It is worth noting that the rate constant is about five times smaller than those observed for similar substituted primary peroxyl radicals, showing that the secondary nature of the radical still tends to decrease the reactivity of this type of mono-substituted radical. However, the multiple substitution in $CF_3CH(O_2)CF_3$ changes the rate constant to values as high as those observed for substituted primary peroxyl radicals, and thus cancels completely the effect of the secondary nature of the peroxyl functionality.

The database for self-reactions of secondary peroxyl radicals is too limited to make reliable recommendations about structure–reactivity relationships. At the present time we can only use the available data for each species for the corresponding class of radicals, i.e. non-cyclic, cyclic, α- or β-, monosubstituted, and multisubstituted radicals, as indicated above.

3.2.3 Tertiary peroxyl radicals

Self-reactions of tertiary peroxyl radicals are even slower than those of the corresponding secondary peroxyl radicals, but the general trends in reactivity are similar. The only tertiary non-substituted alkylperoxyl radical investigated to date is the t-butylperoxyl radical, which exhibits the slowest self-reaction known to date, mainly due to a fairly high activation energy. As observed for secondary radicals, a β-substitution by OH or Br results in a large increase in the rate constant. The database is also too limited to make recommendations about relationships between structure and reactivity. Only rough evaluations can be made, based on the reported room-temperature rate constant values (Table 2).

3.3 CONCLUSIONS CONCERNING SELF-REACTIONS

Self-reactions of peroxyl radicals have been very widely investigated, and yet many uncertainties still remain. This is essentially due to the wide reactivity range (rate constants vary over several orders of magnitude) and to the large sensitivity of rate constants to different factors, i.e. structure, substitution and size. Further progress would require more investigations of radicals bearing particular structures and functional groups. However, experimental difficulties are increasing with the complexity of radicals, difficulties in generating radicals in a clean way, or difficulties arising from the complexity of the chemistry initiated by the self-reaction. It should be emphasized, however, that our knowledge of self-reactions has increased considerably in recent years and that the relationships between structure and reactivity, described above, can nevertheless be used advantageously to construct reaction mechanisms, even though those relationships are still approximate. Note that they are particularly approximate for slow reactions, which are the least important in reaction systems, whereas they seem more reliable for the fastest reactions.

4 REACTIONS OF PEROXYL RADICALS WITH HO_2

Reactions of RO_2 radicals with HO_2 are among the most important combination reactions of peroxyl radicals, both in combustion and in the atmospheric chemistry of hydrocarbons. This is due to the abundance of the HO_2 radical, which can be formed in various ways in hydrocarbon oxidation processes. In addition, these reactions are generally fast, and thus contribute significantly to the chemistry of peroxyl radicals.

The role of $RO_2 + HO_2$ reactions in the atmosphere (at low temperature) and in combustion (at high temperatures) is quite different. Although for most reactions the principal product is the hydroperoxide, ROOH, as discussed below, the reaction is chain-terminating at low temperatures, where the hydroperoxide is stable, whereas at higher temperature (above 600–700 K) the hydroperoxide ROOH generally dissociates into two radicals, RO + OH, and thus the reaction becomes chain-propagating. As a consequence, the kinetics and mechanism of $RO_2 + HO_2$ reactions must be known accurately in order to model the oxidation processes of hydrocarbons under various conditions.

4.1 MECHANISM OF $RO_2 + HO_2$ REACTIONS

The possible reaction pathways are the following:

$$RO_2 + HO_2 \longrightarrow ROOH + O_2 \quad (15a)$$

$$\longrightarrow RO_{-H} + H_2O + O_2 \quad (15b)$$

$$\longrightarrow RO_{-X} + HOX + O_2 \quad (15c)$$

$$\longrightarrow ROH + O_3 \quad (15d)$$

where RO_{-H} and RO_{-X} are the corresponding carbonyl compounds (aldehyde, ketone or acid) and X is a halogen atom. The branching ratio, $\beta = k_{15a}/k_{15}$ of channel (15a), forming the hydroperoxide, is reported in Table 3 for the

Table 3 Recommended rate coefficients of reactions of RO_2 with HO_2

Radical	$k = A \exp(-E/RT)$					Ref./notes
	$\beta_{298\ K}$[a]	$k_{298\ K}$ (cm^3 molecule^{-1} s^{-1})	A (cm^3 molecule^{-1} s^{-1})	(E/R) (K)	T range (K)	
Alkylperoxyl radicals						
CH_3O_2	≈1	5.8×10^{-12}	4.1×10^{-13}	−790	230–680	1, 2[b]
$C_2H_5O_2$	≈1	9.0×10^{-12}	4.4×10^{-13}	−900	210–360	1, 2, 19, 74[c]
neo-$C_5H_{11}O_2$	≈1	1.5×10^{-11}	1.43×10^{-13}	−1380	248–365	75
cyclo-$C_5H_9O_2$	≈1	1.8×10^{-11}	2.1×10^{-13}	−1320	248–364	76
cyclo-$C_6H_{11}O_2$	≈1	1.7×10^{-11}	2.6×10^{-13}	−1245	248–364	76
Peroxyl radicals bearing functional groups						
$HOCH_2O_2$	0.60	1.2×10^{-11}	5.5×10^{-15}	−2300	280–330	1, 2[d]
$HOCH_2CH_2O_2$		1.3×10^{-11}				1, 21, 77[e]
$HOC(CH_3)_2CH_2O_2$		1.4×10^{-11}	5.6×10^{-14}	−1650	306–398	24[f]
$CH_3OCH_2O_2$	0.53					78[g]
$CH_3C(O)CH_2O_2$		9.0×10^{-12}				7
$CH_2{=}CH{-}CH_2O_2$		$\approx 1.0 \times 10^{-11}$				32[h]
$C_6H_5CH_2O_2$		1.0×10^{-11}	3.75×10^{-13}	−980	273–450	33
$CH_3CH(OH)CH(O_2)CH_3$		1.5×10^{-11}				21
$(CH_3)_2C(OH)C(O_2)(CH_3)_2$		$\approx 2 \times 10^{-11}$				21
Halogenated peroxyl radicals						
CH_2FO_2	0.29					79[i]
CH_2ClO_2	0.20	5.2×10^{-12}	3.25×10^{-13}	−820	255–588	34, 80[j]

CH_2BrO_2		6.7×10^{-12}				36
$CHCl_2O_2$	$\approx 0^{k}$	5.8×10^{-12}	5.6×10^{-13}	−700	286–440	16
CCl_3O_2	$\approx 0^{k}$	5.1×10^{-12}	4.8×10^{-13}	−706	298–374	16
CF_3O_2		$<2 \times 10^{-12}$				81
CF_2ClO_2		3.4×10^{-12}				81
$CFCl_2CH_2O_2$		9.2×10^{-12}				81
$CF_2ClCH_2O_2$		6.8×10^{-12}				81
CF_3CFHO_2	>0.95	3.8×10^{-12}	1.8×10^{-13}	−910	210–363	81, 82[l]
$CF_3CF_2O_2$		1.2×10^{-12}				81
$(CH_3)_2CBrC(O_2)(CH_3)_2$		1.25×10^{-11}	1.4×10^{-12}	−680	298–393	50
Acylperoxyl radicals						
$CH_3C(O)O_2$		1.3×10^{-11}	4.3×10^{-13}	−1040	253–368	1, 2, 83[m]
$C_2H_5C(O)O_2$		1.0×10^{-11}				63

[a] Branching ratio of the reaction channel forming the hydroperoxide.
[b] The recommendation is the same as in Ref. 1, where a critical review of the data is given.
[c] The agreement between all measurements is reasonable, after correction for UV absorption cross-sections; the recommended rate expression is an average of the expressions reported in Refs 19 and 74.
[d] See discussion in Refs 1 and 2; other products are $HCOOH + H_2O + O_2$, with a corresponding branching ratio of 0.40.
[e] Average between the two last determinations [21, 77] which are in agreement within experimental uncertainties; the formation of the hydroperoxide has been identified [23] but the yield was not measured.
[f] The formation of the hydroperoxide has been identified [24] but the yield was not measured.
[g] Product analysis only; the branching ratio of the channel forming $CH_3OCHO + H_2O + O_2$ is 0.40.
[h] Extrapolated from measurements at 393–428 K, assuming $E/R \approx -700\ K/T$.
[i] Mechanistic study only; the branching ratio of the channel forming $HCOF + H_2O + O_2$ is equal to 0.71.
[j] The branching ratio of the channel forming $HCOCl + H_2O + O_2$ is equal to 0.80.
[k] See text.
[l] Branching ratio and temperature dependence from Ref. 82; very good agreement obtained between the two studies.
[m] Two reaction channels, forming O_2 and O_3; see discussion in Refs 1 and 2; recent studies [83] have identified the temperature dependence of the branching ratio, $k(O_2)/k(O_3) = 330 \exp(-1430/T)$ (263–333 K), in fairly good agreement with previous studies at room temperature.

reactions where it has been measured. It seems to be nearly one for all alkylperoxyl radicals, as already discussed in detail in preceding reviews [1, 2]. Some evidence has been obtained for a minor contribution of channel (15b) at low pressure ($<20\%$) in the reaction of CH_3O_2 [84], but it would require confirmation and we can assume, at the present time, that the branching ratio for the formation of ROOH is unity for all alkylperoxyl radicals.

Substitution of a hydrogen atom in alkylperoxyl radicals by a halogen atom or an oxygenated group may change substantially the product distribution of the reaction. For example, β is equal to 0.60, 0.29 and 0.20 for $HOCH_2O_2$, CH_2FO_2 and CH_2ClO_2, respectively, while the carbonyl products formed in channel (15b) were clearly characterized (see references in Table 3). However, it should be pointed out that the reaction of the CF_3CFHO_2 radical with HO_2 produces the hydroperoxide with a yield of nearly 100% [82].

In contrast, no hydroperoxide has been observed in the products of the reactions of $CHCl_2O_2$ and CCl_3O_2 with HO_2 [16]. Instead, CHClO and CCl_2O, respectively, have been found as the principal products, indicating that the principal reaction pathway is channel (15c). However, HOCl has not yet been characterized as a product and further evidence is required to confirm the occurrence of this channel.

Channel (15d) is specific to the reactions of acylperoxyl radicals. A yield of ca. 30%, independent of temperature, has been found for the reactions of both the acetylperoxyl [1, 2] and propionylperoxyl radicals [63] and similar branching ratios can probably be assumed for all reactions of acylperoxyl radicals with HO_2.

As for the self-reactions of HO_2 and RO_2 radicals, it is thought that the $RO_2 + HO_2$ cross-reaction also involves a tetroxide intermediate, ROOOOH, which then rearranges and dissociates to the products. Two cyclic structures of the transition state are possible for the tetroxide dissociation of the hydroperoxide:

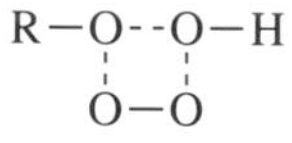

```
R—O--O—H        R—O—O--H
  ¦  ¦               ¦  ¦
  O—O                O—O
  (I)               (II)
```

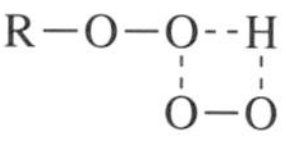

Structure II, involving hydrogen bonding and already proposed for the $HO_2 + HO_2$ reaction [85–87], seems more likely.

Channels (15b) and (15c) require an internal hydrogen- (or halogen-) atom transfer through a six-membered cyclic transition state:

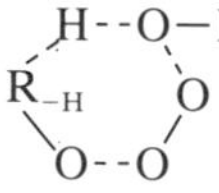

```
   H--O—H            X--O—H
  /      \          /      \
R-H       O   or  R-X       O
  \      /          \      /
   O--O              O--O
```

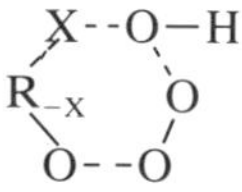

More data are needed to assess the importance of those channels for the principal classes of peroxyl radicals. An internal hydrogen-atom transfer is also necessary for channel (15d), which occurs in acylperoxyl radical reactions. In this case, a seven-membered transition state must be

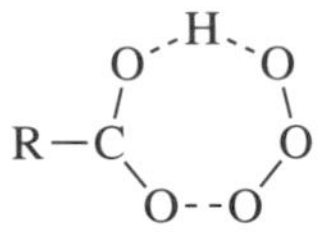

```
        .H.
      O'   'O
      /      \
   R—C        O
      \      /
       O--O
```

considered. Obviously, in this latter case, the proposed mechanism is favoured by both the presence of the carbonyl group, which induces the formation of a hydrogen bond, and by the absence of strain in the seven-membered cycle.

4.2 KINETICS OF $RO_2 + HO_2$ REACTIONS

The recommended kinetic parameters for $RO_2 + HO_2$ reactions are reported in Table 3. It can be seen that the database has been significantly increased since the publication of the previous reviews [1, 2]. As in the case of Table 2, the justification for each recommendation is included in the notes of the table and will not be discussed any further here.

The most striking feature is the much narrower range of rate constants for the various $RO_2 + HO_2$ reactions, compared to the $RO_2 + RO_2$ self-reactions. Most rate constants are similar within a factor of 2–3 and the extreme values hardly differ by more than a factor of 5, when compared to a range of several orders of magnitude for the self-reactions. Consequently, it is much easier to establish relationships between structure and reactivity for $RO_2 + HO_2$ reactions.

Concerning alkylperoxyl radicals, it has already been pointed out that the rate constant increased with the size of the radical, from a value of 5.8×10^{-12} cm^3 molecule^{-1} s^{-1}, accepted for the $CH_3O_2 + HO_2$ reaction, up to 1.5–1.8×10^{-11} cm^3 molecule^{-1} s^{-1} for large radicals ($\geqslant C_4$) [1, 2]. Recent data have confirmed this trend and, in addition, have shown that substitution by various functional groups has only a small influence on the trend, as seen in Table 3. A further significant difference, compared to the self-reactions, is that the structure of the radicals (primary, secondary or tertiary) has apparently no influence on the rate constant. This is particularly clear for the $(CH_3)_2C(OH)C(O_2)(CH_3)_2$ and $(CH_3)C(Br)C(O_2)(CH_3)_2$ tertiary peroxyl radicals, which exhibit slow self-reactions (Table 2) and fast reactions with HO_2 (Table 3). The same behaviour is expected for the t-butylperoxyl radical, but this needs to be confirmed.

As emphasized above, a great deal of attention has recently been given to the oxidation processes of halogenated compounds in the atmosphere. As a result, several halogenated peroxyl radicals have been investigated and the rate constants for their reactions with HO_2 are reported in Table 3. It can be seen that most of the rate constants for C_1 and C_2 radicals are similar, within uncertainties, showing the small influence of halogen substitution, with the exception, apparently, of radicals bearing several fluorine atoms, which tend to have lower rate constants. In contrast, the rate constant is remarkably constant for the series of chloromethylperoxyl radicals: 5.8, 5.2, 5.8, and 5.1×10^{-12} cm^3 molecule^{-1} s^{-1} for CH_3O_2, CH_2ClO_2, $CHCl_2O_2$ and CCl_3O_2, respectively.

4.3 CONCLUSIONS CONCERNING $RO_2 + HO_2$ REACTIONS

The important point to be emphasized about the kinetics of $RO_2 + HO_2$ reactions is the low sensitivity of the rate constant to the structure of the RO_2 radical. In addition, there are no clear influences of substituents or functional groups, with the possible exception of radicals bearing several fluorine atoms (to be confirmed). The main factor which influences the reactivity is apparently the size of the radical (or the number of carbon atoms), as previously pointed out [1, 2], and confirmed by recent measurements. The reaction of the acetylperoxyl radical (and probably those of other acylperoxyl radicals) is fast, like all other combination reactions of this radical.

The database presented in Table 3 allows us to make the following recommendations for the room-temperature rate constants (in units of cm^3 molecule^{-1} s^{-1}):

- C_1 and halogenated C_1–C_2 radicals: $5\text{–}6 \times 10^{-12}$
- Perfluorinated radicals $2\text{–}3 \times 10^{-12}$
- C_2–C_3 radicals $1.0\text{–}1.2 \times 10^{-11}$
- $\geqslant C_4$ and acylperoxyl radicals $1.3\text{–}1.8 \times 10^{-11}$

All reactions exhibit negative temperature dependencies with $E/R \approx -1000$ to -1500 K. As far as the chemistry of polluted atmospheres is concerned, where the terminating $RO_2 + HO_2$ and the propagating $RO_2 + NO$ reactions are in competition, it must be pointed out that these two types of reaction exhibit opposite trends in structure–reactivity relationships. As a result, the ratio of their rate constants, which is a critical parameter in modelling polluted atmospheres, can vary by an order of magnitude with the different structural characteristics given above.

The principal product of the reaction is the hydroperoxide, with a yield of almost 100% for the alkylperoxyl radical. Acylperoxyl reactions produce ozone (~30%), while halogen- and oxygen-substituted radicals may react according to reaction channels 15a and 15b. However, more information is needed to make recommendations about the reaction products for reactions of the latter radicals.

5 CROSS-REACTIONS BETWEEN PEROXYL RADICALS

As in the case of self-reactions, cross-reactions make significant contributions to reaction systems where high peroxyl radical concentrations can build up, such as in combustion, or in hydrocarbon-rich atmospheres under low NO_x concentrations. Under such conditions, the reactions of a particular RO_2 radical preferentially occur with the most abundant peroxyl radicals in the reaction system, provided that the reaction is fast enough. This is why it is important to characterize the kinetics of RO_2 cross-reactions with the most abundant peroxyl radicals.

Although the complexity of the oxidation mechanisms of hydrocarbons means such systems cannot yet be described in detail, it is now well established that large molecules and radical species undergo successive C—C bond splitting to finally end up as CO_2. Thus, all degradation processes converge to produce the simplest radical species, which are, consequently, the most abundant. Such radicals include, in addition to HO_2, CH_3O_2, $C_2H_5O_2$ and $CH_3C(O)O_2$ [88, 89].

Until recently, the only kinetic data available for cross-reactions were obtained from analysis of the reaction mechanism initiated by the propagating channel of certain self-reactions, e.g. as in the case of the acetonylperoxyl (Table 1) and acetylperoxyl radicals [7, 61]. However, direct kinetic studies of typical RO_2 cross-reactions have recently been undertaken [7, 90], in spite of the difficulties of the experimental measurements, which result from both cross- and self-reactions occurring simultaneously. As a result, a limited but nevertheless very valuable database is now available.

To date, there is very little information on the mechanism of cross-reactions and it is assumed that the same mechanism prevails for self- and cross-reactions, with, in the latter case, two possibilities for the terminating channel [6, 61]:

$$RO_2 + R'O_2 \longrightarrow RO + R'O + O_2 \tag{16a}$$

$$\longrightarrow ROH + R'_{-H}O + O_2 \tag{16b}$$

$$\longrightarrow R_{-H}O + R'OH + O_2 \tag{16b}$$

This assumption seems reasonable, and in addition, some support in favour of this mechanism has been obtained from analyses of the reaction mechanism initiated by self-reactions [6, 61]. However, almost no information is available about the branching ratios for the terminating and non-terminating channels. It has only been assumed that the branching ratio α for the non-terminating channel is intermediate between those of the RO_2 and $R'O_2$ self-reactions [88]. All rate coefficients available to date are presented in Table 4, along with the branching ratio α, which is taken as the arithmetic average of the branching ratios for the RO_2 and $R'O_2$ self-reactions.

Table 4 Recommended rate coefficients for cross-reactions between peroxyl radicals at 298 K

RO_2	$k(RO_2 + R'O_2)$[a,b]	$\alpha(RO_2 + R'O_2)$[c]	$k(RO_2 + RO_2)$[a]	Ref./notes
Cross-reactions with $R'O_2 = CH_3O_2$ ($k(CH_3O_2 + CH_3O_2) = 3.7 \times 10^{-13}$ cm^3 molecule^{-1} s^{-1})				
$C_2H_5O_2$	$2.0\ (3.2) \times 10^{-13}$	0.48	7.0×10^{-14}	90
$CH_3C(O)O_2$	$0.95\ (0.5) \times 10^{-11}$	0.66	1.4×10^{-11}	61, 62, 83, 90[d]
$CH_3C(O)CH_2O_2$	$3.8\ (3.4) \times 10^{-12}$	0.54	8.0×10^{-12}	7
neo-$C_5H_{11}O_2$	$1.5\ (1.3) \times 10^{-12}$	0.36	1.2×10^{-12}	90
$CH_2{=}CHCH_2O_2$	$1.7\ (1.0) \times 10^{-12}$	0.47	7.3×10^{-13}	90[f]
$C_6H_5CH_2O_2$	$<2\ (3.3) \times 10^{-12}$	0.37	7.7×10^{-12}	90
cyclo-$C_6H_{11}O_2$	$0.9\ (2.5) \times 10^{-13}$	0.32	4.2×10^{-14}	90
t-$C_4H_9O_2$	$3.1\ (6.6) \times 10^{-15}$	0.66	3.0×10^{-17}	1, 91
CH_2ClO_2	$2.5\ (2.3) \times 10^{-12}$	0.66	3.7×10^{-12}	90
CCl_3O_2	$6.6\ (2.4) \times 10^{-12}$	0.66	4.0×10^{-12}	16
Cross-reactions with $R'O_2 = C_2H_5O_2$ ($k(C_2H_5O_2 + C_2H_5O_2) = 7.0 \times 10^{-14}$ cm^3 molecule^{-1} s^{-1})				
CH_3O_2	$2.0\ (3.2) \times 10^{-13}$	0.48	3.7×10^{-13}	90
$CH_2{=}CHCH_2O_2$	$1.0\ (0.45) \times 10^{-12}$	0.62	7.3×10^{-13}	90
neo-$C_5H_{11}O_2$	$5.6\ (5.8) \times 10^{-13}$	0.51	1.2×10^{-12}	90
cyclo-$C_6H_{11}O_2$	$4.0\ (10) \times 10^{-14}$	0.46	4.2×10^{-14}	90
$CH_3C(O)O_2$	$1.0\ (0.22) \times 10^{-11}$	0.82	1.7×10^{-11}	90
Cross-reactions with $R'O_2 = CH_3C(O)O_2$ ($k(CH_3C(O)O_2 + CH_3C(O)O_2) = 1.4 \times 10^{-11}$ cm^3 molecule^{-1} s^{-1})				
CH_3O_2	$0.95\ (0.5) \times 10^{-11}$	0.47	3.7×10^{-13}	61, 62, 83, 90[d]
$C_2H_5O_2$	$1.0\ (0.22) \times 10^{-11}$	0.82	7.0×10^{-14}	90
cyclo-$C_6H_{11}O_2$	$1.0\ (0.17) \times 10^{-11}$	0.65	4.2×10^{-14}	90
Other reactions				
neo-$C_5H_{11}O_2$ + t-$C_4H_9O_2$	$3.0\ (2.6) \times 10^{-14}$, at 373 K, with $E/R > 0$			1
$CF_3CFHO_2 + CF_3O_2$	$8\ (6.5) \times 10^{-12}$	0.92		56

[a] Units of cm^3 molecule^{-1} s^{-1}.
[b] Figures in parentheses are calculated by using equation (17) (see text).
[c] Branching ratio for the non-terminating channel, assumed to be the average of the self-reaction branching ratios.
[d] The recommendation is the average of the reported values, which are in agreement within experimental uncertainties. Measured values of α are 0.5 [61] and 0.89 [83].
[e] The Arrhenius rate expression for the $CH_2{=}CHCH_2O_2 + CH_3O_2$ reaction is: $k = (3.9 \pm 1.4) \times 10^{-13} \exp[(430 \pm 150)\ K/T]$ cm^3 molecule^{-3} s^{-1} [90].

5.1 CROSS-REACTIONS OF RO_2 RADICALS WITH CH_3O_2 AND $C_2H_5O_2$

The methylperoxyl radical is the most abundant radical formed in hydrocarbon oxidation processes. In clean atmospheres, for example, its concentration can be as high as that of HO_2 [88, 89]. As a consequence it can react with all other peroxyl radicals and the kinetics of the corresponding reactions must be considered.

The self-reaction of CH_3O_2 is among the slowest reactions of primary peroxyl radicals (3.6×10^{-13} cm^3 molecule^{-1} s^{-1}), and it is of interest to compare the rate constant for the cross-reaction to that of the self-reactions of some typical RO_2 radicals. As seen in Table 4 the rate-constant value for cross-reactions is often intermediate between those of the self-reactions, with a few exceptions where the cross-reaction has been found to be faster. In Table 4, the experimental values are compared to the calculated values (in parentheses) by using the following suggested formula [92]:

$$k(RO_2 + R'O_2) = 2\sqrt{k(RO_2 + RO_2) \times k(R'O_2 + R'O_2)} \qquad (17)$$

with $R'O_2$ here being CH_3O_2. It is apparent for most reactions that the experimental and calculated rate constants differ by less than a factor of two, and therefore the above formula is a good approximation for estimating most cross-reaction rate constants, taking into account the large differences observed from one reaction to another and the fairly large uncertainties in measurements (50–60%). It may also be observed that rate constants of cross-reactions with CH_3O_2 are often close to the rate constants of RO_2 self-reactions. This is particularly true for the fastest reactions, which are the most efficient in reaction systems. Thus, an alternative recommendation for CH_3O_2 cross-reactions might be to take a rate constant close to that of the RO_2 self-reaction. This would be incorrect for slow reactions, which generally play a minor role in reaction systems, but would be a good approximation for fast reactions. The only exception to this latter rule is the CH_3O_2 + benzyl(O_2) reaction, which is found to be slower than predicted. This discrepancy may arise from large experimental uncertainties, resulting from the formation of strongly absorbing benzaldehyde, which perturbs the flash photolysis study of this system [33, 90].

As far as cross-reactions with the ethylperoxyl radical are concerned, it seems that similar conclusions can be drawn to those above. The principal exception is the reaction with t-$C_4H_9O_2$, which is much faster than the corresponding reaction of CH_3O_2 (Table 4). This result is rather surprising, and thus the rate coefficient requires confirmation.

5.2 CROSS-REACTIONS OF RO_2 RADICALS WITH $CH_3C(O)O_2$

The reactions of three radicals with the acetylperoxyl radical have been investigated, i.e. two primary radicals and a cyclic secondary peroxyl radicals (Table 4). All reactions are fast, with rate coefficients of ca. 1.0×10^{-11} cm^3 molecule^{-1} s^{-1}, and are obviously independent of the radical structure and of the value of the RO_2 self-reaction rate constant. In spite of the limited number of experimental data, it seems that all cross-reactions of this type are fast. This confirms again the high reactivity of the acetylperoxyl radical. The investigation of a tertiary peroxyl radical, such as the t-butylperoxyl radical, whose self-reaction is very slow (Table 2), would provide a useful complement to this series, to show, as might now be expected, that all such cross-reactions are fast and independent of the radical structure. For the present, a value of 1.0×10^{-11} cm^3 molecule^{-1} s^{-1} can reasonably be recommended for all RO_2 cross-reactions with the $CH_3C(O)O_2$ radical.

5.3 CONCLUSIONS CONCERNING CROSS-REACTIONS BETWEEN PEROXYL RADICALS

The kinetic database for this class of reaction is still limited in spite of the considerable importance of cross-reactions in modelling hydrocarbon oxidation processes. However, the available data already give reasonably good indications as to which reactions are expected to be the most important. The main conclusion is that all cross-reactions with the acetylperoxyl radical are fast, with $k \approx 1.0 \times 10^{-11}$ cm^3 molecule^{-1} s^{-1}, and that cross-reactions of CH_3O_2 or $C_2H_5O_2$ with those RO_2 radicals having fast self-reactions are also fast. In addition, the data show that the kinetics of $RO_2 + R'O_2$ reactions are more sensitive to the radical structure than those of $RO_2 + HO_2$ reactions (see preceding section). The temperature dependences of the rate constants for cross-reactions are expected to be similar to those of RO_2 self-reactions, as can be seen for the cross-reactions of CH_3O_2 with the acetylperoxyl [61] and allylperoxyl [90] radicals.

6 REACTIONS OF PEROXYL RADICALS WITH ATOMS AND OTHER RADICAL SPECIES

Combination reactions of peroxyl radicals with other radical species, including atoms, must be taken into account in certain reaction systems, particularly in hydrocarbon combustion processes, where a fairly large number of such species can reach significant concentrations. Unfortunately, apart from the few reactions of atmospheric interest discussed below, kinetic data are scarce for such reactions and the principal database available at present is the set of rate constants estimated by Tsang and Hampson [93].

The kinetic data available to date are presented in Table 5. The reactions of RO_2 radicals with H atoms, and hydroxyl, alkyl, or alkoxyl radicals are poorly determined and have already been discussed in Ref. 1. The principal new point to make about these reactions is that the reaction of CF_3O_2 with CF_3O was shown to mainly produce the trioxide CF_3OOOCF_3 [53, 97, 98]. However, as shown below, significant new results have been obtained in recent years for the reactions of RO_2 radicals with Cl atoms, and ClO and NO_3 radicals, which are of interest in atmospheric studies.

Table 5 Recommended rate coefficients for peroxyl radical reactions with atoms and other radical species at 298 K

Reaction	k (cm^3 molecule^{-1} s^{-1})	Ref./notes
$CH_3O_2 + O(^3P) \longrightarrow CH_3O + O_2$	4.3×10^{-11}	1
$CH_3O_2 + CH_3 \longrightarrow CH_3O + CH_3O$	8.0×10^{-11}	1[a]
$CCl_3O_2 + CCl_3 \longrightarrow CCl_3O + CCl_3O$	1.0×10^{-12}	1
benzyl(O_2) + benzyl $\longrightarrow$ 2 benzyl(O)	5.6×10^{-11}	94[b]
$CH_3O_2 + CH_3O \longrightarrow$ products	No recommendation	See note[c]
$C_2H_5O_2 + C_2H_5O \longrightarrow$ products	No recommendation	See note[c]
$CH_3O_2 + Cl \longrightarrow CH_3O + ClO$	0.75×10^{-10}	See text
$\longrightarrow CH_2O_2 + HCl$	0.75×10^{-10}	
$C_2H_5O_2 + Cl \longrightarrow$ products	1.5×10^{-10}	10
$CFCl_2O_2 + Cl \longrightarrow CFCl_2O + ClO$	$(2.0 \pm 1.0) \times 10^{-10}$	1
$CCl_3O_2 + Cl \longrightarrow CCl_3O + ClO$	$(2.0 \pm 1.0) \times 10^{-10}$	40
$CH_3O_2 + ClO \longrightarrow CH_3O + ClOO$	$(2.3 \pm 0.5) \times 10^{-12}$	See text
$\longrightarrow CH_3OCl + O_2$		
$CCl_3O_2 + Br \longrightarrow CCl_3O + BrO$	6×10^{-12}	95
$CH_3O_2 + NO_3 \longrightarrow CH_3O + NO_2 + O_2$	$(1.2 \pm 0.5) \times 10^{-12}$	See text
$C_2H_5O_2 + NO_3 \longrightarrow C_2H_5O + NO_2 + O_2$	$(2.4 \pm 0.5) \times 10^{-12}$	See text
$CH_3C(O)O_2 + NO_3 \longrightarrow CH_3C(O)O + NO_2 + O_2$	$(5 \pm 2) \times 10^{-12}$	96

[a] Average of the two values reported in Ref. 1; $E/R = -710$ K.
[b] Independent of temperature from 398 to 525 K [94].
[c] See discussion in Ref. 1.

6.1 REACTIONS OF PEROXYL RADICALS WITH HALOGEN ATOMS

Halogen atoms, particularly chlorine atoms, have often been used to produce peroxyl radicals in the laboratory by hydrogen abstraction from hydrocarbons, addition to unsaturated bonds, or by photodetachment from halides, in the presence of excess oxygen. It has been found that chlorine atoms may react at a very fast rate with RO_2 radicals ($k > 10^{-10}\,cm^{-3}\,molecule^{-1}\,s^{-1}$) if such atoms are not scavenged rapidly enough by the precursor [27, 40, 99, 100]. The results of recent measurements of the kinetics and branching ratios of a few reactions allow us to make the recommendations reported in Table 5.

The principal product of the Cl-atom reaction with perhalogenated peroxyl radicals is apparently ClO, along with the corresponding oxyl radical [27, 40, 99, 100] and the same conclusion has been drawn for the reaction of Cl with the $CFCl_2CH_2O_2$ and CH_3CFClO_2 radicals [101]. In contrast the reaction of CH_3O_2 exhibits two main reaction channels, as shown in four recent studies [10, 11, 63, 102]:

$$CH_3O_2 + Cl \longrightarrow CH_3O + ClO \quad (18a)$$

$$\longrightarrow CH_2O_2 + HCl \quad (18b)$$

Other possible reaction channels have been ruled out. Both channels have been clearly identified but discrepancies exist on the value of the branching ratio. Nevertheless, a value of 0.5 for both channels, which was found in two studies [10, 102], can be recommended. All studies report large values for the rate constant of between 1 and 2×10^{-10} cm^3 molecule^{-1} s^{-1}. The value reported by Maricq *et al.* [10], $k_{18} = (1.5 \pm 0.5) \times 10^{-10}$ cm^3 molecule^{-1} s^{-1}, which overlaps with other values within the uncertainty range, can be recommended. Similar branching ratios and rate coefficients have been found for the reaction of Cl atoms with $C_2H_5O_2$ [10].

The reactions of bromine and iodine atoms with RO_2 radicals apparently proceed through an association process, yielding the adduct as the main products (ROOBr and ROOI, respectively) [9, 47]. However, no kinetic parameters have been reported for this type of reaction. An exception to this behaviour is the reaction of CCl_3O_2 with Br, which forms BrO, with a rate constant of $(6 \pm 3) \times 10^{-12}$ cm^3 molecule^{-1} s^{-1}, and apparently with a high yield [86, 95].

6.2 REACTIONS OF RO_2 RADICALS WITH CLO

The reaction of CH_3O_2 with ClO has been investigated by several groups [11, 13, 14, 103, 104], with the interest in this reaction resulting from its potential importance in the chemistry of the polar atmosphere. All studies agree on the occurrence of two main reaction channels:

$$CH_3O_2 + ClO \longrightarrow CH_3O + ClOO \quad (19a)$$

$$\longrightarrow CH_3OCl + O_2 \quad (19b)$$

Other products have been identified but they have been shown to arise from secondary reactions. The overall rate constant has been measured by two groups [13, 14], and the values obtained are in fairly good agreement, with $k_{19} = (2.3 \pm 0.5) \times 10^{-12}$ cm^3 molecule^{-1} s^{-1}, and almost independent of temperature. However, there exists strong disagreement on the branching ratio of each channel, particularly on k_{19a}/k_{19}, since the reported values vary from ca. 0.3 to 0.7 [11, 103, 104], at room temperature. A better agreement has been obtained for k_{19b}/k_{19} [13, 103] whose value increases significantly on lowering the temperature, approximately from 0.12 to 0.25 between 298 and 220 K. It is apparent that the sum of the branching ratios is less than one, even by taking into account experimental uncertainties, which may indicate that some reaction products have been undetected.

Significantly, higher rate constants were estimated by Wu and Carr [101] for the reactions of $CFCl_2CH_2O_2$ and CH_3CFClO_2 radicals with ClO, with $k_{298\,K} = 6.0$ and $4.5 \times 10^{-12}\,cm^3\,molecule^{-1}\,s^{-1}$, respectively, when assuming a single reaction path corresponding to channel (19a).

6.3 REACTIONS OF RO_2 RADICALS WITH NO_3

The reactions of peroxyl radicals with the NO_3 radical are potentially important in the night-time oxidation of hydrocarbons in the atmosphere. Indeed, NO_3 can play the same role during the night-time as NO by day in propagating oxidation chain reactions:

$$RO_2 + NO_3 \longrightarrow RO + NO_2 + O_2 \quad (20)$$

$$RO + O_2 \longrightarrow R_{-H}O + HO_2 \quad (21)$$

$$HO_2 + NO_3 \longrightarrow OH + NO_2 + O_2 \quad (22)$$

with $R_{-H}O$ being the carbonyl compound corresponding to RO. This process is a night-time source of the OH radical, which then becomes the main chain carrier. In laboratory studies, reaction (20) is necessarily coupled to reaction (23), thus reforming the RO_2 radical:

$$RO + NO_3 \longrightarrow RO_2 + NO_2 \quad (23)$$

which complicates investigation of this system.

Recent investigations of the CH_3O_2 (and CD_3O_2) reaction with NO_3 are in fairly good agreement, with $k = (1.2 \pm 0.5) \times 10^{-12}\,cm^3\,molecule^{-1}\,s^{-1}$ at 298 K [11, 12, 105, 106]. The former value [107] of $k = 2.3 \times 10^{-12}\,cm^3\,molecule^{-1}\,s^{-1}$, was derived from an incomplete mechanism. The reaction of $C_2H_5O_2$ with NO_3 was also investigated in two studies [108, 109], with both being in agreement, and yielding $k = (2.4 \pm 0.5) \times 10^{-12}\,cm^3\,molecule^{-1}\,s^{-1}$. This value is about twice as high as that of the CH_3O_2 reaction and may indicate a trend in that the rate constant increases with the radical size. This should be confirmed, and if the rate constants are higher than those reported above, then the corresponding reactions would be very efficient in the night-time atmosphere. This is certainly the case for the reaction of the acetylperoxyl radical, with a rate constant of $(5 \pm 2) \times 10^{-12}\,cm^3\,molecule^{-1}\,s^{-1}$ [96], which again exhibits a high reactivity in this type of reaction.

7 CONCLUSIONS

It is shown in this brief review that our knowledge of combination reactions of peroxyl radicals has considerably increased during recent years. Both kinetic and product studies have provided a great deal of information on rate coefficients, reaction mechanisms and branching ratios. Given the very large number of peroxyl radicals involved in the various reaction systems, studies have been concentrated on model radicals, with the objective of extrapolating the data to other radicals by using various relationships between structure and reactivity. Such relationships are now available for certain types of reactions, but need to be improved for others.

Cross-reactions between RO_2 and HO_2 radicals are now fairly well described. Rate constants do not vary by more than an order of magnitude and trends in reactivity have been clearly identified. Similarly, it is now clear that all reactions of acylperoxyl radicals, including self-reaction and cross-reactions, are fast, with rate constants of 1.0 to 1.5×10^{-11} cm^3 $molecule^{-1}$ s^{-1}. The situation is not as clear for cross-reactions of RO_2 with $R'O_2$ radicals (CH_3O_2 for example, one of the most abundant radicals), but a few rules have been given for estimating rate coefficients.

The most difficult case remains that of self-reactions. Rate constants can vary by more than five orders of magnitude at room temperature and trends in reactivity are difficult to visualize, apart from the large differences observed between primary, secondary and tertiary alkylperoxyl radicals. Indeed, the trends are not really apparent inside these classes of radicals. Relationships between reactivity and the number of carbon atoms in the radical have been proposed [110] but more data are needed to link the reactivity to the size and the structure of the radicals. The most apparent feature is the fairly small dispersion of rate coefficients observed for primary alkylperoxyl radicals bearing the most common substituents and functional groups. As a consequence, the rate constant for self-reactions can be estimated, within a factor of two, at approximately 4×10^{-12} cm^3 $molecule^{-1}$ s^{-1}.

To end on an optimistic note, it must be pointed out that estimates of rate coefficients are apparently the most reliable for the fastest reactions, which are themselves the most important in reaction systems. This is the case for acylperoxyl radical reactions, cross-reactions with HO_2 and certain reactions of radicals bearing particular substituents, such as the hydroxyl group. This certainly represents the most significant progress, resulting from recent work, in our knowledge of peroxyl radical combination reactions.

AKNOWLEDGEMENTS

The authors wish to thank A.A. Boyd for his help in the preparation of the manuscript and M.E. Jenkin, G.K. Moortgat and G. Poulet for a careful reading of the manuscript prior to publication.

REFERENCES

1. P.D. Lightfoot, R.A. Cox, J.N. Crowley, M. Destriau, G.D. Hayman, M.E. Jenkin, G.K. Moortgat and F. Zabel, *Atmos. Environ.* **26A,** 1805 (1992).
2. T.J. Wallington, P. Dagaut and M.J. Kurylo, *Chem. Rev.* **92,** 667 (1992).
3. R. Atkinson, *Gas-Phase Tropospheric Chemistry of Organic Compounds, J. Phys. Chem. Ref. Data*, Monograph No. 2 (1994).
4. O. Horie, J.N. Crowley and G.K. Moortgat, *J. Phys. Chem.* **94,** 8198 (1990).
5. P.D. Lightfoot, R. Lesclaux and B. Veyret, *J. Phys. Chem.* **94,** 700 (1990).
6. M.E. Jenkin, R.A. Cox, M. Emrich and G.K. Moortgat, *J. Chem. Soc. Faraday Trans.* **89,** 2983 (1993).
7. I. Bridier, B. Veyret, R. Lesclaux and M.E. Jenkin, *J. Chem. Soc. Faraday Trans.* **89,** 2993 (1993).
8. P.D. Lightfoot, R. Lesclaux and B. Veyret, *J. Phys. Chem.* **94,** 700 (1990).
9. M.E. Jenkin and R.A. Cox, *J. Phys. Chem.* **95,** 3229 (1991).
10. M. Maricq, J.J. Szente, E.W. Kaiser and J. Shi, *J. Phys. Chem.* **98,** 2083 (1994).
11. V. Daële, *PhD Thesis*, Université Paris VII (1994); V. Daële and G. Poulet, *J. Chim. Phys.* **93,** 1081 (1996).
12. P. Biggs, C.E. Canosa-Mas, J.M. Fracheboud, D.E. Shallcross and R.P. Wayne, *J. Chem. Soc. Faraday Trans.* **90,** 1205 (1994).
13. A.S. Kukui, T.P.W. Jungkamp and R.N. Schindler, *Ber. Bunsenges. Phys. Chem.* **98,** 1298 (1994).
14. F. Helleis, J.N. Crowley and G.K. Moortgat, *J. Phys. Chem.* **97,** 11464 (1993).
15. F. Danis, F. Caralp, M.T. Rayez and R. Lesclaux, *J. Phys. Chem.* **95,** 7300 (1991).
16. V. Catoire, *PhD Thesis*, Université Bordeaux I (1994); V. Catoire, P.D. Lightfoot and R. Lesclaux, *Air Pollution Research Report 54*, European Commission, (1995) p. 229; V. Catoire, R. Lesclaux, W.F. Schneider and T.J. Wallington, *J. Phys. Chem.* **100,** 14356 (1996).
17. W.B. DeMore, S.P. Sander, D.M. Golden, R.F. Hampson, M.J. Kurylo, C.J. Howard, A.R. Ravishankara, C.E. Kolb and M.J. Molina, *Chemical Kinetics and Photochemical Data for Use in Stratospheric Modelling: Evaluation No. 11*, NASA JPL Publication 94-26, (1994).
18. D. Bauer, J.N. Crowley and G.K. Moortgat, *J. Photochem. Photobiol. A: Chem.* **65A,** 329 (1992).
19. F.F. Fenter, V. Catoire, R. Lesclaux and P.D. Lightfoot, *J. Phys. Chem.* **97,** 3530 (1993).
20. T.J. Wallington, J.M. Andino, A.R. Potts and O.J. Nielsen, *Int. J. Chem. Kinet.* **24,** 649 (1992).
21. M.E. Jenkin and G.D. Hayman, *J. Chem. Soc. Faraday Trans.* **91,**1911 (1995).
22. A.A. Boyd and R. Lesclaux, *Int. J. Chem. Kinet.* in press.
23. I. Barnes, K.H. Becker and L. Ruppert, *Chem. Phys. Lett.* **203,** 295 (1992).
24. A.A. Boyd, R. Lesclaux, M.E. Jenkin and T.J. Wallington, *J. Phys. Chem.*, **100,** 6594 (1996).
25. S. Langer, E. Ljungström, J. Sehested and O.J. Nielsen, *Chem. Phys. Lett.* **226,** 165 (1994).
26. M.E. Jenkin, G.D. Hayman, T.J. Wallington, M.D. Hurley, J.C. Ball, O.J. Nielsen and T. Ellermann, *J. Phys. Chem.* **97,** 11712 (1993).
27. P. Dagaut, T.J. Wallington and M.J. Kurylo, *J. Photochem. Photobiol.* **48,** 187 (1989).
28. O.J. Nielsen, J. Sehested, S. Langer, E. Ljungström and I. Wängberg, *Chem. Phys. Lett.* **238,** 359 (1995).

29. T.J. Wallington, T. Ellermann and O.J. Nielsen, *J. Phys. Chem.* **97,** 8442 (1993).
30. R.A. Cox, J. Munk, O.J. Nielsen, P. Pagsberg and E. Ratajczak, *Chem. Phys. Lett.* **173,** 206 (1990).
31. M.E. Jenkin, T.P. Murrells, S.J. Shalliker and G.D. Hayman, *J. Chem. Soc. Faraday Trans.* **89,** 433 (1993).
32. A.A. Boyd, B. Nozière and R. Lesclaux, *J. Chem. Soc. Faraday Trans.* **92,** 201 (1996).
33. B. Noziere, R. Lesclaux, M.D. Hurley, M.A. Dearth and T.J. Wallington, *J. Phys. Chem.* **98,** 2864 (1994).
34. V. Catoire, R. Lesclaux, P.D. Lightfoot and M.T. Rayez, *J. Phys. Chem.* **98,** 2889 (1994).
35. P. Dagaut, T.J. Wallington and M. Kurylo, *Int. J. Chem. Kinet.* **20,** 815 (1988).
36. E. Villenave and R. Lesclaux, *Chem. Phys. Lett.* **236,** 376 (1995).
37. O.J. Nielsen, J. Munk, G. Locke and T.J. Wallington, *J. Phys. Chem.* **95,** 8714 (1991).
38. J. Sehested, M. Bilde, T.E. Møgelberg, T.J. Wallington and O.J. Nielsen, *J. Phys. Chem.* **100,** 10989 (1996).
39. J. Sehested, T. Ellermann and O.J. Nielsen, *Int. J. Chem. Kinet.* **26,** 259 (1994).
40. J.J. Russel, J.A. Seetula, D. Gutman, F. Danis, F. Caralp, P.D. Lightfoot, R. Lesclaux, C.F. Melius and S.M. Senkan, *J. Phys. Chem.* **94,** 3277 (1990).
41. O.J. Nielsen, T. Ellermann, E. Bartkiewicz, T.J. Wallington and M.D. Hurley, *Chem. Phys. Lett.* **192,** 82 (1992).
42. M. Maricq and J.J. Szente, *J. Phys. Chem.* **96,** 4925 (1992).
43. O.J. Nielsen, T. Ellermann, J. Sehested, E. Bartkiewicz, T.J. Wallington and M.D. Hurley, *Int. J. Chem. Kinet.* **24,** 1009 (1992).
44. P. Dagaut, T.J. Wallington and M.J. Kurylo, *Chem. Phys. Lett.* **146,** 589 (1988).
45. T.J. Wallington, J.M. Andino and S.M. Japar, *Chem. Phys. Lett.* **165,** 189 (1990).
46. T.J. Wallington, M. Bilde, T.E. Møgelberg, J. Sehested and O.J. Nielsen, *J. Phys. Chem.* **100,** 5751 (1996).
47. J.N. Crowley and G.K. Moortgat, *J. Chem. Soc. Faraday Trans.* **88,** 2437 (1992).
48. G. Yarwood, N. Peng and H. Niki, *Int. J. Chem. Kinet.* **24,** 369 (1992).
49. I. Barnes, V. Bastian, K.H. Becker, R. Overath and Zhu Tong, *Int. J. Chem. Kinet.* **21,** 499 (1989).
50. E. Villenave and R. Lesclaux, in *Proceedings of the Joint CEC-EUROTRAC Workshop*, Halipp-Lactoz Working Group, Strasbourg, (1994) p. 407; E. Villenave, PhD Thesis, Université Bordeaux I (1996).
51. O.J. Nielsen, T. Ellermann, J. Sehested and T.J. Wallington, *J. Phys. Chem.* **96,** 10875 (1992).
52. J. Platz, O.J. Nielsen, J. Sehested and T.J. Wallington, *J. Phys. Chem.* **99,** 6570 (1995).
53. J. Sehested, T. Ellermann, O.J. Nielsen, T.J. Wallington and M.D. Hurley, *Int. J. Chem. Kinet.* **25,** 701 (1993).
54. O.J. Nielsen, E. Gamborg, J. Sehested, T.J. Wallington and M.D. Hurley, *J. Phys. Chem.* **98,** 9518 (1994).
55. T.E. Møgelberg, O.J. Nielsen, J. Sehested and T.J. Wallington, *J. Phys. Chem.* **99,** 13437 (1995).
56. M.M. Maricq and J.J. Szente, *J. Phys. Chem.* **96,** 10862 (1992).
57. T.J. Wallington, M.D. Hurley, J.C. Ball and E.W. Kaiser, *Environ. Sci. Technol.* **26,** 1318 (1992).
58. T.J. Wallington, T. Ellermann and O.J. Nielsen, *Res. Chem. Intermed.* **20,** 265 (1994).
59. J. Sehested, *Int. J. Chem. Kinet.* **26,** 1023 (1994).

60. T.E. Møgelberg, O.J. Nielsen, J. Sehested, T.J. Wallington and M.D. Hurley, *J. Phys. Chem.* **99,** 1995 (1995).
61. G.K. Moortgat, B. Veyret and R. Lesclaux, *J. Phys. Chem.* **93,** 2362 (1989).
62. C.M. Roehl, D. Bauer and G.K. Moortgat, *J. Phys. Chem.* **100,** 4038 (1996).
63. I. Bridier, *PhD Thesis*, Université Bordeaux I, (1991).
64. V. Foucher, *Diplome Etude Supérieure*, Université Bordeaux I (1995); V. Foucher, F. Caralp and R. Lesclaux, unpublished results.
65. M.M. Maricq, J.J. Szente, G.A. Khitrov and J.S. Fransisco, *J. Phys. Chem.* **98,** 9522 (1993).
66. T.J. Wallington, T. Ellermann, O.J. Nielsen and J. Sehested, *J. Phys. Chem.* **98,** 2346 (1994).
67. T.J. Wallington, M.D. Hurley and M.M. Maricq, *Chem. Phys. Lett.* **205,** 62 (1993).
68. T.E. Møgelberg, J. Platz, O.J. Nielsen, J. Sehested and T.J. Wallington, *J. Phys. Chem.* **99,** 5373 (1995).
69. E. Furimsky, J.A. Howard and J. Selwyn, *Can. J. Chem.* **58,** 677 (1980).
70. J.E. Bennett, D.M. Brown and B. Mile, *Trans. Faraday Soc.* **66,** 397 (1970).
71. K. Adamic, J.A. Howard and K.U. Ingold, *Can. J. Chem.* **47,** 3803 (1969).
72. P.S. Nangia and S.W. Benson, *Int. J. Chem. Kinet.* **12,** 29 (1980).
73. G.A. Russell, *J. Am. Chem. Soc.* **79,** 3871 (1957).
74. M.M. Maricq and J.J. Szente, *J. Phys. Chem.* **98,** 2078 (1994).
75. D.M. Rowley, R. Lesclaux, P.D. Lightfoot, K. Hughes, M.D. Hurley, S. Rudy and T.J. Wallington, *J. Phys. Chem.* **96,** 7043 (1992).
76. D.M. Rowley, R. Lesclaux, P.D. Lightfoot, B. Nozière, T.J. Wallington and M.D. Hurley, *J. Phys. Chem.* **96,** 4889 (1992).
77. T.P. Murrells, M.E. Jenkin, S.J. Shalliker and G.D. Hayman, *J. Chem. Soc. Faraday Trans.* **87,** 2351 (1991).
78. T.J. Wallington, M.D. Hurley and J.C. Ball, *Chem. Phys. Lett.* **211,** 41 (1993).
79. T.J. Wallington, M.D. Hurley, W.F. Schneider, J. Sehested and O.J. Nielsen, *Chem. Phys. Lett.* **218,** 34 (1994).
80. T.J. Wallington, M.D. Hurley and W.F. Schneider, *Chem. Phys. Lett.* **251,** 164 (1996).
81. G.D. Hayman and F. Battin-Leclerc, *J. Chem. Soc. Faraday Trans.* **91,** 1313 (1995).
82. M. Maricq, J.J. Szente, M.D. Hurley and T.J. Wallington, *J. Phys. Chem.* **98,** 8962 (1994).
83. O. Horie and G.K. Moortgat, *J. Chem. Soc. Faraday Trans.* **88,** 3305 (1992).
84. M.E. Jenkin, R.A. Cox, G.D. Hayman and L.J. Whyte, *J. Chem. Soc. Faraday Trans. 2* **84,** 913 (1988).
85. H. Niki, P.D. Marker, C.M. Savage and L.P. Breitenbach, *Chem. Phys. Lett.* **73,** 43 (1980).
86. R. Patrick, J.R. Barker and D.M. Golden, *J. Phys. Chem.* **88,** 128 (1984).
87. C.C. Kircher and S.P. Sander, *J. Phys. Chem.* **88,** 2082 (1984).
88. S. Madronich and J.G. Calvert, *J. Geophys. Res.* **95,** 5697 (1990).
89. N.M. Donahue and R.G. Prinn, *J. Geophys. Res.* **95,** 18387 (1990).
90. E. Villenave and R. Lesclaux, *J. Phys. Chem.* **100,** 14372 (1996); A.A. Boyd, B. Nozière, E. Villenave and R. Lesclaux, *Air Pollution Research Report 54*, European Commission, (1995), p. 127.
91. D.A. Osborne and D.J. Waddington, *J. Chem. Soc. Perkin. Trans.* **89,** 2993 (1993).
92. J.A. Kerr and A.F. Trotman-Dickenson, *Prog. React. Kinet.* **1,** 105 (1961).
93. W. Tsang, R.F. Hampson, *J. Phys. Chem. Ref. Data* **15,** 1087 (1986).

94. F.F. Fenter, B. Nozière, F. Caralp and R. Lesclaux, *Int. J. Chem. Kinet.* **26,** 171 (1994).
95. F.F. Fenter, P.D. Lightfoot, J.T. Niiranen and D. Gutman, *J. Phys. Chem.* **97,** 5313 (1993).
96. C.E. Canosa-Mas, M.D. King, R. Lopez, C.J. Percival, R.P. Wayne, J. Pyle, D.E. Shallcross and V. Daële, in *Proceedings of the Joint CEC-EUROTRAC Workshop*, Halipp-Lactoz Working Group, Strasbourg, (1994), p. 375.
97. R. Meller and G.K. Moortgat, *J. Photochem. Photobiol. Chem.* **86A,** 15 (1995).
98. E.C. Tuazon and R. Atkinson, *J. Atmos. Chem.* **16,** 301 (1993).
99. H.M. Gillespie, J. Garraway and R.J. Donovan, *J. Photochem.* **7,** 29 (1977).
100. R.W. Carr, D.G. Peterson and F.K. Smith, *J. Phys. Chem.* **90,** 607 (1986).
101. F. Wu and R.W. Carr, *J. Phys. Chem.* **99,** 3128 (1995).
102. T.P.W. Jungkamp, A. Kukui and R.N. Schindler, *Ber. Bunsenges. Phys. Chem.* **99,** 1057 (1995).
103. F. Helleis, J.N. Crowley and G.K. Moortgat, *Geophys. Res. Lett.* **21,** 1795 (1994).
104. P. Biggs, C.E. Canosa-Mas, J.M. Fracheboud, D.E. Shallcross and R.P. Wayne, *Geophys. Res. Lett.* **22,** 1221 (1995).
105. V. Daële, G. Laverdet, G. Le Bras and G. Poulet, *J. Phys. Chem.* **99,** 1470 (1995).
106. G.K. Moortgat, D. Bauer, J.N. Crowley, F. Helleis, O. Horie, S. Koch, S. Limbach, P. Neeb, C. Roehl and F. Sauer, *EUROTRAC Annual Report 1993*, Part 8, (LACTOZ) EUROTRAC ISS, Garmisch Partenkirchen (1994).
107. J.N. Crowley, J.P. Burrows, G.K. Moortgat and G. Poulet, *Int. J. Chem. Kinet.* **22,** 673 (1990).
108. P. Biggs, C.E. Canosa-Mas, J.M. Fracheboud, D.E. Shallcross and R.P. Wayne, *J. Chem. Soc. Faraday Trans.* **91,** 817 (1995).
109. A. Ray, V. Daële, I. Vassali, G. Le Bras and G. Poulet, *J. Phys. Chem.* **100,** 5737 (1996).
110. F. Kirchner and W.R. Stockwell, *J. Geophys. Res.* **101,** 21007 (1996).

7 Reactions of Organic Peroxy Radicals in the Gas Phase

T.J. WALLINGTON AND O.J. NIELSEN
Research Laboratory, Ford Motor Company, Dearborn, USA

and

J. SEHESTED
Risø National Laboratory, Roskilde, Denmark

1 INTRODUCTION

Gas-phase peroxy radical chemistry plays an important role in low-temperature combustion phenomena such as auto-ignition and cool flames. Peroxy radical chemistry is also important in the atmospheric oxidation of the one billion tonnes of organic compounds released into the atmosphere annually [1–3]. Recognition of the adverse environmental impact of the atmospheric release of volatile organic compounds has provided the major impetus for gas-phase peroxy radical research. Accordingly, the bulk of the available kinetic and mechanistic data have been acquired under atmospheric conditions.

Alkylperoxy radicals, RO_2, are important intermediates in the atmospheric oxidation and low-temperature combustion of all organic compounds. Such degradation processes are initiated typically by the reaction of a radical species, such as OH, NO_3, or Cl atoms, with organic compounds via either an abstraction or addition mechanism. Regardless of the reaction mechanism an alkyl radical is formed. Alkyl radicals rapidly add O_2 to give peroxy radicals. The rate constants for the reactions of alkyl radicals with O_2, while dependent on the total pressure, are generally close to, or at the high-pressure limit at pressures above 20 torr. The high-pressure limiting rate constants are all greater than 10^{-12} cm^3 molecule^{-1} s^{-1}. At one atmosphere and 298 K the O_2 concentration in air is 5.2×10^{18} molecules cm^{-3} and the lifetime of alkyl radicals with respect to conversion into peroxy radicals is $< 0.2\ \mu s$. Reaction with O_2 is the sole atmospheric fate of alkyl radicals. The R—O_2 bond strength is typically 30–40 kcal mol^{-1}. At ambient temperatures and below, thermal decomposition of RO_2 radicals into R and O_2 is of no importance. However, as the temperature is increased decomposition becomes important and equilibrium is established. For simple hydrocarbon peroxy radicals the equilibrium is reached at temperatures of 600–800 K

Peroxyl Radicals. Edited by Z. B. Alfassi

(characteristic of cool flames). At higher temperatures, decomposition of the RO_2 radical becomes dominant and peroxy radicals cease to have any significant existence. Also at high temperatures additional channels in the reaction of R radicals with O_2 become important in which products other than RO_2 are formed.

Compared to other radical species such as Cl, F, or O atoms, the OH or NO_3 radicals, perhaps the most striking aspect of peroxy radicals is their relative inactivity. At ambient temperatures peroxy radicals only react with species which carry an unpaired electron. Most, if not all, reactions of peroxy radicals are believed to proceed via the formation of a short-lived adduct which rearranges to give products. The key reactions of peroxy radicals in atmospheric chemistry and low-temperature combustion systems are those with NO, NO_2, and HO_2 radicals, and combination reactions with other peroxy radicals. Combination reactions have been dealt with in a previous chapter. In this present chapter we review the available kinetic and mechanistic data concerning other alkylperoxy radical reactions in the gas phase. Where possible, recommended values are given and discrepancies and uncertainties in the data are highlighted. This chapter is broken down into four sections dealing with the kinetics and mechanisms of organic peroxy radicals with (i) HO_2 radicals, (ii) NO, (iii) NO_2, and (iv) other species. Data published before October 1995 are considered.

2 KINETICS AND MECHANISMS OF THE REACTIONS OF RO_2 AND HO_2 RADICALS

The literature data for the branching ratios and the kinetics of the reactions of peroxy radicals with HO_2 radicals are given in Tables 1 and 2, respectively, and are discussed below.

2.1 THE SELF-REACTION OF HO_2

A substantial body of data exists for the self-reactions of hydroperoxy radicals at temperatures below 500 K, as discussed in the evaluations of kinetic data for atmospheric chemistry (DeMore *et al.* [20], Atkinson *et al.* [21]).At ambient temperature the reaction exhibits both a pressure and water vapour dependence. At higher temperatures, recent data obtained by Lightfoot, Veyret and Lesclaux [23] using the flash photolysis ultraviolet (UV) absorption spectroscopy technique, reconcile the previous low-temperature data with the measurement of Troe [22] at 1100 K by demonstrating that the rate constant increases with temperature above 600 K. These results are interpreted in terms of a direct bimolecular pathway

Table 1 Measured Branching Ratios for the Reactions of Peroxy Radicals with HO_2

k_a/k	k_b/k	Technique[a]	Substrates	Temperature (K)	Pressure (torr)	Ref.
			$CH_3O_2 + HO_2 \longrightarrow CH_3OOH + O_2$ (a)			
			$CH_3O_2 + HO_2 \longrightarrow HCHO + H_2O + O_2$ (b)			
0.92 ± 0.08	–	CP/FTIR	$CH_4/H_2/F_2$	295	700	4
0.92 ± 0.05	–	CP/FTIR	$CH_4/H_2/F_2$	295	15–700	5
			$CD_3O_2 + HO_2 \longrightarrow CD_3OOH + O_2$ (a)			
			$CD_3O_2 + HO_2 \longrightarrow DCDO + HDO + O_2$ (b)			
> 0.8	0.2	MMS	$CD_4/CH_3OH/Cl_2$	300	10.8	6,7
1.00 ± 0.04	–	CP/FTIR	$CD_4/H_2/F_2$	295	10–700	8
			$C_2H_5O_2 + HO_2 \longrightarrow C_2H_5OOH + O_2$ (a)			
1.02 ± 0.06	–	CP/FTIR	$C_2H_6/CH_3OH/Cl_2$	295	700	9
			neo-$C_5H_{11}O_2 + HO_2 \longrightarrow$ neo-$C_5H_{11}OOH + O_2$ (a)			
0.92 ± 0.02	–	CP/FTIR	neo-$C_5H_{12}/CH_3OH/Cl_2$	295	700	10
			cyclo-$C_5H_9O_2 + HO_2 \longrightarrow$ cyclo-$C_5H_9OOH + O_2$ (a)			
0.96 ± 0.03	–	CP/FTIR	cyclo-$C_5H_{10}/CH_3OH/Cl_2$	295	700	11
			cyclo-$C_6H_{11}O_2 + HO_2 \longrightarrow$ cyclo-$C_6H_{11}OOH + O_2$ (a)			
0.99 ± 0.03	–	CP/FTIR	cyclo-$C_6H_{12}/CH_3OH/Cl_2$	295	15–700	11
			$HOCH_2O_2 + HO_2 \longrightarrow HOCH_2OOH + O_2$ (a)			
			$HOCH_2O_2 + HO_2 \longrightarrow HCOOH + H_2O + O_2$ (b)			
0.60	0.40	MMS/UVA	$HCHO/Cl_2$	298	2	12
			$CH_2FO_2 + HO_2 \longrightarrow CH_2FOOH + O_2$ (a)			
			$CH_2FO_2 + HO_2 \longrightarrow HC(O)F + H_2O + O_2$ (b)			
0.29 ± 0.08	0.71 ± 0.11	CP/FTIR	$CH_3F/H_2/Cl_2$	295	10–700	13
			$CH_2ClO_2 + HO_2 \longrightarrow CH_2ClOOH + O_2$ (a)			
			$CH_2ClO_2 + HO_2 \longrightarrow HC(O)Cl + H_2O + O_2$ (b)			
0.20 ± 0.06	0.80 ± 0.14	CP/FTIR	$CH_3Cl/H_2/Cl_2$	295	700	14
			$CF_3CFHO_2 + HO_2 \longrightarrow CF_3CFHOOH + O_2$ (a)			
			$CF_3CFHO_2 + HO_2 \longrightarrow CF_3C(O)F + H_2O + O_2$ (b)			
–	< 0.05	CP/FTIR	$CF_3CFH_2/H_2/Cl_2$	295	700	15
			$CH_3OCH_2O_2 + HO_2 \longrightarrow CH_3OCH_2OOH + O_2$ (a)			
			$CH_3OCH_2O_2 + HO_2 \longrightarrow CH_3OC(O)H + H_2O + O_2$ (b)			
0.53 ± 0.08	0.40 ± 0.04	CP/FTIR	$CH_3OCH_3/CH_3OH/Cl_2$	295	700	16
			$CH_3C(O)O_2 + HO_2 \longrightarrow CH_3C(O)OOH + O_2$ (a)			
			$CH_3C(O)O_2 + HO_2 \longrightarrow CH_3C(O)OH + O_3$ (b)			
0.75	0.25	CP/FTIR	CH_3CHO/Cl_2	298	700	17
0.67	0.33	FP/UVA	$CH_3CHO/CH_3OH/Cl_2$	253–368	600–650	18
–[b]	–[b]	CP/MI-FTIR	$CH_3COCOCH_3$	263–333	730–770	19

[a] CP, continuous photolysis; FTIR, Fourier-transform infrared spectroscopy; MMS, molecular modulation spectroscopy; UVA, ultraviolet absorption spectroscopy; FP, flash photolysis; MI, matrix isolation.
[b] See text.

Table 2 Kinetic data for the Reactions of Peroxy Radicals with HO_2

k (10^{-13} cm³ molecule⁻¹ s⁻¹)	λ (nm)	Temperature (K)	Technique[a]	Ref.
$HO_2 + HO_2$				
		200–300	Review	20
$k = \{2.3 \times 10^{-13} \exp[(600 \pm 200)/T] + 1.7 \times 10^{-33}$ [air] $\exp[(1000 \pm 400)/T]\} \times \{1 + 1.4 \times 10^{-21} [H_2O] \exp(2000)/T]\}$ cm³ molecule⁻¹ s⁻¹				
		230–420	Review	21
$k = \{2.2 \times 10^{-13} \exp[(600/T] + 1.9 \times 10^{-33} [N_2] \exp(980/T\} \times \{1 + 1.4 \times 10^{-21} \exp(2200/T) [H_2O]\}$ cm³ molecule⁻¹ s⁻¹				
1.7–3.3		980–1250	Shock-tube/UVA	22
	210	298–777	FP/UVA ($CH_3OH/Cl_2/O_2$)	23
$k = (2.1 \pm 1.6) \times 10^{-10} \exp(-(5051 \pm 722)/T) + (1.8 \pm 0.2) \times 10^{-13} \exp(885/T)$ cm³ molecule⁻¹ s⁻¹				
	230	750–1120	Shock-tube/UVA (H_2O_2;$(CH_3O)_2/O_2$)	24
$k = 6.97 \times 10^{-10} \exp(-6030/T) + 2.16 \times 10^{-13} \exp(820/T)$ cm³ molecule⁻¹ s⁻¹				
$CH_3O_2 + HO_2$				
8.5 ± 1.2	210, 250	274	MMS/UVA ($CH_4/H_2/Cl_2/O_2$)	25,26
6.5 ± 1.0		298		
3.5 ± 0.5		338		
1.3		298	FTIR (CH_3NNCH_3/O_2)	27
≤6.7 ± 2.2	220, 250	298	MMS/UVA (CH_3CHO/air)	28
4.8 ± 0.2	220, 250	298	MMS/UVA (CH_3CHO/air)	28
6.4 ± 1.0	210, 240	298	FP/UVA ($CH_4/CH_3OH/Cl_2/O_2$)	29
2.9 ± 0.4	215–280	298	FP/UVA ($CH_4/CH_3/OH/Cl_2/O_2$)	30
6.8 ± 0.5	250	228	FP/UVA ($CH_4/CH_3OH/Cl_2/O_2$	31
5.5 ± 0.3		248		
4.1 ± 0.3		273		
2.4 ± 0.5		340		
2.1 ± 0.3		380		
5.4 ± 1.1	260, 1110 cm⁻¹	300	MMS/UVA/IR ($CH_4/H_2O/Cl_2/O_2$)	
6.8 ± 0.9	260	303	MMS/UVA ($CH_4/H_2/Cl_2/O_2$)	
10.37 ± 4.72	210, 260	248	FP/UVA ($CH_4/CH_3OH/Cl_2/O_2$)	32
7.63 ± 1.70	210, 260	248	FP/UVA ($CH_4/CH_3OH/Cl_2/O_2$)	32
7.63 ± 1.70		273		
5.63 ± 1.02		298		
5.22 ± 1.24		323		
2.98 ± 0.84		368		
3.11 ± 0.48		373		
2.39 ± 0.36		473		
1.83 ± 0.38		573		
1.3 ± 0.13	200–300	600	FP/UVA (CH_4/CH_3OH)	33
1.2 ± 0.10		616		
1.1 ± 0.15		648		
0.72 ± 0.25		678		
$CD_3O_2 + HO_2$				
5.4	1378, 594 cm⁻¹	300	CP/IR ($CD_4/CH_3OH/Cl_2/O_2$)	6
$C_2H_5O_2 + HO_2$				
6.3 ± 0.9	210, 260, 1117 cm⁻¹	295	MMS/UVA/IR($C_2H_6/CH_3OH/Cl/O_2$)	34
7.3 ± 1.0	230–280	248	FP/UVA ($C_6H_6/CH_3OH/Cl_2/O_2$)	35
6.0 ± 0.5		273		
5.4 ± 1.2		298		
3.4 ± 1.0		340		
3.1 ± 0.5		380		
24.5 ± 2.7	220, 260	248	FP/UVA ($C_2H_6/CH_3OH/Cl_2/O_2$)	36
15.8 ± 2.8		273		
10.4 ± 1.2		298		

Table 2 – *continued*

k (10^{-13} cm^3 molecule^{-1} s^{-1})	λ (nm)	Temperature (K)	Technique[a]	Ref.
$C_2H_5O_2 + HO_2$ – *continued*				
8.7 ± 1.0		323		
7.6 ± 0.8		332		
4.6 ± 0.5		372		
3.1 ± 0.3		410		
1.9 ± 0.2		480		
19.8	200–300	210	FP/UVA ($C_2H_6/H_2/F_2/O_2$)	37
16.5		233		
11.5		253		
9.5		273		
8.1		295		
6.6		323		
6.1		363		
neo-$C_5H_{11}O_2 + HO_2$				
32.2 ± 8.3	220, 260	248	FP/UVA (neo-$C_5H_{12}/CH_3OH/Cl_2/O_2$)	10
22.9 ± 5.6		273		
16.9 ± 2.3		293		
14.2 ± 1.3		298		
11.3 ± 1.3		326		
8.9 ± 1.4		348		
8.5 ± 2.9		348		
6.1 ± 0.3		365		
cyclo-$C_5H_9O_2 + HO_2$				
37.3 ± 10.6	220, 260	249	FP/UVA (cyclo-$C_5H_{10}/CH_3OH/Cl_2/O_2$)	11
28.5 ± 4.9		274		
18.3 ± 5.7		300		
20.7 ± 3.1		300		
13.1 ± 1.3		324		
10.2 ± 2.8		348		
7.1 ± 364				
cyclo-$C_6H_{11}O_2 + HO_2$				
37.7 ± 7.6	220, 260	249	FP/UVA (cyclo-$C_6H_{12}/CH_3OH/Cl_2/O_2$)	11
32.5 ± 3.8		258		
24.1 ± 7.0		274		
19.8 ± 3.4		300		
16.8 ± 3.8		300		
11.7 ± 1.8		324		
7.8 ± 1.1		364		
Benzylperoxy + HO_2				
15.4 ± 2.8	220–300	273	FP/UVA (toluene/$CH_3OH/Cl_2/O_2$)	38
7.1 ± 3.0		283		
10.2 ± 1.7		296		
10.9 ± 3.2		300		
7.0 ± 1.2		341		
7.0 ± 1.9		355		
5.1 ± 0.9		359		
4.6 ± 0.5		389		
3.2 ± 0.7		411		
5.0 ± 0.9		447		

(continued overleaf)

Table 2 – *continued*

k (10^{-13} cm^3 molecule^{-1} s^{-1})	λ (nm)	Temperature (K)	Technique[a]	Ref.
$CH_2ClO_2 + HO_2$				
9.3 ± 1.4	210, 250	255	FP/UVA ($Cl_2/CH_3Cl/CH_3OH$)	39
7.3 ± 0.8		273		
4.9 ± 0.6		298		
5.0 ± 0.6		307		
3.7 ± 0.4		323		
2.6 ± 0.2		390		
1.9 ± 0.2		460		
1.6 ± 0.2		588		
$CH_2BrO_2 + HO_2$				
6.7 ± 3.8	250–280	255	FP/UVA ($Cl_2/CH_3Br/CH_3OH$)	40
$CF_2ClO_2 + HO_2$				
3.4 ± 1.0	220–240	298	FP/UVA ($H_2O_2/CF_2ClH/O_2$)	1
$HOCH_2O_2 + HO_2$				
12 ± 3	250, 1110 cm^{-1}	298	MMS/UVA/IR ($Cl_2/HCHO/O_2$)	12
25 ± 5	210, 240	275	FP/UVA ($Cl_2/HCHO/O_2$)	42
12 ± 4		295		
12 ± 6		308		
6 ± 4		323		
6 ± 2		333		
$HOCH_2CH_2O_2 + HO_2$				
4.8 ± 0.5	220, 230, 240, 250	298	MMS/UVA ($HOCH_2CH_2I$)	43
$CF_3CFHO_2 + HO_2$				
11 ± 4	180–250	210	FP/UVA (F_2/UVA ($F_2/CF_3CFH_2/H_2$)	15
7.8 ± 4	180–250	210	FP/UVA ($F_2/CF_3CFH_2/H_2$)	15
7.5 ± 2		243		
8.9 ± 3.5		253		
6.8 ± 2.5		273		
4.7 ± 1.7		295		
3.0 ± 1.5		313		
2.8 ± 1.5		323		
2.0 ± 0.6		363		
4.0 ± 2.0	220–240	298	FP/UVA ($H_2O_2/CF_3CFH_2/O_2$)	41
$CF_2ClCH_2O_2 + HO_2$				
6.8 ± 2.0	220–240	298	FP/UVA ($H_2O_2/CF_2ClCH_3/O_2$)	41
$CFCl_2CH_2O_2 + HO_2$				
9.2 ± 2.6	220–240	298	FP/UVA ($H_2O_2/CFCl_2CH_3/O_2$)	41
$CF_3CF_2O_2 + HO_2$				
1.2 ± 0.4	220–240	298	FP/UVA ($H_2O_2/CF_3CF_2H/O_2$)	41
$CH_3COO_2 + HO_2$				
27 ± 5	210	258	FP/UVA ($Cl_2/CH_3CHO/CH_3OH$)	44
13 ± 3	210	298		
7.45 ± 3.0	207	368		
$CH_3C(O)CH_2O_2 + HO_2$				
9.0 ± 1.0	210, 230	298	FP/UVA ($Cl_2/CH_3COCH_3/CH_3OH$)	45

[a] CP, continuous photolysis; FTIR, Fourier-Transform infrared spectroscopy; MMS, molecular modulation spectroscopy; IR, infrared analysis; UVA, ultraviolet absorption spectroscopy; FP, flash photolysis.

accessible at high temperatures (> 600 K):

$$HO_2 + HO_2 \longrightarrow H_2O_2 + O_2$$

competing with the low-temperature mechanism:

$$HO_2 + HO_2 \longrightarrow H_2O_2 + O_2$$

$$\xrightarrow{M} H_2O_2 + O_2$$

Lightfoot, Veyret and Lesclaux [23] fitted their data, together with the existing high-temperature data (298–1100 K, 760 torr), and obtained the following expression:

$$k = (2.1 \pm 6) \times 10^{-10} \exp(-(5051 \pm 722)/T) + (1.8 \pm 0.2) \times 10^{-13} \exp(885/T)$$

in units of cm^3 $molecule^{-1}$ s^{-1}.

Recently, Hippler, Troe and Willner [24] have used a shock-tube–ultraviolet absorption system to measure the kinetics of the self-reaction of HO_2 radicals over the temperature range 720–1120 K. Two sources of HO_2 radicals were used, i.e. thermal dissociation of CH_3OOCH_3 in the presence of excess O_2, and thermal dissociation of H_2O_2. The kinetic data were derived using $\sigma\ (HO_2)_{230\ nm,1000\ K} = 2.56 \times 10^{-18}$ cm^2 $molecule^{-1}$. Hippler and coworkers [24] found no significant difference between their experimental data acquired at a total pressure of 5 bar and data reported by Lightfoot and coworkers [23] at 1 bar, showing that in contrast to the situation at lower temperatures there is no evidence for any pressure effect at high temperatures. Hippler and coworkers [24] combined their high-temperature results with the low-temperature kinetic expression appropriate for 760 torr recommended by Atkinson *et al.* [21] to derive the following expression which is valid over the temperature range 300–1100 K: $k = 6.97 \times 10^{-10} \exp(-6030/T) + 2.16 \times 10^{-13} \exp(820/T)$ cm^3 $molecule^{-1}$ s^{-1}.

Although the kinetic expressions listed in Table 2 appear markedly different, where comparison is possible, they do in fact yield essentially identical results, as shown in Figure 1 where these expressions have been evaluated at a constant number-density of 2.46×10^{19} molecules cm^{-3} of N_2. We recommend use of the expression given by Atkinson *et al.* [21] below 500 K and that of Hippler, Troe and Willner [24] above 500 K.

2.2 THE REACTION OF HO_2 WITH CH_3O_2

The branching ratio of the reaction of CH_3O_2 with HO_2 radicals has been studied by Wallington and Japar [4] and Wallington [5] (see Table 1). These workers used continuous photolysis of $CH_4/H_2/F_2/O_2/N_2$ mixtures at 15–700

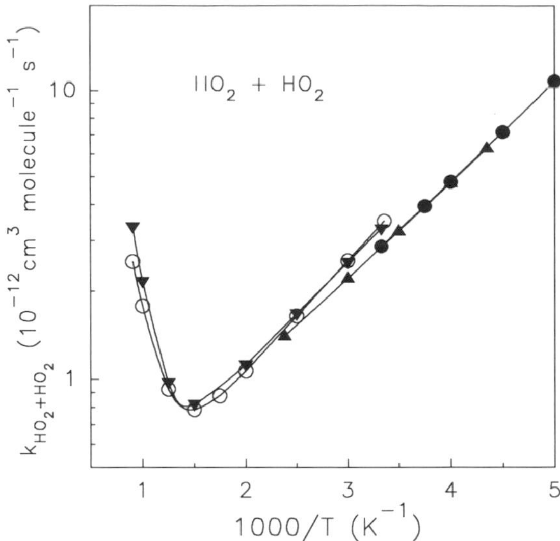

Figure 1 Kinetic data for the self-reaction of HO_2 radicals derived from the expression recommended by DeMore *et al.*[20] (filled circles), Atkinson *et al.*[21] (filled triangles), Lightfoot, Veyret and Lesclaux[23] (open circles), and Hippler, Troe and Willner[24] (filled inverse triangles); the full lines are cubic spline fits over the temperature range for which each recommendation is valid.

torr total pressures and 295 K with FTIR spectroscopy to quantify the resulting CH_3OOH formation, and found $k_a/(k_a + k_b) = 0.92$:

$$HO_2 + CH_3O_2 \longleftrightarrow CH_3O_4H \longrightarrow CH_3OOH + O_2 \quad (a)$$

$$HO_2 + CH_3O_2 \longrightarrow HCHO + H_2O + O_2 \quad (b)$$

In addition there has been several less direct studies of this reaction. In the most recent investigation Lightfoot *et al.* [33] used flash photolysis of molecular oxygen in the presence of methane to produce both CH_3O_2 and HO_2 radicals at temperatures of 600–719 K. The decay of the UV absorption was then fitted to derive kinetic data for the CH_3O_2 + OH reaction. At elevated temperatures CH_3COOH decomposes rapidly to give CH_3O + OH and hence regenerates radical species. HCHO does not undergo significant decomposition. Lightfoot *et al.* [33] were only able to fit the experimentally observed decay of the UV absorption by assuming that the $CH_3O_2 + HO_2$ reaction gave CH_3OOH in essentially 100% yield. This result is consistent with those of Wallington and Japar [4] and Wallington [5] at 295 K, showing that the HCHO-producing channel (b) is unimportant over the temperature range 295–700 K.

An indirect measurement of the branching ratio can be obtained from the study of CH_3CHO photolysis by Moortgat *et al.* [28]. These workers observed that the rate of the reaction channel producing the CH_3OOH product (3.5×10^{-12} cm^3 molecule^{-1} s^{-1}) was slower than the rate of the overall reaction (4.8×10^{-12} cm^3 molecule^{-1} s^{-1}), thus suggesting that $k_a/k = 0.73$ at 300 K and 760 torr. As discussed below, Jenkin *et al.* [6] have reported the observation of a significant yield of HDO following the continuous near-UV photolysis of $CD_4/CH_3OH/O_2/C1_2/N_2$ mixtures at 10.8 torr. By using a detailed modelling approach branching ratios of $k_a/k = 0.6$ and $k_b/k = 0.4$ were derived for the $CD_3O_2 + HO_2$ reaction. Subsequent refinements by Jenkin of the chemical model that was used led to revised branching ratios of $k_a/k = \geqslant 0.8$ and $k_b/k = \leqslant 0.2$ [7]. While the results of Moortgat *et al.* [28] and Jenkin *et al.* [6] suggest that channel (b) may make a small contribution to the overall reaction these studies are considerably less direct than those of Wallington and Japar [4] and Lightfoot *et al.* [33].

Based upon the work of Wallington and coworkers [4,5] and Lightfoot *et al.* [33] we recommend $k_a/k = 1.0$ at 295–700 K, independent of pressure. Experiments to investigate the branching ratio at sub-ambient temperatures are now needed.

The kinetics of the reaction of HO_2 with CH_3O_2 has been studied by Cox and Tyndall [25,26], Kan, Calvert and Shaw [27], Moortgat *et al.* [28,46], McAdam, Veyret and Lesclaux [29], Kurylo *et al.* [30], Dagaut, Wallington and Kurylo [31], Jenkin *et al.* [6], and Lightfoot and co-workers [32,33]. These studies cover the temperature range 228–719 K and reveal a marked negative activation energy for the rate constant of the overall reaction, consistent with the reaction proceeding via a mechanism involving the formation of a short-lived complex. Interestingly, in contrast to the behaviour of the HO_2 self-reaction, no pressure dependence was found between 25 and 760 torr, and no effect of added-water-vapour concentration up to 13 torr [31,32] was observed.

The kinetic data available for this reaction are plotted in Figure 2. There is considerable scatter in the absolute values of the rate constants that have been reported. However, there is general agreement as to the magnitude of the negative temperature dependence of thc reaction. Some of the discrepancies observed may be attributed to differences in the absorption coefficients used by the different authors, but these cannot be easily accounted for due to the complex nature of the kinetic analyses used. The data of Dagaut, Wallington and Kurylo [31] appear to lie systematically lower than the rest of the data. We choose to recommend the Arrhenius expression derived by Lightfoot *et al.* [33] from a fit to the data reported by these workers which was corrected for a recalibration of the UV absorption cross-sections [32,33]:

$$k = (2.9 \pm 0.6) \times 10^{-13} \exp[(860 \pm 90)/T] \text{ cm}^3 \text{ molecule}^{-1} \text{ s}^{-1}.$$

The quoted errors represent 2 standard deviations. This expression is plotted in Figure 2. At 298 K, the expression gives $k = 5.2 \times 10^{-12}$ cm^3 molecule^{-1} s^{-1}.

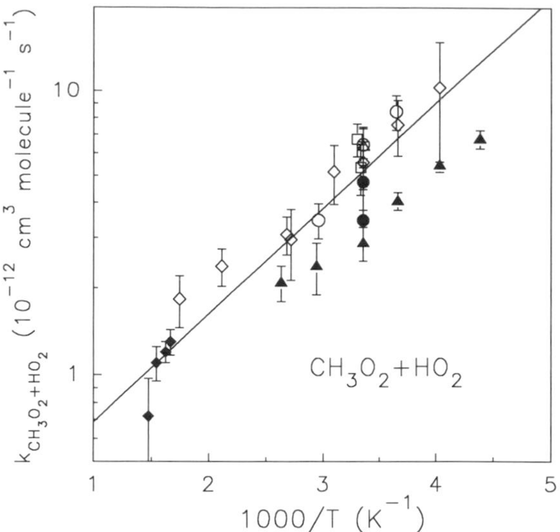

Figure 2 Arrhenius plot for the $CH_3O_2 + HO_2$, with data taken from Cox and Tyndall[25] (open circles), Moortgat *et al.*[28] (filled circles), McAdam, Veyret and Lesclaux[29] (open triangle), Kurylo, Dagaut and Wallington[30] and Dagaut, Wallington and Kurylo[31] (filled triangles), Jenkin *et al.*[6] (open squares), Lightfoot, Veyret and Lesclaux[32] (open diamonds), and Lightfoot *et al.*[33] (filled diamonds); the full line is our recommendation (see text).

2.3 THE REACTION OF HO_2 WITH CD_3O_2

The branching ratio and kinetics of the reaction of HO_2 with CD_3O_2 has been studied by Jenkin *et al.* [6] and Wallington and Hurley [8] using continuous near-UV photolysis of $CD_4/CN_3OH/O_2/C1_2/N_2$ and $CD_4/H_2/O_2/F_2/N_2$ mixtures at total pressures of 10–700 torr:

$$CD_3O_2 + HO_2 \rightarrow CD_3OOH + O_2 \quad (a)$$

$$CD_3O_2 + HO_2 \rightarrow DCDO + HDO + O_2 \quad (b)$$

As discussed above, Jenkin and coworkers monitored the time dependence of HDO production by using infrared tunable diode laser spectroscopy and then used a detailed modelling of the chemical system to extract a value of $k_b/(k_a + k_b) < 0.2$ [7]. Wallington and Hurley [8] used FTIR spectroscopy to quantify the CD_3OOH yield from the $CD_3O_2 + HO_2$ reaction and found that essentially 100% of the reaction proceeds to give CD_3OOH. Consistent with the $CH_3O_2 + HO_2$ reaction, we recommend $k_a/k_a+k_b) \approx 1.0$ at ambient temperature. The kinetic data for the overall reaction of CD_3O_2 radicals with HO_2 reported by Jenkin *et al.* [6], i.e. 5.4×10^{-12} cm^3 molecule^{-1} s^{-1} at 298 K, is essentially the same as the recommended value for $k(CH_3O_2+HO_2)$ given above. This is to be expected if the reaction proceeds via channel (a), with little or no contribution via channel (b).

2.4 THE REACTION OF HO_2 WITH $C_2H_5O_2$

The branching ratio of this reaction has been investigated by Wallington and Japar [9] by photolyzing $C_2H_6/CH_3OH/Cl_2$/air mixtures and analyzing the products using FTIR spectroscopy. It was shown that at 295 K and over the pressure range 20–700 torr, reaction of $C_2H_5O_2$ with HO_2 proceeds by a single reactive route forming C_2H_5OOH and O_2:

$$HO_2 + C_2H_2O_2 \longrightarrow \text{products}$$

The kinetics of the reaction of HO_2 with $C_2H_2O_2$ has been investigated by Cattell *et al.* [34] at room temperature and 2.4 torr, Dagaut, Wallington and Kurylo [35] at 25–400 torr, Fenter *et al.* [36] at 760 torr, and Maricq and Szente [37] at 200–700 torr. No evidence of any pressure dependence on the reaction rate has been reported. The studies span the temperature range 210–480 K, and the results obtained are shown in Figure 3.

Cattell *et al.* [34] produced ethylperoxy and hydroperoxy radicals by modulated photolysis of $C_2H_6/CH_3OH/Cl_2/O_2/N_2$ and azoethane/O_2 mixtures. They monitored $C_2H_5O_2$ and HO_2 by UV absorption at 260 and 210 nm, while

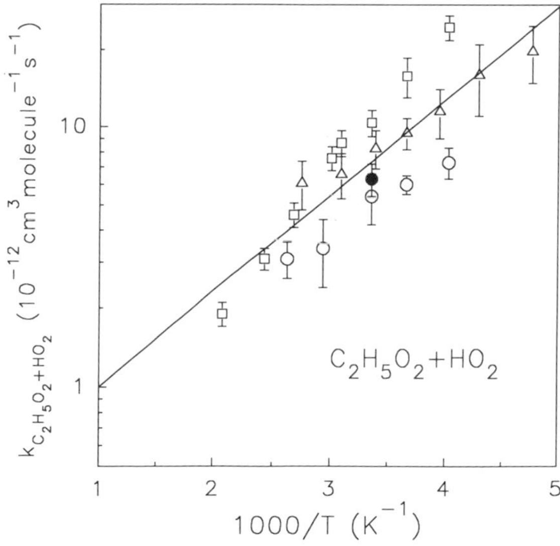

Figure 3 Arrhenius plot for the $C_2H_5O_2 + HO_2$; the full line represents the Arrhenius fit to the data (our recommendation) from Cattell *et al.*[34] (filled circles), Dagaut, Wallington and Kurylo[35] (open circles), Fenter *et al.*[36] (open squares), and Maricq and Szente[37] (open triangles).

HO_2 was also monitored in the infrared at 1117 cm^{-1} by using a tunable diode laser source. Dagaut, Wallington and Kurylo [35] and Fenter *et al.* [36] used flash photolysis UV absorption spectroscopy techniques with radicals produced by the photolysis of $C_2H_6/CH_3OH/Cl_2/O_2/N_2$ mixtures. Maricq and Szente used a laser flash photolysis technique in which F_2 was photolyzed in the presence of C_2H_6 and H_2.

As seen in Figure 3, there is considerable scatter in the available kinetic data. It is difficult to find a reason to prefer the results of any of the studies over the other, and so we have carried out a linear least-squares fit of all of the data in Figure 3 to give:

$$k = (4.4 \pm 2.0) \times 10^{-13} \exp((850 \pm 230)/T) \text{ cm}^3 \text{ molecule}^{-1} \text{ s}^{-1}.$$

At 298 K this expression yields $k = 7.5 \times 10^{-12}$ cm^3 molecule^{-1} s^{-1}.

2.5 THE REACTIONS OF HO_2 WITH NEO-$C_5H_{11}O_2$, CYCLO-$C_5H_9O_2$, AND CYCLO-$C_6H_{11}O_2$

The kinetics and mechanisms of these reactions have been studied by Rowley *et al.* [10,11]. FTIR spectroscopy was used to show that at 295 K and a total pressure of 700 torr hydroperoxides are formed in all three reactions with yields which are, within the experimental uncertainties, indistinguishable from 100%. No evidence for the formation of carbonyl products in these reactions was reported. A flash photolysis UV absorption technique was used to study the kinetics of these reactions over the temperature range 248–365 K, and the results are shown in Figure 4. As with all other

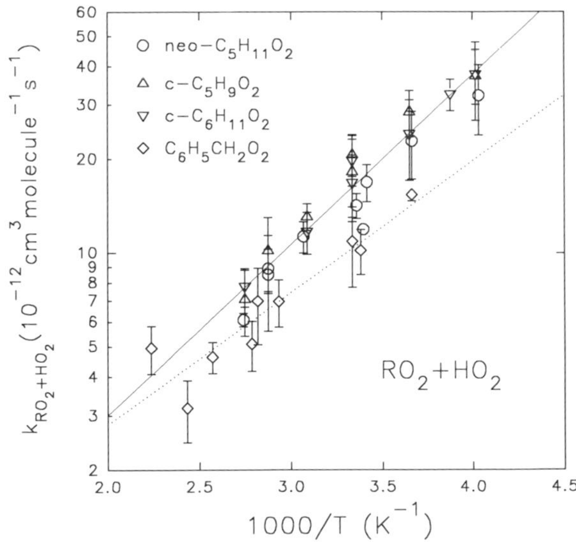

Figure 4 Arrhenius plots for a series of $RO_2 + HO_2$ reactions (see Table 2).

$RO_2 + HO_2$ reactions a substantial negative temperature dependence was observed. As seen from Figure 4, the kinetic data for the three reactions are indistinguishable. The solid line in Figure 4 is a linear least-squares analysis of the neo-$C_5H_{11}O_2$, cyclo-$C_5H_9O_2$, and cyclo-$C_6H_{11}O_2$ data, which gives:

$$k = (2.3 \pm 0.8) \times 10^{-13} \exp(1270 \pm 120/T) \text{ cm}^3 \text{ molecule}^{-1} \text{ s}^{-1}.$$

At 298 K this expression gives $k = 1.6 \times 10^{-11}$ cm^3 molecule^{-1} s^{-1}.

2.6 THE REACTION OF HO_2 WITH $C_6H_5CH_2O_2$

Nozière et al. [38] used a flash photolysis technique to measure the kinetics of this reaction over a range of temperatures. The results are given in Figure 4. The dotted line in this Figure is a linear least-squares analysis which gives:

$$k = (3.8 \pm 0.3) \times 10^{-13} \exp(980 \pm 230/T) \text{ cm}^3 \text{ molecule}^{-1} \text{ s}^{-1}.$$

At 298 K this expression gives $k = 8.6 \times 10^{-12}$ cm^3 molecule^{-1} s^{-1}. There are no mechanistic data available for this reaction.

2.7 THE REACTION OF HO_2 WITH $HOCH_2O_2$

The kinetics of this reaction have been investigated by Burrows *et al.* [12] using molecular modulation spectroscopy at 298 K and by Veyret *et al.* [42] using flash photolysis UV absorption spectroscopy between 275 and 333 K. The radicals were monitored in both the ultraviolet and infrared by Burrows *et al.* [12], but only in the UV by Veyret *et al.* [42]. Hydroxymethylperoxy radicals were produced by photolysis of chlorine in the presence of oxygen and formaldehyde:

$$Cl + HCHO \longrightarrow HCl + HCO$$

$$HCO + O_2 \longrightarrow CO + HO_2$$

$$HO_2 + HCHO \rightleftharpoons HO_2CH_2O \rightleftharpoons HOCH_2O_2$$

$$HO_2 + HOCH_2O_2 \longrightarrow HOCH_2OOH + O_2 \quad (a)$$

$$HO_2 + HOCH_2O_2 \longrightarrow HCOOH + H_2O + O_2 \quad (b)$$

The kinetics of the overall reaction were measured by Veyret *et al.* [42] following a detailed modelling of the transient UV absorption in their system. Burrows *et al.* [12] confirmed the rate of the overall reaction at 298 K and were able to determine a branching ratio of 0.60 and 0.40 for $k_a/k_a+k_b)$ and $k_b/(k_a+k_b)$ at 298 K. The study of Veyret *et al.* [42] provides our recommended Arrhenius expression, valid over the temperature range 275–333 K:

$$k = 5.6 \times 10^{-15} \exp((2300 \pm 1100)/T) \text{ cm}^3 \text{ molecule}^{-1} \text{ s}^{-1}.$$

At 298 K this expression gives $k = 1.3 \times 10^{-11}$ cm^3 molecule^{-1} s^{-1}.

2.8 THE REACTION OF HO_2 WITH CH_2FO_2

Wallington *et al.* [13] used Fourier-transform infrared spectroscopy to identify CH_2FOOH and HC(O)F as products of this reaction. At 700 torr and 295 ± 2 K, branching ratios for the CH_2FOOH-forming channel, i.e. $k_{1a}/k_1 = 0.38 \pm 0.07$, and the HC(O)F-forming channel, i.e. $k_{1b}/k_1 = 0.71 \pm 0.71 \pm 0.11$, were established. There are no kinetic data available for this reaction.

2.9 THE REACTION OF HO_2 WITH CH_2ClO_2

Wallington, Hurley and Schneider [14] used Fourier-transform infrared spectroscopy to identify CH_2ClOOH and HC(O)Cl as products of this reaction. At 700 torr and 295 ± 2 K, branching ratios for the CH_2ClOOH-forming channel, i.e. k_{1a}/k_1, = 0.20 ± 0.06 and the HC(O)Cl-forming channel, $k_{1b}/k_1 = 0.80 \pm 0.14$ were established. Catoire *et al.* [39] have used a flash photolysis technique to determine the kinetics of this reaction over the temperature range 255–588 K at atmospheric pressure. Their data is shown in Figure 5 and is well described by Arrhenius expression $k = (3.26 \pm 0.61) \times 10^{-13} \exp[(822 \pm 63)/T]$ cm^3 molecule^{-1} s^{-1}. At 298 K, this expression gives $k = 5.1 \times 10^{-12}$ cm^3 molecule^{-1} s^{-1}.

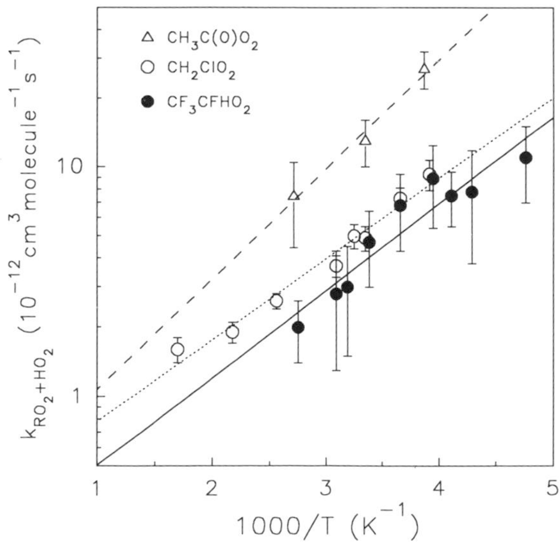

Figure 5 Arrhenius plots for a series of $RO_2 + HO_2$ reactions (see Table 2).

2.10 THE REACTION OF HO_2 WITH CH_2BrO_2

Villenave and Lesclaux [40] used a flash photolysis technique to study this reaction and reported $k = (6.7 \pm 3.8) \times 10^{-12}$ cm^3 molecule^{-1} s^{-1} at ambient temperature. The formation of an unidentified product which absorbs strongly at 250–280 nm was also reported. None of the expected products (CH_2BrOOH, HC(O)Br, or HOBr) are believed to absorb sufficiently strongly in this wavelength region to account for the residual absorption. No recommendation is made for this reaction, and further study is needed.

2.11 THE REACTION OF HO_2 WITH $HOCH_2CH_2O_2$

The kinetics of this reaction have been investigated by Jenkin and Cox [43] by photolyzing $HOCH_2CH_2I/O_2/N_2$ mixtures at 254 nm. Hydroxymethylperoxy radicals were produced following photolysis of 2-iodoethanol in the presence of O_2:

$$HOCH_2CH_2I + h\nu \longrightarrow HOCH_2CH_2 + I$$

$$HOCH_2CH_2 + O_2 + M \longrightarrow HOCH_2CH_2O_2 + M$$

HO_2 radicals were formed in the system by secondary reactions. The authors proposed thc three following production routes:

(a)
$$2\,HOCH_2CH_2O_2 \longrightarrow 2\,HOCH_2CH_2O + O_2$$
$$HOCH_2CH_2O + M \longrightarrow HCHO + CH_2OH + M$$
$$CH_2OH + O_2 \longrightarrow HCHO + HO_2$$

(b)
$$HOCH_2CH_2 + O_2 \longrightarrow HO_2 + HOCHCH_2$$

(c)
$$HOCH_2CH_2I + h\nu \longrightarrow HOCH_2CH_2^{\bullet} + I$$
$$HOCH_2CH_2^{\bullet} \longrightarrow OH + CH_2CH_2$$
$$OH + HOCH_2CH_2I \longrightarrow HOCH_2CH_2 + HOI$$
$$\longrightarrow HOCHCH_2I + H_2O$$
$$HOCHCH_2I + O_2 \longrightarrow HO_2 + ICH_2CHO$$

The rate constant of the reaction of $HOCH_2CH_2O_2$ with HO_2 at 298 K was deduced by modelling the modulated absorption at a variety of wavelengths over the range 220–250 nm at a total pressure of 10 torr. The optimized parameters were as follows: (i) the proportion of $HOCH_2CH_2$ radicals that produced HO_2 via (b); (ii) the branching ratio for the $HOCH_2CH_2O_2$ self-reaction (hence the amount of HO_2 radicals produced via (a)); (iii) $k(HO_2 + HOCH_2CH_2O_2)$. Jenkin and Cox [43] reported a best-fit value of $k(HO_2 + HOCH_2CH_2O_2) = (4.8 \pm 0.5) \times 10^{-12}$ cm^3 molecule^{-1} s^{-1}. Due to the complexity of the system and the large number of parameters necessary for the fitting procedure (including fixed values of σ for HO_2, CH_3O_2 and $HOCH_2CH_2O_2$, known within ca. 10–15%), we recommend an estimated uncertainty of ca. 40%:

$$k = (4.8 \pm 2.0) \times 10^{-12}\ \text{cm}^3\ \text{molecule}^{-1}\ \text{s}^{-1}.$$

2.12 THE REACTION OF HO_2 WITH CF_3CFHO_2

The kinetics and mechanisms of this reaction have been studied by Maricq *et al.* [15]. Using a laser flash photolysis technique the rate constant for this reaction was measured over the temperature range 210–363 K at a total pressure of 200–220 torr. The results are shown in Figure 5 and are fitted by using the expression $k = (1.8^{+24}{}_{-1.0}) \times 10^{-13}$ $\exp[(910 \pm 220)T]$ cm^3 molecule^{-1} s^{-1}, which is recommended. At 298 K this gives $k = 3.8 \times 10^{-12}$ cm^3 molecule^{-1} s^{-1}. The yield of $CF_3C(O)F$ following the Cl-atom initiated oxidation of CF_3CFH_2 at 700 torr of air was unaffected by the presence of H_2, and hence HO_2 radicals. It was concluded that $< 5\%$ of the reaction of CF_3CFHO_2 with HO_2 radicals gives CF_3CFO. The kinetic data derived by Hayman and Battin-Leclerc [41] at 289 K are consistent with that of Maricq *et al.* [15].

2.13 THE REACTIONS OF HO_2 WITH CF_2ClO_2, $CF_2ClCH_2O_2$, $CFCl_2CH_2O_2$, AND $C_2F_5O_2$

Hayman and Battin-Leclerc [41] used a laser flash photolysis technique to study the rates of these reactions at 298 K. It was found that $CF_2ClCH_2O_2$ and $CFCl_2CH_2O_2$ have reactivities towards HO_2 radicals which are indistinguishable from that of $C_2H_5O_2$. In contrast, the reactivity of CF_2ClO_2 and $C_2F_5O_2$ radicals is substantially less than those of the non-halogenated analogues CH_3O_2 and $C_2H_5O_2$. The rate constant measured for the $C_2F_5O_2 + HO_2$ reaction, $(1.2 \pm 0.4) \times 10^{-12}$ cm^3 molecule^{-1} s^{-1}, is the lowest measured for this class of reactions. Electron-withdrawing groups on the carbon atom which bears the O—O(·) group decrease the reactivity of the RO_2 radical towards HO_2 radicals.

2.14 THE REACTION OF HO_2 WITH $CH_3OCH_2O_2$

The products of this reaction have been studied by 700 torr and 295 K by Wallington *et al.* [16]. Two reaction channels were identified:

$$CH_2OCH_2O_2 + HO_2 \longrightarrow CH_3OCH_2OOH + O_2 \quad \text{(a)}$$

$$CH_3OCH_2O_2 + HO_2 \longrightarrow CH_3OCHO + O_2 + H_2O \quad \text{(b)}$$

with $k_{1a}/k_1 = 0.53 \pm 0.08$ and $k_{1b}/k_{1b}/k_1 = 0.40 \pm 0.04$. There are no kinetic data available for this reaction.

2.15 THE REACTION OF HO_2 WITH CH_3COO_2

The mechanism of this reaction has been investigated by Niki *et al.* [17] in an FTIR study of the photolysis of Cl_2 in the presence of CH_3CHO, HCHO and O_2:

$$\begin{aligned} \text{C1} + \text{CH}_3\text{CHO} &\longrightarrow \text{HC1} + \text{CH}_3\text{CO} \\ \text{CH}_3\text{CO} + \text{O}_2 + \text{M} &\longrightarrow \text{CH}_3\text{COO}_2 + \text{M} \\ \text{HCHO} + \text{C1} &\longrightarrow \text{HCO} + \text{HC1} \\ \text{HCO} + \text{O}_2 &\longrightarrow \text{CO} + \text{HO}_2 \end{aligned}$$

The authors demonstrated that acetyl peroxy radicals react with HO_2 by two channels:

$$\begin{aligned} \text{HO}_2 + \text{CH}_3\text{COO}_2 &\longrightarrow \text{CH}_3\text{COO}_2\text{H} + \text{O}_2 \quad \text{(a)} \\ \text{HO}_2 + \text{CH}_3\text{COO}_2 &\longrightarrow \text{CH}_3\text{COOH} + \text{O}_3 \quad \text{(b)} \end{aligned}$$

The relative contributions of (a) and (b) were found to be 0.75 and 0.25, respectively, at 700 torr and 298 K.

Using flash photolysis kinetic spectroscopy, Moortgat, Veyret and Lesclaux [44] studied both the kinetics and mechanisms of this reaction. The mechanistic data [44] was derived in an indirect fashion and have been superseded by the results from a more direct product study reported by Horie and Moortgat. [19]. These latter authors reported their results in the form of a parameter $\beta(= k_a/k_b = 330 \times \exp(-1430/T)$ over the temperature range 263–333 K at 370–770 torr of air as diluent. This expression gives contributions of 0.73 and 0.27 for channels (a) and (b), respectively, at 298 K, in agreement with the results of Niki *et al.* [17].

Moortgat, Veyret and Lesclaux [44] determined the kinetics of the overall reaction of the temperature range 253–368 K at atmospheric pressure (see Table 2). The production of the radicals was achieved by photolysis of $C1_2/CH_3CHO/CH_3OH/O_2/N_2$ mixtures. The Arrhenius expression derived from this work is $k = (4.3 \pm 1.2) \times 10^{-13} \exp((1040 \pm 110)/T)$ cm^3 molecule^{-1} s^{-1}. The reaction of acetyl peroxy with HO_2 radicals shows a strong inverse temperature dependence. Moortgat and coworkers [44] postulated the following mechanism to explain the experimental findings:

$$\begin{aligned} \text{HO}_2 + \text{CH}_3\text{COO}_2 &\longrightarrow \text{CH}_3\text{COO}_4\text{H} \longrightarrow \text{O}_2 + \text{CH}_3\text{COO}_2\text{H} \quad \text{(a)} \\ &\longrightarrow \text{CH}_3\text{C(=O}\cdots\text{H)}-\text{O}-\text{O}-\text{O}-\text{O(H)} \longrightarrow \text{CH}_3\text{COOH} + \text{O}_3 \quad \text{(b)} \end{aligned}$$

with channels (a) and (b) proceeding via four and seven membered transition states. The available kinetic data are shown in Figure 5.

2.14 THE REACTION OF HO_2 WITH $CH_3C(O)CH_2O_2$

Bridier *et al.* [45] used a flash photolysis technique to determine a rate constant of 9.0×10^{-12} cm^3 molecule^{-1} s^{-1} for this reaction at ambient temperature. In contrast to the behaviour of $CH_3C(O)O_2$ radicals, there was no discernible residual absorption that could be attributed to the formation of ozone. It was concluded that the reaction likely proceeds to give the hydroperoxide, $CH_3C(O)CH_2OOH$. The contrasting behaviour of $CH_3C(O)O_2$ and $CH_3C(O)CH_2O_2$ radicals in their reaction with HO_2 can be understood in terms of the proposed mechanisms for ozone formation in the $CH_3C(O)O_2 + HO_2$ reaction, which involves a 7-membered cyclic transition state involving the carbonyl group which is α to the peroxy group. In the $CH_3C(O)CH_2O_2$ radical the carbonyl group is β to the peroxy group, and thus cannot participate in ozone formation.

2.15 DISCUSSION

When considering the available kinetic data presented in Table 2 and Figures 1–5 several interesting points emerge.

First, all reactions for which data are available are rapid, with rate constants at ambient temperature in the range (1–20) $\times 10^{-12}$ cm^3 molecule^{-1} s^{-1}, i.e. only 10–100 times slower than the gas kinetic limiting rate constant of ca. 10^{-10} cm^3 molecule^{-1} s^{-1}.

Secondly, where data is available over a range of temperatures, the reactions display large negative temperature dependencies, with values of $-E_a/R = 800$–1100 being typical. Such behaviour indicates that these reactions proceed via the formation of a short-lived complex which can either dissociate into the reactants, or react to form the products. To account for a negative temperature dependence, the activation energy barrier for the rearrangement of the adduct to give the products must be lower than that of the barrier for dissociation into reactants. The adduct is believed to be the tetroxide, i.e. ROOOOH.

Thirdly, there is no discernible curvature in the Arrhenius plots, and yet at the lowest temperatures studied the reaction rates approach a level which is only a factor of ten less than the gas kinetic limit. As the temperature is decreased further the rate of reaction cannot increase indefinitely, but instead must plateau-out at a value close to the gas kinetic limit. The activation barrier for the formation of the adduct must be small or zero. A schematic potential energy diagram is shown in Figure 6.

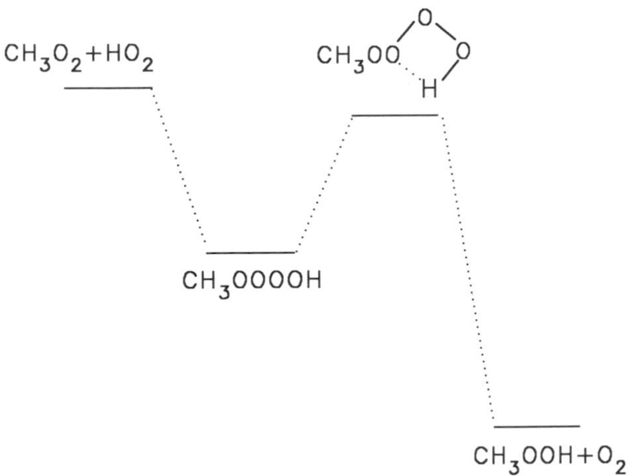

Figure 6 Schematic potential energy diagram for the RO_2 + HO_2 reaction.

Fourth, no complications have been reported in the kinetic analyses of the RO_2 and/or HO_2 decays. The lifetime of the adduct must be less than the typical time-scale of these experiments, i.e. measured in ms or less. The absence of any reported effect of pressure on the rate of these reactions suggests that, under the conditions employed (2–700 torr), the adduct is thermalized. If we arbitrarily assume an adduct lifetime of <1 ms at 250 K, and an A-factor of $10^{16}\ s^{-1}$ (as for $CH_3C(O)O_2NO_2$ and $CH_3O_2NO_2$ [21]), then the bond dissociation energy of the adduct is $>62\ kJ\ mol^{-1}$. This value is consistent with that of 30–60 $kJ\ mol^{-1}$ postulated by Patrick, Barker and Golden [47] for the HO_2—O_2H bond energy in the tetroxide adduct in the self-reaction of HO_2 radicals.

Fifthly, for alkylperoxy radicals, the reactivity increases with increasing size. CH_3O_2 is approximately a factor of two less reactive than the other alkylperoxy radicals. Halogenation decreases the reactivity of the RO_2 radical. These observations are consistent with electron-donating groups facilitating the reaction and electron-withdrawing groups slowing down the reaction.

The limiting amount of mechanistic information regarding the reactions of organic peroxy radicals with HO_2 falls into two groups. For unsubstituted peroxy radicals (CH_3O_2, CD_3O_2, $C_2H_5O_2$, cyclo-$C_5H_9O_2$, cyclo-$C_6H_{11}O_2$, and $(CH_3)_3CCH_2O_2$), only one carbon-containing product is found—the hydroperoxide—in a yield which is indistinguishable from 100%. The yield of carbonyl products in the reaction of unsubstituted peroxy radicals with HO_2 is <10%. Reaction presumably occurs via a four membered transition state.

For example, in the case of $CH_3O_2 + HO_2$

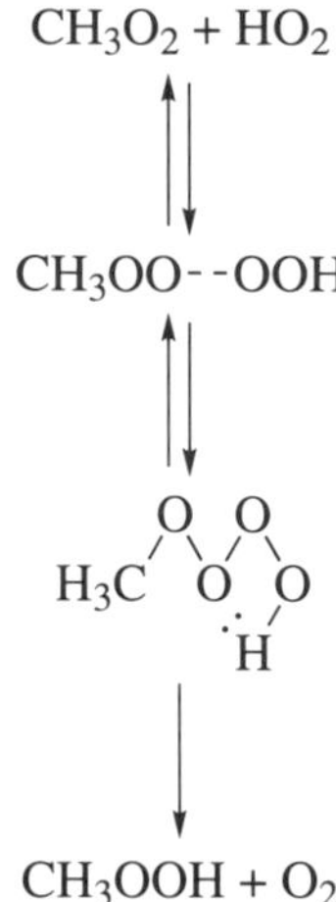

For most substituted peroxy radicals ($HOCH_2O_2$, $CH_3OCH_2O_2$, CH_2FO_2, and CH_2ClO_2), both hydroperoxide and carbonyl products are observed. The substituents are all electron withdrawing, and the importance of the carbonyl channel increases with the electronegativity of the substituent. The mechanism by which the carbonyl products are formed is unknown. There are two possibilities which are illustrated for the case of the $CH_2FO_2 + HO_2$ reaction [13]. The formation of HC(O)F could occur by elimination of HC(O)F, H_2O, and O_2 from the tetroxide intermediate, via a six-membered ring transition state (channel B, below). The six-membered transition state mechanism is the same as that proposed by Russell to explain the formation of molecular products from the self-reactions of RO_2 radicals [48]. Alternatively, the tetroxide could decompose via CH_2FO—OOOH (or CH_2FOOO—OH) bond fission with the resulting radical species forming HC(O)F and HO_2 and OH radicals (channel C below).

In contrast, the reaction of CF_3CFHO_2 radicals with HO_2 does not appear to give much (if any) carbonyl product and is assumed to proceed under ambient conditions to give $>95\%$ yield of the hydroperoxide ($CF_3CFHOOH$). It is puzzling that under ambient conditions CH_2FO_2 radicals react to give HC(O)F in 71% yield [13] while the structurally similar CF_3CFHO_2 radicals give $<5\%$ yield of $CF_3C(O)F$ [15].

The reaction of peroxy radicals with HO_2 radicals is the least understood class of peroxy radical reactions and many important unanswered questions remain. For example, "what is the mechanism, and what factors control carbonyl product formation in the reaction of substituted peroxy radicals ($HOCH_2O_2$, $CH_3OCH_2O_2$, CH_2FO_2, and CH_2ClO_2) with HO_2?" "Why do

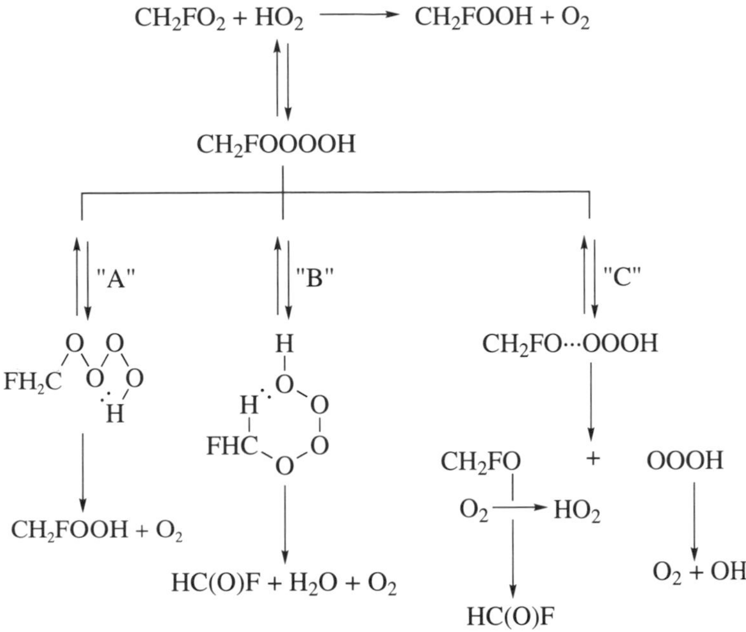

CF_3CFHO_2 and CH_2FO_2 behave so differently?" "Are there any experimental conditions under which unsubstituted peroxy radicals react to give carbonyl products?", and "what influences the rate of reaction?"

Further product and kinetic studies are needed, preferably as a function of both temperature and pressure, for the reaction of structurally diverse peroxy radicals with HO_2 radicals to elucidate the reaction mechanisms. High-level *ab initio* theoretical modelling of these reactions would be particularly helpful.

3 KINETICS AND MECHANISMS OF THE REACTIONS OF RO_2 RADICALS AND NO

The available kinetic data for the reactions of peroxy radicals with NO are given in Table 3. The reaction proceeds via two channels:

$$RO_2 + NO \longrightarrow RO + NO_2 \quad (a)$$

$$RO_2 + NO \longrightarrow RONO_2 \quad (b)$$

As discussed by Atkinson [3] the mechanism of the reaction is believed to involve the formation of a short-lived ROONO complex which can either decompose to give RO and NO_2, or rearrange to give the nitrate $RONO_2$ [49]

Table 3 Kinetic Data for the Reactions of Peroxy Radicals with NO

RO_2	$k \times 10^{12}$	Temperature (K)	Pressure (torr)	Technique[a]	Ref.
HO_2	8.6 (298 K)	232–1271	1–12.5	Review	20
	$k = 3.7 \times 10^{-12} \exp((250 \pm 80)/T)$ cm^3 molecule^{-3} s^{-1}				
CH_3O_2	>1	298	300(N_2)	MMS	50
	8.0 ± 2.0	295	3(He)	DF–MS	51
	3.0 ± 0.2	298	760	FP–UVA	52
	3.2 ± 1.8	296	760	RR[b]	53
	6.5 ± 2.0	298	540	MMS	26
	6.1 ± 0.7	298	75(He)	FP–UVA	54
	6.3 ± 0.9		350(He)		
	8.2 ± 1.1		700(He)		
	8.9 ± 0.7		700(N_2)		
	8.4 ± 1.5	240	40(Ar)	LP–LIF	55
	8.6 ± 1.1	250	40(Ar)		
	9.0 ± 1.1	270	40(Ar)		
	7.8 ± 1.2	298	40–100(Ar)		
	7.8 ± 1.4	339	40(Ar)		
	13 ± 1.4	218	200(CH_4/O_2)	FP–UVA	56
	17 ± 2.2	218	600(CH_4/O_2)		
	7.7 ± 0.9	296	100–600(CH_4/O_2)		
	6.3 ± 1.0	365	200(CH_4O_2)		
	8.6 ± 2.0	295	6(He)	DF–MS	57
	7 ± 2	298	8(N_2)	LP–UVA	58
	9.1 ± 2.0	293	2(He)	DF–MS	59
	8.8 ± 1.4	298	760(SF_6)	PR–UVA	60
	11.9	200	2–6 (He)	FT/MS	61
	9.9	223			
	8.9	248			
	7.4	273			
	7.4	298			
	6.3	329			
	5.9	372			
	5.5	394			
	5.9	410			
	5.8	429			
$C_2H_5O_2$	2.7 ± 0.2	298	350–760	FP–UVA	62
	8.9 ± 3.0	295	5 (He)	DF–MS	63
	8.5 ± 1.2	298	760 (SF_6)	PR–UVA	60
	8.2 ± 1.6	298	1(He)	DF–LIF	64
$(CH_3)_2CHO_2$	3.5 ± 0.3	298	65(Ar)	FP–UVA	65
	5.0 ± 1.2	298	2(He)	DF–MS	66
$(CH_3)_3CO_2$	>1	298	300	MMS	50
	4.0 ± 1.1	298	2	DF–MS	66
$(CH_3)_3CCH_2O_2$	4.7 ± 0.4	298	760(SF_6)	PR–UVA	60
$(CH_3)_3C(CH_3)_2CH_2O_2$	1.8 ± 0.2	298	760(SF_6)	PR–UVA	60
$HOCH_2O_2$	5.6	298	55–265	FP–UVA	42
CH_2FO_2	12.5 ± 1.3	298	760(SF_6)	PR–UVA	60
CHF_2O_2	12.6 ± 1.6	298	760(SF_6)	PR–UVA	60
CF_3O_2	17.8 ± 3.6	295	2–5(He)	DF–MS	67
	14.5 ± 2	298	1–10(N_2)	LP–MS	68
		230–430	1–10	LP–MS	68
	$k = (1.45 \pm 0.2) \times 10^{-11} (T/298)^{-(1.2 \pm 0.2)}$ cm^3 molecule^{-1} s^{-1}				

Table 3 – *continued*

RO_2	$k \times 10^{12}$	Temperature (K)	Pressure (torr)	Technique[a]	Ref.
	16.9 ± 2.6	298	760(SF_6)	PR–UVA	69
	15.3 ± 2.0	297	760	FT–MS	70
	15.0	298	100(He)	FP/LIF	71
	15.7 ± 3.1	293	2–30(N_2)	FP–MS	72
CH_2ClO_2	18.7 ± 2.0	298	760(SF_6)	PR–UVA	60
CH_2BrO_2	10.7 ± 1.1	298	760(SF_6)	PR–UVA	60
$CFCl_2O_2$	16 ± 2	298	1–6(O_2)	LP–MS	73
	14.5 ± 2	298	1–10(N_2)	LP-MS	68
		230–430	1–10	LP–MS	68
	$k = (1.45 \pm 0.2) \times 10^{-11} (T/298)^{-(1.3 \pm 0.2)}$ cm^3 molecule^{-1} s^{-1}				
CF_2ClO_2	16 ± 3	298	1–10(N_2)	LP–MS	68
		230–430	1–10(N_2)	LP–MS	68
	$k = (1.6 \pm 0.3) \times 10^{-11} (T/298)^{-(1.5 \pm 0.4)}$ cm^3 molecule^{-1} s^{-1}				
	13.1 ± 1.2	298	760(SF_6)	PR–UVA	60
CCl_3O_2	18.6 ± 2.8	295	2–8(He/O_2)	DF–MS	74
	17 ± 2	298	1–10(N_2)	LP–MS	68
		230–430	1–10	LP–MS	68
	$k = (1.7 \pm 0.2) \times 10^{-11} (T/298)^{-(1.0 \pm 0.2)}$ cm^3 molecule^{-1} s^{-1}				
CF_3CFHO_2	12.8 ± 3.6	298	760(SF_6)	PR–UVA	75
	13.1 ± 3.0	324	12–25(N_2)	FP–MS	76
	11 ± 3	290	2(He)	DF–MS	77
$FC(O)O_2$	25 ± 8	295	760(SF_6)	PR–UVA	78
$CF_3C(O)O_2$	>9.9	295	760(SF_6)	PR–UVA	79
	55 ± 12	220	104(N_2)	FP–UVA	80
	34 ± 7	254	100(N_2)		
	28 ± 5	295	102(N_2)		
	24 ± 5	324	101(N_2)		
$CF_3CH_2O_2$	12 ± 3	298	760(SF_6)	PR–UVA	81
CH_2FCFHO_2	>8.7	298	760(SF_6)	PR–UVA	82
$CHF_2CF_2O_2$	$>9.7 \pm 1.3$	298	760(SF_6)	PR– UVA	60
$CF_3CF_2O_2$	$>10.7 \pm 1.5$	298	760(SF_6)	PR– UVA	60
$CF_2ClCH_2O_2$	12.8 ± 1.1	298	760(SF_6)	PR–UVA	60
$CFCl_2CH_2O_2$	11.8 ± 1.0	298	760(SF_6)	PR–UVA	60
$CF_3CHO_2(^{\cdot})CF_3$	11 ± 3	295	760(SF_6)	PR–UVA	83
$CH_3CF_2CH_2O_2$	8.5 ± 1.9	295	760(SF_6)	PR–UVA	84
$CCl_3CH_2O_2$	>6.1	295	760(SF_6)	PR–UVA	85
CF_3CClHO_2	10 ± 3	296	760(SF_6)	PR–UVA	86
$CH_3SCH_2O_2$	19 ± 3	295	760(SF_6)	PR–UVA	87
$CH_3OCH_2O_2$	9.1 ± 1.0	296	760(SF_6)	PR–UVA	88
$HO(CH_3)_2CCH_2O_2$	4.9 ± 0.9	298	760(SF_6)	PR-UVA	89

[a] MMS,molecular modulation spectroscopy; UVA, ultraviolet absorption spectroscopy; FP, flash photolysis; FT, flowtube, PR, pulse radiolysis; DF, discharge flow, MS, mass spectrometry; LP, laser photolysis.
[b] Relative to $CH_3O_2 + SO_2 = 8.2 \times 10^{-15}$

(see Figures 7 and 8). Consideration of the mechanism given in Figure 7 shows that the overall rate of reaction is independent of total pressure and only weakly dependent on temperature. In contrast, the relative importance of the RO- and $RONO_2$-forming channels will be pressure and temperature dependent. Increasing pressure favours $RONO_2$ formation, while increasing temperature favours RO formation.

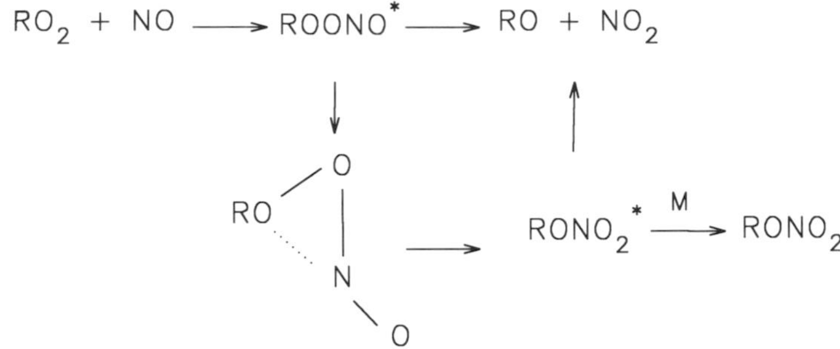

Figure 7 Proposed mechanism for the RO_2 + NO reaction taken from Atkinson [3]

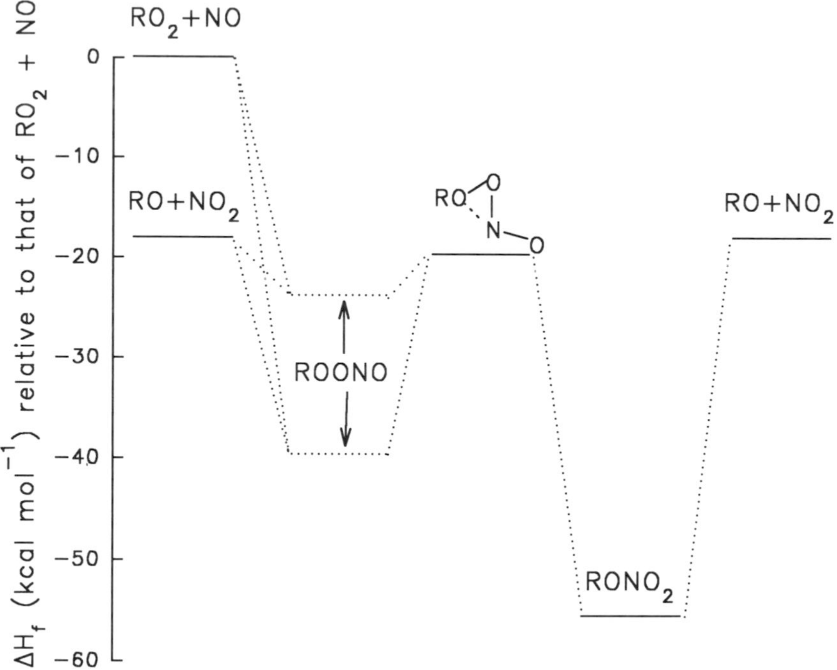

Figure 8 Schematic potential energy diagram for the RO_2 + NO reaction, taken from Atkinson, Carter and Winer[93]; the arrows show the range of likely values for the ROONO species

With increasing size the internal excitation is distributed over an increased number of vibrational degrees of freedom and unimolecular decomposition of the ROONO* and $RONO_2$* species slows down. Hence, with increasing size of the R radical, the importance of the $RONO_2$-producing channel increases steadily. The nitrate yields for selected peroxy radicals are given in Table 4; for a more exhaustive listing of nitrate yields the reader should refer to page 1904 of Lightfoot *et al.* [2]. Atkinson has developed complex expressions that accurately predict the ratio of the nitrate-to alkoxy-radical-forming channels, as a function of temperature and pressure [3].

Table 4 Nitrate yields in Selected Reactions of of Peroxy Radicals with NO

Peroxy radical	Nitrate yield, $k_b/(k_a+k_b)$	Ref.
Methyl	<0.005	90
Ethyl	<0.014	91
1,2-Propyl	0.036	91
t-Butyl	0.18	92
2,3-Pentyl	0.13	91
2,3-Hexyl	0.22	91
2,3,4-Heptyl	0.31	91
2,3,4-Octyl	0.33	91

As seen from an inspection of Table 3, the overall reactions rate decreases with increasing size of the RO_2 radical. The cause of this decrease is not known but may reflect steric factors. Finally, it is evident from Table 3 that in contrast to the behaviour of the $RO_2 + HO_2$ reactions, the rates of reaction to halogenated RO_2 radicals with NO are greater than the unsubstituted species.

3.1 THE REACTION OF NO WITH HO_2

The kinetics and mechanisms of this reaction have been extensively reviewed in the JPL/NASA [20] and IUPAC [21] evaluations. In Table 3 we give the JPL/NASA recommendations which were based upon the results of several studies carried out near room temperature. The data of Howard and Evenson [94], Leu [95], Howard [96], Glaschick-Schimpf *et al.* [97], Hack *et al.* [98] and Thrush and Wilkinson [99] are in good agreement and are used. Other determinations by Burrows *et al.* [100] and Rozenshtein *et al* [101] were disregarded due to problems in interpreting the data. The temperature dependence is that of Howard [96], which agrees with that of Leu [95].

3.2 THE REACTION OF NO WITH CH_3O_2

As seen from Table 3 the reaction of CH_3O_2 with NO has been extensively studied. The early work of Adachi and Basco [52] and Simonaitis and Heicklen [53] has been superseded by later studies by these authors. The agreement between all ten determinations of the room-temperature rate constant is remarkable. Studies have been performed over a wide range of pressures and, with the possible exception of the study of Sander and Watson [54], there is no evidence of any pressure dependence. Even in the study by Sander and Watson the rate constants reported at 75 and 700 torr are indistinguishable within the combined experimental errors. The consistency of the kinetic data obtained in the recent studies by Kenner, Ryan and Plumb [59] and Sehested, Nielsen and Wallington [60] shows that over the range from 2 torr of helium to 760 torr of SF_6 there is no discernible effect of the total pressure on the kinetics of the reaction.

Direct measurements of the formation of NO_2 and CH_3O as products of the reaction of methylperoxy radicals with NO have been made by Ravishankara *et al.* [55] and Zellner and coworkers [58,102] and demonstrate that this reaction proceeds predominately ($>80\%$) via channel (a) (see below). Pate, Finlayson and Pitts Jr [90] photolyzed $CH_3N_2CH_3$ in the presence of O_2 and NO and determined an upper limit of 1% for the molar yield of CH_3ONO_2 (moles of CH_3ONO_2 formed per mole of $CH_3N_2CH_3$ lost) at 296 K under 760 torr of N_2. The experiments of Pate and coworkers [90] show that channel (b) account for $<0.5\%$ of the overall reaction and hence that the reaction proceeds essentially solely via channel (a):

$$CH_3O_2 + NO \longrightarrow CH_3O + NO_2 \quad \text{(a)}$$

$$CH_3O_2 + NO + M \longrightarrow CH_3ONO_2 + M \quad \text{(b)}$$

The temperature dependence of the reaction has been studied by Ravishankara *et al.* [55], Simonaitis and Heicklen [56], and Villalta, Huey and Howard [61]. As seen from Figure 9, the low temperature data reported by Simonaitis and Heicklen [56] appears to be anomalously high. For our recommendation we have performed a linear least squares analysis of all the data shown in Figure 9 (with the exception of the lowest temperature data of Simonaitis and Heicklen [56]) to yield:

$$k = (3.1^{+0.8}{}_{-0.7}) \times 10^{-12} \exp[(260 \pm 70)/T] \text{ cm}^3 \text{ molecule}^{-1} \text{ s}^{-1}$$

which is valid over the temperature range 200–429 K. Quoted errors represent 2σ. At 298 K this expression gives $k = 7.4 \times 10^{-12}$ cm^3 molecule^{-1} s^{-1}.

3.3 THE REACTION OF NO WITH $C_2H_5O_2$

Plumb *et al.* [63] and Atkinson *et al.* [91,103] have studied the products of this reaction. Atkinson and coworkers reported an upper limit of 1.4% for ethyl

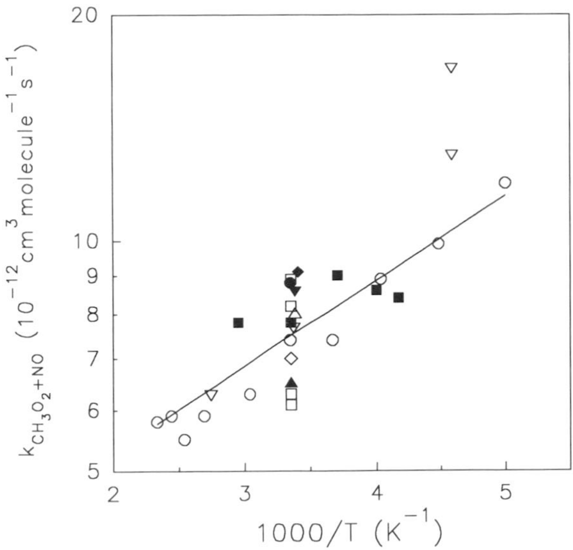

Figure 9 Arrhenius plot for the $CH_3O_2 + NO$, with data taken from Plumb *et al.*[5,57] (open triangle and filled inverse triangle respectively), Cox and Tyndall[26] (filled triangle), Sander and Watson[54] (open squares), Ravishankara *et al.*[55] (filled squares), Simonaitis and Heicklen[56] (open inverse triangles), Zellner, Fritz and Lorenz (open diamond) Kenner, Ryan and Plumb[59] (filled diamond), Sehested, Nielsen and Wallington[60] (filled circle), and Villalta, Huey and Howard[61] (open circles); the full line is our recommendation.

nitrate formation. We recommend the use of a single channel for this reaction:

$$C_2H_5O_2 + NO \longrightarrow C_2H_5O + NO_2$$

There have been four studies of the kinetics of this reaction. The results reported by Adachi and Basco [62] are a factor of three lower than those of Plumb *et al.* [63], Sehested, Nielsen and Wallington [60] and Daële *et al.* [64]. The origin of this discrepancy probably lies in monitoring complications associated with the formation of C_2H_5ONO by the reaction of ethoxy radicals with the excess NO on a time-scale comparable to that of the decay of ethylperoxy radicals. The 250 nm monitoring wavelength used by Adachi and Basco [62] was absorbed appreciably by ethylnitrite. The data of Plumb *et al.* [63], Sehested, Nielsen and Wallington [60], the Daële *et al.* [64] are preferred, and at 298 K we recommend $k = 8.5 \times 10^{-12}$ cm^3 molecule^{-1} s^{-1}, with an estimated uncertainty of $\pm 10\%$. The agreement between the results of Daële *et al.* [64] obtained under 1 torr of helium and Sehested, Nielsen and Wallington obtained under 760 torr of SF_6 demonstrate the absence of any effect of total pressure on this reaction.

3.4 THE REACTION OF NO WITH $(CH_3)_2$ CHO_2

The mechanism of the reaction of isopropylperoxy radicals with NO has been studied by Atkinson *et al.* [91,103] using FTIR analysis of reaction mixtures following the continuous near-UV irradiation of $Cl_2/NO/C_3H_8$ and $CH_3ONO/NO/C_3H_8$ mixtures in air at 735 torr and 298 K. Atkinson and coworkers [91,103] reported a branching ratio of $k_b/(k_a+k_b) = 0.036$, and hence $k_a/(k_a+k_b) = 0.964$:

$$(CH_3)_2CHO_2 + NO \longrightarrow (CH_3)_2CHO + NO_2 \quad (a)$$

$$(CH_3)_2CHO_2 + NO + M \longrightarrow (CH_3)_2CHONO_2 \quad (b)$$

The overall rate constants for this reaction reported by Adachi and Basco [65] and Peeters, Vertommen and Langhans [66] are consistent, and we recommend the average value obtained from these studies, i.e. 4.3×10^{-12} cm^3 $molecule^{-1}$ s^{-1}, with an estimated uncertainty of $\pm 20\%$. As given in Table 4, under ambient conditions this reaction gives 3.6% yield of isopropylnitrate and 97.6% isopropoxy radicals.

3.5 THE REACTION OF NO WITH $(CH_3)_3CO_2$

The rate constant of 4.0×10^{-12} cm^3 $molecule^{-1}$ s^{-1} reported by Peeters, Vertommen and Langhans [66] for this reaction is recommended. This result is consistent with the lower limit of 1×10^{-12} cm^3 $molecule^{-1}$ s^{-1} derived by Anastasi, Smith and Parkes [50]. As given in Table 4, under ambient conditions this reaction gives a 18% yield of tert-butyl-nitrate.

3.6 THE REACTION OF NO WITH $(CH_3)_3C(CH_3)_2CCH_2O_2$ AND $(CH_3)_3CCH_2O_2$

The rate constants reported by Sehested, Nielsen and Wallington [60] (see Table 3) are the only available data for these reactions, and are thus recommended. It is worthy of note that the reaction of $(CH_3)_3C(CH_3)_2CCH_2O_2$ radicals with NO is the slowest of any $RO_2 + NO$ reaction. Close inspection of Table 3 reveals a clear trend of decreasing reactivity of RO_2 radicals with increasing size. Further studies are needed to confirm this trend.

3.7 THE REACTION OF NO WITH $CH_3C(O)O_2$

To date, there have been no direct studies of the kinetics of the reaction of acetylperoxy radicals with NO. However, there have been a number of indirect studies in which the rate of reaction of $CH_3C(O)O_2$ with NO is measured relative to the rate with NO_2:

$$CH_3C(O)O_2 + NO \longrightarrow CH_3C(O)O + NO_2$$

$$CH_3C(O)O_2 + NO_2 + M \longrightarrow CH_3C(O)O_2NO_2 + M$$

Cox *et al.* [104] reported a value for $k_{AcO2+NO}/k_{AcO2+NO2}$ of 1.7 from observations of the effect of[NO]/[NO_2] on the peroxyacetyl nitrate (PAN) product yield on photolysis of HONO/CH_3CHO mixtures in air at atmospheric pressure at 300 K. Further work by Cox and Roffey [105] followed the rate of disappearance of PAN mixtures with NO in air at atmospheric pressure. Over the temperature range 303–328 K there was no significant variation of $k_{AcO2+NO}/k_{AcO2+NO2} = 3.1 \pm 0.5$ at atmospheric pressure, independent of temperature over the range 298–318 K.

More recently, Kirchner, Zabel and Becker [107] and Tuazon, Carter and Atkinson [108] have also monitored the decay rate of PAN in the presence of known concentrations of NO and NO_2 in order to measure the rate constant ratio $k_{AcO2+NO}/k_{AcO2+NO2}$. Tuazon and coworkers [108] reported a ratio of 1.95 ± 0.28 at 740 torr of air, independent of temperature over the range 283–313 K. Kirchner and coworkers [107] reported values for the ratio at atmospheric pressure of air of 2.2 ± 0.3 (304 K), and 2.4 ± 0.1 (321 K). In addition, Kirchner and coworkers [107] measured the rate-constant ratio as a function of pressure over the range 30–1000 mbar of air at a fixed temperature of 313 K, obtaining results of 2.4 ± 0.1 (1000 mbar), 2.6 ± 0.1 (300 mbar), 3.6 ± 0.6 (100 mbar) and 4.0 ± 0.8 (30 mbar). Consistent with previous studies, the results of Kirchner and coworkers show that at atmospheric pressure there is no observable effect of temperature on the ratio over the range 304–321 K.

With the exception of the data reported by Hendry and Kenley [106], all studies of the rate-constant ratio $k_{AcO2+NO}/k_{AcO2+NO2}$ are in agreement. At atmospheric pressure there is no observable effect of temperature over the range 283–328 K, with the average of the values reported by Cox and Roffey [105], Tuazon, Carter and Atkinson [108], and Kirchner, Zabel and Becker [107] being 2.1. This ratio can be combined with our recommended kinetic data for the reaction of $CH_3C(O)O_2$ radicals with NO_2 (see next section) to provide a value for the rate constant of the reaction of $CH_3C(O)O_2$ radicals with NO. At 760 torr our recommended expression for $k_{AcO2+NO2}$ yields effective second-order rate constants of 8.4×10^{-12} and 1.0×10^{-11} cm^3 molecule^{-1} s^{-1} at 328 and 283 K, respectively. Taking an average of $k_{AcO2+NO2} = 9.2 \times 10^{-12}$ cm^3 molecule^{-1} s^{-1} then yields our recommendation of $k_{AcO2+NO} = 1.9 \times 10^{-11}$ cm^3 molecule^{-1} s^{-1} at 760 torr over the temperature range 283–328 K. We estimate the uncertainties associated with this rate constant to be $\pm 50\%$. The rate constant values, $k_{AcO2+NO}/k_{AcO2+NO2}$, reported by Kirchner and coworkers [107] at pressures less than 760 torr can also be combined with values of $k_{AcO2+NO2}$ calculated from our recommended expression (5.1, 7.0 and 8.1 $\times 10^{-12}$ cm^3 molecule^{-1} s^{-1} at 30, 100 and 300 mbar to give values of $k_{AcO2+NO}$ of 2.0 ± 0.4, 2.5 ± 0.4 and 2.1 ± 0.1 $\times 10^{-11}$ cm^3 molecule^{-1} s^{-1} at 30, 100, and 300 mbar, respectively. Consistent with the body of data for reactions of peroxy radicals with NO there is no evidence for any effect of pressure over the range 30–1000 mbar on the kinetics of the reaction of $CH_3C(O)O_2$ radicals with NO. Clearly, there is a need for direct kinetic studies of this reaction.

3.8 THE REACTION OF NO WITH CF_3O_2

The reaction of CF_3O_2 radicals with NO has been the subject of six studies which are all in good agreement, giving an average rate constant of 1.6 $\times 10^{-11}$ cm^3 molecule^{-1} s^{-1} at 298 K, independent of total pressure over the range 1–700 torr. This value is consistent with the kinetics of other reactions of CX_3O_2 (X = F or Cl) with NO. Dognon, Caralp and Lesclaux [68] have studied the effect of temperature on the kinetics of this reaction over the range 230–430 K, and an inverse temperature dependence was observed. The rate expression of Dognon and coworkers [68] given in Table 3 is recommended. No evidence for the existence of reaction channels other than that leading to alkoxy radical (CF_3O) and NO_2 have been detected in any gas-phase study. At 4.2 K in a matrix isolation study, Clemitshaw and Sodeau [109] used FTIR spectroscopy to identify CF_3OONO as an intermediate in the reaction of CF_3O_2 radicals with NO.

3.9 THE REACTION OF NO WITH $CFCl_2O_2$

The kinetics of the reaction of $CFC1_2O_2$ radicals with NO have been studied by Lesclaux and coworkers [68,73] using pulsed laser photolysis of $CFC1_3$ in the presence of O_2 and NO combined with time-resolved mass spectroscopy. Rate constants of (1.6 ± 0.2) [73] and (1.45 ± 0.2) [68] $\times 10^{-11}$ cm^3 molecule^{-1} s^{-1} were determined independent of pressure over the range 1–10 torr of O_2/N_2 at 295 K. No evidence for the existence of reaction channels other than that leading to the alkoxy radical and NO_2 were detected. Experiments performed over the temperature range 230–430 K by Dognon, Caralp and Lesclaux [68] have shown this reaction to have an inverse temperature dependence. The rate expression of Dognon and coworkers [68] given in Table 3 is recommended.

3.10 THE REACTION OF NO WITH CF_2ClO_2

The kinetics and mechanisms of this reaction has been studied by Dognon, Caralp, and Lesclaux [68] using laser photolysis mass spectrometry and by

Sehested, Nielsen and Wallington [60] using a pulse radiolysis UV absorption method. At 298 K, rate constants of (1.6 ± 0.3) and $(1.3 \pm 0.1) \times 10^{-11}$ cm^3 molecule^{-1} s^{-1} were derived. The agreement between these studies demonstrates the absence of any effect of total pressure over the range from 1 torr of N_2 to 760 torr of SF_6. An inverse temperature dependence was observed by Dognon and coworkers [68]. The rate expression obtained by these workers [68] given in Table 3 is recommended. No evidence for the existence of reaction channels other than that leading to the alkoxy radical and NO_2 was found:

$$CF_2ClO_2 + NO \longrightarrow CF_2ClO + NO_2$$

3.11 THE REACTION OF NO WITH CCl_3O_2

The kinetics of this reaction have been studied by Ryan and Plumb [74] using discharge-flow mass spectroscopy at 298 K and by Dognon, Caralp and Lesclaux [68] using laser photolysis mass spectrometry over the temperature range 230–430 K. At 298 K, the results from both studies are in good agreement (see Table 3). Over the temperature range 230–430 K an inverse temperature dependence was observed, with no significant effect of pressure being found. The kinetic data reported by Dognon and coworkers [68] for this reaction are consistent with that of other reactions of CX_3O_2 (X = F or C1) with NO, and the expression calculated by these authors (given in Table 3) is recommended. The major product observed in this reaction was NO_2, suggesting that the reaction proceeds via one channel:

$$CCl_3O_2 + NO \longrightarrow CCl_3O + NO_2$$

3.12 THE REACTION OF NO WITH CF_3CFHO_2

This reaction plays an important role in the atmospheric degradation of CF_3CFH_2, which is a commercially important chlorofluorocarbon replacement. Wallington and Nielsen [75] studied the reaction by using a pulse radiolysis technique with UV absorption at 400 nm being used to monitor the rate of NO_2 formation at 298 K. Bhatnagar and Carr [76] used flash photolysis to create CF_3CFHO_2 radicals in the presence of NO and then followed the resulting NO_2 formation by using mass spectrometry. Peeters and Pultau [77] used a discharge-flow technique with CF_3CFHO_2 radicals being formed via reaction of F atoms with HFC-134a in the presence of O_2; the resulting NO_2 formation was observed using mass spectrometry. In all three studies the rate of NO_2 formation was used to derive a rate constant for the CF_3CFHO + NO reaction. As seen from Table 3, the results of all of the studies are in good agreement. It is expected that this reaction has little or no temperature dependence and we recommend use of the average of the available determinations, i.e. $k = 1.2 \times 10^{-11}$ cm^3 molecule^{-1} s^{-1} at ambient temperature. The yield of NO_2 in all of the studies was consistent with the reaction proceeding essentially completely via one channel to give CF_3CFHO radicals and NO_2.

3.13 THE REACTIONS OF NO WITH $FC(O)O_2$ AND $CF_3C(O)O_2$

These reactions have been studied by Wallington *et al.* [78,79] and Maricq *et al.* [80] using two different techniques. At 298 K the lower limit for $k(CF_3C(O)O_2+NO)$ of 9.9×10^{-12} cm^3 molecule^{-1} s^{-1} from Wallington *et al.* [79] is consistent with the value of $k(CF_3C(O)O_2+NO)$ of $(2.8 \pm 0.5) \times 10^{-11}$ cm^3 molecule^{-1} s^{-1} found by Maricq *et al.* [80]. The Arrhenius expression, $k(CF_3C(O)O_2+NO) = 4.0 \times 10^{-12} \exp(563/T)$ cm^3 molecule^{-1} s^{-1} of Maricq *et al.* [80] is recommended. At 298 K this expression gives $k = 2.6 \times 10^{-11}$ cm^3 molecule^{-1} s^{-1}. For the $FC(O)O_2$ + NO reaction the only available data are from Wallington *et al.* [78] who reported $k = (2.5 \pm 0.8) \times 10^{-11}$ cm^3 molecule^{-1} s^{-1} at 295 K. The similarity in the reactivities of the $FC(O)O_2$ and $CF_3C(O)O_2$ radicals is not surprising in view of the structural similarity of the two molecules, which both possess substituents which have strong electron-withdrawing effects. The reactivities of both fluorinated acylperoxy radicals are greater than that of $CH_3C(O)O_2$, which has a recommended rate constant of 1.9×10^{-11} cm^3 molecule^{-1} s^{-1} (see above).

3.14 REACTIONS OF NO WITH OTHER RO_2 RADICALS

In addition to the reactions discussed above, the research group at Risø National Laboratory has studied the kinetics of the reactions of a number of other peroxy radicals with NO (see Table 3) by using a pulse radiolysis UV absorption technique. We will not discuss these reactions individually but instead we will make some general observations. Points of interest with regard to the data obtained for $CF_3CH_2O_2$, CH_2FCFHO_2, $CFH_2CF_2O_2$, $CF_3CF_2O_2$, $CF_2ClCH_2O_2$, $CFCl_2CH_2O_2$, $CF_3CHO_2(^\bullet)CF_3$, $CH_3CF_2CH_2O_2$, $CCl_3CH_2O_2$, CF_3CClHO_2, $CH_3SCH_2O_2$, $CH_3OCH_2O_2$, and $HO(CH_3)_2CCH_2O_2$ radicals are as follows. First, the reactivity increases with the degree of halogenation on the carbon atom bearing the O—O($^\bullet$) group. There is little or no effect of β-halogenation to the O—O($^\bullet$) group. Secondly, the reactivity of the "large" $HO(CH_3)_2CCH_2O_2$ radical follows the same trend as that observed for the unsubstituted alkylperoxy radicals and is lower than that of the smaller radicals. Thirdly, it is puzzling to note the behaviour of the $CH_3SCH_2O_2$ and $CH_3OCH_2O_2$ radicals. The reactivity of $CH_3OCH_2O_2$ is the same as that of $C_2H_5O_2$, whereas $CH_3SCH_2O_2$ is twice as reactive. It is not clear why there should be such a difference. In view of this discrepancy, and because of the atmospheric importance of $CH_3SCH_2O_2$ as an intermediate in the oxidation of dimethyl sulfide, further work on this reaction is required.

3.15 DISCUSSION

From a consideration of the available data concerning this class of reaction the following points emerge.

First, our understanding of the reactions of peroxy radicals with NO is well advanced. It is well established that the reaction proceeds via two channels:

$$RO_2 + NO \longrightarrow RO + NO_2 \quad (a)$$

$$RO_2 + NO \longrightarrow RONO_2 \quad (b)$$

There is no effect of total pressure on the overall rate of reaction. Where available the data show that the overall rates of reaction tend to display slightly negative temperature dependencies.

Secondly, Atkinson and coworkers have established that the mechanism of reaction involves formation of a ROONO intermediate [93] (see Figures 7 and 8) and developed complex expressions that accurately predict the ratio of the nitrate- to alkoxy-radical-forming channels, k_b/k_a, as a function of temperature and pressure for given alkyperoxy radicals [3]. However, as discussed by Lightfoot *et al.* [2] these expressions are only applicable to alkylperoxy radicals. The available data concerning oxygenated alkylperoxy radicals show that their nitrate yields are substantially less than those observed for the alkylperoxy radicals. Further experimental and theoretical work is needed to understand the formation of nitrates in reactions of the atmospherically important oxygenated peroxy radicals with NO.

Thirdly, Sehested and coworkers [60,89] have shown that the reactivity of peroxy radicals towards NO decreases with increasing size. This is exactly the opposite trend from that observed for the $RO_2 + HO_2$ reactions. In addition, it is now clear that halogenation increases the $RO_2 + NO$ rate, while it decreases the $RO_2 + HO_2$ rate.

4 KINETICS AND MECHANISMS OF THE REACTIONS OF RO_2 RADICALS AND NO_2

Alkyperoxy radicals react with NO_2 to give peroxynitrates:

$$RO_2 + NO_2 + M \longrightarrow ROONO_2 + M$$

As combination reactions they are all expected to be pressure dependent, with the rate increasing with increasing pressure, and to have a negative temperature dependence. While the overall reaction can be written as a simple addition, in reality the situation is more complex, with the initial formation of an excited adduct which can then either decompose back to the reactants or be deactivated by collision with a third body capable of removing some of the vibrational excitation:

$$RO_2 + NO_2 \rightleftharpoons RO_2NO_2^* \quad (a,-a)$$

$$RO_2NO_2^* + M \longrightarrow RO_2NO_2 + M^* \quad (b)$$

At the high-pressure limit, with sufficient diluent gas (M) to deactivate all of the excited adducts, the overall rate of reaction is independent of pressure and the bimolecular rate constant is k_a. At the low-pressure limit, where quenching of the excited adduct does not significantly impact the equilibrium established by reactions (a) and (-a) the overall reaction rate will be linearly dependent on total pressure, with a termolecular rate constant given by $(k_a k_b)/k_a$. Under atmospheric conditions, most addition reactions are not at their high- or low-pressure limits, but instead are somewhere in between. Troe and coworkers [110,111] have performed extensive theoretical studies of addition reactions and have shown that their "effective" bimolecular rate constants in the "fall-off" region between high- and low-pressure limiting behaviour can be described by the three-parameter expression:

$$k([M],T) = k_o(T)[M]/(1+(k_o[M]/k_\infty(T)) \times F_c^{\{1+(\log(k_o(T)[M]/k_\infty(T))^2\}^{-1}}$$

where $k([M],T)$ is the effective bimolecular rate constant at a given third body concentration [M] and temperature T, and k_o and k_∞ are the limiting low- and high-pressure rate constants. F_c is a "broadening factor", which is typically 0.6, but can, in principal, have any value in the range 0–1, and describes the "broadness" of the fall-off (lower values of F_c give more shallow fall-off curves).

The available kinetic data for reactions of peroxy radicals with NO_2 are listed in Table 5. As in all other classes of alkylperoxy radical studies, the best characterized reaction is that of the simplest alkylperoxy radical, CH_3O_2.

4.1 THE REACTION OF NO_2 WITH HO_2

The kinetics and mechanisms of this reaction have been reviewed by Atkinson *et al.*[21] and DeMore *et al.*[20]. Both recommend the use of the kinetic expression derived by Kurylo and Ouellette[125,126] from a fit of their experimental data combined with that of Sander and Peterson[127]. This recommended expression (given in Table 5) is consistent with the previous studies by Howard[96], Simonaitis and Heicklein[128] and Cox and Patrick[129]. Further studies of the temperature dependence of the kinetics of this reaction at pressure above one atmosphere are needed to better characterize the temperature dependence at the high-pressure limit.

Table 5 Kinetic Data for the Reactions of Peroxy Radicals with NO_2

RO_2	$k \times 10^{12}$	Temperature (K)	Pressure (torr)	Technique	Ref.
HO_2	$k_o = 1.8 \times 10^{-31}(N_2)$		298	Review	21
	$k_o = 1.5 \times 10^{-31}(O_2)$		298		
	$k_\infty = 4.7 \times 10^{-12}$		200–300		
	$k_o = 1.8 \times 10^{-31}(T/300)^{-3.2})(N_2)$		220–360	Review	21
	$k_o = 1.5 \times 10^{-31}(T/300)^{-3.2})(O_2)$		220–360		
	$k_o = 1.8 \times 10^{-31}(\text{air})$		298	Review	20
	$k_\infty = 4.7 \times 10^{-12}$		298		
	$k_o = 1.8 \times 10^{-19}(T/300)^{-3.2})(\text{air})$		200–300	Review	20
	$k_\infty = 4.7 \times 10^{-12}(T/300)^{-1.4})$		200–300		
CH_3O_2	1.5 ± 0.1	298	53–580(Ar)	FP–UVA	112
	1.2 ± 0.3	298	50(Ar+CH_4)	MMS	26
	1.6 ± 0.3	298	540(N_2)		
	0.9 ± 0.1	298	50(He)	FP–UVA	54
	1.2 ± 0.2		100(He)		
	1.7 ± 0.1		225(He)		
	2.3 ± 0.3		350(He)		
	2.5 ± 0.3		500(He)		
	2.8 ± 0.4		700(He)		
	1.2 ± 0.1	298	50(N_2)	FP–UVA	54
	1.6 ± 0.2		100(N_2)		
	2.2 ± 0.3		225(N_2)		
	3.0 ± 0.2		350(N_2)		
	3.7 ± 0.2		500(N_2)		
	3.9 ± 0.2		700(N_2)		
	1.3 ± 0.2	298	50(SF_6)	FP–UVA	54
	2.0 ± 0.3		100(SF_6)		
	3.1 ± 0.4		225(SF_6)		
	3.9 ± 0.4		350(SF_6)		
	4.2 ± 0.4		500(SF_6)		
	4.8 ± 0.6		700(SF_6)		
	1.4 ± 0.2	298	76(N_2)	LP–UVA	113
	1.9 ± 0.3		157(N_2)		
	2.6 ± 0.6		258(N_2)		
	2.8 ± 0.3		352(N_2)		
	3.4 ± 0.3		519(N_2)		
	4.1 ± 0.4		722(N_2)		
	1.2 ± 0.2	353	330(N_2)	LP–UVA	113
	1.4 ± 0.2		354(N_2)		
	1.7 ± 0.1		511(N_2)		
	1.9 ± 0.2		696(N_2)		
	2.5 ± 0.3	253	109(N_2)	LP–UVA	113
	3.9 ± 0.3		250(N_2)		
	5.1 ± 0.5		503(N_2)		
	5.8 ± 1.0		519(N_2)		
$C_2H_5O_2$	1.2 ± 0.1	298	44–676(Ar/O_2)	FP–UVA	114
	7.2	256	755(N_2/O_2)	Static photolysis FTIR	115
	4.5	254	76		
	1.8	254	7.8		
	3.8	265	76.8		
$(CH_3)_2CHO_2$	5.6 ± 0.2	298	55–400(Ar)	FP–UVA	65
$(CH_3)_3CO_2$	>0.5	298	300(N_2)	MMS	50
$CH_3C(O)O_2$	2.1 ± 0.1	302	28(N_2)	MMS	116
	6.0 ± 2.0		715(N_2)		
	3.44 ± 0.53	295	76(N_2)	FP–UVA	117
	3.53 ± 0.60		153(N_2)		
	3.64 ± 0.60		306(N_2)		

(continued overleaf)

Table 5 – *continued*

RO_2	$k \times 10^{12}$	Temperature (K)	Pressure (torr)	Technique	Ref.
$CH_3C(O)O_2$ – *continued*					
	3.47 ± 0.66		306(O_2)		
	4.88 ± 0.56		500(N_2)		
	4.77 ± 0.50		612(N_2)		
	10.9	248	90(O_2/air)	FP–UVA	118
	10.8		210		
	10.9		760		
	8.7	273	28		
	9.4		100		
	10.0		190		
	9.9		375		
	10.2		760		
	5.3	298	15		
	6.6		35		
	6.9		42		
	6.7		47		
	7/4		100		
	8.4		160		
	8.3		242		
	9.0		300		
	8.6		450		
	9.6		760		
	5.8	333	36		
	5.6		130		
	6.6		210		
	7.1		335		
	6.3		390		
	6.2		580		
	7.5		760		
	4.5	368	36		
	4.7		98		
	5.4		270		
	5.8		760		
	4.0	393	85		
	4.3		155		
	5.4		335		
	6.3		760		
$CFCl_2O_2$	1.45	298	3.5(O_2)	LP-MS	73
	1.75	233	0.8(O_2)	LP-MS	119
	2.27		1.4		
	2.74		2.0		
	4.31		3.9		
	5.30		5.9		
	5.90		7.8		
	0.72	298	1		
	0.92		1.5		
	0.93		2.1		
	1.17		2.5		
	1.52		3.1		
	1.42		3.5		
	1.85		5.0		
	2.1		6.5		
	2.41		10.0		
	0.43	373	3.1		
	0.72		6.3		
	0.84		12.5		
		≈233–373		Review	20

$k_o = (3.5 \pm 0.5) \times 10^{-29}\ (T/300)^{-(5 \pm 1)}$ cm^6 molecule^{-2} s^{-1} (in air)
$k_\infty = (6.0 \pm 1.0) \times 10^{-12}\ (T/300)^{-(2.5 \pm 1)}$ cm^3 molecule^{-1} s^{-1} (in air)

Table 5 – *continued*

RO_2	$k \times 10^{12}$	Temperature (K)	Pressure (torr)	Technique	Ref.
CF_2ClO_2	0.75 ± 0.13	298	$1(CF_2ClBr)$	FP–MS	120
	0.82 ± 0.33		1.5		
	1.27 ± 0.04		2		
	1.37 ± 0.07		3		
	1.51 ± 0.13		4		
	1.41 ± 0.18		5		
	1.60 ± 0.23		6		
	1.72 ± 0.48		7		
	1.52 ± 0.40		8		
	2.18 ± 0.71		9		
	2.83 ± 0.20		10		
CF_3O_2	1.34	233	$0.8(O_2)$	LP–MS	119
	2.03		1.6		
	3.95		4.7		
	4.55		6.3		
	5.15		7.9		
	0.52	298	1		
	0.92		2		
	1.1		3		
	1.47		6		
	1.76		10		
	0.37	373	2.5		
	0.62		7.5		
	0.69		12.5		
		≈233–373		Review	20
	$k_o = (2.2 \pm 0.5) \times 10^{-29}$ $(T/300)^{-(5 \pm 1)}$ cm^6 molecule^{-2} s^{-1} (in air)				
	$k_\infty = (6.0 \pm 1.0) \times 10^{-12}$ $(T/300)^{-(2.5 \pm 1)}$ cm^3 molecule^{-1} s^{-1} (in air)				
CCl_3O_2	3.25	233	$1.2(O_2)$	LP–MS	119
	3.85		1.6		
	4.87		2.3		
	5.87		3.3		
	1.22	298	2		
	1.58		3		
	3.17		10		
	0.91	373	5		
	1.04		2.6		
	1.32		4.3		
	$k_o = (5.0 \pm 1.0) \times 10^{-29}$ $(T/300)^{-(5 \pm 1)}$ cm^6 molecule^{-2} s^{-1} (in air)				
	$k_\infty = (6.0 \pm 1.0) \times 10^{-12}$ $(T/300)^{-(2.5 \pm 1)}$ cm^3 molecule^{-1} s^{-1} (in air)				
$HO(CH_3)_2CCH_2O_2$	6.7 ± 0.9	298	$760(SF_6)$	PR–UVA	89
$CH_3OCH_2O_2$	7.9 ± 0.4	296	$760(SF_6)$	PR–UVA	88
CF_3CFHO_2	5.0 ± 0.5	298	$760(SF_6)$	PR–UVA	121
$CF_3CH_2O_2$	5.8 ± 1.1	298	$760(SF_6)$	PR–UVA	81
$CF_3C(O)O_2$	6.6 ± 1.3	295	$760(SF_6)$	PR–UVA	122
$FC(O)O_2$	5.5 ± 0.6	295	$380–760(SF_6)$	PR–UVA	123
$CH_3CF_2CH_2O_2$	6.8 ± 0.5	295	$760(SF_6)$	PR–UVA	84
$CCl_3CH_2O_2$	6.5 ± 0.4	295	$760(SF_6)$	PR–UVA	85
$CH_3SCH_2O_2$	9.2 ± 0.9	296	$760(SF_6)$	PR–UVA	124
	7.1 ± 0.9		$228(SF_6)$		
CF_3CClHO_2	6.4 ± 1.5	296	$760(SF_6)$	PR–UVA	86

[a] FTIR, Fourier-transform infrared spectroscopy; MMS, molecular modulation spectroscopy; UVA, ultraviolet absorption spectroscopy; FP, flash photolysis; PR, pulse radiolysis; LP, laser photolysis; MS, mass spectrometry.

[b] Experiments performed at 'room temperature', which was taken to be 295 K.

4.2 THE REACTION OF NO_2 WITH CH_3O_2

The kinetics of this reaction have been studied by Adachi and Basco[112], Cox and Tyndall[26], Sander and Watson[54], and Ravishankara, Eisele and Wine[113]:

$$CH_3O_2 + NO_2 + M \longrightarrow CH_3OONO_2 + M$$

Sander and Watson[54] observed $k(298\,K)$ to increase by factors of three to four on going from 50 to 700 torr of either He, N_2 or SF_6. This dependence was confirmed by the study of Ravishankara and coworkers[113] using nitrogen as the third body (see Figure 10). Ravishankara and coworkers[113] also performed experiments at temperatures other than ambient and observed a negative temperature dependence. The study of Cox and Tyndall[26] at 50 torr of a mixture of methane and argon is consistent with the results of both Sander and Watson[54] and Ravishankara and coworkers[113]. However, the rate constant measured in the presence of 540 torr of N_2 by Cox and Tyndall[26] is considerably lower (by a factor of 2) than that measured by Sander and Watson[54] and Ravishankara and coworkers[114]. As suggested by Sander and Watson[54], the origin of this discrepancy might lie in the relatively long time-scale employed in the molecular modulation technique which may allow significant dissociation of the peroxymethyl nitrate to the reactants and hence led to an underestimation of the rate constant for the association reaction. In the work of Adachi and Basco[112] the rate of

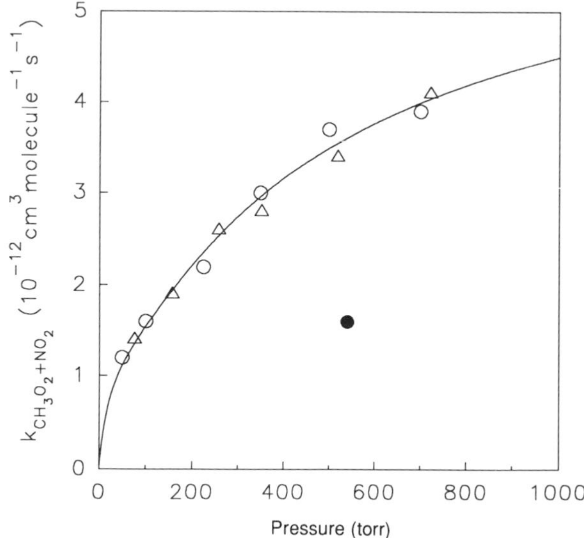

Figure 10 Fall-off plot for the $CH_3O_2 + NO_2$ reaction at 298 K in nitrogen diluent, with data taken from Cox and Tyndall[25] (filled circles), Sander and Watson[54] (open circles), and Ravishankara, Eisele and Wine[113] (triangles); the full line represents our recommendation.

reaction of CH_3O_2 with NO_2 was reported to be independent of pressure over the range 53–580 torr of argon. This result is clearly inconsistent with those obtained from other studies of this reaction and strongly suggests that the results from Reference 112 are in error.

In Figure 11 we show the data obtained at 253 and 353 K (again in N_2) by Ravishankara, Eisele and Wine[113]. Sander and Watson[54] fitted their 298 K kinetic data to the three-parameter expression proposed by Troe and coworkers[110,111]

$$k([M],T = k_o(T)[M]/(1+(k_o[M]/k_\infty(T)) \times F_c^{\{1+(\log(k_o(T)[M]/k_\infty(T))^2\}^{-1}}$$

by using $k_o = 2.3 \times 10^{-30}$ cm^3 molecule^{-2} s^{-1}, $k_\infty = 8 \times 10^{-12}$ cm^3 molecule^{-1} s^{-1} and $F_c = 0.4$. Use of these parameters gives an excellent fit to the data reported by Sander and Watson [54] as well as that measured by Ravishankara and coworkers[113] as shown by the solid line in Figure 10, and is our recommendation for 298 K. For temperatures other than ambient we recommend use of the expressions derived in Ref. 113:

$$k_o(T) = 2.3 \times 10^{-30} (T/298)^{-2.5} \text{ cm}^6 \text{ molecule}^{-2} \text{ s}^{-1}$$

and

$$k_\infty(T) = 8.0 \times 10^{-12} (T/298)^{-3.5} \text{ cm}^3 \text{ molecule}^{-1} \text{ s}^{-1}.$$

The fitting of these expressions to the existing data at 253 K is excellent. At

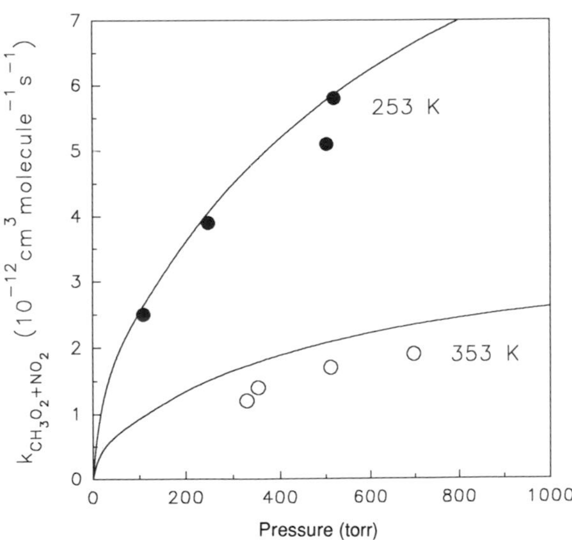

Figure 11 Fall-off plots for the $CH_3O_2 + NO_2$ reaction at 253 and 353 K in nitrogen diluent, with data taken from Ravishankara, Eisele and Wine[113]; the full lines represent our recommendations (see discussion in text).

353 K the fits are only fair, as shown in Figure 11. Our recommendation is thus biased towards the ambient and sub-ambient data, as these are most useful to the atmospheric chemistry community.

4.3 THE REACTION OF NO_2 WITH $C_2H_5O_2$

The kinetics of the reaction of $C_2H_5O_2 + NO_2$ have been studied by Adachi and Basco[114] and Elfers, Zabel and Becker[115]. Adachi and Basco[114] used a flash photolysis UV absorption technique to derive a rate constant of 1.2×10^{-12} cm^3 molecule^{-1} s^{-1} for this reaction, which was independent of pressure over the range 44–676 torr of argon. As noted above the measurements of Adachi and Basco[114] may be subject to large systematic errors. Elfers and coworkers[115] studied the kinetics of the reaction of $C_2H_5O_2$ radicals with NO_2 using a relative rate approach by measuring the rate of reaction of $C_2H_5O_2$ with NO_2 relative to that with NO as ambient temperature and at pressures of ca. 8–755 torr of N_2/O_2 mixtures:

$$C_2H_5O_2 + NO \longrightarrow C_2H_5O + NO_2$$

$$C_2H_5O_2 + NO_2 \longrightarrow C_2H_5O_2NO_2$$

The reacting species were prepared by the photolysis of $Cl_2/C_2H_6/NO_x/O_2/N_2$ mixtures and FTIR spectroscopy was used to monitor changes in the concentrations of NO, NO_2 and $C_2H_5O_2NO_2$. The results were placed upon an absolute basis by using a value of 8.9×10^{-12} cm^3 molecule^{-1} s^{-1} for the rate constant of the reaction with NO[63]. Using the Troe formalism the authors proposed use of the following fall-off parameters: $k_o = 4.8 \times 10^{-29}$ $[N_2]$ cm^{-3} molecule^{-1} s^{-1}, $k_\infty = 1.0 \times 10^{-11}$ cm^3 molecule^{-1} s^{-1}, and $F_c = 0.3$. Further work is needed to better define the kinetics of the reaction of ethylperoxy radicals with NO_2. Nevertheless, we recommend use of the expression derived by Elfers, Zabel and Becker[115].

4.4 THE REACTION OF NO_2 WITH $(CH_3)_2CHO_2$

The kinetics of the reaction of $(CH_3)_2CHO_2$ with NO_2 have been studied by Adachi and Basco[65] using a flash photolysis UV absorption technique at 298 K, with these authors deriving a rate constant of 5.6×10^{-12} cm^3 molecule^{-1} s^{-1} which was independent of pressure over the range 55–400 torr of argon. As noted above, those measurements may be subject to large systematic errors, and in the absence of confirmatory measurements no recommendation is therefore made.

4.5 THE REACTION OF NO_2 WITH $(CH_3)_3CO_2$

A lower limit of $k(298\,\text{K}) > 5 \times 10^{-13}$ cm^3 molecule^{-1} s^{-1} has been reported by Anasasti, Smith and Parkes[50] for this reaction.

4.6 THE REACTION OF NO_2 WITH $CH_3C(O)O_2$

The kinetics of the reaction of acetylperoxy radicals, $CH_3C(O)O_2$, with NO_2 was first studied by Cox and Roffey[105] and Hendry and Kenley[106]. In both of these studies the measured experimental parameter was the rate of the reverse reaction, i.e. the thermal decomposition of PAN. By using estimates of the thermochemistry of the system the equilibrium constant was estimated, and the rate of the association reaction was then calculated:

$$CH_3C(O)O_2 + NO_2 + M \longrightarrow CH_3C(O)O_2NO_2 + M$$

The results from both studies are in good agreement, yielding estimates of the rate constant k of 1.4 and 1.0×10^{-12} cm^3 molecule^{-1} s^{-1} at 1 torr total pressure of either air or N_2, respectively[105,106] with an estimated uncertainty of plus or minus an order of magnitude.

The first absolute study of the kinetics of the reaction of acetylperoxy radicals with NO_2 was performed by Addison *et al.*[116] using the molecular modulation technique. As expected for an association reaction these workers observed a significant effect of pressure, with the rate constant decreasing by a factor of three over the pressure range 715 to 28 torr of added nitrogen. This behaviour was confirmed by Basco and Parmer[117]. In the latest and most comprehensive study of the kinetics of the reaction of acetylperoxy radicals with NO_2, Bridier *et al.*[118] have employed the flash photolysis technique to measure the kinetics of this reaction at total pressures between 15 and 760 torr and temperature in the range 248–393 K. Figure 12 shows the

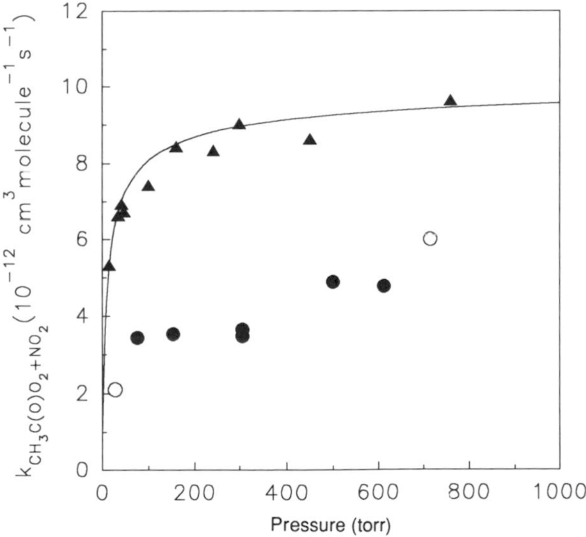

Figure 12 Fall-off plot for the $CH_3C(O)O_2 + NO_2$ reaction at 298 K in nitrogen diluent, with data taken from Addison *et al* [116] (open circles), Basco and Parmer[117] (filled circles), and Bridier *et al.*[118] (triangles); the full line is our recommendation.

kinetic data reported by Addison and coworkers[116], Basco and Parmer[117], and Bridier and coworkers[118] as a function of pressure at 298 K. As seen from this figure the latest data from Bridier and coworkers[118] is a factor of two higher than that previously reported by Addison *et al.*[116] and Basco and Parmer[117]. As discussed by Bridier and coworkers[118] the origin of this discrepancy probably lies in the different chemical mechanism used to simulate the observed behaviour of the UV absorption. The chemical mechanism adopted in the later work[118] is more complete than those used in the earlier studies and so the results of Bridier *et al.*[118] are to be preferred. For consistency with the other three-body association reactions we have fitted the 298 K data of Bridier *et al.*[118] to the Troe expression[110,111]:

$$k([\mathrm{M}],T) = k_o(T)[\mathrm{M}]/(1+(k_o[\mathrm{M}]/k_\infty(T)) \times F_c^{\{1+(\log(k_o(T)[\mathrm{M}]/k_\infty(T))^2\}^{-1}}$$

with F_c fixed at 0.3 (as recommended by Bridier *et al.*[118] to obtain our recommendation of $k_o = (2.0 \pm 0.5) \times 10^{-28}$ $(T/298)^{\{-7.1 \pm 1.7\}}$ cm^6 $molecule^{-2}$ s^{-1}, and $k_\infty = (1.1 \pm 0.1) \pm 10^{-12}$ $(T/298)^{\{-0.9 \pm 0.15\}}$ cm^3 $molecule^{-1}$ s^{-1}. Temperature dependencies were taken directly from Bridier *et al.*[118]. At 760 torr of air this expression yields $k = 9.4 \times 10^{-12}$ cm^3 $molecule^{-1}$ s^{-1}, with uncertainties estimated to be $\pm$ 40%.

There is a clear need for further studies of kinetics of this atmospherically important reaction to check the data reported by Bridier *et al.*[118] over as wide a range of temperatures and pressures as possible. The large impact of uncertainties in the temperature dependence of this reaction on our understanding of trospospheric ozone formation has been discussed by Dodge[130]. In addition to the direct experimental studies reported above there have been a number of indirect studies in which relative rate techniques have been used. In these studies the competitive between reaction of acetylperoxy radicals with either NO or NO_2 has been measured, and the results from these studies are discussed in the section dealing with the reaction of acetylperoxy radicals and NO.

4.7 THE REACTION OF NO_2 WITH $CFCl_2O_2$

The kinetics of the reaction of $CFCl_2O_2$ with NO_2 has been studied by Lesclaux and Caralp[73] and Caralp *et al.*[119] using laser flash photolysis of $CFC1_3$ in the presence of O_2 and NO_2. Rate-constant data were measured over the pressure range 1–12 torr of oxygen at either 298[3] or 233, 298 and 373 K[19]. At 298 K the results from both studies by Caralp and coworkers are in agreement. At the pressures investigated this reaction is in the fall-off region between second- and third-order kinetic behaviour. Caralp *et al.*[73] fitted their kinetic data to the Troe formalism with a temperature dependent F_c value given by $F_c = \exp(-T/342)$. DeMore *et al.*[20] have shown that the data of Lesclaux and Caralp *et al.*[73] can equally well be fitted by the Troe formalization with F_c held constant at a value of 0.6. For the sake of simplicity and consistency we recommend use of the expression given by DeMore *et al.*[20]:

$$k([\mathrm{M}],T) = k_o(T)[\mathrm{M}]/(1+(k_o[\mathrm{M}]/k_\infty(T)) \times F_c^{\{1+(\log(k_o(T)[\mathrm{M}]/k_\infty(T))^2\}^{-1}}$$

with $k_o = (3.5 \pm 0.5) \times 10^{-29}\ (T/300)^{\{-5 \pm 1\}}$ cm^6 molecule^{-2} s^{-1}, $k_\infty = (6.0 \pm 1.0) \times 10^{-12}\ (T/300)^{\{-2.5 \pm 1\}}$ cm^3 molecule^{-1} s^{-1} and $F_c = 0.6$. At 760 torr and 300 K this expression gives $k = 5.4 \times 10^{-12}$ cm^3 molecule^{-1} s^{-1}.

4.8 THE REACTION OF NO_2 WITH CF_2ClO_2

The kinetics of the reaction of CF_2ClO_2 with NO_2 have been studied by Moore and Carr[120] using flash photolysis coupled with time-resolved mass spectrometry over the pressure range 1–10 torr in CF_2ClBr at 298 K. At the pressures investigated the reaction is in the fall-off region between second- and third-order kinetic behaviour. Moore and Carr[120] fitted their kinetic data to the Troe expression:

$$k([\mathrm{M}],T) = k_o(T)[\mathrm{M}]/(1+(k_o[\mathrm{M}]/k_\infty(T)) \times F_c^{\{1+(\log(k_o(T)[\mathrm{M}]/k_\infty(T))^2\}^{-1}}$$

using $k_o = (3.5 \pm 1.8) \times 10^{-29}$ cm^6 molecule^{-2} s^{-1}, $k_\infty = 5.2 \times 10^{-12}$ cm^3 molecule^{-1} s^{-1} and $F_c = 0.6$.

Measurement of the kinetics of the reaction of CF_2ClO_2 with NO_2 as a function of temperature at high pressures are necessary to better define the high-pressure limit and its temperature dependence. We recommend use of the expression of Moore and Carr[120] for this reaction at 298 K.

4.9 THE REACTION OF NO_2 WITH CF_3O_2

The kinetics of the reaction of CF_3O_2 with NO_2 has been studied by Caralp *et al.*[119] using pulsed laser photolysis and time-resolved mass spectroscopy in the pressure range 1–10 torr and over the temperature range 233–373 K. These workers fitted their kinetic data to the Troe formalism with a temperature-dependent F_c value given by $F_c = \exp(-T/416)$. DeMore *et al.*[20] have shown that the data of Caralp *et al.*[119] can equally well be fitted by the Troe formalism with F_c held constant at 0.6. For the sake of simplicity and consistency with the bulk of the data reported in this present review we recommend use of the expression given by DeMore *et al.*[20]:

$$k([\mathrm{M}],T) = k_o(T)[\mathrm{M}]/(1+(k_o[\mathrm{M}]/k_\infty(T)) \times F_c^{\{1+(\log(k_o(T)[\mathrm{M}]/k_\infty(T))^2\}^{-1}}$$

with $k_o = (2.2 \pm 0.5) \times 10^{-29}\ (T/300)^{\{-5 \pm 1\}}$ cm^6 molecule^{-2} s^{-1}, $k_\infty = (6.0 \pm 1.0) \times 10^{-12}\ (T/300)^{\{-2.5 \pm 1\}}$ cm^3 molecule^{-1} s^{-1} and $F_c = 0.6$.

4.10 THE REACTION OF NO_2 WITH CCl_3O_2

The kinetics of the reaction of CCl_3O_2 with NO_2 has been studied by Caralp *et al.*[119] using pulsed laser photolysis and time-resolved mass spectrometry in the pressure range 1–10 torr and over the temperature range 233–373 K. These workers fitted their kinetic data to the Troe formalism with a temperature-dependent F_c value given by $F_c = \exp(-T/260)$. DeMore *et al.*[20] have shown that the data of Caralp *et al.*[119] can equally well be fitted by the Troe formalism with F_c held constant at 0.6. For the sake of simplicity and consistency with the bulk of the data reported in this present review we recommend use of the expression given by DeMore *et al.*[20]:

$$k([M],T) = k_o(T)[M]/(1+(k_o[M]/k_\infty(T)) \times F_c^{\{1+(\log(k_o(T)[M]/k_\infty(T))^2\}^{-1}}$$

with $k_o = (5.0 \pm 1.0) \times 10^{-29}\ (T/300)^{\{-5 \pm 1\}}$ cm^6 molecule^{-2} s^{-1}, $k_\infty = (6.0 \pm 1.0) \times 10^{-12}\ (T/300)^{\{-2.5 \pm 1\}}$ cm^3 molecule^{-1} and $F_c = 0.6$.

4.11 REACTIONS OF NO_2 WITH OTHER RO_2 RADICALS

The kinetics of the reactions of $HO(CH_3)_2CCH_2O_2$, $CH_3OCH_2O_2$, CF_3CFHO_2, $CF_3CH_2O_2$, $CF_3C(O)O_2$, $FC(O)O_2$, $CH_3CF_2CH_2O_2$, $CCl_2CH_2O_2$, $CH_3SCH_2O_2$, and CF_3CClHO_2 radicals with NO_2 have been studied by using pulse radiolysis coupled with time-resolved UV absorption spectroscopy at 760 torr of SF_6 and ambient temperature. The results are given in Table 5. The reactions of $FC(O)O_2$ and $CH_3SCH_2O_2$ radicals with NO_2 were also studied at total pressures of 380 and 228 torr. The reaction of $FC(O)O_2$ radicals with NO_2 was independent of the total pressure from 380 to 760 torr, while the reaction of $CH_3SCH_2O_2$ with NO_2 showed a small dependence on the total pressure over the pressure range that was investigated. In the case of the reaction of CH_3O_2 radicals with NO_2 the third-body efficiency of SF_6 is approximately 1.5 times the third-body efficiency of N_2[54]. Considering this point and also that the reactions of CH_3O_2, $CF_XCl_{3-X}O_2$, and $CH_3C(O)O_2$ radicals with NO_2 are near their high-pressure limits at 760 torr of N_2, the rate constants measured at 760 torr of SF_6 are also expected to be close to their high-pressure limits. The high pressure limit for the ten reactions measured using SF_6 as the third body are in the range $(5\text{–}9.2) \times 10^{-12}$ cm^3 molecule^{-1} s^{-1}. These values are consistent with the range of k_∞ values reported for other reactions of this type, and are therefore recommended.

4.12 DISCUSSION

The reactions of peroxy radicals with NO_2 are pressure dependent up to 760 torr. With the exception of HO_2 and CH_3O_2, the reaction proceed at ambient temperature and a pressure of 760 torr with rate constants which are within 20% of their high-pressure limits. The reaction rates are well represented by the expression proposed by Troe and coworkers[110,11]:

$$k([M],T) = k_o(T)[M]/(1+(k_o[M]/k_\infty(T)) \times F_c^{\{1+(\log(k_o(T)[M]/k_\infty(T))^2\}^{-1}}$$

with appropriate values of k_o, k_∞, and F_c. The pressure dependence at 298 K is shown in Figure 13, while the recommended values of k_o(298 K), k_∞(298 K), and F_c are listed in Table 9 (see below). The recommended temperature dependencies of k_o and k_∞ are also given in Table 9.

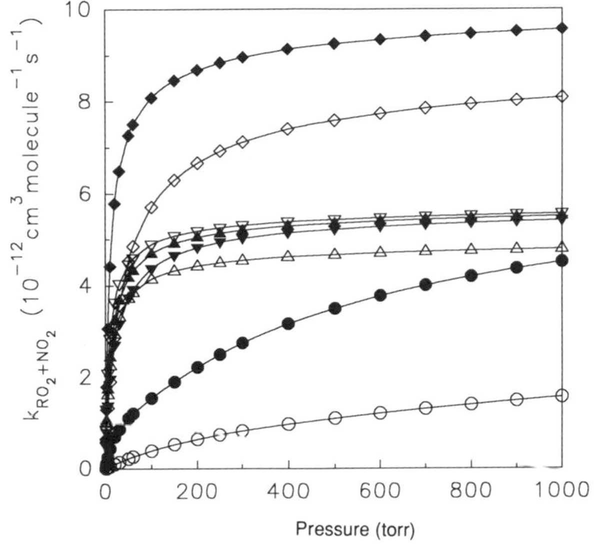

Figure 13 Recommended kinetic data for the reaction of peroxy radicals with NO_2 at 298 K as a function of pressure, with data points calculated by using the rate expressions given in the text: HO_2 (open circles); CH_3O_2 (filled circles); CF_2ClO_2 (open triangles); $CFCl_2O_2$ (filled triangles); $C_2H_5O_2$ (open diamonds); $CH_3C(O)O_2$ (filled diamonds); CCl_3O_2 (open inverse triangles); CF_3O_2 (filled inverse triangles).

Destriau and Troe [131] suggested that the high-pressure limit for the reactions of all peroxy radicals with NO_2 should be 7.5×10^{-12} cm^3 molecule^{-1} s^{-1}. The recommended values of k_∞ in Table 9 all lie in the range (5–10) $\times 10^{-12}$ cm^3 molecule^{-1} s^{-1} which is consistent with the suggestion of Destriau and Troe [131]. In contrast to k_∞, the values of k_o vary greatly for different peroxy radicals. For R = H, $k_o = 1.8 \times 10^{-31}$ cm^6 molecule^{-2} s^{-1}, while $k_o = 2.0 \times 10^{-28}$ cm^6 molecule^{-2} s^{-1} for R = CH_3CO. This difference in k_o reflects the effectiveness of the stabilization and thereby the lifetime of the highly excited reaction complex, RO_2NO_2, which is formed following recombination of RO_2 and NO_2. Larger RO_2 radicals have more degrees of freedom in which to distribute the excess energy and hence have greater values of k_o.

The recommendations given in Table 9 are derived by extrapolations from data over limited pressure and temperature ranges. The uncertainties of the temperature dependence and of the high-pressure limits for the rate constants for reactions of peroxy radicals which are important in atmospheric chemistry studies, such as CH_3O_2, HO_2 and $C_2H_5O_2$ with NO_2, are large. More work is therefore needed to establish high-pressure limits and the temperature dependencies of both high- and low-pressure limits for the reactions of peroxy radicals with NO_2.

5 KINETICS AND MECHANISMS OF THE REACTIONS OF RO_2 RADICALS WITH MISCELLANEOUS SPECIES

As noted above, peroxy radicals are relatively unreactive species. At ambient temperatures they only react with species which carry an unpaired electron. While the majority of experimental studies have centred on the self-reactions with NO, NO_2, and HO_2 radicals, some studies have focused on the reactions of peroxy radicals with other species. The available data for such reactions is gathered in Table 6, which has been adapted from Lightfoot *et al.*[2].

5.1 THE REACTION OF $O(^3P)$ WITH CH_3O_2

Zellner *et al.*[133] studied the 193 nm laser photolysis of $N_2O/CH_3N_2CH_3/O_2$ mixtures at 298 K, using laser-induced fluorescence to monitor the formation of CH_3O radicals. From a detailed modelling of the reactions occurring in this chemical system it was deduced that this reaction proceeds to give CH_3O and O_2 products with a rate constant of $(4.3 \pm 2.0) \times 10^{-11}$ cm^3 molecule^{-1} s^{-1}. This result is consistent with the lower limit of 3×10^{-11} cm^3 molecule^{-1} s^{-1}, derived by Washida and Bayes[132].

5.2 THE REACTION OF Cl WITH CH_3O_2 AND $C_2H_5O_2$

Maricq *et al.*[134] used laser flash photolysis combined with UV absorption and tunable diode laser spectroscopy to study the reaction of Cl atoms with CH_3O_2 and $C_2H_5O_2$ at 298 K. Both reactions have an overall rate constant of $(1.5 \pm 0.2) \times 10^{-10}$ cm^3 molecule^{-1} s^{-1}. In the CH_3O_2 reaction, branching ratios of $k_a/(k_a+k_b) = 0.51 \pm 0.05$ and $k_b/(k_a+k_b)$ $0.49 \pm 0.49 \pm 0.05$ were determined:

$$CH_3O_2 + Cl \longrightarrow CH_3O + ClO \quad (a)$$

$$CH_3O_2 + Cl \longrightarrow CH_2O_2 + HCl \quad (b)$$

For the $C_2H_5O_2$ + Cl reaction with ClO yield is 0.49 ± 0.05. The observation of a rapid reaction between Cl and CH_3O_2 and $C_2H_5O_2$ to give ClO radicals is consistent with the available data concerning the reaction of Cl atoms with CCl_3O_2 and CF_2ClO_2 radicals, as discussed below. Channel (b) is interesting,

Table 6 Kinetic Data for the Reactions of Peroxy Radicals with Miscellaneous Species

Reaction	k (cm^3 molecule^{-1} s^{-1})	Temperature (K)	Ref.
$CH_3O_2 + O(^3P) \longrightarrow CH_3O + O_2$	$>3 \times 10^{-11}$	298	132
	$(4.3 \pm 2.0) \times 10^{-11}$	298	133
$CH_3O_2 + Cl \longrightarrow$ products	$(1.5 \pm 0.2) \times 10^{-10}$	298	134
$C_2H_5O_2 + Cl \longrightarrow$ products	$(1.5 \pm 0.2) \times 10^{-10}$	298	134
$CH_3O_2 + H \longrightarrow$ products	1.6×10^{-10}	298	135
$CH_3O_2 + ClO \longrightarrow$ products	$(3.1 \pm 1.7) \times 10^{-12}$	298	136
$CH_3O_2 + ClO \longrightarrow CH_3O + ClOO$	$<4 \times 10^{-12}$	200	137
$CH_3O_2 + ClO \longrightarrow CH_3O + OClO$	$<1 \times 10^{-15}$	200	137
$CH_3O_2 + ClO \longrightarrow$ products	$(1.9 \pm 0.4) \times 10^{-12}$	298	59
$CH_3O_2 + ClO \longrightarrow$ products	$(2.0 \pm 0.1) \times 10^{-12}$	225	138
	$(2.1 \pm 0.1) \times 10^{-12}$	255	
	$(2.2 \pm 0.1) \times 10^{-12}$	295	
	$(2.3 \pm 0.2) \times 10^{-12}$	355	
$CH_3O_2 + NO_3 \longrightarrow CH_3O + NO_2 + O_2$	$(2.3 \pm 0.7) \times 10^{-12}$	298	139
	$(1.0 \pm 0.6) \times 10^{-12}$	298	140
	$(1.2 \pm 0.6) \times 10^{-12}$	298	141
$CH_3O_2 + CH_3 \longrightarrow CH_3O + CH_3O + O_2$	$(6.0 \pm 1.0) \times 10^{-11}$	298	142
	9.1×10^{-11}	298	143
	$8.4 \times 10^{-12} \exp(710/T)$	298–582	
$CF_3O_2 + CF_3O \longrightarrow CF_3O_3CF_3$	$>2.5 \times 10^{-11}$	298	144
$CH_3O_2 + CH_3O_2 \longrightarrow$ products	1.0×10^{-11}	298	145
$CH_3O_2 + O_3 \longrightarrow$ products	$<2.4 \times 10^{-17}$	298	146
$CH_3O_2 + SO_2 \longrightarrow$ products	$<5 \times 10^{-17}$	298	147
$CH_3O_2 + CO \longrightarrow$ products	$<7 \times 10^{-18}$	298	54
$CCl_3O_2 + Cl \longrightarrow CCl_3O + ClO$	$(2 \pm 1) \times 10^{-10}$	298	148
$CF_2ClO_2 + Cl \longrightarrow CCl_3O + ClO$	$(2 \pm 1) \times 10^{-10}$	298	149
$CCl_3O_2 + CCl_3 \longrightarrow CCl_3O + CCl_3O$	$(1^{+1}_{-0.5}) \times 10^{-12}$	298	148

as it offers a means to produce the 'Criegee' biradical, CH_2O_2. Maricq *et al.*[134] proposed that reaction of Cl atoms with CH_3O_2 proceeds via the formation of an excited CH_3OOCl adduct which decomposes via either ClO or HCl elimination.

5.3 THE REACTION OF H WITH CH_3O_2

As discussed by Lightfoot *et al.*[2], the rate of mechanism of this reaction are highly uncertain. The only experimental data arises indirectly from a study of the H + CH_3OOH reaction by Slemr and Warneck[150]. From the observed H_2O yield, Slemr and Warneck suggested that at 298 K the CH_3O_2 + H reaction proceeds predominantly (87 ± 13%) via channel (b) with a minor or zero (13 ± 13%) contribution from channel (a) (see below). Tsang and Hampson [135] have estimated that the rate constant for this reaction is close to the gas kinetic limit at 1.6×10^{-10} cm^3 molecule^{-1} s^{-1} and noted that the H_2O product observed by Slemr and Warneck[150] was likely due to decomposition of CH_3OOH on the walls of the reaction vessel. It seem reasonable to suppose that like other peroxy radical reactions the CH_3O_2 + H reaction proceeds via the formation of a short-lived adduct (CH_3OOH^*) which decomposes in to the products. Decomposition of CH_3OOH^* via O—O bond scission is more exothermic (by 25 kcal mol^{-1}) than via C—O bond scission. Thus it is reasonable to expect that reaction proceeds mainly via channel (a). No recommendation is made for this reaction, and further experimental studies are needed.:

$$CH_3O_2 + H \longrightarrow CH_3O + OH \quad (a)$$

$$CH_3O_2 + H \longrightarrow HCHO + H_2 \quad (b)$$

$$CH_3O_2 + H \longrightarrow CH_3 + HO_2 \quad (c)$$

5.4 REACTION OF ClO WITH CH_3O_2

This reaction has been studied by Simon *et al.*[136], DeMore[137], Kenner, Ryan and Plumb[59], and Helleis, Crowley and Moortgat[138]. Helleis and coworkers[138] have shown that over the temperature range 225–355 K the reaction proceeds via two pathways:

$$CH_3O_2 + ClO \longrightarrow CH_3O + ClOO \quad (a)$$

$$CH_3O_2 + ClO \longrightarrow CH_3OCl + O_2 \quad (b)$$

With increasing temperature, channel (a) increases in importance at the expense of channel (b). No effect of total pressure on either the branching ratio or kinetic was observed. Over the temperature range 225–355 K the branching ratio $k_a/(k_a+k_b)$ is $(1.51 \pm 0.56)\exp((-218 \pm 93)/T)$, and the overall rate constant is $(3.25 \pm 0.50) \times 10^{-12} \exp((-114 \pm 38)/T)$. At 298 K, $k_a/(k_a+k_b) = 0.73$, $k_b/(k_a+k_b) = 0.27$ and $k_a+k_b = 2.2 \times 10^{-12}$ cm^3 molecule^{-1} s^{-1}. We recommend this result, which is consistent with data obtained by Simon *et al.*[136], Kenner, Ryan and Plumb[59], and DeMore[137].

5.4 THE REACTION OF NO_3 WITH CH_3O_2

There have been three studies of this reaction, all of which show that the reaction proceeds to give CH_3O and NO_2:

$$CH_3O_2 + NO_3 \longrightarrow CH_3O + NO_2 + O_2$$

We recommend a rate constant of 1.5×10^{-12} cm^3 molecule^{-1} s^{-1} at 298 K, derived by averaging the three determinations given in Table 6. We estimate the uncertainty of this rate constant to be ± 50%.

5.5 THE REACTION OF CH_3 WITH CH_3O_2

This is a secondary reaction in experimental studies of CH_3O_2, and can be important under conditions of low-O_2 or high-CH_3 radical concentrations:

$$CH_3 + O_2 + M \longrightarrow CH_3O_2 + M$$

$$CH_3O_2 + CH_3 \longrightarrow 2CH_3O$$

In the pioneering work of Parkes[142] it was noted that the CH_3O_2 yield in the photolysis of azomethane/O_2 mixtures decreased as the O_2 concentration was lowered. This observation was ascribed to the reaction of CH_3O_2 with CH_3 radicals and a rate constant of $(6.0 \pm 1.0) \times 10^{-11}$ cm^3 molecule^{-1} s^{-1} was deduced at 298 K. More recently, Pilling and coworkers [143,145] have used laser flash photolysis techniques to study this reaction over the temperature range 298–582 K, deriving $k = 8.4 \times 10^{-12} \exp(170/T)$ cm^3 molecule^{-1} s^{-1} [143], which is somewhat higher than the value reported Parkes [142].

5.6 THE REACTION OF CF_3O WITH CF_3O_2

Numerous recent product studies of the simulated atmospheric oxidation of hydrofluorocarbons containing a CF_3 group (e.g. CF_3H, CF_3CFH_2, and CF_3CH_3) have reported the formation of the trioxide, $CF_3O_3CF_3$ [151–154]. The only reasonable source of this trioxide in these experiments is the association reaction of CF_3O_2 with CF_3O radicals. The observation of 100% conversion of CF_3H into $CF_3O_3CF_3$ following irradiation of $CF_3H/F_2/O_2$ mixtures, reported by Nielsen *et al.* [144], shows that only one channel is important:

$$CF_3O_2 + CF_3O + M \longrightarrow CF_3O_3CF_3 + M$$

Nielsen *et al.* [144] modelled the formation of $CF_3O_3CF_3$ and derived a lower limit for the rate constant of 2.15×10^{-11} cm^3 molecule^{-1} s^{-1}.

5.7 THE REACTION OF CF_3O WITH CF_3CFHO_2

In a study of the Cl-atom initiated oxidation of CF_3CFH_2 in air Sehested and Wallington [155] reported the observation of a short-lived product which has infrared spectroscopic features consistent with the trioxide, $CF_3CFHO_3CF_3$. This species decomposed in the dark with a lifetime of ca. 10 min. By analogy with the reaction of CF_3O_2 and CF_3O radicals it was proposed that CF_3CFHO_2 radicals add to CF_3O to form a trioxide which is unstable with respect to decomposition at 298 K. No kinetic data were obtained.

5.8 THE REACTION OF CH_3O WITH CH_3O_2

As discussed by Lightfoot *et al.* [2], there are many erroneous reports in the literature concerning this reaction. It was originally assumed to be the source hydroperoxide observed following the CH_3O_2 self-reaction:

$$CH_3O_2 + CH_3O_2 \longrightarrow CH_3O + CH_3 + O_2$$

$$CH_3O + CH_3O_2 \longrightarrow CH_3OOH + HCHO$$

It is now well established that the source of CH_3OOH in this system is the reaction of CH_3O_2 with HO_2 radicals:

$$CH_3O_2 + CH_3O_2 \longrightarrow CH_3O + CH_3O + O_2$$

$$CH_3O + O_2 \longrightarrow HCHO + HO_2$$

$$CH_3O_2 + HO_2 \longrightarrow CH_3OOH + O_2$$

The only reliable evidence for the occurrence of the $CH_3O_2 + CH_3O$ reaction comes from the laser flash photolysis study of Pilling and Smith [145], in which it was found that a rapid reaction of CH_3O_2 with CH_3O was necessary to explain the observed rate of loss of CH_3O_2 radicals. Pilling and Smith [145] obtained a rate constant of 1.0×10^{-11} cm^3 molecule^{-1} s^{-1} at 298 K and suggested that the reaction could proceed via two channels:

$$CH_3O_2 + CH_3O + M \longrightarrow CH_3O_3CH_3 + M \quad (a)$$

$$CH_3O_2 + CH_3O \longrightarrow CH_2O_2 + CH_3OH \quad (b)$$

Based upon the behaviour of the reactions of CF_3O_2 and CF_3CFHO_3 with CF_3O radicals it seems likely that channel (a) dominates. It should be noted that unlike $CF_3O_3CF_3$, the trioxide $CH_3O_3CH_3$ is unstable under ambient conditions and decomposes to reform CH_3O_2 and CH_3O radicals.

5.9 THE REACTION OF O_3 WITH CH_3O_2

Simonaitus and Heicklen [146] studied the rate of ozone loss following the 253.7 nm irradiation of O_3/O_2 mixtures both with and without added CH_4. The loss of ozone was unaffected by the presence of CH_4 and an upper limit of 2.4×10^{-17} cm^3 molecule^{-1} s^{-1} was established at 298 K. This upper limit does not preclude this reaction from playing a minor role in atmospheric chemistry. Further studies are needed.

5.10 THE REACTION OF SO_2 WITH CH_3O_2

Early studies of the kinetics of this reaction which suggested a rate constant of 1×10^{-14} cm^3 molecule^{-1} s^{-1} [53,156] have been superseded by more direct investigations which observe no reaction between these species. Based upon the work of Sander and Watson [147] it is now accepted that $k(CH_3O_2 + SO_2) < 5 \times 10^{-17}$ cm^3 molecule^{-1} s^{-1} (i.e. there is no discernible reaction).

5.11 THE REACTION OF CO WITH CH_3O_2

Sander and Watson [54] used a flash photolysis UV absorption technique to study the self-reaction of CH_3O_2 radicals. Addition of CO to the reaction mixtures *lowered* the rate of CH_3O_2 loss. From this observation, Sander and Watson were able to derive an upper limit of k of 7×10^{-18} cm^3 molecule^{-1} s^{-1} for the CH_3O_2 + CO reaction. The decrease in the observed CH_3O_2 rate loss when CO is present can be explained if CO causes regeneration of CH_3O_2 radicals. The most likely regeneration mechanism is a reaction of CO with CH_3O radicals to given CH_3 radicals, which will then add O_2 to give CH_3O_2:

$$CH_3O_2 + CH_3O_2 \longrightarrow CH_3O + CH_3O$$

$$CH_3O + CO \longrightarrow CH_3 + CO_2$$

$$CH_3 + O_2 + M \longrightarrow CH_3O_2 + M$$

The mechanism by which the reaction of CH_3O with CO gives CH_3 could either be a direct O-atom transfer or could involve formation of the CH_3OCO intermediate. It is known that CF_3O radicals add to CO to give CF_3OCO radicals [157,158].

5.12 THE REACTION OF Cl WITH CCl_3O_2

Hautecloque [159] was the first to suggest the importance of this reaction as a source of ClO radicals in experimental studies of the Cl-atom initiated oxidation $CHCl_3$. In a study of the formation of CCl_3O_2 and its thermal decomposition, Russell *et al.*[148] derived an estimate of $k = (2 \pm 1) \times 10^{-10}$ cm^3 molecule^{-1} s^{-1} at 298 K.

5.13 THE REACTION OF Cl WITH CF_2ClO_2

Dagaut, Wallington and Kurylo [149] conducted a study of the UV spectrum of the CF_2ClO_2 + Cl reaction by using flash photolysis of $CF_2ClH/Cl_2/O_2$ mixtures. In their study a broad UV absorption spectrum matching that of ClO radicals was observed which was attributed to a fast reaction of Cl with

CF_2ClO_2. It was estimated that $k = (2 \pm 1) \times 10^{-10}$ cm^3 molecule^{-1} s^{-1} at 298 K.

5.14 THE REACTION OF CCl_3 WITH CCl_3O_2

Russell *et al.* [148] have reported $k = (1.0^{+1.0}_{-0.5}) \times 10^{-12}$ cm^3 molecule^{-1} s^{-1} for the reaction:

$$CCl_3O_2 + CCl_3 \longrightarrow CCl_3O + CCl_3O$$

5.15 THE REACTION OF I WITH CF_3O_2

Clemitshaw and Sodeau have shown that CF_3I is converted into COF_2 in the 253.7 nm irradiation of CF_3I/O_2 mixtures and suggested that this occurs via reaction of CF_3O_2 radicals with I atoms to give CF_3OOI, which then decomposes to give COF_2 and FOI [160]. These authors have used low-temperature matrix isolation experiments with FTIR spectroscopy to directly identify CF_3OOI as an intermediate in this system [161].

DISCUSSION

As shown in Table 6 peroxy radicals react with a variety of other radical species, but no reactions have been reported between peroxy radicals and closed-shell species at ambient temperatures. For the majority of reactions only one study has been performed and clearly further studies are required. The reaction of peroxy radicals with alkoxy radicals (e.g. $CH_3O_2 + CH_3O$) has been given relatively little attention and should be studied further as these reactions may be important in kinetic studies of peroxy radical self-reactions. The reaction of CH_3O_2 radicals with O_3 is potentially of atmospheric importance and deserves further study.

6 CONCLUSIONS

There now exists a substantial database concerning the reactions of organic peroxy radicals in the gas phase. Tables 7–9 list our recommended kinetic data for reactions of peroxy radicals with HO_2 radicals, NO and NO_2. The data obtained for reactions with miscellaneous species are given in Table 6. Despite the progress made over the past few years, there are still significant gaps in our understanding of peroxy radical reactions. Of particular relevance to computer modelling of hydrocarbon oxidation in the atmosphere are uncertainties in the kinetics and mechanisms of the reactions with HO_2 radicals and NO_X. Additional experiments coupled with *ab initio* computational studies are needed in several areas. Product data and hence

Table 7 Recommended Kinetic Parameters for the Reactions of RO_2 and HO_2 Radicals

Reaction	k (cm^3 $molecule^{-1}$ s^{-1})	Temperature (K)
$HO_2 + HO_2 \longrightarrow H_2O_2 + O_2$	$\{2.2 \times 10^{-13} \exp(600/T) + 1.9 \times 10^{-33}[N_2]\exp(980/T)\} \times \{1 + 1.4 \times 10^{-21}\exp(2200/T)[H_2O]\}$	200–500
	$6.97 \times 10^{-10} \exp(-6030/T) + 2.16 \times 10^{-13} \exp(820/T)$	500–1100
$CH_3O_2 + HO_2 \longrightarrow CH_3OOH + O_2$	$2.9 \times 10^{-13} \exp(860/T)$	248–678
$CD_3O_2 + HO_2 \longrightarrow CD_3OOH + O_2$	5.4×10^{-12}	300
$C_2H_5O_2 + HO_2 \longrightarrow C_2H_5OOH + O_2$	$4.3 \times 10^{-13} \exp(850/T)$	210–480
neo-$C_5H_{11}O_2 + HO_2 \longrightarrow$ products	$2.3 \times 10^{-13} \exp(1270/T)$	248–364
cyclo-$C_5H_9O_2 + HO_2 \longrightarrow$ products	$2.3 \times 10^{-13} \exp(1270/T)$	248–364
cyclo-$C_6H_{11}O_2 + HO_2 \longrightarrow$ products	$2.3 \times 10^{-13} \exp(1270/T)$	248–364
$C_6H_5CH_2O_2 + HO_2 \longrightarrow$ products	$3.8 \times 10^{-13} \exp(980/T)$	273–447
$HO_2 + HOCH_2O_2 \longrightarrow$ products	$5.6 \times 10^{-15} \exp(2300/T)$	275–333
$CH_2ClO_2 + HO_2 \longrightarrow$ products	$3.3 \times 10^{-13} \exp(822/T)$	255–588
$CF_3CFHO_2 + HO_2 \longrightarrow$ products	$1.8 \times 10^{-13} \exp(910/T)$	210–363
$CH_3C(O)O_2 + HO_2 \longrightarrow$ products	$4.3 \times 10^{-13} \exp(1040/T)$	253–368

Table 8 Recommended Kinetic Parameters for the Reactions of RO_2 Radicals with NO

Peroxy radicals	k (cm^3 molecule^{-1} s^{-1} s^{-1})	Temperature (K)
HO_2	3.7×10^{-12} exp(240/T)	232–1271
CH_3O_2	3.1×10^{-12} exp(260/T)	240–370
$C_2H_5O_2$	8.5×10^{-12}	298
$(CH_3)_2CHO_2$	4.3×10^{-12}	298
$(CH_3)_2CO_2$	4.0×10^{-12}	298
$(CH_3)_3CCH_2O_2$	4.7×10^{-12}	298
$(CH_3)_3C(CH_3)_2CCH_2O_2$	1.8×10^{-12}	298
$CH_3C(O)O_2$	1.9×10^{-11}	283–328
CF_2ClO_2	1.6×10^{-11} (T/298)$^{-1.5}$	230–430
$CFCl_2O_2$	1.5×10^{-11} (T/298)$^{-1.3}$	230–430
CCl_3O_2	1.7×10^{-11} (T/298)$^{-1.0}$	230–430
CF_3O_2	1.5×10^{-11} (T/298)$^{-1.2}$	230–430
CF_3CFHO_2	1.2×10^{-11} (T/298)$^{-1.2}$	290–324
$FC(O)O_2$	2.5×10^{-11}	298
$CF_3C(O)O_2$	4.0×10^{-12} exp(563/T)	220–324
$CF_3CH_2O_2$	1.2×10^{-11}	298
CH_2FCFHO_2	$> 8.7 \times 10^{-12}$	298
$CHF_2CF_2O_2$	$> 8.4 \times 10^{-12}$	298
$CF_3CF_2O_2$	$>9.2 \times 10^{-12}$	298
$CF_2ClCH_2O_2$	1.3×10^{-11}	298
$CFCl_2CH_2O_2$	1.2×10^{-11}	298
$CF_3CHO_2(^{\cdot})CF_3$	1.1×10^{-11}	298
$CH_3CF_2CH_2O_2$	8.5×10^{-12}	298
$CCl_3CH_2O_2$	$< 6.1 \times 10^{-12}$	295
CF_3CClHO_2	1.0×10^{-11}	298
$CH_3SCH_2O_2$	1.9×10^{-11}	298
$CH_3OCH_2O_2$	9.1×10^{-12}	298
$HO(CH_3)_2CCH_2O_2$	4.9×10^{-12}	298

mechanism information concerning reactions with HO_2 radicals are needed as a function of temperature for structurally diverse peroxy radicals. Kinetic experiments are needed to confirm the results from the limited experiments performed thus far, which suggest that the reactivity of peroxy radicals with HO_2 is increased by electron-donating and decreased by electron-withdrawing substituents. Work is needed to confirm the observed decrease in reactivity of peroxy radicals towards NO with increasing size of the peroxy radical. Further experiments are needed to better define the temperature dependence of reactions with NO. Direct studies of the atmospherically important $CH_3C(O)O_2 + NO$ reaction are needed. Nitrate yields in the

Table 9 Recommended Kinetic Parameters for Reactions of RO_2 Radicals and NO_2

Peroxy radicals	k_o (cm^6 molecule^{-2} s^{-2})	k_∞ (cm^3 molecule^{-1} s^{-1})	F_c
HO_2	$1.8 \times 10^{-31} (T/300)^{-3.2}$	$4.7 \times 10^{-12} (T/300)^{-1.4}$	0.6
CH_3O_2	$2.3 \times 10^{-30} (T/300)^{-2.5}$	$8.0 \times 10^{-12} (T/300)^{-3.5}$	0.4
$C_2H_5O_2$	4.8×10^{-29}	1.0×10^{-11}	0.3
$CH_3C(O)O_2$	$2.0 \times 10^{-28} (T/298)^{-7.1}$	$1.1 \times 10^{-11} (T/298)^{-0.9}$	0.3
$CF_2Cl_2O_2$	3.5×10^{-29}	5.2×10^{-12}	0.6
$CFCl_2O_2$	$3.5 \times 10^{-29} (T/300)^{-5}$	$6.0 \times 10^{-12} (T/300)^{-2.5}$	0.6
CCl_3O_2	$5.0 \times 10^{-29} (T/300)^{-5}$	$6.0 \times 10^{-12} (T/300)^{-2.5}$	0.6
CF_3O_2	$2.2 \times 10^{-29} (T/300)^{-5}$	$6.0 \times 10^{-12} (T/300)^{-2.5}$	0.6
$HO(CH_3)_2CCH_2O_2$		6.7×10^{-12}	
$CH_3OCH_2O_2$		7.9×10^{-12}	
CF_3CFHO_2		5.0×10^{-12}	
$CF_3CH_2O_2$		5.8×10^{-12}	
$CF_3C(O)O_2$		6.6×10^{-12}	
$FC(O)O_2$		5.5×10^{-12}	
$CH_3CF_2CH_2O_2$		6.8×10^{-12}	
$CCl_3CH_2O_2$		6.5×10^{-12}	
$CH_3SCH_2O_2$		9.2×10^{-12}	
CF_3CClHO_2		6.4×10^{-12}	

reactions of oxygenated peroxy radicals with NO need to be quantified as a function of temperature and pressure. Finally, additional experiments would be useful to better define the pressure and temperature dependence of reactions with NO_2.

REFERENCES

1. T.J. Wallington, P. Dagaut and M.J. Kurylo, *Chem. Rev.* **92,** 667 (1992).
2. P.D. Lightfoot, R.A. Cox, J.N. Crowley, M. Destriau, G.D. Hayman, M.E. Jenkin, G.K. Moortgat and F. Zabel, *Atmos. Environ.* **26A,** 1805 (1992).
3. R. Atkinson, *J. Phys. Chem. Ref. Data*, Monograph No. 2, (1994).
4. T.J. Wallington and S.M. Japar, *Chem. Phys. Lett.* **167,** 513 (1990).
5. T.J. Wallington, *J. Chem. Soc. Faraday Trans.* **87,** 2379 (1991).
6. M.E. Jenkin, R.A. Cox, G.D. Hayman and L.J. Whyte, *J. Chem. Soc. Faraday Trans 2* **84**, 913 (1988).
7. M.E. Jenkin, private communication.
8. T.J. Wallington and M.D. Hurley, *Chem. Phys. Lett.* **193**, 84 (1992).
9. T.J. Wallington and S.M. Japar, *Chem. Phys. Lett.* **166**, 495 (1990).
10. D.M. Rowley, R. Lesclaux, P.D. Lightfoot, K. Hughes, M.D. Hurley, S. Rudy and T.J. Wallington, *J. Phys. Chem.* **96,** 7043 (1992).
11. D.M. Rowley, R. Lesclaux, P.D. Lightfoot, B. Nozière, T.J. Wallington and M.D. Hurley, *J. Phys. Chem.* **96,** 4889 (1992).
12. J.P. Burrows, G.K. Moortgat, G.S. Tyndall, R.A. Cox, M.E., Jenkin, G.D., Hayman and B. Veyret, *Phys. Chem.* **93,** 2375 (1989).
13. T.J. Wallington, M.D. Hurley, W.F. Schneider, J. Sehested and O.J. Nielsen,

Chem. Phys. Lett. **218,** 34 (1994).
14. T.J Wallington, M.D. Hurley and W.F. Schneider, *Chem. Phys. Lett.* (1996).
15. M.M. Maricq, J.J. Szente, M.D. Hurley and T.J. Wallington, *J. Phys. Chem.* **98,** 8962 (1994).
16. T.J. Wallington, M.D. Hurley, J.C. Ball and M.E. Jenkin, *Chem. Phys. Lett.* **211,** 41 (1993).
17. H. Niki, P.D. Maker, C.M. Savage and L.P. Breitenbach, *J. Phys. Chem.* **89,** 588 (1985).
18. G.K. Moortgat, B. Veyret and R. Lesclaux, *Chem. Phys. Lett.* **160,** 443 (1989).
19. O. Horie and G.K Moortgat, *J. Chem. Soc. Faraday Trans.* **88,** 3305 (1992).
20. W.B. DeMore, S.P. Sander, D.M. Golden, R.F. Hampson, M.J. Kurylo, C.J. Howard, A.R. Ravishankara and C.E. Kolb, *Chemical Kinetics and Photochemical Data for Use in Stratospheric Modelling Evaluation Number 11*, NASA-JPL Publication 94–26, (1994).
21. R. Atkinson, D.L. Baulch, R.A. Cox, R.F. Hampson, J.A. Kerr and J. Troe, *J. Phys. Chem. Ref.* Data **881,** (1989).
22. J. Troe, *Ber. Bunsenges. Phys. Chem.*, **73,** 946 (1969).
23. P.D. Lightfoot, B. Veyret and R. Lesclaux, *Chem. Phys. Lett.* **150,** 120 (1988).
24. H. Hippler, J. Troe, J. Willner, *J. Chem. Phys.*, **93,** 1755 (1990).
25. R.A. Cox and G.S. Tyndall, *Chem. Phys. Lett.* **65,** 357 (1979).
26. R.A. Cox and G.S. Tyndall, *J. Chem. Soc. Faraday Trans.* **76,** 153 (1980).
27. C.S. Kan, J.G. Calvert and J.H. Shaw, *J. Phys. Chem.* **84,** 3411 (1980).
28. G.K. Moortgat, R.A. Cox, G. Schuster, J.P. Burrows and G.S. Tyndall, *J. Chem. Soc. Faraday Trans.* **85,** 809 (1989).
29. K. McAdam, B. Veyret and R. Lesclaux, *Chem. Phys. Lett.* **133,** 39 (1987).
30. M.J. Kurylo, P. Dagaut, T.J. Wallington and D.M. Neuman, *Chem. Phys. Lett.* **139,** 513 (1987).
31. P. Dagaut, T.J. Wallington and M.J. Kurylo, *J. Phys. Chem.* **92,** 3833 (1988).
32. P.D. Lightfoot, B. Veyret and R. Lesclaux, *J. Phys. Chem.* **94,** 708 (1990).
33. P.D. Lightfoot, P. Roussel, F. Caralp, R. Lesclaux, *J. Chem. Soc. Faraday Trans.* **87,** 3213 (1991).
34. F.C. Cattell, J. Cavanagh, R.A. Cox and M.E. Jenkin, *J. Chem. Soc. Faraday Trans. 2*, **82,** 1999 (1986).
35. P. Dagaut, T.J. Wallington and M.J. Kurylo, *J. Phys Chem.* **92,** 3836 (1988).
36. F.F. Fenter, V. Catoire, R. Lesclaux and P.D. Lightfoot, *J. Phys. Chem.* **97,** 3530 (1993).
37. M.M. Maricq and J.J. Szente, *J. Phys. Chem.* **98,** 2078 (1994).
38. B. Nozière, R. Lesclaux, M.D. Hurley, M.A. Dearth and T.J. Wallington, *J. Phys. Chem.* **98,** 2864 (1994).
39. V. Catoire, R. Lesclaux, P.D. Lightfoot and M.T. Rayez, *J. Phys. Chem.* **98,** 2889 (1994).
40. E.Villenave and R. Lesclaux, *Chem. Phys. Lett.* **236,** 376 (1995).
41. G.D. Hayman and F. Battin-Leclerc, *J. Chem. Soc. Faraday Trans.* **91,** 1313 (1995).
42. B. Veyret, R. Lesclaux, M.-T. Rayez, J.-C. Rayez, R.A. Cox and G.K. Moortgat, *J. Phys. Chem.* **93,** 2368 (1989).
43. M.E. Jenkin and R.A. Cox, *J. Phys. Chem.* **95,** 3229 (1990).
44. G.K. Moortgat, B. Veyret and R. Lesclaux, *Chem. Phys. Lett.* **160,** 443 (1989).
45. I. Bridier, B. Veyret, R. Lesclaux and M.E. Jenkin, *J. Chem. Soc. Faraday Trans.* **89,** 2993 (1993).
46. G.K. Moortgat, J.P. Burrows, W. Schneider, G.S. Tyndall and R.A. Cox, *Proceedings of the 4th European Symposium on the Physical and Chemical*

Behaviour of Atmospheric Pollutants, edited by G. Angeletti and G. Restelli, Reidel, Dordrecht, The Netherlands (1987).
47. R. Patrick, J.R. Barker and D.M. Golden, *J. Phys. Chem.* **88,** 128 (1984).
48. G.A. Russell, *J. Am. Chem. Soc.* **79,** 3871 (1957).
49. R. Atkinson, W.P.L. Carter and A.M. Winer, *J. Phys. Chem.* **87,** 2012 (1983).
50. C. Anastasi, I.W.M. Smith and D.A. Parkes, *J. Chem. Soc. Faraday Trans. 1* **74,** 1693 (1978).
51. I.C. Plumb, K.R. Ryan, J.R. Steven and M.F.R. Mulcahy, *Chem. Phys. Lett.* **63,** 255 (1979).
52. H. Adachi and N. Basco, *Chem. Phys. Lett.* **63,** 490 (1979).
53. R. Simonaitis and J. Heicklen, *Chem. Phys. Lett.* **65,** 361 (1979).
54. S.P. Sander and R.T. Watson, *J. Phys. Chem.* **84,** 1664 (1980).
55. A.R. Ravishankara, F.L. Eisele, N.M. Kreutter and P.H. Wine, *J. Chem. Phys.* **74,** 2267 (1981).
56. R. Simonaitis and J. Heicklen, *J. Phys. Chem.* **85,** 2946 (1981).
57. I.C. Plumb, K.R. Ryan, J.R Steven and M.F.R. Mulcahy, *J. Phys. Chem.* **85,** 3136 (1981).
58. R. Zellner, B. Fritz and K. Lorenz, *J. Atmos. Chem.* **4,** 241 (1986).
59. R.D. Kenner, K.R. Ryan and I.C. Plumb, *Geophys. Res. Lett.* **20,** 1571 (1993).
60. J. Sehested, O.J. Nielsen and T.J. Wallington, *Chem. Phys. Lett.* **213,** 457 (1993).
61. P.W. Villalta, G. Huey and C.J. Howard, *J. Phys. Chem.* **99,** 12829 (1995).
62. H. Adachi and N. Basco, *Chem. Phys. Lett.* **64,** 431 (1979).
63. I.C. Plumb, K.R. Ryan, J.R. Stevens and M.F.R. Mulcahy, *Int. J. Chem. Kinet.* **14,** 183 (1982).
64. V. Däele, A. Ray, I. Vassalli, G. Poulet and G. Le Bras, *Int. J. Chem. Kinet.* **28,** 1121 (1995).
65. H. Adachi and N. Basco, *Int. J. Chem. Kinet.* **14,** 1243 (1982).
66. J. Peeters, J. Vertommen and I. Langhans, *Ber. Bunsenges, Phys. Chem.* **96,** 431 (1993).
67. I.C. Plumb and K.R. Ryan, *Chem. Phys. Lett.* **92,** 236 (1992).
68. A.M. Dognon, F. Caralp and R. Lesclaux, *J. Chim. Phys.* **82,** 349 (1985).
69. J. Sehested and O.J. Nielsen, *Chem. Phys. Lett.* **206,** 369 (1993).
70. T.J. Bevilacqua, D.R. Hanson and C.J. Howard, *J. Phys. Chem.* **97,** 3750 (1993).
71. A.R. Ravishankara, A.A. Turnipseed, N.R. Jensen, S. Barone, M. Mills, C.J. Howard and S. Solomon, *Science* **263,** 75 (1994).
72. A. Bhatnagar and R.W. Carr, *Chem. Phys. Lett.* **231,** 454 (1994).
73. R. Lesclaux and F. Caralp, *Int. J. Chem. Kinet.* **16,** 1117 (1984).
74. K.R. Ryan and I.C. Plumb, *Int. J. Chem. Kinet.* **16,** 591 (1984).
75. T.J. Wallington and O.J. Nielsen, *Chem. Phys. Lett.* **187,** 33 (1991).
76. A. Bhatnagar and R.W. Carr, *Chem. Phys. Lett.* **238,** 9 (1995).
77. J. Peeters and V. Pultau, in *Proceedings of CEC/EUROTRAC Workshop on Chemical Mechanisms Describing Tropospheric Processes*, Edited by J. Peeters, (1992).
78. T.J. Wallington, T. Ellerman, O.J. Nielsen and J. Sehested, *J. Phys Chem.* **98,** 2346 (1994).
79. T.J. Wallington, M. Hurley, O.J. Nielsen and J. Sehested, *J. Phys Chem.* **98,** 5686 (1994).
80. M.M. Maricq, J.J. Szente, G.A. Khitrov and J.S. Francisco, *J. Phys. Chem.* **100,** 4514 (1996).
81 O.J. Nielsen, E. Gamborg, J. Sehested, T.J. Wallington and M.D. Hurley, *J. Phys. Chem.* **98,** 9518 (1994).
82. T.J. Wallington, M.D. Hurley, J.C. Ball, T. Ellerman, O.J. Nielsen and J.

Sehested, *J. Phys. Chem.* **98,** 5435 (1994).
83. T.E. Møgelberg, J. Platz, O.J. Nielsen, J. Sehested and T.J. Wallington, *J. Phys. Chem.* **99,** 5373 (1995).
84. T.E. Møgelberg, O.J. Nielsen, J. Sehested, T.J. Wallington and M.D. Hurley, *J. Phys. Chem.* **99,** 1995 (1995).
85. J. Platz, O.J. Nielsen, J. Sehested and T.J. Wallington, *J. Phys. Chem.* **99,** 6570 (1995).
86. T.E. Møgelberg, O.J. Nielsen, J. Sehested and T.J. Wallington, *J. Phys. Chem.* **99,** 13437 (1995).
87. T.J. Wallington, T. Ellermann and O.J. Nielsen, *J. Phys. Chem.* **97,** 8442 (1993).
88. S. Langer, E. Ljungstrom, T. Ellermann, O.J. Nielsen and J. Sehested, *Chem. Phys. Lett.* **240,** 53 (1995).
89. S. Langer, E. Ljungström, J. Sehested and O.J. Nielsen, *Chem. Phys. Lett.* **226,** 165 (1994).
90. C.T. Pate, B.J. Finlayson and J.N. Pitts Jr, *J. Am. Chem. Soc.* **96,** 6554 (1974).
91. R. Atkinson, S.M. Aschmann, W.P.L. Carter, A.M. Winer and J.N. Pitts Jr., *J. Phys. Chem.* **86,** 4563 (1982).
92. K.H. Becker, H. Geiger and P. Wiesen, *Chem. Phys. Lett.* **184,** 256 (1991).
93. R. Atkinson, W.P.L. Carter and A.M. Winer, *J. Phys. Chem.* **87,** 2012 (1983).
94. C.J. Howard and K.M. Evenson, *Geophys. Res. Lett.* **4,** 437 (1977).
95. M.T. Leu, *J. Chem. Phys.* **70,** 1662 (1979).
96. C.J. Howard, *J. Chem. Phys.* **67,** 5258 (1977).
97. I. Glaschick-Schimpf, A. Leiss, P.B. Monkhouse, U. Schurath, K.H. Becker and E.H. Fink, *Chem. Phys. Lett.* **67,** 318 (1979).
98. W. Hack, A.W. Preuss, F. Temps, H. Wagner Gg and K. Hoyermann, *Int. J. Chem. Kinet.* **12,** 851 (1980).
99. B.A. Thrush and J.P.T. Wilkinson, *Chem. Phys. Lett.* **81,** 1 (1981).
100. J.P. Burrows, D.I. Cliff, B.A. Thrush and J.P.T. Wilkinson, *Proc. Soc. (London) A368*, 463 (1979).
101. V.B. Rozenshtein, Y.M. Gershenzon, S.O. Il'in and O.P. Kishkovitch, *Chem. Phys. Lett.* **112,** 473 (1984).
102. R. Zellner, *J. Chim. Phys. Chim. Biol.* **84,** 403 (1987).
103. R. Atkinson, S.M. Aschmann, W.P.L. Carter, A.M. Winer and J.N. Pitts, Jr, *In. J. Chem. Kinet.* **16,** 1085 (1984).
104. R.A Cox, R.G. Derwent, P.M. Holt and J.A. Kerr, *J. Chem. Soc. Faraday Trans. 1* **72,** 106 (1976).
105. R.A. Cox and M.J. Roffey, *Environ. Sci. Technol.* **11,** 900 (1977).
106. D.G. Hendry and R.A. Kenley, *J. Am. Chem. Soc.* **99,** 3198 (1977).
107. F. Kirchner, F. Zabel and K.H. Becker, *Ber. Bunsen Ges. Phys. Chem.* **94,** 1379 (1990).
108. E.C.Tuazon, W.P.L. Carter and R. Atkinson, *J. Phys. Chem.* **95,** 2434 (1991).
109. K.C. Clemitshaw and J.R. Sodeau, *J. Phys. Chem.* **91,** 3650 (1987).
110. J. Troe, *Ber. Bunsenges. Phys. Chem.* **87,** 161 (1983).
111. R.G. Gilbert, K. Luther and J. Troe, *Ber. Bunsenges. Phys. Chem.* **87,** 169 (1983).
112. H. Adachi and N. Basco, *Int. J. Chem. Kinet.* **12,** 1 (1980).
113. A.R. Ravishankara, F.L. Eiesele and P.H. Wine, *J. Chem. Phys.* **73,** 3743 (1980).
114. H. Adachi and N. Basco, *Chem. Phys. Lett.* **67,** 324 (1979).
115. G. Elfers, F. Zabel and K.H. Becker, *Chem. Phys. Lett.* **168,** 14 (1980).
116. M.C. Addison, J.P. Burrows, R.A. Cox and R. Patrick, *Chem. Phys. Lett.* **73,** 283 (1980).
117. N. Basco and S.S. Parmer, *Int. J. Chem. Kinet.* **19,** 115 (1987).

118. I. Bridier, F. Caralp, H. Loirat, R. Lesclaux, B. Veyret, K.H. Becker, A. Reimer and F. Zabel, *J. Phys. Chem.* **95,** 3594 (1991).
119. F. Caralp, R. Lesclaux, M.T. Rayez, J.C. Rayez and W. Forst, *J. Chem. Soc. Faraday Trans.* 2 **84,** 569 (1988).
120. S.B. Moore and R.W. Carr, *J. Phys. Chem.* **94,** 1393 (1980).
121. T.E. Møgelberg, O.J. Nielsen, J. Sehested, T.J. Wallington, M.D. Hurley and W.F. Schneider, *Chem. Phys. Lett.* **225,** 375 (1994).
122. T.J. Wallington, J. Sehested and O.J. Nielsen, *Chem. Phys. Lett.* **225,** 563 (1994).
123. T.J. Wallington, W.F. Schneider, T.E. Møgelberg, O.J. Nielsen and J. Sehested, *Int. J. Chem. Kinet.* **27,** 391 (1995).
124. O.J. Nielsen, J. Sehested and T.J. Wallington, *Chem. Phys. Lett.* **236,** 385 (1995).
125. M.J. Kurylo and P.A. Ouellette, *J. Phys. Chem.* **90,** 441 (1986).
126. M.J. Kurylo and P.A. Ouellette, *J. Phys. Chem.* **91,** 3365 (1987).
127. S.P. Sander and M. Peterson, *J. Phys. Chem.* **88,** 1566 (1984).
128. R. Simonaitis and J. Heicklen, *Int. J. Chem. Kinet.* **10,** 67 (1978).
129. R.A. Cox and R. Patrick, *Int. J. Chem. Kinet.* **11,** 635 (1979).
130. M.C. Dodge, *J. Geophys. Res.* **94,** 5121 (1989).
131. M. Destriau and J. Troe, *Int. J. Chem. Kinet.* **22,** 915 (1990).
132. N. Washida and K.D. Bayes, *Int. J. Chem. Kinet.* **8,** 777 (1976).
133. R. Zellner, D. Hartmann, J. Karthauser, D. Rhasa and G. Weibring, *J. Chem. Soc. Faraday Trans.* **84,** 549 (1988).
134. M.M. Maricq, J.J. Szente, E.W. Kaiser and J. Shi, *J. Phys. Chem. 98,* 2083 (1994).
135. W. Tsang and R.F. Hampson, *J. Phys. Chem. Ref. Data,* **15,** 1087 (1986).
136. F.G. Simon, J.P. Burrows, W. Schneider, G.K. Moortgat and P. Crutzen, *J. Phys. Chem.* **93,** 7807 (1989).
137. W.B. DeMore, *J. Geophys. Res.* **96,** 4995 (1991).
138. F. Helleis, J.N. Crowley and G.K. Moortgat, *J. Phys. Chem.* **97,** 11464 (1993).
139. J. Crowley, J.P. Burrows, G.K. Moortgat, G. Poulet and G. LeBras, *Int. J. Chem. Kinet.* **22,** 673 (1990).
140. P. Biggs, C.E. Canosa-Mas, J.M. Fracheboud, D.E. Shallcross and R.P. Wayne, *J. Chem. Soc. Faraday Trans.* **90,** 1205 (1994).
141. F. Däele, G. Laverdet, G. Le Bras and G. Poulet, *J. Phys. Chem.* **99,** 1470 (1995).
142. D.A. Parkes, *Int. J. Chem. Kinet.* **9,** 451 (1977).
143. M. Keiffer, M.J. Pilling and M.J.C. Smith, *J. Phys. Chem.* **91,** 6028 (1987).
144. O.J. Nielsen, T. Ellerman, J. Sehested, E. Bartkiewicz, T.J. Wallington and M.D. Hurley, *Int. J. Chem. Kinet.* **24,** 1009 (1992).
145. M.J. Pilling and M.J.C. Smith, *J. Phys. Chem.* **89,** 4713 (1985).
146. R. Simonaitis and J. Heicklen, *J. Phys. Chem.* **79,** 298 (1975).
147. S.P. Sander and R.T. Watson, *Chem. Phys. Lett.* **77,** 473 (1981).
148. J.J. Russell, J.A. Seetula, D. Gutman, F. Danis, F. Caralp, P.D. Lightfoot, R. Lesclaux, C. Melius and S.M. Senkan, *J. Phys Chem.* **94,** 3277 (1990).
149. P. Dagaut, T.J. Wallington and M.J. Kurylo, *J. Photochem. Photobiol. Chem.* **48,** 187 (1989).
150. F. Slemr and P. Warneck, *Int. J. Chem. Kinet.* **9,** 267 (1977).
151. T.J. Wallington, M.D. Hurley, J.C. Ball and E.W. Kaiser, *Environ. Sci. Technol.* **26,** 1318 (1992).
152. E.O. Edney and D.J. Driscoll, *Int. J. Chem. Kinet.* **24,** 1067 (1992).
153. E.C. Tuazon and R. Atkinson, *J. Atmos. Chem.* **17,** 179 (1993).
154. J. Sehested, T. Ellermann, O.J. Nielsen, T.J. Wallington and M.D. Hurley, *Int. J. Chem. Kinet,* **25,** 701 (1993).

155. J. Sehested and T.J. Wallington, *Environ. Sci. Technol.* **27,** 146 (1993).
156. E. Sanhueza, R. Simonaitus and J. Heicklen, *Int. J. Chem. Kinet.* **11,** 907 (1979).
157. T.J. Wallington and J.C. Ball, *J. Phys. Chem.* **99,** 3201 (1995).
158. A.A. Turnipseed, S.B. Barone, N.R. Jensen, D.R. Hanson, C.J. Howard and A.R. Ravishankara, *J. Phys. Chem.* **99,** 6000 (1995).
159. S. Hautecloque, *J. Photochem.* **14,** 157 (1980).
160. K.C. Clemitshaw and J.R. Sodeau, *J. Photochem. Photobiol. Chem.* **A86**, 9 (1995).
161. K.C. Clemitshaw and J.R. Sodeau, *J. Phys. Chem.* **93,** 3552 (1989).

8 Peroxyl Radicals in Aqueous Solutions

CLEMENS VON SONNTAG
Max-Planck-Institut für Strahlenchemie, Mülheim, Germany

and

HEINZ-PETER SCHUCHMANN
Max-Planck-Institut für Strahlenchemie, Mülheim, Germany

1 INTRODUCTION

The reactions of peroxyl radicals in aqueous solutions are attracting increasing interest in the context of environmental studies of aquatic systems, including the reactions in water droplets in atmospheric free-radical chemistry. Peroxyl radicals result from the reaction of free radicals ($R^{\bullet}$) with molecular oxygen (reaction (1)). Although organic free radicals may be formed directly by photolytic processes (reaction (2)), they usually have other, more reactive species as precursors, e.g. the OH radical (see below). For example, in surface waters which practically always contain nitrate ions, solar radiation produces OH radicals via reaction (3) (for mechanistic details, see Refs 1 and 2):

$$R^{\bullet} + O_2 \longrightarrow RO_2^{\bullet} \quad (1)$$

$$R{-}X + h\nu \longrightarrow R^{\bullet} + X^{\bullet} \quad (2)$$

$$NO_3^- + h\nu + H_2O \longrightarrow NO_2^{\bullet} + {}^{\bullet}OH + OH^- \quad (3)$$

OH radicals react at high specific rates with almost any organic compound (perhalogenated alkanes are an exception; for a compilation of rate constants see Ref. 3) by either addition to C=C double bonds or by H-abstraction (e.g. reactions (4) and (5)).

$$CH_2{=}CH_2 + {}^{\bullet}OH \longrightarrow HOCH_2{-}CH_2^{\bullet} \quad (4)$$

$$CH_4 + {}^{\bullet}OH \longrightarrow {}^{\bullet}CH_3 + H_2O \quad (5)$$

At present, the most commonly used oxidant in drinking water and waste water processing is ozone. While many pollutants are ozone-refractory (for ozone reaction rate constants, see Ref. 4), the high reactivity of the OH radical enables it to degrade such materials as well. There are a number of

Peroxyl Radicals. Edited by Z. B. Alfassi

pathways that furnish OH radicals: the photolysis of hydrogen peroxide (reaction (6)), the reaction of ozone with hydrogen peroxide (reaction (7)), the photolysis of ozone (reaction (8)), the reaction of hydrogen peroxide with Fe(ii) in acid media (the Fenton reaction (9)), and ionizing radiation (reaction (10)). These OH-radical-based technologies are often summarized under the term *advanced oxidation processes*. For a detailed discussion of the primary processes see Refs 5 and 6:

$$H_2O_2 + h\nu \longrightarrow 2^{\bullet}OH \tag{6}$$

$$H_2O_2 + 2O_3 \longrightarrow 2^{\bullet}OH + 3O_2 \tag{7}$$

$$H_2O_2 \rightleftharpoons HO_2^- + H^+ \ (pK_a = 11.7) \tag{7a}$$

$$HO_2^- + O_3 \longrightarrow HO_2^{\bullet} + O_3^{\bullet -} \tag{7b}$$

$$O_3^{\bullet -} + H^+ \rightleftharpoons HO_3^{\bullet} \ (pK_a = 6.15) \tag{7c}$$

$$HO_3^{\bullet} \longrightarrow {}^{\bullet}OH + O_2 \tag{7d}$$

$$HO_2^{\bullet} \rightleftharpoons O_2^{\bullet -} + H^+ \ (pK_a = 4.8) \tag{7e}$$

$$O_2^{\bullet -} + O_3 \longrightarrow O_2 + O_3^{\bullet -} \tag{7f}$$

$$3O_3 + h\nu + H_2O \longrightarrow 2^{\bullet}OH + 4O_2 \tag{8}$$

$$O_3 + h\nu \longrightarrow O + O_2 \tag{8a}$$

$$O + H_2O \longrightarrow H_2O_2 \tag{8b}$$

$$H_2O_2 + Fe^{2+} + H^+ \longrightarrow {}^{\bullet}OH + Fe^{3+} + H_2O \tag{9}$$

$$H_2O + \text{ionizing radiation} \longrightarrow {}^{\bullet}OH, e^-_{aq}, H^{\bullet}, H^+, H_2, H_2O_2 \tag{10}$$

Radiation techniques provide the most powerful tool for studying these aqueous-solution oxidation processes, since they allow the generation of OH radicals without the use of reactive additives, and most of our present knowledge of peroxyl radical reactions in aqueous solutions is based on radiolytic studies [7]. For this reason it is appropriate here to recall the principal reactions. The solvated electron formed in reaction (10) can be converted by N_2O into further OH radicals (reaction (11)). Saturation of a solution of a given solute which is present at low concentrations (typically 10^{-3} mol dm^{-3}) with a 4:1 mixture of N_2O and O_2 at atmospheric pressure converts the solvated electron almost completely into OH radicals and at the same time provides oxygen at a concentration sufficient to convert the OH-radical-induced solute radicals and the H atoms (from reaction (10)) into the corresponding peroxyl radicals (reactions (1) and (12)). The $HO_2^{\bullet}$ radicals are in equilibrium with their conjugated base, the superoxide radical $O_2^{\bullet -}$(cf. reaction (7e); pK_a($HO_2^{\bullet}$) = 4.8) [8]:

$$e^-_{aq} + N_2O \longrightarrow {}^{\bullet}OH + N_2 + OH^- \quad (11)$$

$$H^{\bullet} + O_2 \longrightarrow HO_2^{\bullet} \quad (12)$$

Under such conditions the radiation-chemical yield $G(RO_2^{\bullet})$ [9] of solute peroxyl radicals is 5.8×10^{-7} mol J^{-1}, while $HO_2^{\bullet}/O_2^{\bullet -}$ radicals are formed with a G value of 0.6×10^{-7} mol J^{-1}. Product studies are usually carried out at the low dose rates of γ-radiolysis, while for kinetic studies the ionizing radiation can be delivered at times $\leqslant 1\ \mu s$ (pulse radiolysis; for reviews, see Refs 10–12). The peroxyl radicals can also be generated by using pulse photolysis techniques (cf. Ref. 13).

The chemistry of peroxyl radicals in aqueous solution may be considerably different from that encountered in the gas phase or in organic solvents. This is not only reflected in the very different behaviour of e.g. $O_2^{\bullet -}$ [14–18] (see also Chapter 13 of this book), but also in the fact that the polar and hydrogen-bonding solvent water can mediate reaction channels different from those followed by the peroxyl radicals in the gas phase or in organic solvents; these aqueous pathways are often not as clearly understood (cf. the bimolecular decay of cyclopentylperoxyl radicals [19, 20], see below). In addition, oxyl radicals which are typical products of peroxyl radical decay often show different reactions and different specific rates in these different environments (see below). It is worth emphasizing that the exceptional strength of the H–OH bond makes water uniquely inert against free-radical attack, and therefore it is a very useful solvent for studying the chemistry of solute radicals.

This present chapter is devoted to a discussion of the principal general reaction paths of the formation and decay of peroxyl radicals. Some individual systems are also presented to exemplify certain mechanistic aspects in more detail.

2 FORMATION OF PEROXYL RADICALS AND THEIR OBSERVATION

Spectroscopic detection of reactive intermediates is the most desirable monitoring technique when carrying out fast kinetic studies. The UV/VIS absorption spectra of a given radical and its corresponding peroxyl radical are often quite different. The situation frequently encountered is that the parent radical has a well-characterized absorption in the UV/VIS region while the corresponding peroxyl radical absorbs more weakly in the UV with a lack of significant spectral structure. A case in point is the radical derived from

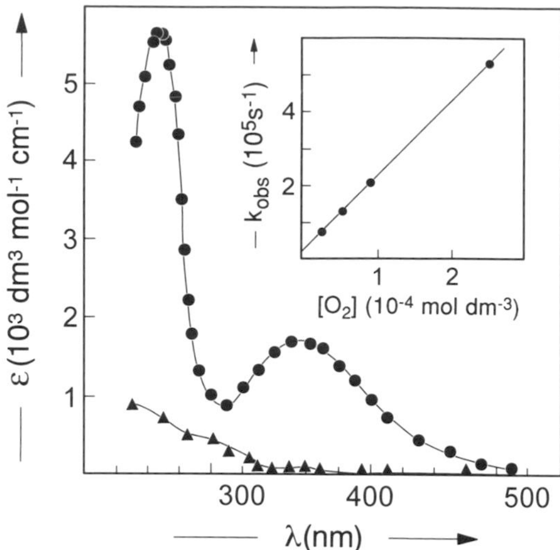

Figure 1 Pulse radiolysis of glycine anhydride in aqueous solution (10^{-3} mol dm^{-3}, pH 6.5): (●) UV/VIS spectrum of the glycine anhydride radical **1**, 6.5 μs after the pulse (N_2O-saturation); (▲) spectrum of the glycine anhydride peroxyl radical **2**, 15 μs after the pulse (N_2O/O_2 (4/1)-saturation). Inset: observed rate constant k_{13} of the disappearance of **1** as a function of the oxygen concentration, measured at 360 nm [7]. Reproduced by permission of VCH Verlagsgesellschaft mbH.

glycine anhydride **1** and its peroxyl radical **2** (Figure 1). This dramatic change in absorption is due to the loss of mesomery in the capto-dative [21, 22] glycine anhydride radical **1** when the peroxyl radical is formed (reaction (13)).

$$\mathbf{1} \xrightarrow[(13)]{O_2} \mathbf{2}$$

With radicals that are less conjugated and hence only absorb in the UV region, the difference between their absorption and that of their corresponding peroxyl radical is not quite so pronounced (e.g. Ref. 23).

The inverse case where the parent radical exhibits practically no absorption in the accessible wavelength region while the peroxyl radical absorbs in the near-UV or even in the visible, is rare. This situation has so far only been encountered in vinyl and aryl radicals (e.g. **3**) and their corresponding peroxyl radicals (e.g. **4**, cf. reaction (14)), and for the thiyl/thiylperoxyl system (for the latter, see Chapter 14 of this book).

$$\underset{\mathbf{3}}{C_6H_5^{\bullet}} \xrightarrow[(14)]{O_2} \underset{\mathbf{4}}{C_6H_5{-}O{-}O^{\bullet}}$$

Vinyl and phenyl are σ-radicals and hence the unpaired electron barely interacts with the π-system. This causes their absorption bands to largely coincide with those of the parent compounds and the absorption of the latter precludes easy detection of these intermediates. On the other hand, the corresponding peroxyl radicals undergo a charge-transfer interaction between the unpaired electron of the peroxyl radical function and the π-system, which then gives rise to a long-wavelength absorption band [24]. An example is shown in Figure 2. The absorption maxima and molar absorption coefficients of several such peroxyl radicals are compiled in Table 1.

The spectral differences between the parent radical and the corresponding peroxyl radical allows the determination of the rate constant of the oxygen addition reaction in a laser flash or pulse radiolysis experiment. In the case where the addition of oxygen is practically irreversible and the oxygen concentration is much higher than the radical

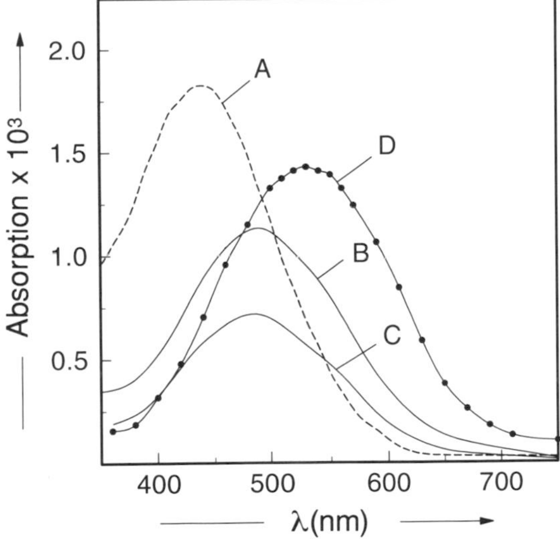

Figure 2 Absorption spectra of halogenated vinylperoxyl radicals in the visible region: (A) vinylperoxyl; (B) 2-chlorovinylperoxyl; (C) 1-chlorovinylperoxyl; (D) 1,2-dichlorovinylperoxyl/1,1-dichlorovinylperoxyl (4/1) [25]. Reproduced by permission of VCH Verlagsgesellschaft mbH.

Table 1 Absorption Maxima and Molar Absorption Coefficients of Some Vinylperoxyl and Arylperoxyl Radicals in Aqueous Solution [25–31]

Peroxyl radial	λ_{max} (nm)	ϵ ($dm^3\,mol^{-1}\,cm^{-1}$)
Vinyl	440	1100
1-Chlorovinyl	480	530
2-Chlorovinyl	480	890[a], 900[b]
Dichlorovinyl[c]	540	1100
Trichlorovinyl	580	1300
Phenyl	490	1600
4-Methoxyphenyl	600	2100
4-Cyanophenyl	490	1300
2-Carboxyphenyl	510, 520	154, 1600
3-Hydroxyphenyl	520	1400
4-Bromophenyl	530	650
3-Methoxyphenyl	540	1000
4-Methylphenyl	560	600
4-Aminophenyl	590	2000
4-*N*,*N*-Dimethylaminophenyl	600	1200
5-Uracilyl	570	950

[a] From *cis*-1,2-dichloroethene.
[b] From *trans*-1,2-dichloroethene.
[c] 1,2-Dichlorovinylperoxyl/1,1-dichlorovinylperoxyl (4/1).

concentration, the kinetics of the process are first-order and a plot of k_{obs} vs. the oxygen concentration is linear. An example is given in the inset of Figure 1. It can be shown that the slope represents the bimolecular rate constant of $R^{\bullet} + O_2$. Many of these rate constants centre on $2 \times 10^9\,dm^3\,mol^{-1}\,s^{-1}$, i.e. they are close to, but not fully diffusion controlled. Examples of smaller rate constants are discussed below. For a compilation of rate constants see Ref. 32.

When the C—$O_2^{\bullet}$ bond strength is weak, the $R^{\bullet} + O_2/RO_2^{\bullet}$ system becomes reversible. In the gas phase, the allylperoxyl radical [33, 34], as well as the trichloromethyperoxyl radical [35] already show dissociation into their various components under conditions of low oxygen pressure at temperatures of ca. 80°C, so that this effect might also begin to materialize in aqueous solutions under certain conditions. To date, examples of reversibility at room temperature have however only been found in some types of dienyl radicals [36–41]. In aqueous solutions, the phenomenon has been extensively studied in a series of substituted hydroxycyclohexadienyl radicals **5** [38–42]. These are formed when OH radicals add to benzene and its derivatives (cf. reactions (15) and (16)).

OH
H
5
O_2
(15)
(−15)
OH
H
H
O−O•
6
(16)
(−16)
H OH
H O−O•
7

Hydroxycyclohexadienyl radicals absorb strongly near 310 nm, while the corresponding peroxyl radicals absorb only weakly at this wavelength [39]. Due to the reversibility of the reaction, the 310 nm absorbance does not decay to the low level that would be given by the absorption of the remaining hydroxycyclohexadienylperoxyl radicals alone, but settles at a higher plateau value, with the height of this plateau depending on the oxygen concentration (Figure 3) [39, 40, 42]. Subsequent to the rapid attainment of this near-equilibrium situation (and to some extent also within this time-span) the

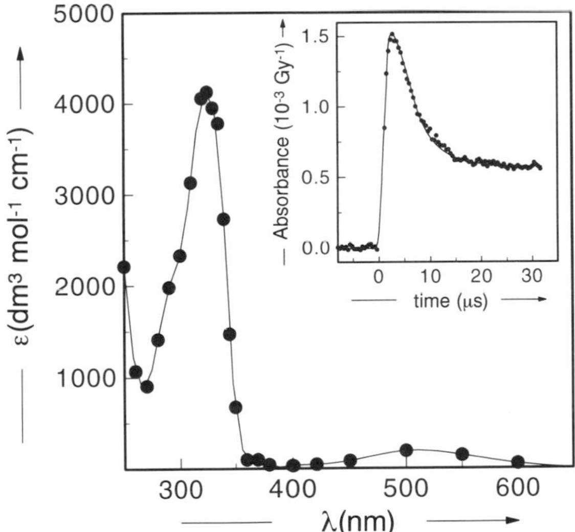

Figure 3 Spectrum of the sum of OH- and H-adduct radicals from pulse radiolysis of N_2O-saturated aqueous solutions of anisole (10^{-3} mol dm^{-3}). Inset shows the decay of the anisole hydroxycyclohexadienyl radical at 325 nm in the presence of 15% O_2 and formation of the 'plateau' [40]. Reproduced by permission of VCH Verlagsgesellschaft mbH.

radicals decay more slowly by both bimolecular and unimolecular processes which will be discussed below. For the simple case:

$$R^{\cdot} + O_2 \underset{\text{reverse}}{\overset{\text{forward}}{\rightleftharpoons}} RO_2^{\cdot} \xrightarrow{\text{decay}} P$$

it can be shown that when the decay process is much slower than the rates of the forward and reverse reactions, the rate constant, k_{obs}, for the disappearance of $R^{\cdot}$ is given by the expression $k_{obs} = k_{forward}[O_2] + k_{reverse}$, i.e. when k_{obs} of the decay of the hydroxycyclohexadienyl radical absorption near its maximum in the region just above 300 nm is plotted as a function of the oxygen concentration, the slope represents the rate constant of the forward reaction, and the intercept that of the reverse reaction. An example is shown in Figure 4.

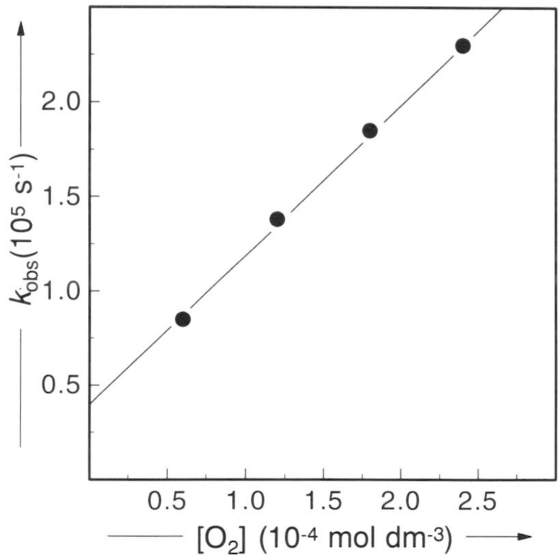

Figure 4 Plot of the fast decay at 325 nm as a function of the oxygen concentration from pulse radiolysis of N_2O/O_2-saturated aqueous solutions of anisole (10^{-3} mol dm^{-3}) [40]. Reproduced by permission of VCH Verlagsgesellschaft mbH.

It has been demonstrated that this approach can be generalized and applied to systems with parallel reactions of this kind involving several species ($^{\cdot}R_i$, $R_iO_2^{\cdot}$), provided their reactivities are not very dissimilar [40]. In this case, a diagram such as Figure 4 yields the 'effective' forward and reverse rate constants.

The ratio of the forward and reverse rate constants gives the stability constant of the peroxyl radical. This quantity can also be obtained from the height of the 'plateau' values as a function of the oxygen concentration. In terms of concentrations, the stability constant is defined by equation (17):

$$K = [RO_2^\cdot]_{eq}/([R^\cdot]_{eq}[O_2]) = ([R^\cdot]_0 - [R^\cdot]_{eq})/([R^\cdot]_{eq}[O_2]) \qquad (17)$$

where the subscripts 'eq' and '0' signify 'equilibrium' and 'initial'. The initial absorbance is $A_0 = [R^\cdot]\epsilon(R^\cdot)$. In the case of Ref. 39, $\epsilon(R^\cdot) = 4000\,dm^3\,mol^{-1}\,cm^{-1}$. The absorbance at equilibrium is given by equation (18). Equations (17) and (18) lead to equation (19):

$$A_{eq} = [R^\cdot]_{eq}\epsilon(R^\cdot) + [RO_2^\cdot]_{eq}\epsilon(RO_2^\cdot) = [R^\cdot]_{eq}\epsilon(R^\cdot) + \epsilon(RO_2^\cdot)([R^\cdot]_0 - [R^\cdot]_{eq}) \qquad (18)$$

$$1 + K[O_2] = [\epsilon(R^\cdot) - \epsilon(RO_2^\cdot)]A_0/[\epsilon(R^\cdot)A_{eq} - \epsilon(RO_2^\cdot)A_0] \qquad (19)$$

Of the quantities that constitute the right-hand side of expression (19), only $\epsilon(RO_2^\cdot)$ cannot be readily obtained under the usual experimental conditions (atmospheric pressure) where the equilibrium does not yet lie sufficiently close to the side of $RO_2^\cdot$, as $\epsilon(RO_2^\cdot)$ is considerably smaller than $\epsilon(R^\cdot)$ at the monitoring wavelength. However, it is possible by trial and error to solve equation (19): when the parameter $\epsilon(RO_2^\cdot)$ is chosen correctly, the plot of $(1 + K[O_2])$ vs. $[O_2]$ is linear (the slope is then equal to K). If $\epsilon(RO_2^\cdot)$ is chosen to have too high a value, the curve bends upward, and if chosen too low, it bends downward. In the case of the unsubstituted hydroxy-cyclohexadienyl radical system such an analysis has shown that $\epsilon(RO_2^\cdot) = 500\,dm^3\,mol^{-1}\,cm^{-1}$, i.e. ca. 13% of $\epsilon(R^\cdot)$ (Figure 5).

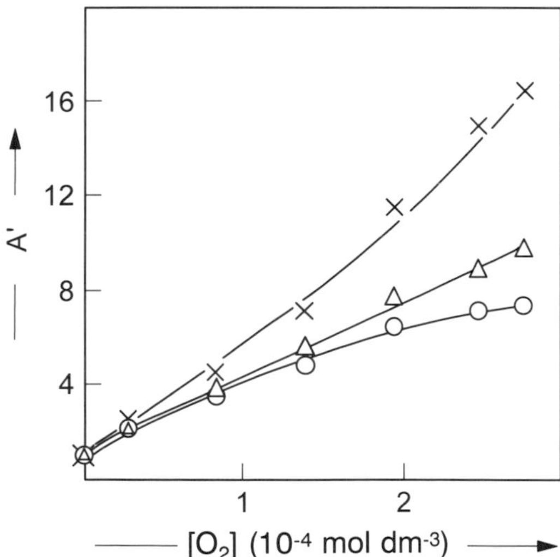

Figure 5 Plots of $(1 + K[O_2])$ vs. $[O_2]$: (×) $\epsilon(RO_2^\cdot) = 650$; (△) $\epsilon(RO_2^\cdot) = 500$; (○) $\epsilon(RO_2^\cdot) = 400\,dm^3\,mol^{-1}\,cm^{-1}$. $RO_2^\cdot$ = unsubstituted hydroxycyclohexadienylperoxyl radical [39]. Reproduced by permission of Zeitschrift für Naturforschung.

Table 2 Reactions of Substituted Hydroxycyclohexadienyl Radicals with Oxygen, Showing Rate Constants of Forward (k_f) and Reverse (k_r) Reactions, the Stability Constant, $K(=k_f/k_r)$, and Rate Constants of Peroxyl Radical Decomposition, k_p [40]

Substituent	k_f (10^8 dm^3 mol^{-1} s^{-1})	k_r (10^4 s^{-1})	K (10^4 dm^3 mol^{-1})	k_p (10^3 s^{-1})
OCH_3	8.0	3.9	2.1	5.5
CH_3	4.8	7.5	0.64	1.0
F	4.6	5.5	0.84	0.59
H[a]	3.1	1.2	2.6	0.80
Cl	2.6	5.5	0.47	1.1
CH_2Cl	2.2	6.4	0.34	1.1
CO_2^-	2.0	1.3	1.5	0.34
Alanyl[b]	1.7[c]; 1.8	5.2[c]; 5.4	0.33	1.5
1,4-$(CO_2)_2$	0.16	0.34	0.47	0.39
CO_2H	0.03	n.d.	n.d.	n.d.
CO_2Et	0.07	n.d.	n.d.	n.d.
CN	0.05	n.d.	n.d.	n.d.
NO_2	$\leq$0.02	n.d.	n.d.	n.d.

n.d. = not determined.
[a] From Ref. 39.
[b] From Ref. 38.
[c] Measured at 15°C [40].

Kinetic parameters have been determined for a number of substituted hydroxycyclohexadienyl radicals. These are compiled in Table 2, from where it can be seen that in all of the cases studied the rate constant for oxygen addition is significantly below the value of 2×10^9 dm^3 mol^{-1} s^{-1} that is typical for other carbon-centred radicals. In addition, the data show that the substituents affect the equilibrium constant by influencing both the forward and the reverse reactions. In some cases the rate constant of oxygen addition is so slow that in the pulse-radiolytic experiment, bimolecular decay processes of the radicals obscure to a significant extent the kinetics of the forward, reverse and unimolecular-decay reactions. A case in point is the hydroxycyclohexadienyl radical formed by the addition of the OH radical to benzonitrile. Nevertheless, the first-order component of the decay can still be isolated. This is achieved by plotting the inverse of the first half-life of the spectral decay of the cyanohydroxycyclohexadienyl radical vs. the pulse dose at various oxygen concentrations. Extrapolation to zero dose yields a series of intercepts (Figure 6). These in turn are plotted vs. $[O_2]$ (Figure 6, inset); it has been shown that the slope represents the quantity $k_{forward}/\ln 2$ and that k_{decay} in the benzonitrile case must be very small [40]. The absence of a detectable intercept in the latter plot indicates that $k_{reverse}$ must also be quite small. When the rate of oxygen addition drops below a value of ca. 2×10^6 dm^3 mol^{-1} s^{-1} (e.g. with the hydroxynitrocyclohexadienyl radical) it can no longer be determined by pulse radiolysis

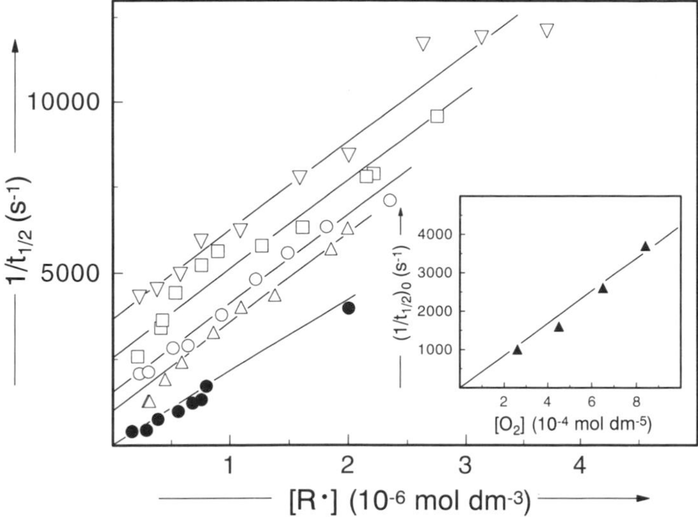

Figure 6 Inverse of the first half-life of the (mainly bimolecular) decay of the cyanohydroxycyclohexadienyl radical as a function of the dose per pulse following pulse radiolysis of N_2O- (●) and N_2O/O_2-saturated (O_2: (△) 20; (○) 35; (□) 50; (▽) 65%) aqueous solutions of benzonitrile (10^{-4} mol dm^{-3}). Inset shows the dependence of the intercept on O_2 concentration [40]. Reproduced by permission of VCH Verlagsgesellschaft mbH.

even at the lowest possible pulse doses. However, the reaction can still be monitored by product studies at the much lower dose rates such as are typical for ^{60}Co γ-irradiations [40].

While the substituted hydroxycyclohexadienyl radicals *do* react with oxygen, even though often with low rate constants, with the reversibility of the addition indicating a low C–$O_2^{\cdot}$ bond strength (this has been determined at 22.5 kJ mol^{-1} in one case) [40], phenoxyl-type radicals apparently do not react with oxygen at all, even under conditions of very low steady-state radical concentrations, despite having a considerable spin density at carbon. A case in point is the tyrosine phenoxyl radical **8**. In the absence of oxygen this radical mainly recombines (>90%), giving rise to 2,2′-bityrosyl **10** (reactions (20) and (21)). Thus in this respect it *does* act like most other kinds of carbon-centred radicals [43]. The rate constant of bimolecular decay ($k = 4.6 \times 10^8$ dm^3 mol^{-1} s^{-1}) is not far from that of other radicals of similar size.

From the data presented in Refs 43 and 44 it has been concluded that if oxygen adds to these phenoxyl radicals, the rate constant must be less than 10^3 dm^3 mol^{-1} s^{-1}. The low reactivity of oxygen towards phenoxyl-type radicals has also been observed for the radical derived from phloroglucine **11** ($k < 4 \times 10^5$ dm^3 mol^{-1} s^{-1}) [45]. However, when a phenolic group is deprotonated, the reaction of oxygen with the radical anion **12** becomes noticeable ($k = 2.1 \times 10^8 \, dm^3 \, mol^{-1} \, s^{-1}$); this also holds for the monoanion in the keto-form (**13**) ($k = 1.4 \times 10^8 \, dm^3 \, mol^{-1} \, s^{-1}$) [45].

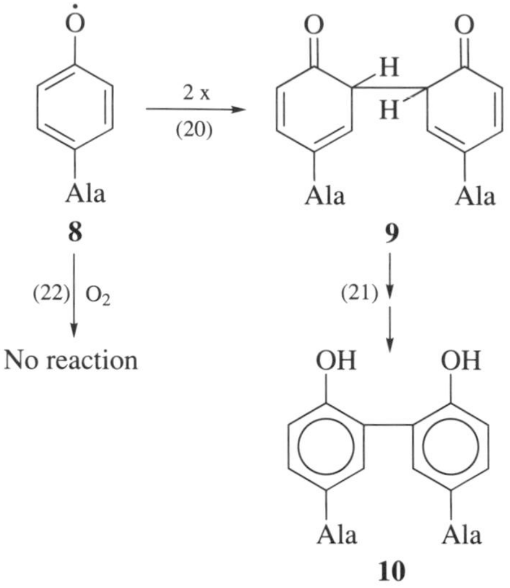

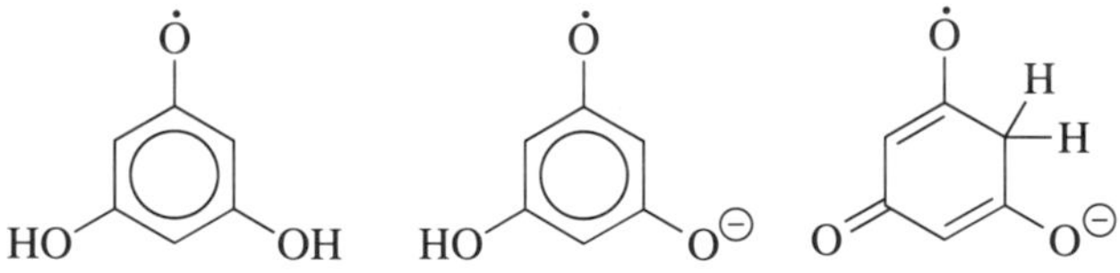

This observation, and the fact that in a number of substituted hydroxycyclohexadienyl radicals the rate of oxygen addition is relatively fast only with electron-donating substituents, both indicate that electron-rich carbon-centres are a requirement for the formation of peroxyl radicals. This explains why radical cations and radicals with a tendency to shift electron and spin densities to a heteroatom, such as **14** and **15**, exhibit either a reduced reactivity towards oxygen or none at all [46, 47]. In fact, in the 2′-deoxyadenosine case the reactivity toward O_2 is so low that it is feasible to measure the reactivity of radical **15** toward superoxide $O_2^{\bullet -}$ [47].

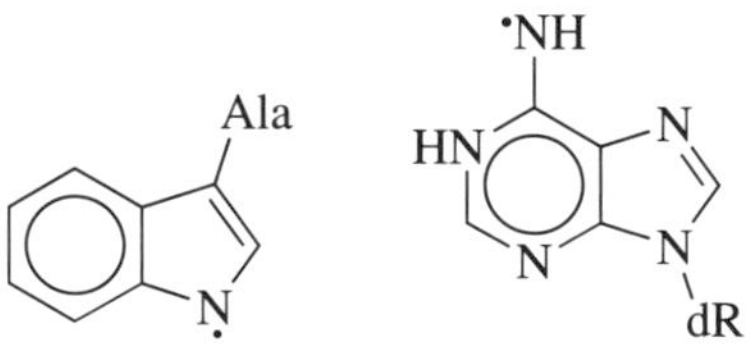

The existence of the foregoing peroxyl radicals as well as of others, such as alkylperoxyl, is of course well established. Nevertheless, sometimes the intermediacy of the peroxyl-radical stage in the oxidation of a radical by O_2 is not obvious. Several types of peroxyl radical are capable of eliminating $HO_2^{\bullet}$ or $O_2^{\bullet-}$ quite rapidly (see below). Thus it is not surprising that in some cases (e.g. **16** and **17**) a peroxyl intermediate has not been observed at all (cf. reactions (23) and (24)), and that $O_2^{\bullet-}$ appears to be formed instantly.

$$\underset{\mathbf{16}}{{}^{\bullet}C(=O)O^{\ominus}} \xrightarrow[(23)]{O_2} CO_2 + O_2^{\bullet\ominus}$$

$$\underset{\mathbf{17}}{(CH_3)_2N{-}CH_2^{\bullet}} \xrightarrow[(24)]{O_2} (CH_3)_2\overset{\oplus}{N}{=}CH_2 + O_2^{\bullet\ominus}$$

This does not prove, however, that the reaction really proceeds by genuine electron transfer. Typically, the rise-times of peroxyl radical formation in air- or oxygen-saturated solutions are of the order of a few μs. Thus a very short-lived peroxyl intermediate ($k(O_2^{\bullet-}$ release$) > 10^6\,s^{-1}$) may easily escape detection. The experiment would therefore have to be carried out at elevated oxygen pressures in order to place a narrower limit on the lifetime of potential peroxyl intermediates.

The 2- and 4-hydroxyphenoxyl radical anions (e.g. **18**) fall into this group as well. Again, in their reactions with oxygen no peroxyl intermediates have been observed (e.g. equilibrium (25)). Therefore, these reactions have been treated as genuine cases of electron transfer, with the equilibrium constant being determined by the reduction potential of the species involved [48].

$$\underset{\mathbf{18}}{{}^{\bullet}O{-}C_6H_4{-}O^{\ominus}} + O_2 \underset{(-25)}{\overset{(25)}{\rightleftharpoons}} O{=}C_6H_4{=}O + O_2^{\bullet\ominus}$$

The foregoing examples involve $O_2^{\bullet-}$ production, quite possibly via very short-lived peroxyl intermediates. There is now an indication that the reverse reaction may actually be a pathway for the formation of certain peroxyl radicals (**20**), e.g. nucleophilic addition of $O_2^{\bullet-}$ to ketomalonic acid **19** (reaction (26)) [49–51].

$$\underset{\textbf{19}}{\begin{matrix} CO_2^{\ominus} \\ | \\ C{=}O \\ | \\ CO_2^{\ominus} \end{matrix}} + O_2^{\cdot\ominus} \xrightarrow{(26)} \underset{\textbf{20}}{{}^{\ominus}O{-}\begin{matrix} CO_2^{\ominus} \\ | \\ C \\ | \\ CO_2^{\ominus} \end{matrix}{-}O{-}O^{\cdot}}$$

In non-aqueous media a specific photochemical method exists for the production of acylperoxyl radicals by excitation of an α-diketone in the presence of O_2 [52]. Whether this procedure would be as successful in water has not yet been investigated. Here one might expect one of the carbonyl functions of the diketone to be hydrated, and as a consequence of this a system with more than one single kind of peroxyl radical would be obtained.

3 $HO_2^{\cdot}$ ELIMINATION REACTIONS

Peroxyl radicals may undergo a number of unimolecular processes. The most ubiquitous ones are $HO_2^{\cdot}$ or $O_2^{\cdot-}$ elimination, but intramolecular addition to a C=C double bond or H-abstraction have also been demonstrated (see below).

The driving force of the $HO_2^{\cdot}$ elimination consists in the formation of a double bond, such as in reactions (27)–(29) [38–40, 53–59].

$$\underset{\textbf{21}}{HO{-}\begin{matrix} CH_3 \\ | \\ C \\ | \\ CH_3 \end{matrix}{-}O{-}O^{\cdot}} \xrightarrow{(27)} \begin{matrix} CH_3 \\ | \\ C{=}O \\ | \\ CH_3 \end{matrix} + HO_2^{\cdot}$$

(28): **22** (cyclohexadienyl peroxyl radical: ring carbon bearing H, H; adjacent carbon bearing O–O$^{\cdot}$ and H) $\longrightarrow$ benzene $+ HO_2^{\cdot}$

(29): **6** (cyclohexadienyl peroxyl radical: ring carbon bearing HO, H; adjacent carbon bearing O–O$^{\cdot}$ and H) $\longrightarrow$ phenol (OH) $+ HO_2^{\cdot}$

It has been suggested that this elimination occurs via a five-membered transition state [53]. This is supported by the fact that in the series of cyclohexadienylperoxyl radicals only those that carry the peroxyl radical function in an *ortho*-position to the H atom to be eliminated (e.g. **22** and **6**), undergo this reaction (reactions (28) and (29)), while those that carry the radical function in the *para*-position (e.g. **23**) cannot eliminate $HO_2^{\bullet}$ (reaction (30)) [59]. This may be largely due to steric reasons. The C^1—O—$O^{\bullet} \cdots$ H—C^2 distance for **22** is calculated at 1.8 ± 0.4 Å, while **23** has a C^1—O—$O^{\bullet} \cdots$ H—C^4 distance of 3.85 ± 0.35 Å [59].

H H

H O−O•
23

$\xrightarrow{(30)}$ no $HO_2^{\bullet}$ elimination

The rate of the $HO_2^{\bullet}$ elimination reaction from α-hydroxyalkylperoxyl radicals varies considerably with the nature of the flanking substituents. For the purpose of illustration, some examples are listed in Table 3. It can be seen that in this series of simple α-hydroxyalkylperoxyl radicals, the rate of $HO_2^{\bullet}$ elimination increases with increasing degree of substitution by methyl groups. The trend reflects the increasing ease of the oxygen-bound carbon atom to assume the sp^2 configuration and parallels the decreasing stability of the corresponding hydrates: in water, $R_2C(OH)_2/R_2C{=}O$ is ca. 2000 for formaldehyde, ca. 1.3 for acetaldehyde, and 0.002 for acetone [60]. The C=O bond strength of the final product also increases in the same direction. This adds weight to the above suggestion that the driving force of the reaction is connected with the emergence of the resulting double bond. In

Table 3 Rate Constants for Spontaneous $HO_2^{\bullet}$ Elimination and OH^--Induced $O_2^{\bullet-}$ Elimination from some α-Hydroxyalkylperoxyl Radicals

Peroxyl radical	k($HO_2^{\bullet}$ elimination) (s^{-1})	k(OH^--induced $O_2^{\bullet-}$ elimination) ($dm^3\,mol^{-1}\,s^{-1}$)	Ref.
$HOCH_2O_2^{\bullet}$	<10	1.5×10^{10}	53
	<3	1.8×10^{10}	55
$HOCH(CH_3)O_2^{\bullet}$	50	4×10^9	56
$HOC(CH_3)_2O_2^{\bullet}$	665	5×10^9	53
	700	5.2×10^9	57
$HOCH_2$–$CH(OH)O_2^{\bullet}$	190	–	54
$HOC(OH)_2O_2^{\bullet}$	$>10^5$		58

this context it is of interest to note that the peroxyl radical derived from glycine anhydride **2** does not (or merely very slowly, $k < 1\,s^{-1}$) eliminate $HO_2^{\bullet}$ [61]. This would require the formation of a C=N double bond. A larger driving force than this seems to be required here, and is supplied by N—H deprotonation, whereupon $O_2^{\bullet -}$ is eliminated (see below).

In the series of cyclohexadienylperoxyl radicals shown above, the rate of $HO_2^{\bullet}$ elimination also varies by several orders of magnitude (cf. Table 2) [40]. This process is especially fast in the case of the unsubstituted peroxyl radical (reaction (28)), where it is actually too fast to be observed as it overlaps with the process of O_2 addition [59].

4 $O_2^{\bullet -}$ ELIMINATION REACTIONS

The $O_2^{\bullet -}$ elimination reactions may be divided into three groups. Those peroxyl radicals that have an —OH or —NHR function in the α-position make up the first group. The peroxyl radical anion is formed upon deprotonation at the heteroatom by OH^-, with the acidity of the functions α to the peroxyl group being enhanced compared to the situation in the original molecule, as this group is electron withdrawing; for instance, the pK_a of the acetic-acid-derived peroxyl radical is 2.1, compared to 4.8 for the parent compound [62]. As before, the driving force for the elimination reaction is the formation of a C=O or C=NR double bond (in addition to the energy gained by the formation of the mesomery-stabilized $O_2^{\bullet -}$ radical). Examples are provided by reactions (31)–(36).

$$\underset{\mathbf{21}}{HO{-}C(CH_3)_2{-}O{-}O^{\bullet}} \underset{(31/-31)}{\overset{OH^{\ominus}/-H_2O}{\rightleftharpoons}} \underset{\mathbf{24}}{{}^{\ominus}O{-}C(CH_3)_2{-}O{-}O^{\bullet}} \xrightarrow{(32)} (CH_3)_2C{=}O + O_2^{\bullet\ominus}$$

$$\underset{\mathbf{25}}{HO{-}C(CO_2^{\ominus})_2{-}O{-}O^{\bullet}} \underset{(33/-33)}{\overset{OH^{\ominus}/-H_2O}{\rightleftharpoons}} \underset{\mathbf{26}}{{}^{\ominus}O{-}C(CO_2^{\ominus})_2{-}O{-}O^{\bullet}} \xrightarrow{(34)} \underset{\mathbf{19}}{(CO_2^{\ominus})_2C{=}O} + O_2^{\bullet\ominus}$$

$$\mathbf{2} \underset{(35/-35)}{\overset{OH^{\ominus}/-H_2O}{\rightleftharpoons}} \mathbf{27} \xrightarrow[(36)]{-O_2^{\bullet\ominus}} \mathbf{28}$$

(**2**: glycine anhydride peroxyl radical, ring with H, H, NH, HN, H, O–O•, two C=O; **27**: the corresponding N$^{\ominus}$ anion; **28**: ring with C=N double bond, H.)

The peroxyl radical anion **24** formed in reaction (31) has an immeasurably short ($\ll 10^{-6}$ s) lifetime, i.e. k_{32} is much larger than $k_{-31}[H_2O]$, and even at high $[OH^-]$ the rate of acetone formation is essentially given by $k_{31}[OH^-]$ [53]. The situation is similar for peroxyl radical anions derived from methanol and ethanol; however, α-hydroxyalkylperoxyl radical anions do not necessarily have lifetimes much below the limit of resolution (here, $<10^{-6}$ s). For example, the elimination of $O_2^{\bullet -}$ from the peroxyl radical anion **26** derived from hydroxymalonic acid (reaction (34)), is not much faster ($k_{34} = 8.9 \times 10^4\ s^{-1}$) than the elimination of $HO_2^{\bullet}$ from the protonated form **25** which has a rate constant of $1.1 \times 10^4\ s^{-1}$ [51]. Even longer lifetimes with respect to superoxide elimination are shown by the peroxyl radical anions derived from the glycine and alanine anhydrides [59]. As seen in Table 4, substitution by a methyl group (alanine anhydride) shifts the pK_a to higher values compared to the prototype radical and peroxyl radical, owing to the electron-donating properties of the methyl group. For the same reason, $O_2^{\bullet -}$ release is facilitated by the methyl substituent.

Table 4 pK_a Values of Radicals Derived from Glycine Anhydride and Alanine Anhydride, plus Rate Constants for the $O_2^{\bullet -}$ Elimination Reactions from their Peroxyl Radical Anions [61, 63]

Parameter	Glycine anhydride	Alanine anhydride
pK_a of parent radical	9.8	10.6
pK_a of peroxyl radical	10.7	11.2
$k(O_2^{\bullet -}$ elimination) (s^{-1})	1.6×10^5	3.7×10^6

The second group, represented by the 1-hydroxycyclohexadienyl-4-peroxyl radicals (e.g. **7**), is related to the first group in that, for both of these, double-bond formation is instrumental. However, here, in order to induce $O_2^{\bullet -}$ elimination, a distant *carbon*-bound hydrogen must be removed (reactions (37) and (38)) [41].

$$\mathbf{7} \underset{(37/-37)}{\overset{OH^{\ominus}/-H_2O}{\rightleftharpoons}} \mathbf{29} \xrightarrow{(38)} C_6H_5OH + O_2^{\bullet\ominus}$$

Whereas base-induced deprotonation at a heteroatom is very fast (practically diffusion controlled, cf. Ref 64), deprotonation at carbon is generally much slower (cf. Refs 65 and 66). Thus this type of $O_2^{\bullet -}$ elimination is observed over a range of higher pH values compared to the reactions discussed before. It should be emphasized that while the elimination of $HO_2^{\bullet}$ is subject to steric restrictions, the OH^--induced $O_2^{\bullet-}$-elimination is not. Thus, under strongly basic conditions, the peroxyl radical **7/29** *does* eliminate $O_2^{\bullet-}$, bringing the phenol yield close to 100% as the overall rate of $HO_2^{\bullet}$ plus $O_2^{\bullet-}$ elimination is enhanced, and competing reactions thereby suppressed (see below) [41]. It is noted that because of the faster deprotonation at oxygen, peroxyl radicals **6** and **7** are preferentially deprotonated at the oxygen. Such a deprotonation is, however, without consequence. Only when the deprotonation occurs at carbon (e.g. reaction (37)) can $O_2^{\bullet-}$ elimination ensue. Whether under these conditions the OH group is also deprotonated (cf. Ref. 67), or whether, after O-deprotonation, a (water-assisted) 1,2-H-shift occurs which then facilitates the $O_2^{\bullet-}$ elimination cannot be decided on the basis of the existing experimental data.

The third class of peroxyl radicals is characterized by the dissociation into a carbocation (e.g. **31**) and $O_2^{\bullet-}$. One example has been unequivocally established to date (reaction (39)) [68].

$$H_3C{-}C(OCH_3)_2{-}O{-}O^{\bullet}\ (\mathbf{30}) \xrightarrow{(39)} H_3C{-}C^{\oplus}(OCH_3)_2\ (\mathbf{31}) + O_2^{\bullet\ominus}$$

Here the driving force lies in the stabilization of the cation by the adjacent electron-donating groups. In this case, the rate constant is $6.5 \times 10^4\,s^{-1}$. It is typical for ionic dissociation processes that their rates depend strongly on the stabilization energies of the ions formed. Thus it is not surprising that the peroxyl radicals derived from diisopropyl ether [69] or 1,3-dioxane [70] do not display a similarly fast $O_2^{\bullet-}$ elimination, as the former lacks the second alkoxyl function, and the latter an alkyl substituent.

It has been mentioned above that in reactions (23) and (24) no peroxyl-radical intermediates have been observed. If they existed, they would count among this group.

5 ADDITION TO THE C=C DOUBLE BOND

In the series of hydroxycyclohexadienylperoxyl radicals one encounters the competition between the $HO_2^{\bullet}$ elimination leading to phenol, and fragmentation of the ring (see below). The latter has been attributed to an intramolecular addition of the peroxyl radical function to a diene double bond (reaction (40)) [41, 71]. This reaction seems to be reversible, but when oxygen adds to the newly created carbon-centred radical **32** (reaction (41)) the endoperoxidic function is locked in (cf. **33**).

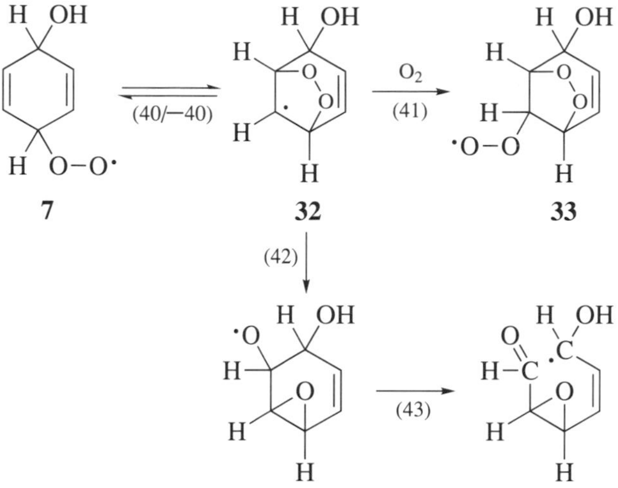

In competition with O_2 addition, β-alkylperoxide species such as **32** may undergo radical-induced cleavage of the peroxide function (reaction (42)) [71–73] and eventually fragmentation of the resulting alkoxyl radical **34** (reaction (43)). Similarly, the oxidation of indols in the presence of oxygen has been initiated by the addition of the trichloromethylperoxyl radical to the indol $C^2{=}C^3$ double bond [74].

Peroxyl radicals of multi-unsaturated compounds (e.g. **36**) may undergo chain-like peroxidation if the C=C double bonds are suitably disposed (reaction (44)) [75, 76].

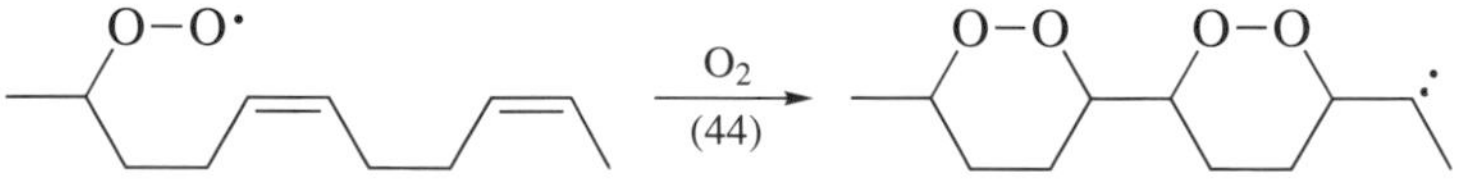

The auto-oxidation of polyunsaturated fatty acids (cf. Ref. 77) is usually monitored by the formation of 'malondialdehyde' using the 2-thiobarbituric acid assay. This is carried out under rather severe conditions (pH 2, 30 min at 90°C) which decompose its precursor, product **41**. This malonaldehyde-like product **41**, which is a major one, is obviously formed via a cyclization reaction of a peroxyl radical (reaction (45)), followed by other processes, e.g. further cyclization (reaction (46)) and hydroperoxide formation (reaction

(47)). Compounds such as **41** may eliminate malonaldehyde upon induction of endoperoxide homolytic cleavage.

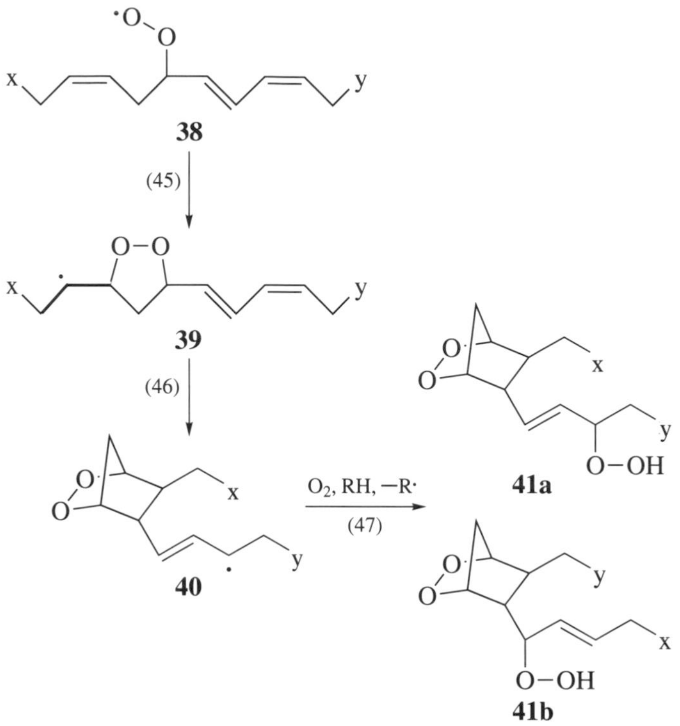

38

(45)

39

(46)

40

O_2, RH, −R·

(47)

41a

41b

In allylperoxyl radicals **42**/**43**, allylic rearrangement (reaction (48)) leads to a 1,3-migration of the peroxyl function, with a corresponding shift of the double bond [78].

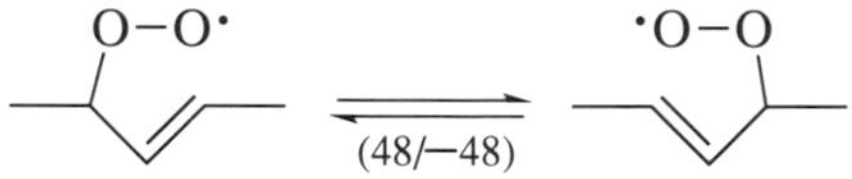

42 (48/−48) **43**

Evidence has been adduced that for many such systems, the apparently obvious cyclic intermediate 1,2-dioxolan-4-yl is *not* realized in the course of this rearrangement; a dioxygen–allyl radical pair is postulated instead [79, 80]. Nevertheless, this cyclic structure has been invoked in the gas phase reaction [81, 82], and an exothermicity of 96 kJ mol^{-1} been computed for the formation of the cyclic intermediate relative to the level of allyl plus O_2 [82].

6 HYDROGEN-ABSTRACTION REACTIONS

Hydrogen-abstraction reactions by peroxyl radicals are common. An important example is the autoxidation of polyunsaturated lipids [77, 83–87]. These involve *intra*molecular as well as *inter*molecular H-transfer (e.g. reaction (49)). When this process is studied in dilute aqueous solutions of these, micelle-forming, compounds, it starts to become noticeable only above the critical micelle concentration as the reaction acquires chain character [88, 89]. The ROO—H bond energy in hydroperoxides derived from weakly oxidizing peroxyl radicals is ca. 360–370 kJ mol^{-1} [90, 91]. Thus for this reaction to occur at an appreciable rate, the C—H bond energy of the donor must be sufficiently low. This is easily the case for bis-allylic hydrogen in polyunsaturated fatty acids (e.g. **44**) which is as low as 320 kJ mol^{-1} [92]. Thus this H-abstraction is exothermic by ca. 50 kJ mol^{-1}.

R−O−O• + [bis-allylic diene with CH₂ bearing H, H] **44**

(49) ↓

R−O−OH + [pentadienyl radical, H] **45**

In other cases, H-abstraction by peroxyl occurs mostly intramolecularly. Evidence has been reported that reactions such as (50) contribute to the chain character of certain autoxidation processes, with the OH radical as chain carrier (cf. Refs 93–95).

$$\underset{\mathbf{46}}{(H_3C)_2C(OO^\bullet)-CH_2-CH(CH_3)_2} \xrightarrow{(50)} (CH_3)_2C{=}O + H_2C{=}C(CH_3)_2 + {}^\bullet OH$$

It can be estimated, by e.g. using the method of group increments [96, 97], that this kind of reaction is exothermic; in the example given here, by 33 kJ mol^{-1}. It apparently occurs at elevated temperatures; there is no indication as yet that it plays any role in aqueous solutions at room temperature. Instead, an intramolecular H-abstraction reaction *without* fragmentation has been observed with diethylether in dilute aqueous solution, at low steady-state concentrations of peroxyl radicals **47** (reaction

(51)) where the bimolecular decay is slow; its rate constant is ca. 1 s^{-1} [98].

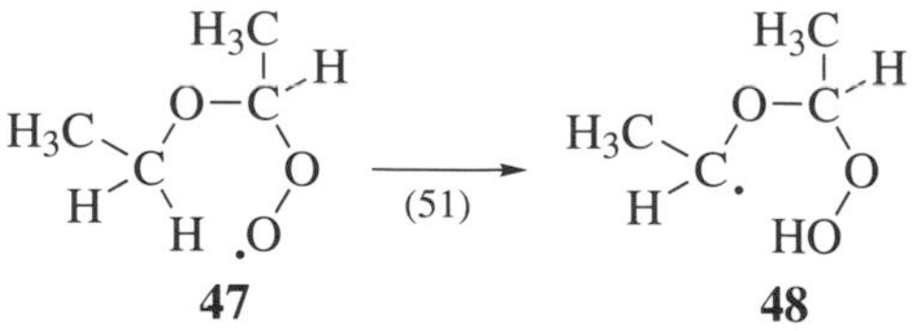

(51) **47** → **48**

It is possible that this H-abstraction reaction is slightly endothermic considering that in tetrahydrofuran the C—H binding energy of the CH_2 group in the α-position to the oxygen is reported to be 381 kJ mol^{-1} [92]. In the case of the peroxyl radicals derived from 2,4-dimethylglutaric acid [23] and poly(acrylic acid) [99] such H-transfer is also observed. Kinetic parameters for the reactions of a series of peroxyl radicals with various alcohols, with pre-exponential factors near 10^8 and activation energies varying from 30–60 kJ mol^{-1} have been reported [91].

In poly(uridylic acid), nucleobase peroxyl radicals which are formed upon oxygen addition to the OH-adduct radicals are implicated in the formation of strand breaks and release of unaltered uracil via H-abstraction from the ribose moiety [100, 101]. This leads to an enhanced O_2 uptake [102].

Hydrogen atom abstraction by peroxyl radicals is often an essential link in autoxidation chain reactions. These can be inhibited by H-atom donors RH, where R$^\bullet$ is practically inert to peroxidation, e.g. sterically hindered and other phenols (cf. the function of vitamin E in protecting biomembranes against autoxidation). It has been shown [103] that depending on the nature of the attacking peroxyl radical (reaction (52)), the dependence of the rate constant on the nature of the sterically hindered phenol is described by the Hammett parameters of the phenols (e.g., when R$^\bullet$ is $^\bullet CH(CH_3)Ph$), or is governed by the size of the dipole moments of the phenols (e.g. when ROO$^\bullet$ is 1,4-dioxanyl-2-peroxyl). At room temperature, rate constants for the 1,4-dioxanyl-2-peroxyl radical are in the order of 10^3–10^4 $dm^3\ mol^{-1}\ s^{-1}$, depending on the phenol [103].

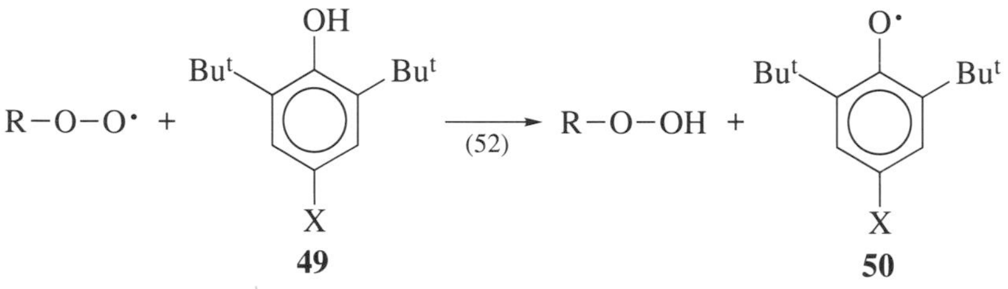

$$R{-}O{-}O^\bullet + \mathbf{49} \xrightarrow{(52)} R{-}O{-}OH + \mathbf{50}$$

These experiments were done in non-aqueous media. In aqueous solutions one may expect a smaller influence of the dipole moment on the size of the rate constants.

7 ELECTRON-TRANSFER REACTIONS

Peroxyl radicals act readily as electron acceptors, and rate constants are often high. A series of alkylperoxyl radicals have been reacted with *N,N,N′,N′*-tetramethylphenylenediamine **51**. This compound is a popular probe for oxidizing intermediates. It is very easily oxidized to the radical cation **52** which is characterized by its strong absorption in the visible region; this is how it is monitored. Rate constants of its reaction with peroxyl radicals range from 1.1×10^6 ($(CH_3)_2CHO_2^{\bullet}$) to $1.7 \times 10^9\, dm^3\, mol^{-1}\, s^{-1}$ ($CCl_3O_2^{\bullet}$) [104] and $1.9 \times 10^9\, dm^3\, mol^{-1}\, s^{-1}$ ($CH_3C(O)O_2^{\bullet}$) [105] (reaction (53)). Thus, $CH_3C(O)O_2^{\bullet}$ and $CCl_3O_2^{\bullet}$ are among the most strongly oxidizing peroxyl radicals.

$$H_3C-C(=O)-O-O^{\bullet} + \textbf{51} \xrightarrow{(53)} H_3C-C(=O)-O-O^{\ominus} + \textbf{52}$$

(**51**: $(H_3C)_2N-C_6H_4-N(CH_3)_2$; **52**: the corresponding radical cation, ring marked $\oplus$ and $\bullet$)

The rate constants for the oxidation of ascorbate show a variation over a similar range; again $CCl_3O_2^{\bullet}$ is the most reactive [104, 106–108]. Surprisingly, vinylperoxyl radicals are of a comparable reactivity [25]. A similar gradation is observed when $O_2^{\bullet-}$ is the reductant. The rate of the reaction of these highly reactive radicals with $O_2^{\bullet-}$ is, as one would expect, close to being diffusion-controlled (reaction (54), $k_{54} \cong 10^9\, dm^3\, mol^{-1}\, s^{-1}$) [105], while the hydroxyethylperoxyl radical reacts with $O_2^{\bullet-}$ merely with a rate constant of ca. $10^7\, dm^3\, mol^{-1}\, s^{-1}$ [56].

$$H_3C-C(=O)-O-O^{\bullet} + O_2^{\bullet\ominus} \xrightarrow{(54)} H_3C-C(=O)-O-O^{\ominus} + O_2$$

It has been observed that the oxidizing power of halogenated peroxyl radicals generally increases with increasing degree of halogenation [29, 109–112]. Other electron-withdrawing functions such as carbonyl oxygen have a similarly strong reactivity-promoting effect [105]. The high reactivity of the trichloromethylperoxyl radical is reflected in its relatively high reduction potential of 1.15 V (vs. 0.77 V for non-halogenated alkylperoxyl radicals) [113]. Arylperoxyl radicals are more reactive than alkylperoxyl ones, increasingly so as the number of aromatic rings increases [28], and also as the electron-drawing capacity of the aromatic ring increases [28]. Electron transfer from organic reductants to chlorinated methylperoxyl radicals may take place via a transient adduct of the peroxyl radical to the reductant [114]. Electron transfer to the peroxyl radical is thought to be concerted with the transfer of the proton from the solvent to the incipient hydroperoxide anion [115]. The pronounced preference shown by the trichloromethylperoxyl radical to react via electron transfer rather than hydrogen-atom transfer (or via another pathway) is nicely demonstrated by its behaviour towards 4-methoxyphenylthiol, where it is shown that electron transfer from the thiolate **53** (reaction (55)) is many times faster ($k_{55} = 8 \times 10^8\,dm^3\,mol^{-1}\,s^{-1}$) than its reaction with the protonated thiol [116].

$$CCl_3\text{-}O\text{-}O^{\bullet} + {}^{\ominus}S\text{-}C_6H_4\text{-}OCH_3\ (\mathbf{53}) \xrightarrow{(55)} CCl_3\text{-}O\text{-}O^{\ominus} + {}^{\bullet}S\text{-}C_6H_4\text{-}OCH_3\ (\mathbf{54})$$

Haloalkylperoxyl radicals effect one-electron oxidation of dialkyl sulfides via an adduct **55** which is converted into the tetraalkyldisulfide radical cation **56** by a second thioether molecule (reaction (56)) [117].

$$CCl_3\text{-}O\text{-}O\text{-}S^{\bullet}(CH_3)_2\ (\mathbf{55}) \xrightarrow[(56)]{S(CH_3)_2} CCl_3\text{-}O\text{-}O^{\ominus} + (H_3C)_2S\overset{\bullet}{\therefore}S(CH_3)_2^{\oplus}\ (\mathbf{56})$$

A peculiar form of unimolecular transformation of certain peroxyl radicals, **57/58**, through solvent-assisted charge transfer (equilibrium (57)) has been implicated in the oxidative radiolytic decomposition of hydroxymalonic [51] and ketomalonic [49] acids.

$${}^{\ominus}O\text{-}C(CO_2^{\ominus})_2\text{-}O\text{-}O^{\bullet}\ (\mathbf{57}) \underset{(57/-57)}{\rightleftharpoons} {}^{\bullet}O\text{-}C(CO_2^{\ominus})_2\text{-}O\text{-}O^{\ominus}\ (\mathbf{58})$$

Even though the forward reaction in such a couple appears to be energetically uphill, it might nevertheless become product-forming if the resulting oxyl radical was particularly prone to fragmentation. A similar intramolecular charge transfer has been postulated for the anion of the 4-hydroxyphenylperoxyl radical [31].

8 OXYGEN ATOM TRANSFER

These reactions of peroxyl radicals are sometimes referred to as two-electron reductions [113, 117, 118], in analogy to the one-electron reduction discussed above, although the reaction type is quite different.

Oxygen-atom transfer requires the addition of the alkylperoxyl radical to an electron-rich center. In some cases, this complex may simply decay into an oxide such as **60** and the alkoxyl radical, as in the oxidation of diaryltelluride **59** (reaction (58)) [119].

$$HO{-}C(CH_3)_2{-}CH_2{-}O{-}O^{\bullet} + (R{-}C_6H_4)_2Te \xrightarrow{(58)} HO{-}C(CH_3)_2{-}CH_2{-}O^{\bullet} + O{=}Te(C_6H_4{-}R)_2$$

59 **60**

Interestingly, the behaviour of the trichloromethylperoxyl radical contrasts with this, in that the latter radical reacts predominantly (70%) by electron transfer [119]. Similarly, one of the pathways of thioether oxidation by haloalkylperoxyl radicals in aqueous solution gives rise to sulfoxide (reaction (58a)) [117]. Confusingly, however, it has simultaneously been shown by isotopic labelling [117] that the sulfoxide oxygen must derive from the solvent water and not from the peroxyl radical:

$$R{-}O{-}O^{\bullet} + S(CH_3)_2 \longrightarrow R{-}O^{\bullet} + (CH_3)_2S{=}O \qquad (58a)$$

Trivalent phosphorus compounds (**61**) are known to act as reductants of peroxyl radicals. One of the pathways (reactions (59) and (60)) is oxygen-atom transfer leading to the phosphine oxide **63** [120]. It is known that phosphoranyl radicals such as **62**, besides undergoing α-cleavage (reaction (60)) may add O_2 to form the heteroatom peroxyl radical such as **63** (reaction (61)) (see Chapter 14 of this book).

$$R{-}O{-}O^{\bullet} + P(OR)_3 \ (\mathbf{61}) \xrightarrow{(59)} R{-}O{-}O{-}\dot{P}(OR)_3 \ (\mathbf{62})$$

$$\mathbf{62} \xrightarrow{(60)} R{-}O^{\bullet} + O{=}P(OR)_3 \ (\mathbf{63})$$

$$\mathbf{62} \xrightarrow[(61)]{O_2} R{-}O{-}O{-}P(OR)_3{-}O{-}O^{\bullet} \ (\mathbf{63})$$

The trichloromethylperoxyl radical adds to the iodide ion (reaction (62)) [118]. The adduct **64** is expected to decompose into the trichloromethoxyl radical **65** and hypoiodide (which in the presence of excess iodide is converted into the triiodide anion; for an alternative mechanism see Ref. 118). The subsequent reactions (63) and (64) lead to trichloromethanol, whose lifetime is very short (reaction (65), $k_{65} \geqslant 8 \times 10^4\,s^{-1}$) [121a]. The hydrolysis of the ensuing phosgene (reaction (66)) has been followed in this and another system by pulse conductometry (k_{66} at 25°C = $9\,s^{-1}$, $E_a = 53\,kJ\,mol^{-1}$) [121a].

$$Cl_3C{-}O{-}O^{\bullet} + I^{\ominus} \xrightarrow{(62)} Cl_3C{-}O{-}O{\cdot\cdot}I^{\ominus} \ (\mathbf{64}) \xrightarrow{(63)} Cl_3C{-}O^{\bullet} \ (\mathbf{65}) + I{-}O^{\ominus}$$

$$\mathbf{65} \xrightarrow[(64)]{I^{\ominus},\,H^{\oplus}} Cl_3C{-}OH + I^{\bullet}$$

$$Cl_3C{-}OH \xrightarrow[(65)]{-HCl} Cl_2C{=}O \xrightarrow[(66)]{H_2O} 2\,HCl + CO_2$$

The reaction of peroxyl radicals with alkenes may give rise to epoxides such as **67** via the intermediate **66** (reactions (67) and (68)) (cf. Refs 52 and 122).

$$R{-}O{-}O^{\bullet} + X_2C{=}CX_2 \xrightarrow{(67)} X{-}C(X)(O{-}O{-}R){-}\dot{C}(X){-}X \ (\mathbf{66}) \xrightarrow[(68)]{-RO^{\bullet}} X_2C{-}CX_2 \text{ (epoxide, } \mathbf{67})$$

The rate constant of reaction (67) increases with increasing electron-donating capacity of the C=C double bond [121b] and the oxidative power of the peroxyl radical [52]. In benzene, the addition of acylperoxyl radicals to olefins has been found to be ca. 10^5 times faster than that of alkylperoxyl [52]; in the gas phase for the propyl-2-peroxyl radical $k_{67} \approx 6\,\mathrm{dm^3\,mol^{-1}\,s^{-1}}$ at room temperature [123]. With rate constants as low as this latter one, reaction (67) is expected to represent the rate determining step of epoxide formation, since reactions such as (68) are first-order, with rate constants ranging from 10^2 [73] to $10^6\,\mathrm{s^{-1}}$ [72].

9 BIMOLECULAR DECAY OF PEROXYL RADICALS

9.1 RATE CONSTANTS

Most of the rate constants of the bimolecular decay of peroxyl radicals in aqueous solutions have been determined by pulse radiolysis (cf. Refs 10 and 11) where the peroxyl radical is produced by reacting a given substrate with OH radicals in the presence of oxygen. Under such conditions, organic peroxyl radicals are formed in 90% yield, while the remaining 10% are $HO_2^{\bullet}/O_2^{\bullet -}$. The organic peroxyl radicals usually react faster with one another than with $O_2^{\bullet -}$ (see below), and the self-termination of the $HO_2^{\bullet}/O_2^{\bullet -}$ radicals is also relatively slow (see Chapter 13 of this book). Hence their decay, at least in the early stage, is largely governed by their bimolecular termination. For the determination of the bimolecular rate constant, the initial radical concentration must be known and is provided by dosimetry. The rate of the decay is usually monitored by following the evolution of the optical absorbance of the radicals involved. The inverse of the first half-life of the decay is plotted vs. the initial radical concentration, or the pulse dose D. From its slope the bimolecular rate constant of their disappearance can be derived according to equations (69) and (70):

$$1/t_{1/2} = 2k[RO_2^{\bullet}]_0 \quad (69)$$

$$[RO_2^{\bullet}]_0 = G(RO_2^{\bullet}) \times D \quad (70)$$

Peroxyl radicals which do not decay in one of the unimolecular processes discussed above must disappear bimolecularly. In contrast to many other radicals, they cannot undergo disproportionation. Hence they are left to decay via a recombination process (reaction (71); an exception may be their reaction with the superoxide radical, $O_2^{\bullet -}$; see above), the immediate result of which is a tetroxide:

$$2RO{-}O^{\bullet} \rightleftharpoons R{-}O{-}O{-}O{-}O{-}R \quad (71, -71)$$

$$R{-}O{-}O{-}O{-}O{-}R \longrightarrow \text{Products} \quad (72)$$

Such an intermediate is well established in organic solvents at low temperatures (see Chapter 10 of this book). However, at the temperatures accessible in liquid aqueous solutions the tetroxide, owing to its low ROO–OOR bond strength, estimated at 21 to 33 kJ mol^{-1} for R = alkyl [124–127], can attain a steady-state concentration which is so low that even at the high radical concentrations achievable in a pulse radiolysis experiment it has not yet been detected. Two unimolecular processes of the tetroxide limit its steady-state concentration: the reverse reaction (–71) and the various decay routes into the products (see below), summarized here as reaction (72). For a considerable proportion of bimolecular peroxyl radical decay processes (i.e. the most primary ones), the observed rate constants, $2k_{obs}$, can be up to a few times 10^9 dm^3 mol^{-1} s^{-1} in water (cf. Ref. 32), i.e. they can be close to being diffusion-controlled. This means that the rate of the reverse reaction (–71) in those cases cannot be much faster than that of the sum of the product-forming processes (reaction (72)), as can easily be seen from the algebraic expression (73) derived for k_{obs} under the steady-state approximation $d(RO_4R)/dt = 0$, which should hold quite well as the concentration of the tetroxide is always much smaller than that of $RO_2^{\bullet}$, i.e. always close to zero:

$$2k_{obs} = 2\frac{k_{71}k_{72}}{k_{-71} + k_{72}} \tag{73}$$

Conversely, the finding that the rates of the bimolecular decay of the (primary) peroxyl radicals derived from t-butanol and deuterated t-butanol are practically the same, even though reactions involving the transfer of one or two deuterium atoms (see below) are generally disfavoured relative to the situation with ^{1}H, making k_{72} smaller, indicates that in this case of a primary peroxyl radical k_{72} is at least as large as k_{-71} [128]. Indeed, deuteration causes a considerable shift within the different product-forming decay channels [128].

There is a variation in the rate constants of the bimolecular decay of peroxyl radicals in aqueous solutions, depending on the degree of substitution; in general, tertiary peroxyl radicals decay considerably more slowly than secondary and primary ones. Systematic surveys of the behaviour of tertiary [129, 130], as well as secondary and primary peroxyl radicals [131] in non-aqueous media demonstrate, interestingly, that their reactivity can be quite well described by a Taft relationship (74), where the parameters $2k^0_{obs}$ and ρ^* assume the values displayed in Table 5.

$$\log(2k_{obs}) = \log(2k^0_{obs}) + \rho^*\sigma^* \tag{74}$$

A similar survey regarding aqueous solutions is at present unavailable. The makeup of $2k_{obs}$ (equation (73)) suggests that the substituent effects are

Table 5 Taft-Relationship Parameters for Peroxyl Radical Termination in Non-Aqueous Media at Room Temperature [129, 131]

$RO_2^{\bullet} + RO_2^{\bullet}$	$\log(2k^0_{obs})$	ρ*
Primary/primary	8.5, 8.5[a]	0.0, 0[a]
Secondary/secondary	6.8	2.6
Tertiary/tertiary	4.6, 5.0[a]	5.6, 2.5[a]

[a] From Ref. 130.

introduced into $2k_{obs}$ in part through k_{-71}, since it may be argued on the basis of the behaviour of certain perfluorinated peroxy compounds that electron-drawing substituents tend to stabilize the O—O linkage [127], which would reduce the value of k_{-71}. Perfluoroalkyl is an extreme example of an electron-drawing group. Indeed, such peroxides and trioxides [132] are known to be remarkably stable compared to the corresponding alkyl compounds (cf. Refs 124, 133 and 134). Moreover, certain product-forming reaction channels are closed to tertiary peroxyl radicals (see below), i.e. k_{72} tends to become smaller. This may also contribute to a lowering of $2k_{obs}$ compared to what is observed for secondary and primary peroxyl radicals.

9.2 MECHANISMS OF BIMOLECULAR DECAY

The study of a prodigious number of peroxyl radical reaction systems has shown the existence of several channels:

$$R_2CH{-}O_4{-}CHR_2 \quad (R = \text{alkyl, H})$$

$$\xrightarrow{(75)} R_2C{=}O + R_2CHOH + O_2$$

$$\xrightarrow{(76)} 2\,R_2C{=}O + H_2O_2$$

$$\xrightarrow{(77)} 2\,R_2CHO^{\bullet} + O_2$$

$$\xrightarrow{(78)} R_2CHOOCHR_2 + O_2$$

Depending on the identity of the peroxyl radicals involved, as well as on the properties of the medium (including the gas phase), these channels may be accessed in differing proportions. In particular, the product-forming self-reaction of tertiary peroxyl radicals is restricted to paths (77) and (78), since paths (75) and (76) require the existence of C—H α to the peroxyl function. The exact mechanism of these reactions is still controversial. They have in common the eventual cleavage of the lateral O—O bond of the tetroxide intermediate. There is a dispute concerning whether the product-forming processes are sequential or concerted. Reaction (75) has been described by Russell as a concerted process [135], and this process bears his name. It is formulated as having a six-membered transition state, **68**.

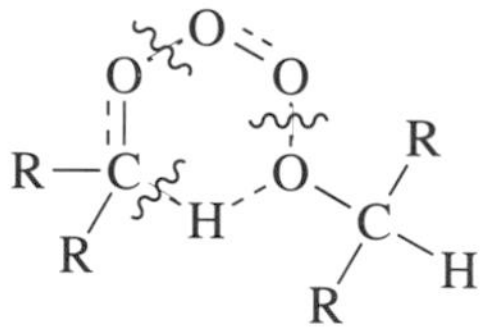

68

Starting from a singlet ground state, the tetroxide gives rise to a carbonyl compound, a hydroxyl compound (usually an alcohol), and oxygen, perhaps as singlet oxygen ($^1O_2$) (cf. Ref. 136). Alternatively, oxygen can be formed in its triplet ground state, and the carbonyl compound in its triplet excited state, which is the cause of the chemiluminescence observed in these reactions (cf. Refs 137–139).

The concertedness of reaction (75) has been questioned on energetic and several other grounds [140, 141]. It has been proposed instead [140] that the carbonyl oxide $RCH{=}O^{+}{-}O^{-}$ (the 'Criegee intermediate'), and the alkoxyl radical $R_2CH{-}O^{\bullet}$ play a central role. For aqueous media, this hypothesis which implies a chain reaction [140] must be ruled out, since on account of the swift rearrangement of primary and secondary alkoxyl radicals under these conditions (reaction (79)) [142–145], the alkoxyl radical cannot fulfil its function [140] as a chain carrier:

$$R_2CH{-}O^{\bullet} \longrightarrow {}^{\bullet}C(OH)R_2 \qquad (79)$$

Moreover, in the case of the methoxyl radical it could be shown that in aqueous solution the corresponding carbonyl oxide may at best play a minor role [146]. Asymmetric O–O bond homolysis of the tetroxide as a first step to product formation has been invoked (cf. Ref. 141), and the idea of the Russell concerted proccess replaced by a three-step sequence, (80)–(82):

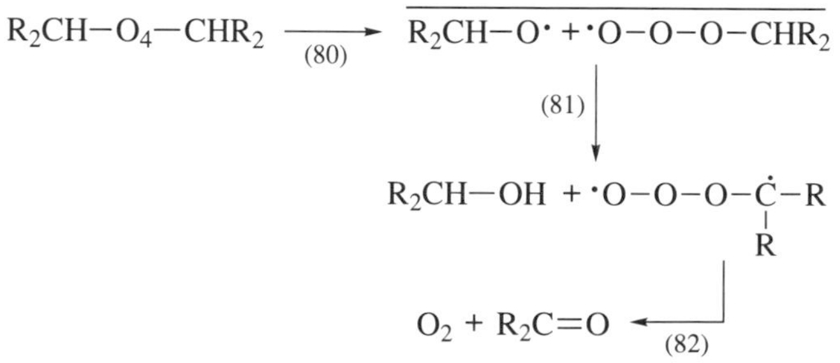

$$R_2CH{-}O_4{-}CHR_2 \xrightarrow{(80)} \overline{R_2CH{-}O^{\bullet} + {}^{\bullet}O{-}O{-}O{-}CHR_2}$$

$$\xrightarrow{(81)} R_2CH{-}OH + {}^{\bullet}O{-}O{-}O{-}\dot{C}R{-}R$$

$$\xrightarrow{(82)} O_2 + R_2C{=}O$$

Process (76) yields hydrogen peroxide and two molar equivalents of the carbonyl compound. This reaction too has often been thought of as being concerted, proceeding via a transition state involving two five-membered rings, **69** (cf. Refs 147 and 148, sometimes termed the Bennett mechanism), or two six-membered rings with the participation of two water molecules, **70**.

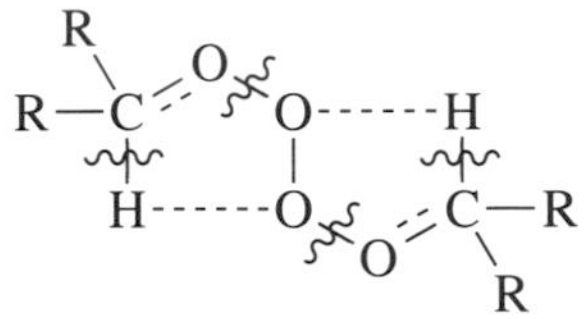

69 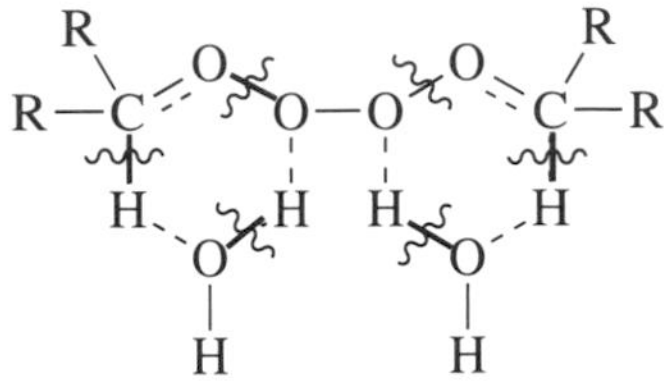 **70**

This hypothesis has been criticized on account of the excessive entropic requirements of the bicyclic transition state [141]. Similar to the case of the Russell mechanism, it has been proposed instead that process (76) consists of several steps, again starting with the homolytic cleavage of the lateral O—O bond (sequence (80)), i.e. (83)–(85) [141]:

$$R_2CHO^{\bullet} + {}^{\bullet}OOOCHR_2 \longrightarrow R_2C{=}O + HOOOCHR_2 \qquad (83)$$

$$HOOOCHR_2 \longrightarrow HOO^{\bullet} + {}^{\bullet}OCHR_2 \qquad (84)$$

$$HOO^{\bullet} + {}^{\bullet}OCHR_2 \longrightarrow H_2O_2 + O{=}CHR_2 \qquad (85)$$

The hypothesis of initial lateral O—O bond homolysis is based on thermochemical arguments regarding polyoxides and polyoxyl radicals [90]. From these data, one estimates bond dissociation energies, BDE(RO—OOOR), ranging from 13 kJ mol^{-1} (R = CH_3) to 35 kJ mol^{-1} (R = cyclo-C_6H_{11}), which would easily enable asymmetric O—O bond homolysis at room temperature to initiate irreversible tetroxide decomposition. However, other thermochemical data on polyoxides and polyoxyls [91, 127] give considerably higher values for the strength of this bond (with R = CH_3), namely BDE = 98 [149], 94 [97], and 99 [127] kJ mol^{-1} from group increments, and BDE = 47 kJ mol^{-1} based on quantum-chemically computed heats of formation [127]. Of these, only the latter value (which is, however, questionable, see below) could be reconciled with the above hypothesis, if the frequency factor of the homolysis was very large. A final decision as to the viability of the assumption of asymmetric O—O bond homolysis cannot be made until the thermochemical basis regarding polyoxides/polyoxyls has been further consolidated, but nevertheless the hypothesis now looks unlikely. As an alternative, in this chapter which deals with peroxyl radical chemistry in aqueous solution, we prefer to formulate a heterolytic water-assisted process (reaction (86)) as the first step toward both the Russell and Bennett mix of products. The fate of the hydrotrioxide formed in reaction (86) will be discussed below.

$$\mathrm{R_2C(H)(O{-}O{-}O{-}O{-}CR_2H)(OH)} \xrightarrow{(86)} \mathrm{R_2C{=}O} + \mathrm{H_2O} + \mathrm{H{-}O{-}O{-}O{-}CR_2{-}H}$$

While process (77) seems of minor importance in primary and secondary peroxyl radicals, compared to those reactions where a hydrogen atom from the position α to the peroxyl group is transferred (processes (75) and (76)) [136], it is essentially the only path for tertiary peroxyl radicals. It has been concluded that tertiary peroxyl radicals do not produce singlet oxygen, as the $^1O_2$ sometimes observed upon their decay in non-aqueous media derives from the reactions of non-tertiary peroxyls formed upon fragmentation (see below) of the tertiary alkoxyl [136]. Similar to what has been discussed before with regard to process (75), it is not clear whether process (77) is a one-step, concerted reaction, where the two lateral O—O bonds are broken at once [149], or whether it is a two-step process (reactions (80) and (87)) [141]:

$$\mathrm{R{-}O{-}O{-}O^{\bullet}} \longrightarrow \mathrm{R{-}O^{\bullet}} + \mathrm{O_2} \qquad (87)$$

Again, the ultimate mechanistic choice will depend on the size of BDE(RO–OOOR). Process (78) appears subordinate to (77), since it is best regarded as a cage recombination of the alkoxyl radicals from process (77). Evidently for non-tertiary peroxyl radicals, products otherwise ascribable to process (75) can be formed as a result of alkoxyl radical cage disproportionation.

The problems regarding the concept of tetroxide asymmetric O—O bond homolysis have been discussed above; the analogous reaction of a hydrotrioxide (reaction (84)) has been thought additionally to intervene in the last step (disproportionation) of the pseudo-Bennett pathway leading to hydrogen peroxide and a further molecule of the carbonyl compound [90, 141]. However, the concept of hydrotrioxide O—O bond homolysis seems even more problematic than that of the tetroxide. On the basis of the thermochemical data reported by Khursan and Martem'yanov [90], a bond strength, BDE(RO—OOH), of 147 kJ mol^{-1} is estimated, which is far too high to allow this reaction to take place at room temperature. It is true that other estimates are lower: BDE = 106 [127] or 99 [97] kJ mol^{-1}, on the basis of group increments; BDE = 45 kJ mol^{-1}, based on thermochemical data obtained by quantum-chemical computation [127]. The quality of the quantum-chemically obtained thermochemical data appears questionable, judged by the fact that the computation of the heat of formation of the relatively simple species, $HO_2^{\bullet}$, within the same frame of reference gave -42.7 kJ mol^{-1}, instead of the generally accepted range of values from above zero up to +21 kJ mol^{-1} [125] (+12.6 kJ mol^{-1} [127]) obtained by group additivity. Therefore, the value of 45 kJ mol^{-1}, which would easily permit hydrotrioxide homolysis to occur at room temperature to the exclusion of other pathways, must be viewed with suspicion, in a similar way to that of the value of 47 kJ mol^{-1} given for the tetroxide lateral O—O bond homolysis (see above).

Indeed, it has been observed that the decomposition of some hydrotrioxides occurs mainly by a non-homolytic pathway [150]; a lifetime of the order of a few seconds at room temperature can be inferred from such data. Decay by O—O bond homolysis makes a relatively small contribution in organic solvents; in aqueous medium it is undetectable. Interestingly, the relative importance of the homolytic pathway increases with the temperature: an activation energy of 105 kJ mol^{-1} has been found in the case of $(CH_3)_2C(OH)OOOH$ [150], which is approximately the same as the values estimated on the basis of group increments (see above).

For aqueous systems, we envisage water-assisted heterolytic fragmentation pathways for the hydrotrioxide, such as reactions (88) and (89), instead of the homolytic routes.

$$R_2CH{-}O{-}O{-}O{-}H + H{-}O{-}H \xrightarrow{(88)} R_2CH{-}OH + O_2 + H_2O$$

$$R_2C(H){-}O{-}O{-}OH + H{-}O{-}H \xrightarrow{(89)} R_2C{=}O + H_2O + H_2O_2$$

Reaction (86), together with reaction (88), mimic the Russell process, and the same reaction (86), with reaction (89), the Bennett process, i.e. both have a common root. As two-step processes, they avoid the objections that have been raised against the one-step Russell and Bennett routes.

The likelihood of a concerted fragmentation of the tetroxide into oxygen and two oxyl radicals (reaction (77)) has been referred to above. There is evidence that fragmentation may go a stage further, e.g. in the aqueous cyclopentylperoxyl system [19], where it is inferred from the products that the 'unzipping' extends to the cleavage of the cyclopentyl ring (reaction (90)).

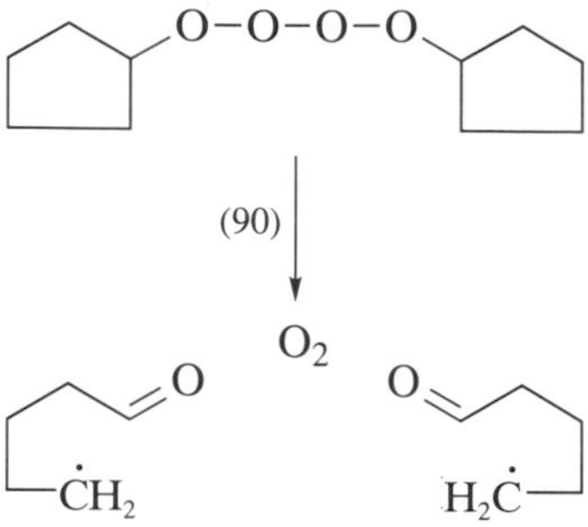

(90)

O_2

$\dot{C}H_2$ $H_2\dot{C}$

A similarly extensive fragmentation process has been observed in the acetate system [151]. It produces formaldehyde, carbon dioxide, and hydrogen peroxide, perhaps via the transition state **71**.

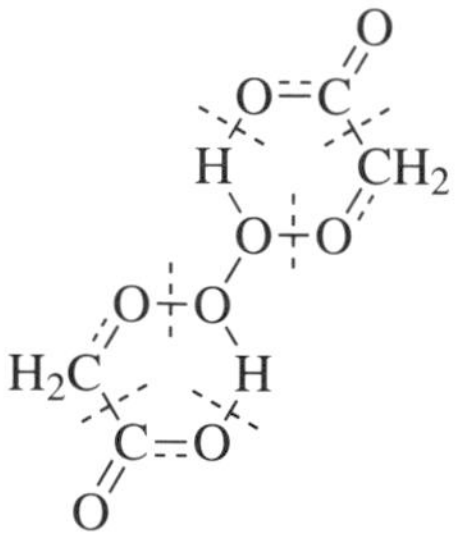

71

9.3 ALKOXYL RADICALS IN PEROXYL RADICAL SYSTEMS

It has been seen that in their bimolecular decay nearly all organic peroxyl radicals give rise to oxyl radicals (usually in competition with other processes). The appearance of alkoxyl radicals complicates the study of peroxyl radical reactions, since in these systems they usually give rise to the formation of further peroxyl radicals which are different from the initial ones. Alkoxyl radicals can undergo β-fragmentation (reaction (91)); those which carry a hydrogen atom in the α-position may additionally undergo 1,2-H shift (reaction (79)) [142–145]:

$$R_3C{-}O^{\bullet} \longrightarrow R^{\bullet} + R_2C{=}O \qquad (91)$$

Compared to the gas phase and non-polar media, β-fragmentation is speeded up considerably with increasing polarity [152], and is fastest in water (cf.

$k[(CH_3)_3CO^{\bullet} \longrightarrow {}^{\bullet}CH_3 + (CH_3)_2CO)] = 10^3\,s^{-1}$ in the gas phase [153] and $10^6\,s^{-1}$ in water [154]).

The effect of water on the 1,2-H shift, reaction (79), is even more drastic. While it has not been observed in the gas phase [155] (this is in agreement with theoretical predictions which indicate it to have a high activation energy of up to 200 kJ mol^{-1} [156]), it proceeds in water with rate constants $>10^6\,s^{-1}$ [143]. Thus, it must be a water-assisted process. Its mechanism, however, has not yet been elucidated.

Alkoxyl radicals are thus often precursors of $O_2^{\bullet -}$ (cf. reaction sequences (79), (1) and (32)). The formation of the superoxide radical can be monitored by its fast reaction with tetranitromethane which yields the strongly absorbing (ϵ(350 nm) = 15 000 dm^3 mol^{-1} cm^{-1}) nitroform anion (reaction (92)) [157].

$$O_2^{\bullet -} + C(NO_2)_4 \longrightarrow O_2 + C(NO_2)_3^- + {}^{\bullet}NO_2 \qquad (92)$$

In contrast to the $HO_2^{\bullet}$ and $O_2^{\bullet -}$ elimination described above where the build-up of the nitroform anion is kinetically first-order, here it is of second-order since it reflects the original encounter of two peroxyl radicals. Apart from using the tetranitromethane method, the formation of H^+ plus $O_2^{\bullet -}$ can also be determined by conductometry [11]. In this case, special dosimetry is required [158].

Fragmentation, such as in reaction (91), creates a new alkyl radical which is peroxylated and eventually gives rise to characteristic products. However, in view of possible 'unzipping', such as that exemplified by reaction (90), the observation of fragment products is not necessarily proof for the intermediacy of oxyl radicals.

10 SELECTED SYSTEMS: PRODUCTS AND THEIR PRECURSORS

In the preceding sections, the types of reaction that have emerged from a large number of detailed product and kinetic studies have been discussed. In this present section, the results obtained with some special systems are reported. For those systems where detailed product studies are available, the products and their yields have been tabulated. However, the main thrust is aimed at providing additional mechanistic insight. This will also concern the fate of short-lived, non-radical intermediates produced by the peroxyl radicals.

The most detailed information has come from radiolytic studies. As has been discussed above, saturation with N_2O/O_2 (4/1) provides a system which contains initially 90% OH radicals and 10% $HO_2^{\bullet}/O_2^{\bullet -}$ radicals (saturation with O_2 leads to 45% OH radicals and 55% $HO_2^{\bullet}/O_2^{\bullet -}$ radicals) as primary intermediates. In the reaction of OH radicals with substrates, more than one kind of radical is often generated. However, one of these might be formed in a predominant yield, and in favourable cases practically only one single kind of radical is produced. Yet, even though a system may start out with a single kind of peroxyl radical, it may develop into a multiple-radical one, owing to fragmentation reactions such as reaction (91). This may then preclude the construction of a mechanistic scheme that would be entirely satisfactory in quantitative terms.

The following subsections are grouped on the basis of types of substrate precursors according to a common peroxyl radical chemistry.

10.1 ALKANES

In the case of methane [146], cyclopentane [19], and cyclohexane [19], a single kind of alkylperoxyl radical is generated (reactions (93)–(98)). The products formed in their decay are compiled in Tables 6–8.

$$CH_4 \xrightarrow[(93)]{^{\bullet}OH} {}^{\bullet}CH_3 \xrightarrow[(94)]{O_2} CH_3\text{-O-O}^{\bullet}$$

$$c\text{-}C_5H_{10} \xrightarrow[(95)]{^{\bullet}OH} c\text{-}C_5H_9^{\bullet} \xrightarrow[(96)]{O_2} c\text{-}C_5H_9\text{-O-O}^{\bullet}$$

$$c\text{-}C_6H_{12} \xrightarrow[(97)]{^{\bullet}OH} c\text{-}C_6H_{11}^{\bullet} \xrightarrow[(98)]{O_2} c\text{-}C_6H_{11}\text{-O-O}^{\bullet}$$

The methylperoxyl radical system is the only one where it has been possible to quantify the formation of the dialkyl peroxide in aqueous solutions (cf. reaction (99)). Because methoxyl rearranges very rapidly into hydroxymethyl (reaction (79)), it must be concluded that dimethylperoxide is formed as a cage product, in agreement with what has been discussed above:

$$2CH_3O^{\bullet} \longrightarrow CH_3\text{–O–O–}CH_3 \qquad (99)$$

Since the dimethylperoxide yield was found to be only 2% of the peroxyl radical yield (Table 6) in the present system, it is assumed that peroxide formation should also be a minor process in other peroxyl radical systems. Although this has never been proven by actually identifying the peroxides

Table 6 Products and Relative Yields from the Radiolytic Oxidation of Methane in N_2O/O_2-Saturated Aqueous Solutions [146]

Product	Yield (% of ·OH)	Product	Yield (% of ·OH)
Formaldehyde	50	Formic acid	5
Methanol	27	Dimethyl peroxide	2
Methyl hydroperoxide	14	Hydrogen peroxide	23

and measuring their yields, material-balance considerations support this view for practically all systems thus far.

With respect to the cyclopentyl- and cyclohexylperoxyl radicals (Tables 7 and 8), it is noted that a considerable amount of ring fragmentation takes place. This is more pronounced in the case of the cyclopentyl than of the cyclohexyl system, and it has been argued that this may be due to the effect of strain in the cyclopentyl ring (cf. Ref. 159) compared to the cyclohexyl ring [19]. The ring-fragmentation products have been attributed to the 'unzipping'

Table 7 Products and Relative Yields from the Radiolytic Oxidation of Cyclopentane in N_2O/O_2-Saturated Aqueous Solutions [19]

Product	Yield (% of ·OH)	Product	Yield (% of ·OH)
Cyclopentanone	31	Formaldehyde	Absent
Cyclopentyl alcohol	15	Organic hydroperoxides	9
Glutaraldehyde	22	Hydrogen peroxide	18
5-Hydroxypentanal	15	Oxygen uptake	93

Table 8 Products and Relative Yields from the Radiolytic Oxidation of Cyclohexane in N_2O/O_2-Saturated Aqueous Solutions [19]

Product	Yield (% of ·OH)	Product	Yield (% of ·OH)
Cyclohexanone	46	Organic hydroperoxides	17
Cyclohexanol	17	Hydrogen peroxide	16
Total acyclic aldehydes	154	Oxygen uptake	74
Formaldehyde	Absent		

reaction (90), with free cyclopentoxyl radicals being ruled out as precursors of the glutaraldehyde and 5-hydroxypentanal, based on the observation [144] that in water the cyclopentoxyl radical undergoes a 1,2-H shift (reaction (100)), even though it is particularly prone to fragmentation (reaction (101)) (cf. Ref. 159).

H O• — (100) → OH; — (101) → H O C ĊH$_2$

There is an interesting difference between the reactions of cyclopentylperoxyl radicals in aqueous [19] and cyclopentane solutions [20]. While in the former the cyclopentanone/cyclopentanol ratio is 2/1, it is close to unity in cyclopentane solutions.

10.2 ALKENES AND TERTIARY ALCOHOLS: β-HYDROXYALKYLPEROXYL RADICALS

When OH radicals react with olefins at room temperature (cf. Ref. 160), they preferentially add to the C–C double bond (e.g. ethene [161], reaction (102)), rather than abstract a carbon-bound hydrogen. An exception to this rule of thumb is 1,4-cyclohexadiene. In this compound, the four bisallylic hydrogen atoms are only loosely bound (see above), and hence addition and H-abstraction occur with comparable probability (reactions (103) and (104)) [162]. Addition furnishes β-hydroxyalkyl radicals. The same type of radical arises when OH radicals abstract a hydrogen atom from a tertiary alcohol, such as tert-butanol (reaction (105)). In the reaction with ethene, practically no abstraction of vinylic hydrogens is observed [163], and in the case of tert-butanol only 5% of the OH radicals abstract the oxygen-bound hydrogen [164]. Thus these reactions are rather clean sources for β-hydroxyalkyl radicals.

In β-hydroxyalkylperoxyl radicals, the peroxyl function is practically unaffected by the presence of the hydroxyl group. This contrasts with the situation in their α-homologues where the hydroxyl group exerts a labilizing influence, expressed by the tendency to eliminate $HO_2^{\bullet}/O_2^{\bullet-}$.

Detailed product studies have been carried out for the three systems mentioned above. The products and their yields are compiled in Tables 9–11.

$$H_2C{=}CH_2 \xrightarrow[(102)]{^{\bullet}OH} HO{-}CH_2{-}\dot{C}H_2$$

$$\text{1,4-cyclohexadiene} \xrightarrow[(103)]{^{\bullet}OH} \text{6-hydroxycyclohex-3-enyl radical}$$

$$\text{1,4-cyclohexadiene} \xrightarrow[(104)]{^{\bullet}OH} \text{cyclohexadienyl radical} + H_2O$$

$$H_3C{-}C(CH_3)_2{-}OH \xrightarrow[(105)]{^{\bullet}OH} H_2\dot{C}{-}C(CH_3)_2{-}OH + H_2O$$

Table 9 Products and Relative Yields from the Radiolytic Oxidation of Ethene in N_2O/O_2-Saturated Aqueous Solutions [161]

Product	Yield (% of $^{\bullet}OH$)	Product	Yield (% of $^{\bullet}OH$)
Glycolaldehyde	59	Organic hydroperoxide	9
Formaldehyde	29	Hydrogen peroxide	29
Ethylene glycol	14	Oxygen uptake	87
Acetaldehyde[a]	9		

[a]Produced via ethylperoxyl radicals which originate from the H atoms (10%, see text) through addition to ethene.

Table 10 Products and Relative Yields (in the Presence of Superoxide Dismutase) from the Radiolytic Oxidation of 1,4-Cyclohexadiene in N_2O/O_2-Saturated Aqueous Solutions [59]

Product	Yield (% of $^{\bullet}OH$)	Product	Yield (% of $^{\bullet}OH$)
Benzene	30	6-Hydroxycyclohex-3-enyl hydroperoxides	5
6-Hydroxycyclohex-3-enone	23	Formaldehyde	30
trans-4,5-Dihydroxycyclohexene	13	Total organic hydroperoxides	11
cis-4,5-Dihydroxycyclohexene	7	Oxygen uptake	100

Table 11 Products and Relative Yields from the Radiolytic Oxidation of tert-Butanol in N_2O/O_2-Saturated Aqueous Solutions [128]

Product	Yield (% of ·OH)	Product	Yield (% of ·OH)
Formaldehyde	23	2-Methyl-2,3-propanediol	9
Acetone	29	Hydrogen peroxide	25
2-Hydroxy-2-methyl-propionaldehyde	36	Bis(2-hydroxy-2-methylpropyl)peroxide	12

10.3 AROMATIC COMPOUNDS

In a similar fashion to the alkenes, aromatic compounds form the OH-adduct radicals very easily. Thus when peroxyl radical formation in these compounds is induced by means of the OH radical, the hydroxyl function is always introduced in addition to the peroxyl one. It seems, however, that this does not detract from the generality of the conclusions drawn with respect to the behaviour of the peroxyl radicals, since in these systems the hydroxyl function does not strongly influence this behaviour when compared to non-hydroxylated functions. The reason seems to be that the double bond formed upon $HO_2^{\bullet}$ elimination is not >C=O, as in the case of α-hydroxyalkylperoxyl radiclas, but >C=C< (rearomatization).

In the discussion of the fate of hydroxycyclohexadienylperoxyl radicals (cf. reactions (28)–(30), (37)–(39)) it has been mentioned that besides the elimination of $HO_2^{\bullet}$, which yields phenolic products and which is driven by the rearomatization tendency (cf. reactions (28) and (29)), a considerable number of fragmentation products are also formed, via reactions such as (40)–(43) which take placc relatively easily on account of the proximity of unsaturation to the peroxyl function. Thus far, only the benzene [41] and chlorobenzene systems [71] have been studied in detail. The products and their yields are compiled in Tables 12 and 13. Some information exists also with respect to the toluene system (Table 14) [165].

Many natural antioxidants have aromatic structures, e.g. the catechol derivatives. These are powerful reactants and can reduce even superoxide (cf. Refs 50, and 166–168). With $O_2^{\bullet-}$ the reaction might proceed via a superoxide–catechol adduct [166]. Lipid and other peroxyl radicals react with these compounds under aroxyl radical formation. Rate constants between 10^7 and $10^9\ dm^3\ mol^{-1}\ s^{-1}$ have been determined [169].

10.4 ETHERS AND ACETALS

Hydrogen-abstraction occurs most easily at the position α to the ether function (cf. Refs 69 and 98). In the bimolecular decay of peroxyl radicals

Table 12 Products and Relative Yields from the Oxidation of Benzene by OH Radicals in Oxygenated Aqueous Solutions [41]

Product	Yield (% of ·OH)
Phenol	55
Cyclohexa-2,5-diene-1,4-diol	2
Hydroquinone	2
Catechol	1
Mucondialdehyde	Absent
5,6-Epoxy-4-hydroxycyclohex-2-enone	0.5
CHO–CH=CH–CO–CO–CH_2OH	3
CHO–CHOH–CH=CH–CHOH–CHO	9
5,6-Dihydroxycyclohex-2-ene-1,4-dione	0.2
CHO–CH_2–CH=CH–CO–CHO	0.5
CHO–CH=CH–CH_2–CHOH–CHO	2
CHO–CHOH–CHOH–CHOH–CHO	3.5
CHO–CO–CH=CH–CO_2H	0.7
CHO–CH=CH–CO–CH_2OH	9
CHO–CH_2–CH_2–CO–CHO	2
CHO–CH=CH–CHO	3.5
CHO–CHOH–CO–CH_2OH	2
CH_3–CHOH–CO_2H	3.5
CHO–CO–CHO	5.5
CH_2OH–CO_2H	3.5
CHO–CHO	3.5
CH_3–CHO	5.5
CH_2O	12.5
HCO_2H	16
Oxygen consumption	100
Hydrogen peroxide	17
Organic hydroperoxides	Absent
$HO_2^{\cdot}/O_2^{\cdot-}$	62

Table 13 Products and Relative Yields from the Radiolytic Oxidation of Chlorobenzene in N_2O/O_2-Saturated Aqueous Solutions [71]

Product	Yield (% of ·OH)	Product	Yield (% of ·OH)
2-Chlorophenol	13	Cl-free compounds containing 2 carbon atoms	6
3-Chlorophenol	9	Cl-free compounds containing 3 carbon atoms	14
4-Chlorophenol	16	Cl-free compounds containing 4 carbon atoms	18
Formic acid	45	Cl-free compounds containing 5 carbon atoms	18
Glyoxylic acid	2	Cl-free compounds containing 6 carbon atoms	6
Glycolic acid	2	Vinylic-chlorine-containing compounds	5

Table 14 Products and Relative Yields from the Radiolytic Oxidation of Toluene in Air-Saturated Aqueous Solutions [165]

Product	Yield (% of ·OH)	Product	Yield (% of ·OH)
o-Cresol	12	Benzyl alcohol	1
m-Cresol	9	Acids	39
p-Cresol	12		

derived from primary ethers, oxyl radicals are formed which undergo either a 1,2-H shift (e.g. reactions (106) and (108)) or, in competition, β-fragmentation (e.g. reactions (107) and (109)).

$$H_3C{-}CH(O^{\bullet}){-}O{-}CH_2{-}CH_3 \xrightarrow{(106)} H_3C{-}\dot{C}(OH){-}O{-}CH_2{-}CH_3$$

$$H_3C{-}CH(O^{\bullet}){-}O{-}CH_2{-}CH_3 \xrightarrow{(107)} {}^{\bullet}CH_3 + H{-}C({=}O){-}O{-}CH_2{-}CH_3$$

$${-}O{-}CH_2{-}CH(O^{\bullet}){-}O{-}CH_2{-} \xrightarrow{(108)} {-}O{-}CH_2{-}\dot{C}(OH){-}O{-}CH_2{-}$$

$${-}O{-}CH_2{-}CH(O^{\bullet}){-}O{-}CH_2{-} \xrightarrow{(109)} {-}O{-}CH_2^{\bullet} + H{-}C({=}O){-}O{-}CH_2{-}$$

A product study has shown that in the case of diethyl ether [98] the 1,2-H shift dominates over the fragmentation reaction (k_{106}/k_{107} estimated at 3–4 [98]), while in poly(ethylene glycol) practically only fragmentation is observed [98] (cf. also Ref. 170). This marked difference is plausible from an energetic point of view because in reaction (107) a methyl radical is liberated while reaction (109) leads to an α-alkoxy-stabilized radical. For diethyl ether [98], diisopropyl ether [69] and 1,4-dioxane [70], detailed product studies exist. The results are complicated in Tables 15–17.

Acetals combine their characteristic structural element, —O—CHR—O—, with that of an ether, R′—O—CHR—. The most interesting feature of the peroxyl radical chemistry of acetals is the rapid loss of $O_2^{\bullet-}$ from the tertiary peroxyl radical (e.g. acetaldehyde dimethyl acetal, reaction (39)). This reaction eventually leads to the formation of methyl acetate and methanol (cf. Table 18). Since considerable OH-radical attack also occurs at the two methoxy groups, other products, notably acetaldehyde, formaldehyde and formic acid, result from this attack.

Table 15 Products and Relative Yields from the Radiolytic Oxidation of Diethyl Ether in N_2O/O_2-Saturated Aqueous Solutions [98]

Product	Yield (% of ·OH)	Product	Yield (% of ·OH)
Acetaldehyde	11	Hydrogen peroxide	23
Ethanol	11	1-Ethoxyethyl hydroperoxide	21
Ethyl acetate	50	Bis(1-ethoxyethyl) peroxide	Absent (<1)
Ethyl formate	13	2-Ethoxyacetaldehyde	9
Formaldehyde	7	Oxygen uptake	80

Table 16 Products and Relative Yields from the Radiolytic Oxidation of Diisopropyl Ether in N_2O/O_2-Saturated Aqueous Solutions [69]

Product	Yield (% of ·OH)	Product	Yield (% of ·OH)
Isopropyl acetate	32	Organic hydroperoxides	16
Acetone	39	Hydrogen peroxide	29
2-Propanol	21	Acids	9
Formaldehyde	32	Oxygen consumption	91
2-Isopropoxypropanal	25		

Table 17 Products and Relative Yields from the Radiolytic Oxidation of 1,4-Dioxane in N_2O/O_2-Saturated Aqueous Solutions [70]

Product	Yield (% of ·OH)	Product	Yield (% of ·OH)
Formaldehyde	11	1,2-Ethanediol diformate	50
1,4-Dioxan-2-one	7	Organic hydroperoxides	20
2-Hydroxy-1,4-dioxane	7	Hydrogen peroxide	2
1,2-Ethanediol monoformate	11	Oxygen uptake	111

Table 18 Products and Relative Yields from the Radiolytic Oxidation of Acetaldehyde Dimethyl Acetal in N_2O/O_2-Saturated Aqueous Solutiosn [68]

Product	Yield (% of ·OH)	Product	Yield (% of ·OH)
Methanol	53	Formic acid	29
Formaldehyde	29	Hydrogen peroxide	11
Acetaldehyde	27	Organic hydroperoxide(s)	31
Methyl acetate	38	Oxygen uptake	95

10.5 PRIMARY, SECONDARY AND POLYHYDRIC ALCOHOLS; CARBOHYDRATES

The $HO_2^{\bullet}$-elimination (cf. reactions (27), (31) and (32)) dominates the chemistry of α-hydroxyalkylperoxyl radicals. The OH radical attacks these compounds preferentially in the α-position to a hydroxyl function (cf. Ref. 164). At the O_2 concentrations that prevail under conditions of air- or N_2O/O_2-saturation (at atmospheric pressure), oxygen addition to the radicals is usually much faster than the competing reactions, e.g. the water elimination from vicinal diols (reaction (110)) [171]:

$$^{\bullet}CHOH{-}CH_2OH \longrightarrow H_2O + CHO{-}CH_2^{\bullet} \quad (110)$$

There are some exceptions to this. Sugar phosphates undergo the analogous elimination of the phosphate groups [91, 172, 173] so fast that it competes unsuccessfully with the O_2-addition. The same is true for β-chloro-α-hydroxyalkyl [174, 175], but not for β-acetoxy-α-hydroxyalkyl [175, 176]. In polyhydric alcohols and carbohydrates, the rate of $HO_2^{\bullet}$-elimination depends strongly on the position of the peroxyl function within the molecule [54]. Details are not clearly understood; from the data reported on D-glucose [54] it is clear that the C^1 (hemiacetal-type) peroxyl radical, **72**, eliminates fast (reaction (111)); the rate constant exceeding $10^5\,s^{-1}$), while the radical at C^5, **73**, eliminates only slowly. Here, $HO_2^{\bullet}$-elimination can proceed only when the ring is opened as well (e.g. reactions (112) and (113)).

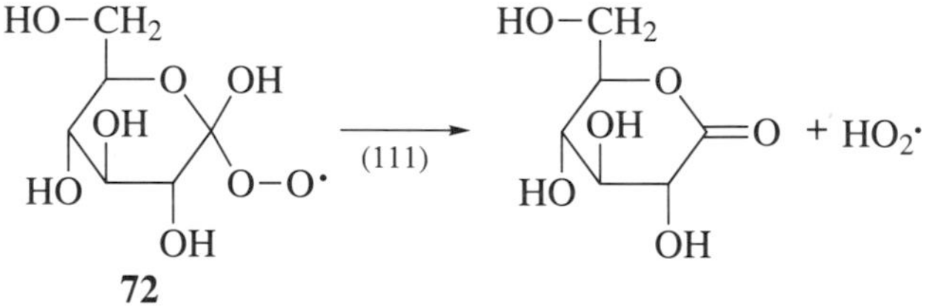

72

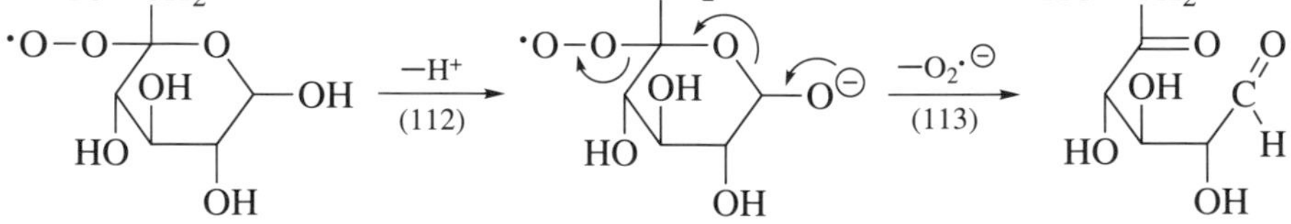

73

A detailed product analysis (results shown in Table 19) shows that the products resulting from the $HO_2^{\bullet}$-elimination reaction do indeed dominate. (The fragmentation products present in low yields are likely to arise for the most part from the fragmentation of oxyl radicals formed by OH-radical attack at an alcoholic hydroxyl group (cf. Ref. 164). A similar product study is available for D-ribose (data not shown) [177]. Near pH 7 the $HO_2^{\bullet}$ radicals are already largely dissociated, and in contrast to other peroxyl radicals, $O_2^{\bullet -}$ radicals do not undergo H-abstraction reactions. This is whys in the case of aqueous D-glucose solutions that no chain auto-oxidation is observed, while 2-methyltetrahydrofuran is autoxidized effectively under otherwise equal conditions [178]. The behaviour found with D-glucose can be generalized to other carbohydrates, where it becomes obvious that the absence of an auto-oxidation chain reaction is of tremendous importance for the biosphere as well as economically. The natural lifetime of cellulose-based materials such as wood, cotton, and linen would be measured in years, rather than decades or centuries, if they were subject to this kind of autoxidation.

Table 19 Products and Relative Yields from the Radiolytic Oxidation of D-Glucose by OH Radicals in Oxygenated Aqueous Solutions [179]

Product	Yield (% of $^{\bullet}$OH)	Product	Yield (% of $^{\bullet}$OH)
D-Gluconic acid	17	L-*threo*-Tetrodialdose	4
D-*arabino*-Hexosulose	17	D-Erythrose	0.2
D-*ribo*-Hexos-3-ulose	11	D-Erythronic acid	0.2
D-*xylo*-Hexos-4-ulose	9	D-Gyceraldehyde	1
D-*xylo*-Hexos-5-ulose	11	D-Glyceric acid	1
D-*gluco*-hexodialdose	29	Glyoxal	2
D-Glucuronic acid	1	Glyoxylic and glycolic acid	7
D-Arabinose	1	Formaldehyde	2
D-Arabinonic acid	0.5	Formic acid	11
xylo-Pentodialdose	1	Hydrogen peroxide	42
D-Xylose	0.2	D-*arabino*-Hexulonic acid	Secondary product

It has been observed with disaccharides and polysaccharides under anoxic conditions that several different free-radical reaction paths eventually lead to the scission of the glycosidic linkage [180–184]. These reactions are sufficiently slow [184] to be readily suppressed by O_2. As a consequence, *G* (scission) is reduced [184, 185]. This does not mean, of course, that there should not be any specifically oxygen-induced scission of this bond at all. For example, one expects superoxide elimination similar to reaction (39) at the acetalic C^1 to lead to the C^1-carbocation and subsequent hydrolysis. Attempts have been made to unravel the underlying reactions using cellobiose as a model system [185]. However, mechanistic details of this 'protection' by O_2 are not clearly understood.

10.6 CARBOXYLIC ACIDS

Detailed product studies are at present available for the acetic [151], 2,4-dimethylglutaric [23], hydroxymalonic [51] and ketomalonic [49] acids (Tables 20–22). An interesting feature of the acetate system is the formation of carbon dioxide; this might proceed via the transition state **71**, as proposed above. 2,4-Dimethylglutaric acid has been investigated as a low-molecular-weight model for poly(acrylic acid) [23]. In its peroxyl radical chemistry the

Table 20 Products and Relative Yields from the Radiolytic Oxidation of Acetate Ions by OH Radicals in Oxygenated Aqueous Solutions [151]

Product	Yield (% of ·OH)	Product	Yield (% of ·OH)
Glyoxylic acid	50	Organic hydroperoxide(s)	13
Glycolic acid	13	Hydrogen peroxide	32
Formaldehyde	26	Oxygen uptake	98
Carbon dioxide	26		

Table 21 Products and Relative Yields from the Radiolytic Oxidation of 2,4-Dimethylglutaric Acid by OH Radicals in Oxygenated Aqueous Solutions [23]

Product	Yield (mol%)
2-Formyl-4-methyl glutaric acid	7–9
2,4-Dimethyl-3-oxoglutaric acid	13
2-Hydroxymethyl-4-methylglutaric acid	4
2- and 3-Hydroxy-2,4-dimethylglutaric acid	4
2-Methyl-4-oxovaleric acid	25
2-Methylglutaraldehydic acid	4
5-Hydroxy-2-methyl-4-oxovaleric acid	3
Acetylacetone	9
Unidentified products containing 7 carbon atoms	4
2-Methyl-malonaldehydic acid	4
Acetaldehyde	4
Formaldehyde	2
Minor unidentified fragment products	7
Carbon dioxide	46
Hydrogen peroxide	20
Organic hydroperoxides	5

Table 22 Products and Relative Yields from the Radiolytic Oxidation of Hydroxymalonic Acid by OH Radicals in Oxygenated Aqueous Solutions [51][a]

Products	Yield (% of $^{\cdot}OH$)		
	pH 3	pH 7	pH 10
Ketomalonic acid	91	57	23
Oxalic monoperacid	4	20	50
Ketomalonic acid—H_2O_2	7	3	1.5
Carbon dioxide	n.d.	n.d.	43

[a] n.d. = not determined.

formation of acetylacetone is noteworthy (Table 21). This product shows oxidation at two positions, i.e. C^2 and C^4, compared to the substrate molecule, which indicates that the initial peroxyl radical **74** undergoes an intramolecular H-transfer (reaction (114), via a six-membered transition state), giving rise to peroxyl radical **76** (reaction (115)) which (among other reactions) can be reduced by $O_2^{\cdot -}$, yielding the dihydroperoxidic product **77** (reaction (116)). This can decarboxylate into acetylacetone (reaction (117)). In poly(acrylic acid) this type of reaction sequence also occurs and leads to considerable oxidation to give a product which retains polymeric character [99, 186].

$$^{\cdot}O{-}O{-}C(CH_3)(CO_2H){-}CH_2{-}CH(CH_3)(CO_2H)\ (\mathbf{74}) \xrightarrow{(114)} HO{-}O{-}C(CH_3)(CO_2H){-}CH_2{-}{}^{\cdot}C(CH_3)(CO_2H)\ (\mathbf{75}) \xrightarrow[(115)]{O_2} HO{-}O{-}C(CH_3)(CO_2H){-}CH_2{-}C(CH_3)(CO_2H){-}O{-}O^{\cdot}\ (\mathbf{76})$$

$$\mathbf{76} \xrightarrow[(116)]{O_2^{\cdot\ominus},\ H^{\oplus},\ -O_2} HO{-}O{-}C(CH_3)(CO_2H){-}CH_2{-}C(CH_3)(CO_2H){-}O{-}OH\ (\mathbf{77})$$

$$\mathbf{77} \xrightarrow{(117)} H_3C{-}\overset{O}{\overset{\|}{C}}{-}CH_2{-}\overset{O}{\overset{\|}{C}}{-}CH_3 + 2\,CO_2 + 2\,H_2O$$

An interesting example of the complexity of some of these systems is presented by the peroxyl radical chemistry of hydroxymalonic acid [51] (beginning with reactions (119) and (120); for products see Table 22), and in combination with this, that of its product, ketomalonic acid, **19** [49]. The major aspects have been discussed above (cf. reactions (26), (33) and (34)); here, attention is drawn to the fact that the intermediate **58** undergoes fragmentation, leading to oxalic monoperacid and the $CO_2^{\cdot-}$ radical (reaction (118)). The latter reacts with O_2, giving rise to the superoxide radical (reaction (23)). This induces one of the few chain reaction known to occur with the superoxide radical acting as a chain carrier in protic solvents [50].

$O_2^{\cdot-}$ (26) / (−26) — $^-O_2C{-}C(=O){-}CO_2^-$ **19** — $O_2^{\cdot-}$ (34) / (−34)

$^-O{-}O{-}C(CO_2^-)_2{-}O^\cdot$ **58** ⇌ (57) / (−57) $^\cdot O{-}O{-}C(CO_2^-)_2{-}O^-$ **57**

58 → (118) $+ H^+$ → $HO{-}O{-}C(=O){-}CO_2^- + CO_2^{\cdot-}$

$CO_2^{\cdot-}$ → O_2 (23) → $CO_2 + O_2^{\cdot-}$

57 ⇌ $OH^-/-H_2O$ (33/−33) $^\cdot O{-}O{-}C(CO_2^-)_2{-}OH$ **25**

$H{-}C(CO_2^-)_2{-}OH$ → $^\cdot OH/-H_2O$ (119) → $^\cdot C(CO_2^-)_2{-}OH$ → O_2 (120) → **25**

The rate constants for the reaction of O_2 with carboxylic-acid-derived radicals are typically close to being diffusion-controlled ($k \geqslant 10^9\,dm^3\,mol^{-1}\,s^{-1}$). Therefore, the recently reported value of only $1.7 \times 10^7\,dm^3\,mol^{-1}\,s^{-1}$ [187] for the rate constant of the reaction of the malonic-acid-derived radical with oxygen, determined from electron spin resonance (ESR) experiments, came as a surprise. This reaction is very important for modelling the Belousov–Zhabotinsky reaction. It has therefore been remeasured more directly by pulse radiolysis [188], and it has been found that its value is as large as expected ($10^9\,dm^3\,mol^{-1}\,s^{-1}$) on the basis of the above homologues.

10.7 HALOGENATED HYDROCARBONS

Halogenated compounds play a major part as environmental pollutants, especially the solvents tetrachloroethene and trichloroethene that have seaped into the ground in many places through negligence or because of industrial accidents. In the soil they are degradable by bacteria to *cis*-1,2-dichloroethene and vinyl chloride, and these secondary pollutants often predominate over the primary ones. Eventually they reach the ground water and so threaten drinking-water resources. The more volatile pollutants such as the carcinogenic vinyl chloride can re-emanate and may diffuse into basements, subterranean garages, or subway systems. Besides localized contamination hot-spots, there is a more general threat: transported by the atmosphere, chlorinatd hydrocarbons enter the groundwater generally and in remote locations, as they are washed down by rain. In water remediation, the OH-radical-generating *advanded oxidation processes* are gaining in importance. Hence, detailed information on the elementary steps in these processes is urgently needed, and an impressive body of data on the products and the mechanistic aspects of the peroxyl radical chemistry of halogenated (mainly chlorinated) compounds has already been assembled.

With chlorinated olefins as with other olefins, OH radicals react by addition (e.g. reaction (121)). The resulting radicals, **78**, are geminal chlorohydrins which eliminate HCl very rapidly (reaction (122), $k_{122} \geqslant 8 \times 10^5\,\mathrm{s}^{-1}$) [174, 189]. In the absence of O_2, the radicals produced (**79**) combine to yield tetrachlorosuccinic dichloride [174]. It is noted that on the pulse radiolysis time-scale, i.e. in the submillisecond to low millisecond range, the acylchloride function of radicals **79** does not yet hydrolyse [174]. In the presence of oxygen these radicals are converted into the corresponding peroxyl radicals (**80**) (reaction (123)). Again, on the pulse radiolysis time-scale these radicals seem to decay mainly by second-order processes (e.g. reaction (124)) [190]. Their rate of hydrolysis (reaction (125)) is not yet known. The knowledge of this rate constant would be of some importance for the assessment of the contribution of the peroxyl radicals **81** in the bimolecular decay of such peroxyl radicals at much lower concentrations, i.e. at much longer overall lifetimes than prevail under pulse radiolysis conditions.

$$^{\bullet}OH + Cl_2C{=}CCl_2 \longrightarrow Cl_2(OH)C{-}CCl_2^{\bullet}\ (\mathbf{78}) \qquad (121)$$

$$Cl_2(OH)C{-}CCl_2^{\bullet}\ (\mathbf{78}) \longrightarrow HCl + Cl(O)C{-}CCl_2^{\bullet}\ (\mathbf{79}) \qquad (122)$$

$$Cl(O)C{-}CCl_2^{\bullet}\ (\mathbf{79}) + O_2 \longrightarrow Cl(O)C{-}CCl_2O_2^{\bullet}\ (\mathbf{80}) \qquad (123)$$

$$2Cl(O)C{-}CCl_2O_2^{\bullet}\ (\mathbf{80}) \longrightarrow 2Cl(O)C{-}C(O)Cl + O_2 + 2Cl^{\bullet} \qquad (124)$$

$$Cl(O)C{-}CCl_2O_2^{\bullet}\ (\mathbf{80}) + H_2O \longrightarrow HO_2C{-}CCl_2O_2^{\bullet}\ (\mathbf{81}) + HCl \qquad (125)$$

Chlorine atoms appear to be set free in the bimolecular decay of peroxyl radical **80** (and similarly of **81**), and are scavenged by the chlorinated olefin (e.g. reaction (126)). As a consequence of this, the peroxyl radical **83** (cf. reaction (127)) also plays a considerable role in the subsequent bimolecular decay processes [191]. For example, its decay can lead to the formation of trichloroacetyl chloride via the oxyl radical **84** (cf. reactions (128) and (129)) which hydrolyses into the major product, trichloroacetic acid (Table 23).

$$^{\bullet}Cl + Cl_2C{=}CCl_2 \longrightarrow Cl_3C{-}CCl_2^{\bullet}\ (\mathbf{82}) \quad (126)$$

$$Cl_3C{-}CCl_2^{\bullet}\ (\mathbf{82}) + O_2 \longrightarrow CCl_3{-}CCl_2O_2^{\bullet}\ (\mathbf{83}) \quad (127)$$

$$CCl_3{-}CCl_2O_2^{\bullet}\ (\mathbf{83}) + RO_2^{\bullet} \longrightarrow Cl_3C{-}CCl_2O^{\bullet}\ (\mathbf{84}) + O_2 + RO^{\bullet} \quad (128)$$

$$Cl_3C{-}CCl_2O^{\bullet}\ (\mathbf{84}) \longrightarrow Cl_3C{-}C(O)Cl + {}^{\bullet}Cl \quad (129)$$

$$CCl_3{-}CCl_2O^{\bullet}\ (\mathbf{84}) \longrightarrow Cl_3C^{\bullet} + Cl_2C{=}O \quad (130)$$

Besides being capable of undergoing fragmentation of an α C–Cl bond, the likely intermediate **84** may alternatively suffer C–C bond cleavage, e.g. reaction (130). This leads to the formation of phosgene. Phosgene hydrolyses only slowly ($9\,s^{-1}$ at 25°C [121a]) and has been detected as an intermediate in the pulse radiolysis of N_2O/O_2-saturated solutions of tetrachloroethene [190]. It is a major precursor of carbon dioxide (Table 23).

Table 23 Products and Relative Yields from the Radiolytic Oxidation of Tetrachloroethene by OH Radicals in Oxygenated Aqueous Solutions [190]

Product	Yield (% of ·OH)	Product	Yield (% of ·OH)
Chloride ion	590	Carbon monoxide	27
Carbon dioxide	190	Trichloroacetic acid	37

In reactions such as (124) or (129) (and in reactions of the trichloromethyl radicals formed in reaction (130); cf. Ref. 192), chlorine atoms are set free, and these behave as chain carriers (reaction (126)). In water, the chain length (ca. 6 chloride ions are formed per OH radical, cf. Table 23) is relatively short. The chain-terminating processes are at present not fully understood. Interestingly, the TiO_2/UV-induced degradation of chlorinated hydrocarbons [193–195], although often discussed as being OH-radical-induced, seems to proceed by different pathways.

The bimolecular decay of peroxyl radicals **80/83** gives rise to acidity [190]. Trichloroacetyl chloride has been shown by stopped-flow techniques to hydrolyse on the submillisecond time-scale [196]. Oxalyl dichloride (from reaction (124)), perhaps via a very short-lived oxyl radical analogous to **84**) in water is known to decompose into the gaseous products, carbon monoxide and carbon dioxide, together with two molecules of HCl [197]. Owing to effervescence which disturbs the measurement, the rate of its hydrolysis could not be determined with any accuracy [196], but it seems that its hydrolysis/decomposition must be similarly fast. Carbon monoxide is among the products (Table 23), and it may be suggested that these reactions contribute to the very rapid formation of acid in these systems.

In the OH-radical-induced degradation of *cis*-1,2-dichloroethene (cf. reactions (131) and (132) peroxyl radical **87** is formed which carries a hydrogen atom α to the peroxyl function. This could give wide scope to reactions such as (133) and (134), as in the case of other systems carrying a hydrogen agom in the α-position (cf. reactions (75) and (76)):

$$\mathrm{HOCHCl{-}CHCl^{\bullet}}\ (\mathbf{85}) \longrightarrow \mathrm{HCl + HC(O){-}CHCl^{\bullet}}\ (\mathbf{86}) \qquad (131)$$

$$\mathrm{HC(O){-}CHCl^{\bullet}}\ (\mathbf{86}) + \mathrm{O_2} \longrightarrow \mathrm{HC(O){-}CHClO_2^{\bullet}}\ (\mathbf{87}) \qquad (132)$$

$$2\ \mathrm{HC(O){-}CH(Cl){-}O{-}O^{\bullet}}\ (\mathbf{87}) \xrightarrow{(133)} \mathrm{HC(O){-}C(O)Cl} + \mathrm{O_2} + \mathrm{HC(O){-}CH(Cl){-}OH}$$

$$2\ \mathrm{HC(O){-}CH(Cl){-}O{-}O^{\bullet}}\ (\mathbf{87}) \xrightarrow{(134)} 2\ \mathrm{HC(O){-}C(O)Cl} + \mathrm{H_2O_2}$$

$$2\ \mathrm{HC(O){-}CH(Cl){-}O{-}O^{\bullet}}\ (\mathbf{87}) \xrightarrow{(135)} 2\ \mathrm{HC(O){-}CH(Cl){-}O^{\bullet}} + \mathrm{O_2}$$

$$\mathrm{HC(O){-}CH(Cl){-}O^{\bullet}} \xrightarrow{(136)} \mathrm{HC^{\bullet}(O)} + \mathrm{H{-}C(O)Cl}$$

The product of these reactions are glyoxylic chloride and the geminal chlorohydrin 2-chloro-2-hydroxyacetaldehyde. The latter gives rise to glyoxal, while the former hydrolyses to glyoxylic acid and HCl. As can be seen from Table 24 neither one of these products predominates. Instead, major amounts of carbon dioxide, formic acid and carbon monoxide are formed. The carbon monoxide arises from the formyl chloride formed in reaction (136). The latter is known to predominantly decompose into HCl and carbon monoxide (reaction 137) [198], and only a small fraction suffers hydrolysis in neutral and weakly acidic, or even basic solutions [199]. Its rate of decomposition is very fast ($k_{137} = 10^4\,s^{-1}$ [199]) and hence can compete successfully with the hydrolysis. Only at very high pH levels does the latter process acquire some importance:

$$HC(OCl) \longrightarrow HCl + CO \qquad (137)$$

Table 24 Products and Relative Yields from the Radiolytic Oxidation of *cis*-1,2-Dichloroethene by OH Radicals in Oxygenated Aqueous Solutions [200]

Product	Yield (% of ·OH)	Product	Yield (% of ·OH)
Chloride ion	200	Carbon dioxide	93
Formate ion	102	Glyoxal	11
Carbon monoxide	32	Glyoxylic acid	4

The precursor of the formic acid is not yet equally obvious. In separate experiments we have observed that glycolate reacts very rapidly with hydrogen peroxide and the ensuing hydroxyhydroperoxide decomposes into formic acid and carbon dioxide (reactions (138) and (139)).

$$\mathrm{HC(=O)-C(=O)OH} \xrightarrow[(138)]{H_2O_2} \mathrm{HO-O-CH(OH)-C(=O)OH} \xrightarrow[(139)]{-CO_2,\,-H_2O} \mathrm{H-C(=O)OH}$$

Data on 1,1-dichloroethene [201] and dichloromethane [202] also exist. They are compiled in Tables 25 and 26.

Table 25 Products and Relative Yields from the Radiolytic Oxidation of 1,1-Dichloroethene by OH Radicals in Oxygenated Aqueous Solutions [201]

Product	Yield (% of ·OH)	Product Product	Yield (% of ·OH)
Carbon dioxide	67	Chloride ion	217
Formic acid	30	Glycolic acid	38
Carbon monoxide	7	Chloracetic acid	8
Formaldehyde	59	Hydrogen peroxide	21

Table 26 Products and Relative Yields from the Radiolytic Oxidation of Dichloromethane by OH Radicals in Oxygenated Aqueous Solutions [202]

Product	Yield (% of ·OH)	Product	Yield (% of ·OH)
Carbon dioxide	61	Formaldehyde[a]	4
Formic acid	16	Chloride ion	190
Carbon monoxide	21	Hydrogen peroxide	23

[a] Induced by a contribution of the solvated electron.

10.8 DNA PEROXYL RADICALS AND THEIR MODEL SYSTEMS

The importance of DNA-derived peroxyl radicals in radiation-induced cell killing (cf. radiotherapy of cancer) is obvious from the strong enhancement of this effect by O_2 (cf. Ref 203). Such intermediates are also formed by some anti-cancer drugs such as bleomycin (e.g. Refs 204–206). Considerable effort has been spent in order to determine the products formed under such conditions, mainly at the model level of DNA in aqueous solution, and its sub-units, the nucleobases. For a survey of the current knowledge in this area, the reader is referred to recent reviews [9, 207 and 208]. From a mechanistic point of view, DNA peroxyl radical chemistry still remains to a large extent speculative. The reason for this is obvious: OH-radical attack leads to a multitude of different kinds of radicals, both base-derived and sugar-derived. Some of these radicals are readily converted into the corresponding peroxyl radicals, while others are not (cf. **15** [47]). These radicals may interact with one another, or may eliminate $HO_2^{\bullet}/O_2^{\bullet -}$ which could also interact with the DNA-derived radicals/peroxyl radicals. Owing to its high molecular weights and its high charge density, the rate of bimolecular decay of these polymer-sited peroxyl radicals is slow and the kinetics are necessarily complex (cf. also simpler polymer systems [99, 209 and 210]). Therefore, we will only present here some studies of model systems where both product and kinetic data exist. A case in point is the OH-radical-induced degeneration of uracil in the presence of oxygen, which shows a considerable change of product distribution as a function of pH (Table 27) [211]. Upon OH-radical attack, the C^5- and C^6-OH-adduct radicals are formed in a ratio of 4.5/1 [212]. In the presence of O_2 these are converted into the corresponding peroxyl radicals. The dominating radical, **88**, can undergo OH^--induced $O_2^{\bullet -}$ elimination (reactions (140) and (141)), similar to the sequence discussed above for the peptide-derived peroxyl radicals (cf. reactions (35) and (36)). In reaction (141) the ‘isopyrimidine’ **90** is formed, which then undergoes subsequent rearrangement into the corresponding pyrimidine and/or hydrates (cf. Refs 67 and 213).

Table 27 Products and Relative Yields from the Radiolytic Oxidation of Uracil by OH Radicals in Oxygenated Aqueous Solutions [211]

Product	Yield (% of ·OH)		
	pH 3	pH 6.5	pH 10
5,6-*cis*-Dihydroxy-5,6-dihydrouracil	11	16	25
5,6-*trans*-Dihydroxy-5,6-dihydrouracil	9	20	18
Isobarbituric acid	0	4	21
1-*N*-formyl-5-hydantoin	29	25	4
Dialuric acid	16	7	4
Isodialuric acid	2	4	2
5-Hydroxyhydantoin	7	7	5
Unidentified products	16	11	16

$$\mathbf{88} \xrightarrow[(140/-140)]{OH^{\ominus},\ -H_2O} \mathbf{89} \xrightarrow[(141)]{-O_2^{\bullet\ominus}} \mathbf{90}$$

The marked pH-dependence of the product distribution (Table 27) arises through the suppression of reactions (140) and (141) at low pH, which forces the system along the bimolecular-decay route [211].

In DNA free-radical chemistry, oxygen usually acts as an enhancer of damage. A curious example of the opposite effect has been found at the model compound level, e.g. the electron adduct of uracil and thymine (**91**). This adduct and its O-protonated conjugate acid (**92**) react with O_2 by 'electron transfer' (probably via intermediate adducts which are too short-lived to be detected), thereby re-forming the nucleobase (reactions (144) and (146)) [214]. In the presence of phosphate buffer, this 'protective' effect is to some extent reversed. This is because the buffer accelerates the protonation of the electron adduct at the carbon, C^6, which is very much slower than the protonation reaction, C^4—O^- [215, 216]. The rate constants of protonation at oxygen anions and at carbanions can differ by many orders of magnitude [217]. Moreover, the pK_a of **93** is expected to be above that of **92**, in analogy to the situation of the barbituric acid system (pK_a(enol form) = 2.3; pK_a(keto form) = 4.0 [217]). At high buffer concentrations, reaction (143) becomes fast enough for **93** to play a role, whereby O_2 increasingly reacts with **93** and so gives rise to a destruction of the pyrimidine chromophore (reaction (145)).

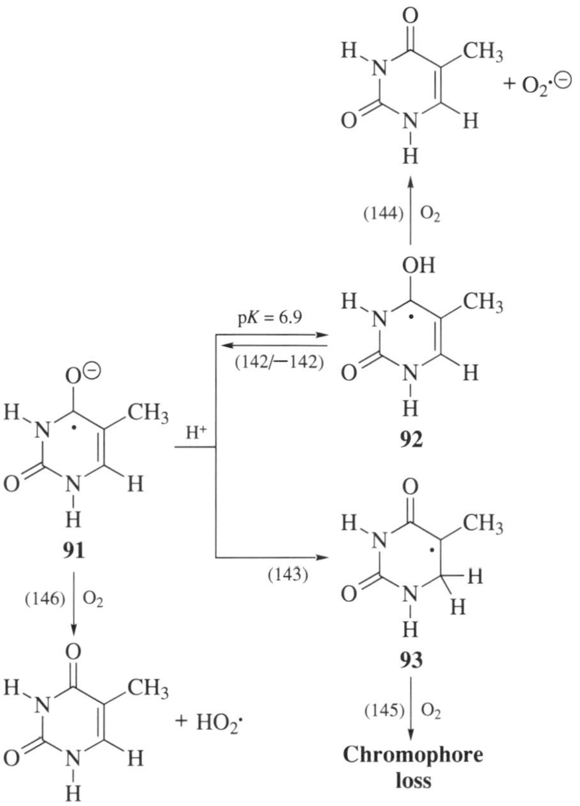

DNA peroxyl radicals already release $O_2^{\cdot-}$ in low yields on the time-scale of pulse radiolysis [218]. It is tempting to correlate this, at least in part, with the known [219] formation of DNA-bound 2′-deoxyribonolactone, **97** (reactions (147)–(149)).

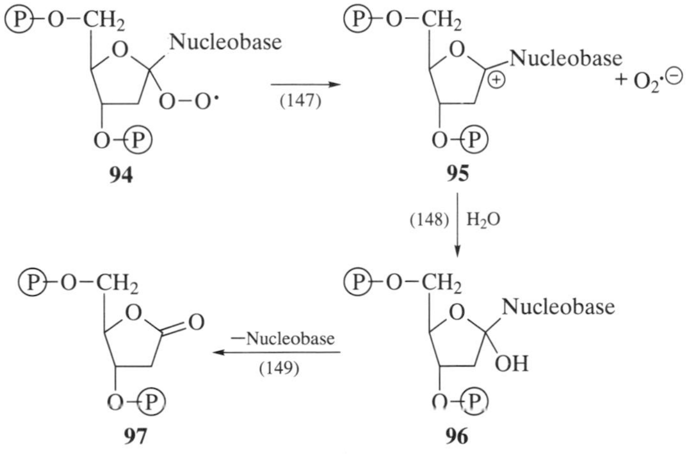

This would be in analogy to reaction (39) discussed above, with the intermediate carbocation **95** being stabilized by –OR and $-NR_2$ functions, rather than by two –OR groups as in the case of **31**.

REFERENCES

1. G. Mark, H.-G. Korth, H.-P. Schuchmann and C. von Sonntag, *J. Photochem. Photobiol. A: Chem.* in press (1996).
2. R.G. Zepp, J. Hoigné and H. Bader, *Environ. Sci. Technol.* **21,** 443 (1987).
3. G.V. Buxton, C.L. Greenstock, W.P. Helman and A.B. Ross, *J. Phys. Chem. Ref. Data* **17,** 513 (1988).
4. P. Neta, R.E. Huie and A.B. Ross, *J. Phys. Chem. Ref. Data* **17,** 1027 (1988).
5. C. von Sonntag, G. Mark, R. Mertens, M.N. Schuchmann and H.-P. Schuchmann, *J. Water Supply Res. Technol.—Aqua* **42,** 201 (1993).
6. C. von Sonntag, *J. Water Supply Res. Technol.—Aqua* **45,** 84 (1996).
7. C. von Sonntag and H.-P. Schuchmann, *Angew. Chem. Int. Ed. Engl.* **30,** 1229 (1991).
8. B.H.J. Bielski, D.E. Cabelli, R.L. Arudi and A.B. Ross, *J. Phys. Chem. Ref. Data* **14,** 1041 (1985).
9. C. von Sonntag, *The Chemical Basis of Radiation Biology*, Taylor and Francis, London (1987).
10. K.-D. Asmus, *Meth. Enzymol.* **105,** 167 (1984).
11. C. von Sonntag and H.-P. Schuchmann, *Meth. Enzymol.* **233,** 3 (1994).
12. Y. Tabata (Ed.), *Pulse Radiolysis*, CRC Press, Boca Raton (1991).
13. S.L. Khursan and A.I. Nikolaev, *Khim. Fiz.* **9,** 914 (1990).
14. B.H.J. Bielski and D.E. Cabelli, *Int. J. Radiat. Biol.* **59,** 291 (1991).
15. A.A. Frimer, in *Superoxide Dismutase*, edited by L.W. Oberley, CRC Press, Boca Raton (1982), p. 83.
16. J.A. Fee and J.S. Valentine, in *Superoxide and Superoxide Dismutases*, edited by M. Michelson, J.M. McCord and I. Fridovich, Academic Press, New York (1977), p. 19.
17. A.O. Allen and B.H.J. Bielski, in *Superoxide Dismutase*, edited by L.W. Oberley, CRC Press, Boca Raton (1982), p. 125.
18. E. Lee-Ruff, *Chem. Soc. Rev.* **6,** 195 (1977).
19. H. Zegota, M.N. Schuchmann and C. von Sonntag, *J. Phys. Chem.* **88,** 5589 (1984).
20. H.-P. Schuchmann, C. von Sonntag and R. Srinivasan, *J. Photochem. Photobiol. A: Chem.* **45,** 49 (1988).
21. H.G. Viehe, Z. Janousek, R. Merenyi and L. Stella, *Acc. Chem. Res.* **18,** 148 (1985).
22. F. Welle, S.P. Verevkin, M. Keller, H.-D. Beckhaus and C. Rüchardt, *Chem. Ber.* **127,** 697 (1994).
23. P. Ulanski, E. Bothe, J.M. Rosiak and C. von Sonntag, *J. Chem. Soc. Perkin Trans. 2,* 5 (1996).
24. M. Krauss and R. Osman, *J. Phys. Chem.* **99,** 11387 (1995).
25. R. Mertens and C. von Sonntag, *Angew. Chem. Int. Ed. Engl.* **33,** 1262 (1994).
26. X. Fang, R. Mertens and C. von Sonntag, *J. Chem. Soc. Perkin Trans.* 1033 (1995).
27. M. Sommeling, P. Mulder, R. Louw, D.V. Avila, J. Lusztyk and K.U. Ingold, *J. Phys. Chem.* **97,** 8361 (1993).
28. Z.B. Alfassi, G.I. Khaikin and P. Neta, *J. Phys. Chem.* **99,** 265 (1995).
29. G.I. Khaikin, Z.B. Alfassi and P. Neta, *J. Phys. Chem.* **99,** 11447 (1995).
30. Z.B. Alfassi, S. Marguet and P. Neta, *J. Phys. Chem.* **98,** 8019 (1994).
31. G.I. Khaikin, Z.B. Alfassi and P. Neta, *J. Phys. Chem.* **99,** 16722 (1995).

32. P. Neta, R.E. Huie and A.B. Ross, *J. Phys. Chem. Ref. Data* **19,** 413 (1990).
33. C.A. Morgan, M.J. Pilling, J.M. Tulloch, R.P. Ruiz and K.D. Bayes, *J. Chem. Soc. Faraday Trans. 2* **78,** 1323 (1982).
34. R.P. Ruiz, K.D. Bayes, M.T. MacPherson and M.J. Pilling, *J. Phys. Chem.* **85,** 1622 (1981).
35. J.J. Russell, J.A. Seetula, D. Gutman, F. Danis, F. Caralp, P.D. Lightfoot, R. Lesclaux, C.F. Melius and S.M. Senkan, *J. Phys. Chem.* **94,** 3277 (19909).
36. N.A. Porter, *Acc. Chem. Res.* **19,** 262 (1986).
37. N.A. Porter, in *Membrane Lipid Oxidation*, edited by C. Vigo-Pelfrey, CRC Press, Boca Raton (1990), p. 33.
38. D. Wang, H.-P. Schuchmann and C. von Sonntag, *Z. Naturforsch.* **48B,** 761 (1993).
39. X.-M. Pan and C. von Sonntag, *Z. Naturforsch.* **45B,** 1337 (1990).
40. X. Fang, X. Pan, A. Rahmann, H.-P. Schuchmann and C. von Sonntag, *Chem. Eur. J.* **1,** 423 (1995).
41. X.-M. Pan, M.N. Schuchmann and C. von Sonntag, *J. Chem. Soc. Perkin Trans. 2* 289 (1993).
42. X. Fang, G. Mark and C. von Sonntag, *Ultrasonics—Sonochem.* **3,** 57 (1996).
43. F. Jin, J. Leitich and C. von Sonntag, *J. Chem. Soc. Perkin Trans. 2* 1583 (1993).
44. E.P.L. Hunter, M.F. Desrosiers and M.G. Simic, *Free Radical Biol. Med.* **6,** 581 (1989).
45. D. Wang, G. György, K. Hildenbrand and C. von Sonntag, *J. Chem. Soc. Perkin Trans. 2* 45 (1994).
46. S.V. Jovanovic and M.G. Simic, *Free Radical Biol. Med.* **1,** 125 (1985).
47. C. von Sonntag, *Int. J. Radiat. Biol.* **66,** 485 (1994).
48. P. Wardman, *J. Phys. Chem. Ref. Data* **18,** 1637 (1989).
49. M.N. Schuchmann, H.-P. Schuchmann, M. Hess and C. von Sonntag, *J. Am. Chem. Soc.* **113,** 6934 (1991).
50. C. von Sonntag, D.J. Deeble, M. Hess, H.-P. Schuchmann and M.N. Schuchmann, in *Active Oxygens, Lipid Peroxides and Antioxidants*, edited by K. Yagi, Japan Scientific Societies Press, Tokyo (1993), p. 127.
51. M.N. Schuchmann, H.-P. Schuchmann and C. von Sonntag, *J. Phys. Chem.* **99,** 9122 (1995).
52. Y. Sawaki and Y. Ogata, *J. Org. Chem.* **49,** 3344 (1984).
53. E. Bothe, G. Behrens and D. Schulte-Frohlinde, *Z. Naturforsch.* **32B,** 886 (1977).
54. E. Bothe, D. Schulte-Frohlinde and C. von Sonntag, *J. Chem. Soc. Perkin Trans. 2* 416 (1978).
55. J. Rabani, D. Klug-Roth and A. Henglein, *J. Phys. Chem.* **78,** 2089 (1974).
56. E. Bothe, M.N. Schuchmann, D. Schulte-Frohlinde and C. von Sonntag, *Z. Naturforsch.* **38B,** 212 (1983).
57. Y. Ilan, J. Rabani and A. Henglein, *J. Phys. Chem.* **80,** 1558 (1976).
58. E. Bothe and D. Schulte-Frohlinde, *Z. Naturforsch.* **35B,** 1035 (1980).
59. X.-M. Pan, M.N. Schuchmann and C. von Sonntag, *J. Chem. Soc. Perkin Trans 2* 1021 (1993).
60. R.P. Bell, in *Advances in Physical Organic Chemistry*, edited by V. Gold, Academic Press, London (1966), p. 1.
61. O.J. Mieden, M.N. Schuchmann and C. von Sonntag, *J. Phys. Chem.* **97,** 3783 (1993).
62. M.N. Schuchmann, H.-P. Schuchmann and C. von Sonntag, *J. Phys. Chem.* **93,** 5320 (1989).
63. O.J. Mieden and C. von Sonntag, *Z. Naturforsch.* **44B,** 959 (1989).
64. M. Eigen, W. Kruse, G. Maass and L. De Maeyer, *Progr. React. Kinet.* **2,** 285 (1964).
65. M. Eigen, G. Ilgenfritz and W. Kruse, *Chem. Ber.* **98,** 1623 (1965).
66. M.N. Schuchmann and C. von Sonntag, *Z. Naturforsch.* **37B,** 1184 (1982).

67. M.N. Schuchmann, M. Al-Sheikhly, C. von Sonntag, A. Garner and G. Scholes, *J. Chem. Soc. Perkin Trans. 2* 1777 (1984).
68. M.N. Schuchmann, H.-P. Schuchmann and C. von Sonntag, *J. Am. Chem. Soc.* **112,** 403 (1990).
69. M.N. Schuchmann and C. von Sonntag, *Z. Naturforsch.* **42B,** 495 (1987).
70. C. Nese, M.N. Schuchmann, S. Steenken and C. von Sonntag, *J. Chem. Soc. Perkin Trans. 2* 1037 (1995).
71. G. Merga, H.-P. Schuchmann, B.S.M. Rao and C. von Sonntag, *J. Chem. Soc. Perkin Trans. 2*, 1097 (1996).
72. A.J. Bloodworth, J.L. Courtneidge and A.G. Davies, *J. Chem. Soc. Perkin Trans. 2* 523 (1984).
73. S. Phulkar, B.S.M. Rao, H.-P. Schuchmann and C. von Sonntag, *Z. Naturforsch.* **45B,** 1425 (1990).
74. X. Shen, J. Lind, T.E. Eriksen and G. Merenyi, *J. Chem. Soc. Perkin Trans. 2* 555 (1989).
75. N.A. Porter, A.N. Roe and A.T. McPhail, *J. Am. Chem. Soc.* **102,** 7474 (1980).
76. A.N. Roe, A.T. McPhail and N.A. Porter, *J. Am. Chem. Soc.* **105,** 1199 (1983).
77. N.A. Porter, L.S. Lehman, B.A. Weber and K.J. Smith, *J. Am. Chem. Soc.* **103,** 6447 (1981).
78. G.O. Schenck, O.-A. Neumüller and W. Eisfeld, *Liebigs Ann. Chem.* **618,** 202 (1958).
79. A.L.J. Beckwith, A.G. Davies, I.G.E. Davison, A. Maccoll and M.H. Mruzek, *J. Chem. Soc. Perkin Trans. 2* 815 (1989).
80. N.A. Porter, K.A. Mills, S.E. Caldwell and G.R. Dubay, *J. Am. Chem. Soc.* **116,** 6697 (1994).
81. Z.H. Lodhi and R.W. Walker, *J. Chem. Soc. Faraday Trans.* **87,** 2361 (1991).
82. J.W. Bozzelli and A.M. Dean, *J. Phys. Chem.* **97,** 4427 (1993).
83. K. Hasegawa and L.K. Patterson, *Photochem. Photobiol.* **28,** 817 (1978).
84. L.K. Patterson and K. Hasegawa, *Ber. Bunsenges. Phys. Chem.* **82,** 951 (1978).
85. L.K. Patterson, in *Oxygen and Oxy-Radicals in Chemistry and Biology*, edited by M.A.J. Rodgers and E.L. Powers, Academic Press, New York (1981), p. 89.
86. M.G. Simic, S.V. Jovanovic and E. Niki, *ACS Symp. Ser.* **500,** 14 (1992).
87. L.R.C. Barclay, K.A. Baskin, S.J. Locke and M.R. Vinquist, *Can. J. Chem.* **67,** 1366 (1989).
88. J.M. Gebicki and A.O. Allen, *J. Phys. Chem.* **73,** 2443 (1969).
89. M. Al-Sheikhly and M.G. Simic, *J. Phys. Chem.* **93,** 3103 (1989).
90. S.L. Khursan and V.S. Martem'yanov, *Russ. J. Phys. Chem.* **65,** 321 (1991).
91. E.T. Denisov and T.G. Denisova, *Kinet. Catal.* (*Engl. Transl.*) **34,** 738 (1993).
92. D.F. McMillen and D.M. Golden, *Annu. Rev. Phys. Chem.* **33,** 493 (1982).
93. A. Fish, in *Organic Peroxides*, edited by D. Swern, Vol. 1, John Wiley & Sons, New York (1970), p. 181.
94. N.F. Trofimova, V.V. Kharitonov and E.T. Denisov, *Dokl. Akad. Nauk SSSR* **241,** 416 (1978).
95. N.F. Trofimova, V.V. Kharitonov and E.T. Denisov, *Dokl. Akad. Nauk SSSR* **253,** 651 (1980).
96. S.W. Benson, *Thermochemical Kinetics*, John Wiley & Sons, New York (1976).
97. A.C. Baldwin, in *The Chemistry of Functional Groups: Peroxides*, edited by S. Patai, John Wiley & Sons, Chichester (1983), p. 97.
98. M.N. Schuchmann and C. von Sonntag, *J. Phys. Chem.* **86,** 1995 (1982).
99. P. Ulanski, E. Bothe, K. Hildenbrand, J.M. Rosiak and C. von Sonntag, *J. Chem. Soc. Perkin Trans. 2* 23 (1996).
100. D.J. Deeble and C. von Sonntag, *Int. J. Radiat. Biol.* **49,** 927 (1986).
101. E. Bothe, G. Behrens, E. Böhm, B. Sethuram and D. Schulte-Frohlinde, *Int. J. Radiat. Biol.* **49,** 57 (1986).
102. M. Isildar, M.N. Schuchmann, D. Schulte-Frohlinde and C. von Sonntag, *Int. J.*

Radiat. Biol. **41,** 525 (1982).
103. S.L. Khursan, A.N. Teregulova, Y.C. Zimin, E.I. Faizova and A.Y. Gerchikov, *Khim. Fiz.* **14,** 41 (1995).
104. P. Neta, R.E. Huie, S. Mosseri, L.V. Shastri, J.P. Mittal, P. Maruthamuthu and S. Steenken, *J. Phys. Chem.* **93,** 4099 (1989).
105. M.N. Schuchmann and C. von Sonntag, *J. Am. Chem. Soc.* **110,** 5698 (1988).
106. J.E. Packer, R.L. Willson, D. Bahnemann and K.-D. Asmus, *J. Chem. Soc. Perkin Trans. 2* 296 (1980).
107. K.-D. Asmus, M. Lal, J. Mönig and C. Schöneich, in *Oxygen Radicals in Biology and Medicine*, edited by M.G. Simic, K.A. Taylor, C. von Sonntag and J.F. Ward, Plenum, New York (1988), p. 67.
108. J. Mönig, K.-D. Asmus, M. Schaeffer, T.F. Slater and R.L. Willson, *J. Chem. Soc. Perkin Trans. 2* 1133 (1983).
109. G.S. Nahor, P. Neta and Z.B. Alfassi, *J. Phys. Chem.* **95,** 4419 (1991).
110. Z.B. Alfassi, S. Mosseri and P. Neta, *J. Phys. Chem.* **91,** 3383 (1987).
111. M. Lal, C. Schöneich, J. Mönig and K.-D. Asmus, *Int. J. Radiat. Biol.* **54,** 773 (1988).
112. S.K. Kapoor and C. Gopinathan, *Int. J. Chem. Kinet.* **24,** 1035 (1992).
113. G. Merenyi, J. Lind and L. Engman, *J. Chem. Soc. Perkin Trans. 2* 2551 (1994).
114. Z.B. Alfassi, R.E. Huie, M. Kumar and P. Neta, *J. Phys. Chem.* **96,** 767 (1992).
115. P. Neta, R.E. Huie, P. Maruthamuthu and S. Steenken, *J. Phys. Chem.* **93,** 7654 (1989).
116. M.G. Simic and E.P.L. Hunter, *Free Radical Biol. Med.* **2,** 227 (1986).
117. C. Schöneich, A. Aced and K.-D. Asmus, *J. Am. Chem. Soc.* **113,** 375 (1991).
118. M. Bonifacic, C. Schöneich and K.-D. Asmus, *J. Chem. Soc. Chem. Commun.* 1117 (1991).
119. L. Engman, J. Persson, G. Merenyi and J. Lind, *Organometallics* **14,** 3641 (1995).
120. E. Furimsky and J.A. Howard, *J. Am. Chem. Soc.* **95,** 369 (1973).
121. (a) R. Mertens, C. von Sonntag, J. Lind and G. Merenyi, *Angew. Chem. Int. Ed. Engl.* **33,** 1259 (1994); (b) L.C.T. Shoute, Z.B. Alfassi, P. Neta and R.E. Huie, *J. Phys. Chem.* **98,** 5701 (1994).
122. M.E. Morgan, D.A. Osborne and D.J. Waddington, *J. Chem. Soc. Perkin* Trans. 2 1869 (1984).
123. M.I. Sway and D.J. Waddington, *J. Chem. Soc. Perkin Trans. 2* 999 (1982).
124. S.W. Benson and R. Shaw, in *Organic Peroxides*, edited by D. Swern, Vol. 1, John Wiley & Sons, New York (1970), p. 105.
125. P.S. Nangia and S.W. Benson, *J. Phys. Chem.* **83,** 1138 (1979).
126. J.E. Bennett, G. Brunton, J.R.L. Smith, T.M. Salmon and D.J. Waddington, *J. Chem. Soc. Faraday Trans. 1* **83,** 2421 (1987).
127. J.S. Francisco and I.H. Williams, *Int. J. Chem. Kinet.* **20,** 455 (1988).
128. M.N. Schuchmann and C. von Sonntag, *J. Phys. Chem.* **83,** 780 (1979).
129. S.L. Khursan and V.S. Martemyanov, *React. Kinet. Catal. Lett.* **40,** 253 (1989).
130. L.A. Tavadian, *Khim. Fiz.* **10,** 655 (1991).
131. S.L. Khursan and V.S. Martemyanov, *React. Kinet. Catal. Lett.* **40,** 269 (1989).
132. P.G. Thompson, *J. Am. Chem. Soc.* **89,** 4316 (1967).
133. R. Hiatt, in *Organic Peroxides*, edited by D. Swern, Vol. 3, John Wiley & Sons, Chichester (1972), p. 1.
134. B. Plesnicar, in *The Chemistry of Functional Groups: Peroxides*, edited by S. Patai, John Wiley & Sons, Chichester (1983), p. 483.
135. G.A. Russell, *J. Am. Chem. Soc.* **79,** 3871 (1957).
136. Q.J. Niu and G.D. Mendenhall, *J. Am. Chem. Soc.* **114,** 165 (1992).
137. S.-H. Lee and G.D. Mendenhall, *J. Am. Chem. Soc.* **110,** 4318 (1988).
138. G.D. Mendenhall, X.C. Sheng and T. Wilson, *J. Am. Chem. Soc.* **113,** 8976 (1991).
139. G. Vasvary and D. Gal, *Ber. Bunsenges. Phys. Chem.* **97,** 22 (1993).

140. P.S. Nangia and S.W. Benson, *Int. J. Chem. Kinet.* **12,** 43 (1980).
141. S.L. Khursan, V.S. Martem'yanov and E.T. Denisov, *Kinet. Catal.* (*Engl. Transl.*) **31,** 899 (1990).
142. V.M. Berdnikov, N.M. Bazhin, V.K. Fedorov and O.V. Polyakov, *Kinet. Catal.* (*Engl. Transl.*) **13,** 986 (1972).
143. B.C. Gilbert, R.G.G. Holmes, H.A.H. Laue and R.O.C. Norman, *J. Chem. Soc. Perkin Trans. 2* 1047 (1976).
144. B.C. Gilbert, R.G.G. Holmes and R.O.C. Norman, *J. Chem. Res.* (*S*) 1 (1977).
145. H.-P. Schuchmann and C. von Sonntag, *J. Photochem.* **16,** 289 (1981).
146. H.-P. Schuchmann and C. von Sonntag, *Z. Naturforsch.* **39B,** 217 (1984).
147. J.E. Bennett and R. Summers, *Can. J. Chem.* **52,** 1377 (1974).
148. E. Bothe and D. Schulte-Frohlinde, *Z. Naturforsch.* **33B,** 786 (1978).
149. P.S. Nangia and S.W. Benson, *Int. J. Chem. Kinet.* **12,** 29 (1980).
150. V.V. Shereshovets, F.A. Galieva, R.A. Sadykov, V.D. Komissarov and G.A. Tolstikov, *Izv. Akad. Nauk SSSR Ser. Khim.* **10,** 2208 (1989).
151. M.N. Schuchmann, H. Zegota and C. von Sonntag, *Z. Naturforsch.* **40B,** 215 (1985).
152. C. Walling and P.J. Wagner, *J. Am. Chem. Soc.* **86,** 3368 (1964).
153. K.Y. Choo and S.W. Benson, *Int. J. Chem. Kinet.* **13,** 833 (1981).
154. B.C. Gilbert, P.D.R. Marshall, R.O.C. Norman, N. Pineda and P.S. Williams, *J. Chem. Soc. Perkin Trans. 2* 1392 (1981).
155. J. Heicklen, *Adv. Photochem.* **14,** 177 (1988).
156. S.M. Colwell and N.C. Handy, *J. Chem. Phys.* **82,** 1281 (1985).
157. K.-D. Asmus, A. Henglein, M. Ebert and J.P. Keene, *Ber. Bunsenges. Phys. Chem.* **68,** 657 (1964).
158. H.-P. Schuchmann, D.J. Deeble, G.O. Phillips and C. von Sonntag, *Radiat. Phys. Chem.* **37,** 157 (1991).
159. C. Walling and R.T. Clark, *J. Am. Chem. Soc.* **96,** 4530 (1974).
160. Z.B. Alfassi, in *Supplement A: The Chemistry of Double-Bonded Functional Groups*, edited by S. Patai, John Wiley & Sons, Chichester (1989), p. 527.
161. A. Piesiak, M.N. Schuchmann, H. Zegota and C. von Sonntag, *Z. Naturforsch.* **39B,** 1262 (1984).
162. X.-M. Pan, E. Bastian and C. von Sonntag, *Z. Naturforsch.* **43B,** 1201 (1988).
163. T. Söylemez and C. von Sonntag, *J. Chem. Soc. Perkin Trans 2* 391 (1980).
164. K.-D. Asmus, H. Möckel and A. Henglein, *J. Phys. Chem.* **77,** 1218 (1973).
165. H. Christensen and R. Gustafsson, *Acta Chem. Scand.* **26,** 937 (1972).
166. D.J. Deeble, B.J. Parsons, G.O. Phillips, H.-P. Schuchmann and C. von Sonntag, *Int. J. Radiat. Biol.* **54,** 179 (1988).
167. S.V. Jovanovic, Y. Hara, S. Steenken and M.G. Simic, *J. Am. Chem. Soc.* **117,** 9881 (1995).
168. W. Bors, M. Saran and C. Michel, *J. Phys. Chem.* **83,** 3084 (1979).
169. W. Bors, C. Michel and M. Saran, *Meth. Enzymol.* **234,** 420 (1994).
170. U. Gröllmann and W. Schnabel, *Makromol. Chem.* **181,** 1215 (1980).
171. C. von Sonntag, *Adv. Carbohydr. Chem. Biochem.* **37,** 7 (1980).
172. A. Samuni and P. Neta, *J. Phys. Chem.* **77,** 2425 (1973).
173. S. Steenken, G. Behrens and D. Schulte-Frohlinde, *Int. J. Radiat. Biol.* **25,** 205 (1974).
174. R. Mertens and C. von Sonntag, *J. Chem. Soc. Perkin Trans. 2* 2181 (1994).
175. G. Behrens and G. Koltzenburg, *Z. Naturforsch.* **40C,** 785 (1985).
176. G. Koltzenburg and T. Matsushige, *Z. Naturforsch.* **31B,** 960 (1976).
177. C. von Sonntag and M. Dizdaroglu, *Carbohydr. Res.* **58,** 21 (1977).
178. M.N. Schuchmann and C. von Sonntag, *Z. Naturforsch.* **33B,** 329 (1978).
179. M.N. Schuchmann and C. von Sonntag, *J. Chem. Soc. Perkin Trans. 2* 1958 (1977).
180. M. Dizdaroglu and C. von Sonntag, *Z. Naturforsch.* **28B,** 635 (1973).

181. C. von Sonntag, M. Dizdaroglu and D. Schulte-Frohlinde, *Z. Naturforsch.* **31B,** 857 (1976).
182. H. Zegota and C. von Sonntag, *Z. Naturforsch.* **32B,** 1060 (1977).
183. S. Al-Assaf, G.O. Phillips, D.J. Deeble, B. Parsons, H. Starnes and C. von Sonntag, *Radiat. Phys. Chem.* **46,** 207 (1995).
184. D.J. Deeble, E. Bothe, H.-P. Schuchmann, B.J. Parsons, G.O. Phillips and C. von Sonntag, *Z. Naturforsch.* **45C,** 1031 (1990).
185. M.N. Schuchmann and C. von Sonntag, *Int. J. Radiat. Biol.* **34,** 397 (1978).
186. P. Ulanski, E. Bothe, K. Hildenbrand, J.M. Rosiak and C. von Sonntag, *Radiat. Phys. Chem.* **46,** 909 (1995).
187. B. Neumann, S.C. Müller, M.J.B. Hauser, O. Steinbock, R.H. Simoyi and N.S. Dalal, *J. Am. Chem. Soc.* **117,** 6372 (1995).
188. R. Rao, M.N. Schuchmann and C. von Sonntag, unpublished results.
189. R. Köster and K.-D. Asmus, *Z. Naturforsch.* **26B,** 1108 (1971).
190. R. Mertens and C. von Sonntag, unpublished results.
191. R. Mertens and C. von Sonntag, *J. Photochem. Photobiol. A: Chem.* **85,** 1 (1995).
192. J. Mönig, D. Bahnemann and K.-D. Asmus, *Chem.-Biol. Interact.* **45,** 15 (1983).
193. Y. Mao, C. Schöneich and K.-D. Asmus, *J. Phys. Chem.* **96,** 8522 (1992).
194. Y. Mao, C. Schöneich and K.-D. Asmus, *J. Phys. Chem.* **95,** 10080 (1991).
195. Y. Mao, C. Schöneich and K.-D. Asmus, in *Photocatalytic Purification and Treatment of Water and Air*, edited by H. Al-Ekabi and D.F. Ollis, Elsevier, Amsterdam (1993), p. 49
196. P. Dowideit and C. von Sonntag, unpublished results.
197. H. Staudinger, *Chem. Ber.* **41,** 3558 (1908).
198. K.B. Krauskopf and G.K. Rollefson, *J. Am. Chem. Soc.* **56,** 2542 (1934).
199. P. Dowideit, R. Mertens and C. von Sonntag, *J. Am. Chem. Soc.* **118,** 11288 (1996).
200. R. Mertens and C. von Sonntag, unpublished results.
201. R. Mertens and C. von Sonntag, unpublished results.
202. R. Mertens and C. von Sonntag, unpublished results.
203. C. von Sonntag and H.-P. Schuchmann, in *Sulfur-Centered Reactive Intermediates in Chemistry and Biology*, edited by C. Chatgilialoglu and K.-D. Asmus, Plenum, New York (1990), p. 409.
204. R.M. Burger, K. Drlica and B. Birdsall, *J. Biol. Chem.* **269,** 25978 (1994).
205. R.M. Burger, J. Peisach and S.B. Horwitz, *Life Sci.* **78,** 715 (1981).
206. L. Giloni, M. Takeshita, F. Johnson, C. Iden and A.P. Grollman, *J. Biol. Chem.* **256,** 8608 (1981).
207. C. von Sonntag and H.-P. Schuchmann, *Meth. Enzymol.* **186,** 511 (1990).
208. S. Steenken, *Chem. Rev.* **89,** 503 (1989).
209. P. Ulanski, E. Bothe, J.M. Rosiak and C. von Sonntag, *Makromol. Chem.* **195,** 1443 (1994).
210. P. Ulanski, E. Bothe, K. Hildenbrand and C. von Sonntag, *J. Chem. Soc. Perkin Trans. 2* 13 (1996).
211. M.N. Schuchmann and C. von Sonntag, *J. Chem. Soc. Perkin Trans. 2* 1525 (1983).
212. S. Fujita and S. Steenken, *J. Am. Chem. Soc.* **103,** 2540 (1981).
213. M.I. Al-Sheikhly, A. Hissung, H.-P. Schuchmann, M.N. Schuchmann, C. von Sonntag, A. Garner and G. Scholes, *J. Chem. Soc. Perkin Trans. 2* 601 (1984).
214. D.J. Deeble and C. von Sonntag, *Int. J. Radiat. Biol.* **51,** 791 (1987).
215. S. Das, D.J. Deeble, M.N. Schuchmann and C. von Sonntag, *Int. J. Radiat. Biol.* **46,** 7 (1984).
216. D.J. Deeble, S. Das and C. von Sonntag, *J. Phys. Chem.* **89,** 5784 (1985).
217. H. Koffer, *J. Chem. Soc. Perkin Trans. 2* 819 (1975).
218. M.N. Schuchmann and E. Bothe, unpublished results.
219. M. Dizdaroglu, D. Schulte-Frohlinde and C. von Sonntag, *Int. J. Radiat. Biol.* **32,** 481 (1977).

NOTE ADDED IN PROOF

A general mechanism of the OH-radical-induced oxidation of several alkylbenzenes, based on the quantum-chemical computation of the energies of intermediate radical species, including endoperoxidic structures, has been developed [220] and appeared while the present review was in preparation. It is in general support of the mechanisms of the OH-radical-induced oxidation of benzene [41] and chlorobenzene [71] mentioned above (Section 5, p. 191).

The bond dissociation energies of hydrogen trioxide and the methyl and ethyl hydrotrioxides have been computed *ab initio*. For reaction (84a), values situated just above 130 kJ mol^{-1} (R=H) and near 125 kJ mol^{-1} (R=CH_3, C_2H_5) have been obtained at different theoretical levels [221], in further support of the contention made above (Section 9.2, p. 204) that homolytic cleavage of this bond should not play any role in the process of product formation following the termination reaction of peroxyl radicals.

$$\mathrm{RO{-}OOH} \longrightarrow \mathrm{RO^{\bullet} + {}^{\bullet}OOH} \qquad (84a)$$

A recent theoretical study of the mechanism of hydrogen trioxide decomposition (reaction (88a)) [222] also sustains our postulate (Section 9.2, p. 205) of water-induced electrocyclic cleavage, reaction (88).

$$\mathrm{HOOOH + H_2O} \longrightarrow \mathrm{2\,H_2O + O_2} \qquad (88a)$$

The results of ongoing work on the OH-radical-induced oxidation of chlorohydrocarbons suggest [223] (as do those of the photolytic study of tetrachloroethene [191]) the possibility of explicit heterolysis upon termination, summarized in reactions (124a) and (150).

$$\mathrm{2\,Cl(O)C\text{-}CCl_2O_2^{\bullet}\,(\mathbf{80}) + H_2O} \longrightarrow \mathrm{Cl(O)C\text{-}C(O)Cl + O_2 + 2\,HCl + CCl_2O + CO_2} \qquad (124a)$$

$$\mathrm{2\,Cl_3C\text{-}CCl_2O_2^{\bullet}\,(\mathbf{83}) + H_2O} \longrightarrow \mathrm{2\,CCl_3\text{-}C(O)Cl + O_2 + HCl + HOCl} \qquad (150)$$

The charge distribution in polyoxides is such that the lateral oxygen atoms are more negative than the central ones [224] which may enable a good leaving group such as Cl^- to be expelled from the α position of the intermediate tetroxide as the first step in its decomposition into the final products.

REFERENCES

220. J.M. Andino, J.N. Smith, R.C. Flagan, W.A. Goddard III and J.H. Seinfeld, *J. Phys. Chem.* **100,** 10967 (1996).
221. P.T.W. Jungkamp and J.H. Seinfeld, *Chem. Phys. Lett.* **257,** 15 (1996).
222. J. Koller and B. Plesnicar, *J. Am. Chem. Soc.* **118,** 2470 (1996).
223. P. Dowideit, R. Mertens, C. von Sonntag and H.-P. Schuchmann, unpublished results.
224. B. Plesnicar, D. Kocjan, S. Murovec, and A. Azman, *J. Am. Chem. Soc.* **98,** 3143 (1976).

9 Kinetic Studies of Organic Peroxyl Radicals in Aqueous Solutions and Mixed Solvents

Z.B. ALFASSI
Ben Gurion University of the Negev, Israel

and

R.E. HUIE and P. NETA
National Institute of Standards and Technology, Gaithersburg, USA

1 INTRODUCTION

Peroxyl radicals are key intermediates in the oxidation of organic compounds. The abstraction of a hydrogen atom from an organic compound, or the addition of a radical to a double bond, leads to the formation of a carbon-centred radical. In the presence of O_2, a peroxyl radical is formed:

$$R^{\cdot} + O_2 \longrightarrow RO_2^{\cdot} \qquad (1)$$

These reactions are typically fast ($k > 10^9\,L\,mol^{-1}\,s^{-1}$) [1, 2] and irreversible at room temperature, although in a few important cases the reactions are reversible and this has a profound effect on the subsequent behaviour of the chemical system. It is the subsequent reactions of peroxyl radicals which lead to chain propagation, chain branching, or chain termination. Organic peroxyl radicals play an important role in physiological chemistry, particularly in the damage induced by ionizing radiation, possibly in ageing, and in ischemia. In the condensed phase, a number of different types of reactions of peroxyl radicals are known to be important, including electron transfer, addition to double bonds, and hydrogen abstraction. Due to their relatively unreactive nature, reactions of organic peroxyl radicals with other organic peroxyl radicals and with the hydroperoxyl radical, $HO_2^{\cdot}/O_2^{-\cdot}$, are also likely to be important. In addition, the very fast reactions of peroxyl radicals with nitric oxide are potentially of great importance.

The kinetics and mechanisms of the reactions of peroxyl radicals in solution have been studied extensively and have been the subject of a number of reviews [1, 3]. Rate constants for peroxyl radical reactions have been compiled in several publications [2, 4]. Compared to other free radicals, alkylperoxyl radicals are quite unreactive towards hydrogen abstraction and

Peroxyl Radicals. Edited by Z.B. Alfassi

addition to double bonds, and only moderately reactive towards electron transfer. The reactivity increases strongly upon the substitution of halogens on the peroxyl carbon. Halogenated alkylperoxyl radicals are of increased interest due to the replacement of the totally halogenated halocarbons in industry with new hydrohalocarbon compounds. In the atmosphere, the hydrohalocarbons react by hydrogen abstraction with hydroxyl radicals and the carbon-centred radicals that are produced then form halogenated alkylperoxyl radicals. The subsequent reactions of these radicals lead to the removal of the halogens from the atmosphere. Also of interest is the toxicity of these replacement compounds. Previous work suggested that the toxicity of CCl_4 arises from its one-electron reduction in the body to $CCl_3^{\bullet}$, which then forms the reactive $CCl_3O_2^{\bullet}$ radical [5]. The physiological behaviour of the replacement hydrohalocarbons might also be expected to relate to the chemical behaviour of the peroxyl radicals derived from them. In this present chapter we will discuss the kinetics and mechanisms of organic peroxyl radicals in aqueous solutions and in mixed solvents, and shall concentrate on those radicals that are generally more reactive, mainly due to the presence of halogens or unsaturated bonds in the immediate vicinity of the oxyl group.

2 PRODUCTION OF PEROXYL RADICALS

Peroxyl radicals are generally produced by the reaction of O_2 with alkyl radicals. Since alkyl radicals can be produced via many different reactions, we will discuss first the procedures by which different types of alkyl radicals can be produced. We concentrate here on radiolytic methods since these were the main methods used to obtain the results summarized in this chapter.

Substituted alkyl radicals are produced in irradiated solutions by the reaction of primary radiolytic species with the solvent or with solutes. In aqueous solutions the radicals are generated by the reactions of e_{aq}^- or $^{\bullet}OH$ radicals with various organic compounds. If the $^{\bullet}OH$ radical is to be used, the solutions typically contain N_2O to convert the e_{aq}^- into $^{\bullet}OH$, thus increasing the yield of the desired alkyl radical and reducing possible complications arising from other reactions of the electron:

$$H_2O \rightsquigarrow e_{aq}^- + H^{\bullet} + {}^{\bullet}OH + H_2 + H_2O_2 \tag{2}$$

$$e_{aq}^- + N_2O \longrightarrow N_2 + OH^- + {}^{\bullet}OH \tag{3}$$

$${}^{\bullet}OH + RH \longrightarrow H_2O + R^{\bullet} \tag{4}$$

If e_{aq}^- is to be used to produce the desired alkyl radical, the $^{\bullet}OH$ radical cannot simply be converted to the hydrated electron, but often can be converted to a reducing radical capable of producing more of the desired alkyl radical. Otherwise, secondary products arising from the $^{\bullet}OH$ must be taken into account in the kinetic analysis.

Many alkyl radicals have been formed by hydrogen abstraction from a solvent or a solute by primary or secondary radiolytic products. This can result in a mixture of several alkyl radicals, depending upon the selectivity of the reactant radical. The $^{\cdot}OH$ radical is very non-selective and is used primarily to produce radicals from simple precursors, e.g. methanol, acetone, or acetonitrile, where essentially all of the alkyl radicals will be identical. For 2-propanol (2-PrOH), which is slightly more complex, 86% of the alkyl radicals will be secondary while 13% will be primary radicals. When $^{\cdot}OH$ is allowed to react with precursors containing aromatic moieties, the situation may be even more complex, with the formation of OH-adduct radicals, along with radicals formed by abstraction from side chains. Rate constants for a large number of hydroxyl radical reactions have been summarized in a recent compilation [6].

Many simple alkyl radicals can be formed cleanly by the reaction of $^{\cdot}OH$ with alkyl sulfoxides via an addition/fragmentation mechanism [7]:

$$^{\cdot}OH + R_2SO \longrightarrow [R_2SO(OH)]^{\cdot} \rightarrow RSO_2H + R^{\cdot} \tag{5}$$

Since the rate constants for the reaction of $^{\cdot}OH$ with the sulfoxides are $>10^9\,L\,mol^{-1}\,s^{-1}$ and the lifetimes of the OH-adducts are $<200\,ns$, in $0.01\,mol\,L^{-1}$ sulfoxide solutions the radicals $R^{\cdot}$ are produced in a time $<1\,\mu s$ after the pulse.

Alkyl radicals are also produced by the reaction of e_{aq}^- with halogenated organic compounds, leading to reductive elimination of a halide ion. The $^{\cdot}OH$ radicals can be removed by scavenging them with 2-PrOH or with formate ions:

$$e_{aq}^- + RX \longrightarrow R^{\cdot} + X^- \tag{6}$$

$$^{\cdot}OH + (CH_3)_2CHOH \longrightarrow H_2O + (CH_3)_2\dot{C}OH \tag{7}$$

$$^{\cdot}OH + HCO_2^- \longrightarrow H_2O + CO_2^{-\cdot} \tag{8}$$

The radical from 2-PrOH will reduce some halogenated compounds (such as CCl_4 or CH_3I) to produce more of the desired radical [8]:

$$(CH_3)_2\dot{C}OH + RX \longrightarrow (CH_3)_2CO + H^+ + X^- + R^{\cdot} \tag{9}$$

Many halogenated compounds, however, do not react rapidly with $(CH_3)_2\dot{C}OH$, which will instead react with O_2 to produce a peroxyl radical [9, 10]. This may need to be taken into account in the subsequent kinetic analysis.

The $CO_2^{-\cdot}$ radical formed by reaction (8) reacts rapidly with O_2 to yield $O_2^{-\cdot}$ and CO_2 [11]. The $O_2^{-\cdot}$ is a very weak oxidant [12], and in most cases does not interfere with the oxidation of organic reductants by the peroxyl radicals. Further details of the reactions involved in the preparation of the various substituted alkyl radicals have been given elsewhere [8, 13–15].

Alkyl radicals have also been produced by reaction of e_{aq}^- with alkylammonium derivatives, a reaction that proceeds by reductive elimination of ammonia [16]:

$$e_{aq}^- + C_6H_5CH_2NH_3^+ \longrightarrow NH_3 + C_6H_5CH_2^{\cdot} \tag{10}$$

Elimination of ammonia occurs only when the amino group is protonated and the α-carbon bears an electron-withdrawing group such as carbonyl or phenyl. This process is almost quantitative for diphenylmethylammonium, which deaminates 95% of the time, but is less quantitative for benzylammonium, for which deamination occurs for 70% of the reactions. The rest of the hydrated-electron reactions occur via ring addition and protonation to give a hydrogen adduct.

Peroxyl radicals have also been produced by photochemical methods, generally by forming an alkyl radical in the presence of O_2, although direct photolytic production from a precursor such as di-t-butyl diperoxycarbonate has also been utilized. These methods were not used in the studies discussed in this present chapter.

3 REACTIONS OF PEROXYL RADICALS

Peroxyl radicals are detected by their optical absorption or electron spin resonance (ESR) spectra. They exhibit weak absorptions in the UV region, generally with a peak at or below 250 nm, with molar absorptivities typically of ca. 1000 L mol^{-1} cm^{-1}. Although the UV absorptions may be used to follow the self-reaction of a peroxyl radical, they have not proven as useful in monitoring the reactions of peroxyl radicals with other substrates, particularly when the substrates absorb in the UV. Therefore, the rates of reaction of peroxyl radicals with inorganic and organic compounds have been determined in most cases by following the decay of the ESR signal of peroxyl radicals or the build-up of optical absorption of the product radical. When the latter exhibits no intense absorption at $\lambda > 250$ nm, as in the case of alkenes and fatty acids, the rate constants have been determined by competition kinetics using compounds such as chlorpromazine [2-chloro-10-(3-dimethylaminopropyl)phenothiazine hydrochloride] (ClPz), ABTS [2,2′-azinobis(3-ethylbenzthiazoline-6-sulfonate ion)], or porphyrins as reference reactants.

Peroxyl radicals react with various compounds via different mechanisms. With saturated aliphatic compounds they may react by hydrogen abstraction:

$$RO_2^{\cdot} + R'H \longrightarrow RO_2H + R'^{\cdot} \tag{11}$$

This reaction plays a major role when the compound contains weak C–H bonds or other weakly bonded hydrogens, such as those in thiols:

$$RO_2^{\cdot} + RSH \longrightarrow RO_2H + RS^{\cdot} \quad (12)$$

With unsaturated compounds the reaction may be via addition to the unsaturated bond:

$$RO_2^{\cdot} + \;>C{=}C< \;\longrightarrow\; RO_2{-}\overset{|}{\underset{|}{C}}{-}\dot{C}< \quad (13)$$

or by abstraction from allylic or bisallylic C–H bonds. With compounds that are readily oxidized the reaction is generally via electron transfer, e.g. oxidation of phenolate ions to phenoxyl radicals:

$$RO_2^{\cdot} + PhO^- \longrightarrow RO_2^- + PhO^{\cdot} \quad (14)$$

or phenothiazine (Pz) to its radical cation:

$$RO_2^{\cdot} + Pz \longrightarrow RO_2^- + Pz^{+\cdot} \quad (15)$$

In the following sections we will discuss the results presented for each of these mechanisms separately.

4 REACTION WITH NITRIC OXIDE

The fastest reaction of an organic peroxyl radical is likely to be its reaction with nitric oxide. The reaction of $RO_2^{\cdot}$ with NO is fast, with rate constants of $1\text{–}3 \times 10^9\,L\,mol^{-1}\,s^{-1}$ for the peroxyl radicals derived from MeOH, 2-PrOH, and t-BuOH [17]. These reactions lead to the formation of an intermediate species which absorbs light at ca. 300 nm, similar to the peroxynitrite species, $^{\cdot}OONO^-$, formed in the reaction of $O_2^{-\cdot}$ with NO [18]:

$$RO_2^{\cdot} + NO \longrightarrow ROONO \quad (16)$$

This intermediate species decays with a rate constant of $0.1\text{–}0.3\,L\,mol^{-1}\,s^{-1}$ in water, and approximately 200 times faster in alcoholic solutions. Although rate constants have not been reported for the reaction of any other peroxyl radicals with NO in solution, a large number of these reactions have been investigated in the gas phase [19]. The reactions are all fast and, typically, involve oxygen-atom transfer, probably through a peroxynitrite intermediate:

$$RO_2^{\cdot} + NO \longrightarrow RO + NO_2 \quad (17)$$

In the liquid phase, the peroxynitrite may also decompose to radical products, in which case NO will act as a potent pro-oxidant, or it may rearrange to an organic nitrate:

$$ROONO \longrightarrow RONO_2 \qquad (18)$$

and NO would be an anti-oxidant. This behaviour might be a function of the solvent, as is the decomposition rate. Some evidence points to an anti-oxidant behaviour of NO in lipid peroxidation [20].

5 ABSRACTION OF HYDROGEN ATOMS

The hydrogen-abstraction reactions of alkylperoxyl radicals are slow and highly selective. For abstraction from saturated alkanes, the rate constants are typically less than $1\,L\,mol^{-1}\,s^{-1}$ [4]. The abstraction rates for allylic hydrogens are faster, with a rate constant of less than $10^4\,L\,mol^{-1}\,s^{-1}$ for abstraction of the bisallylic hydrogen in cyclohexadiene or linoleic acid. Although these reactions are relatively slow, they constitute the chain-propagation steps in the autoxidation of unsaturated organic compounds [21]. Indeed, the rate of cellular lipid peroxidation has been shown to increase exponentially with the number of bisallylic C—H bonds [22]. This behaviour reflects the much lower C—H bond strengths of allylic bonds (ca. $372\,kJ\,mol^{-1}$) as compared to primary, secondary, or tertiary C—H bonds (422, 413, and $402\,kJ\,mol^{-1}$, respectively) [23]. The bond strength for the O—H bond in tert-butyl hydroperoxide was determined to be $359\,kJ\,mol^{-1}$ [23], and from this value and a comparison of the activation energies for abstraction by secondary and tertiary peroxyl radicals, the bond strength in secondary hydroperoxides was estimated to be $365\,kJ\,mol^{-1}$ [24]. Halogenation at the α-carbon increases the reactivity of peroxyl radicals toward hydrogen abstraction enormously. By considering the increased reactivities of $CCl_3CCl_2COO^{\bullet}$ and $CHCl_2CCl_2COO^{\bullet}$ toward n-C_8H_{18}, the O—H bond dissociation energy of chlorinated hydroperoxides was estimated to be $407\,kJ\,mol^{-1}$ [24].

Rate constants for hydrogen abstraction by $CCl_3O_2^{\bullet}$ radicals were measured [25] by competition kinetics, using as a reference reaction the oxidation of a metalloporphyrin and monitoring the optical absorption of the porphyrin π-radical cation. The rate constant for cyclohexane was found to be $1.0 \times 10^3\,L\,mol^{-1}\,s^{-1}$, indicating a rate constant of $1.6 \times 10^2\,L\,mol^{-1}\,s^{-1}$ per CH_2 group. The rate constant for the reaction of $CCl_3O_2^{\bullet}$ with cyclohexene was found to be $1.0 \times 10^5\,L\,mol^{-1}\,s^{-1}$. Since the rate constant for the two CH_2 groups that are remote from the double bond is only $3.2 \times 10^2\,L\,mol^{-1}\,s^{-1}$, the main contribution to the rate constant for cyclohexene must be the abstraction of the activated allylic hydrogens of the other two CH_2 groups and/or the addition to the double bond. The rate constant for hexamethylbenzene was found to be $7.5 \times 10^4\,L\,mol^{-1}\,s^{-1}$ and was ascribed to abstraction of the benzylic hydrogens, i.e. $1.2 \times 10^4\,L\,mol^{-1}\,s^{-1}$ per methyl group. If we assume that the reactivity of the allylic CH_2 group in cyclohexene is similar to or slightly higher than that of the benzylic CH_3 group, we find that the rate constant for the addition to the double bond in cyclohexene is $\leq 8 \times 10^4\,L\,mol^{-1}\,s^{-1}$. The addition rate constant becomes much higher in activated double bonds, as discussed below.

The same ambiguity exists in the reactions of halogenated peroxyl radicals with fatty acids, where both the addition to double bonds and abstraction of allylic and doubly allylic hydrogen atoms may contribute to the overall reactivity. (As mentioned above, for non-halogenated alkylperoxyl radicals, abstraction appears to be the dominant mechanism.) Rate constants for the reactions of $CCl_3O_2^{\cdot}$ with oleic, linoleic, and linolenic acids were measured by two groups [26, 27] by the competition kinetics method under somewhat different conditions. The absolute rate constants reported by the two groups (Table 1) vary by a factor of 5–10, possibly due to differences in solvent (50% t-BuOH vs. 50% 2-PrOH) and pH (natural pH vs. pH 1). The order of reactivity, however, was always oleic < linoleic < linolenic (1/2.3/4.1 or 1/1.2/2.7, respectively). The increasing reactivity may be ascribed to the increasing number of allylic and doubly allylic hydrogens and to the increasing number of double bonds, and thus the reaction may involve hydrogen abstraction and/or addition.

Table 1 Rate constants for reactions of peroxyl radicals with fatty acids

Radical	Fatty acid	k^a ($L\,mol^{-1}\,s^{-1}$)	k^b ($L\,mol^{-1}\,s^{-1}$)
$CCl_3O_2^{\cdot}$	Oleic	4×10^5	1.7×10^6
	Linoleic	5×10^5	3.9×10^6
	Linolenic	1.1×10^6	7×10^6
	Arachidonic		7.3×10^6
$CF_3CHClO_2^{\cdot}$	Oleic	2×10^4	3×10^5
	Linoleic	8×10^4	8×10^5
	Linolenic	3×10^5	1.3×10^6
	Arachidonic		1.5×10^6
$CF_3O_2^{\cdot}$	Linolenic	6.9×10^6	
$CBr_3O_2^{\cdot}$	Linolenic	1.2×10^6	
$(CH_3)_2C(OH)O_2^{\cdot}$	Oleic	$<3 \times 10^3$	
	Linoleic	$\approx 6 \times 10^3$	

[a] From Ref. 27.
[b] From Ref. 26.

Other hydrogens that undergo abstraction readily are those in phenols and thiols. In these cases, however, the same products may be produced by a different mechanism, namely oxidation via electron transfer, and it is not always clear which mechanism prevails. It has been suggested that the reactions of peroxyl radicals that are weak oxidants, in low-polarity solvents, may occur via hydrogen abstraction, whereas strongly oxidizing radicals, particularly in more polar solvents, react by electron transfer (see below).

6 ADDITION TO DOUBLE BONDS

The addition of alkylperoxyl radicals to carbon–carbon double bonds is also generally very slow and strongly dependent upon the substituents about the double bond [4]. Often, both addition and abstraction reactions can take place. For example, the rate constant for addition of $(CH_3)_3CO_2^{\cdot}$ to 2,3-dimethyl-2-butene was reported to be $22\,L\,mol^{-1}\,s^{-1}$ in benzyl chloride and the abstraction rate constant was reported to be $2.9\,L\,mol^{-1}\,s^{-1}$ [28]. Without product analysis, it is frequently difficult to ascertain if a reported rate constant refers to an addition or an abstraction reaction. Rate constants for the reactions of several peroxyl radicals with a series of unsaturated hydrocarbons and alcohols were determined by competition kinetics. For compounds of low reactivity, it was difficult to distinguish between hydrogen abstraction and addition. For more complex compounds, however, as the reactivity increased, it became clear that this increase was ascribable to more reactive double bonds rather than to an increase in the number or the reactivity of abstractable hydrogen atoms.

The reactions of halogenated peroxyl radicals with alkenes are much faster than the corresponding reactions of alkylperoxyl radicals. The rate constants for reactions of alkenes with $C_4F_9O_2^{\cdot}$ [29], $CCl_3O_2^{\cdot}$ and $CBr_3O_2^{\cdot}$ [30], and other perhaloalkylperoxyl radicals [31] in methanolic solutions are summarized in Table 2. For each radical, the rate constants for the different alkenes vary by about two orders of magnitude. These variations do not follow the pattern expected for hydrogen abstraction, i.e. diallyl > allyl > alkyl and, within each type, tertiary > secondary > primary [32], and do not increase with the number of reactive hydrogens. Therefore, hydrogen abstraction is ruled out as a major route for these reactions. The rate constant for the addition reaction should depend on the substituents about the double bond. In fact, a good correlation was found between the measured rate constants and Taft's σ^* substituent constants [33], which reflect the inductive effect of the substituents. A plot of k for the reactions of $C_4F_9O_2^{\cdot}$ with various alkenes vs. the sum of the σ^* values for the four substituents on the double bond gave a good straight line with a slope of $\rho^* = -1.37$, suggesting that the mechanism involves an electrophilic addition to the double bond. Similar straight lines, with practically identical slopes, were obtained for $CCl_3O_2^{\cdot}$ and other polyhalogenated peroxyl radicals.

Table 2 Rate constants ($L\,mol^{-1}\,s^{-1}$) for reactions of halogenated peroxyl radicals with alkenes in methanolic solutions

Alkene	Radical								
	$CBr_3O_2^{\cdot}$	$CCl_3O_2^{\cdot}$	$CF_3O_2^{\cdot}$	$CF_2ClO_2^{\cdot}$	$CF_2BrO_2^{\cdot}$	$CF_3CCl_2O_2^{\cdot}$	$C_2F_5O_2^{\cdot}$	$(CF_3)_2CFO_2^{\cdot}$	$C_4F_9O_2^{\cdot}$
$CH_2CH_2CH{=}CHCH_2CH_2$ (cyclic)	9.1×10^4	9.5×10^4	7.5×10^5	4.5×10^5	4.9×10^5	2.1×10^5	1.7×10^6	7.7×10^5	1.5×10^6
$CH_2{=}CHCH_2CH{=}CH_2$									1.0×10^6
$CH_2{=}C(CH_3)CH_2CH{=}CH_2$		4.4×10^4							2.6×10^6
$CH_2CH_2C(CH_3){=}CHCH_2CH_2$ (cyclic)		1.3×10^6	1.7×10^7	9.1×10^6	7.7×10^6	3.2×10^6	2.6×10^7	6.4×10^6	
$CH{=}CHCH_2CH{=}CHCH_2$ (cyclic)									7.5×10^6
$(CH_3)_2C{=}CHCH_3$		1.3×10^6	1.8×10^7	9.4×10^6	7.9×10^6	2.8×10^6	1.9×10^7	6.8×10^6	3.0×10^7
$C(CH_3){=}CHCH_2CH{=}CHCH_2$ (cyclic)		1.8×10^6							4.0×10^7
$(CH_3)_2C{=}C(C_2H_5)_2$	2.5×10^6	3.2×10^6				9.7×10^6			9.7×10^7
$(CH_3)_2C{=}C(CH_3)_2$	1.0×10^7	1.4×10^7	2.2×10^8	8.3×10^7	7.7×10^7	3.1×10^7	1.3×10^8	5.0×10^7	1.7×10^8
$CH_2CH_2C(CH_3){=}C(CH_3)CH_2CH_2$ (cyclic)	3.3×10^6	3.4×10^6	7.5×10^7	3.2×10^7	2.7×10^7	8.0×10^6	6.7×10^7	2.5×10^7	6×10^8
$C_6H_5CH{=}CH_2$	2.1×10^5	3.2×10^5				5.7×10^5			1.1×10^7
$C_6H_5CH{=}CHC_6H_5$									2.3×10^7
$CH_2{=}CHCH_2OH$									2.5×10^5
$CH_3CH{=}CHCH_2OH$	2.3×10^4	2.6×10^4				4.9×10^4			9.3×10^5
$CH_2{=}C(CH_3)CH_2CH_2OH$	5.0×10^4	5.0×10^4				9.3×10^4			
$(CH_3)_2C{=}CHCH_2OH$	1.2×10^5	1.6×10^5				3.3×10^5			

All of the above plots of $\log k$, vs. σ^* exhibited a certain degree of scatter of the data points about the straight line. However, when $\log k$ for any haloalkylperoxyl radical with a series of alkenes was plotted against $\log k$ for $CCl_3O_2^{\cdot}$ with the same compounds, the correlation was generally very good ($R^2 \geqslant 0.97$) and the slope of all the lines was 1.00 ± 0.15. This indicates that the scatter in the $\log k$ vs. σ^* plots are due to inaccuracies in the values of σ^*, rather than in the measured rate constants. The plot of $\log k$ for $C_4F_9O_2^{\cdot}$ vs. $\log k$ for $CCl_3O_2^{\cdot}$ indicates a deviation of the point representing 1,2-dimethylcyclohexane, whereas such deviations were not observed for other radicals, and suggests that the rate constant reported for the reaction of this alkene with $C_4F_9O_2^{\cdot}$ is somewhat higher than it should be.

The reactivity of the alkylperoxyl radicals is also affected by the substituents attached to the alkyl group, mainly those at the α-carbon. It is clear from Table 2 that the trihaloalkylperoxyl radicals exhibit increased reactivity with increasing electron affinity of the halogen, i.e. $CBr_3O_2^{\cdot} < CCl_3O_2^{\cdot} \ll CF_3O_2^{\cdot}$. For the mixed-halogen radical, it can be seen that, within the experimental uncertainty (±15%), $k(CF_2ClO_2^{\cdot}) = [k(CF_3O_2^{\cdot})^2 \times k(CCl_3O_2^{\cdot})]^{1/3}$ and $k(CF_2BrO_2^{\cdot}) = [k(CF_3O_2^{\cdot})^2 \times k(CBr_3O_2^{\cdot})]^{1/3}$. The good agreement between the experimental and calculated values in Table 3 indicates an additive effect of the substituents on $\log k$. Comparison of the rate constants of $CF_3CF_2O_2^{\cdot}$ with those of $CF_3CCl_2O_2^{\cdot}$, and $CF_3O_2^{\cdot}$ with those of $CCl_3O_2^{\cdot}$, provides another indication for the additivity of the substituent effect on $\log k$. For the same five alkenes in Table 3, we find the ratios: $k(CF_3CF_2O_2^{\cdot})/k(CF_3CCl_2O_2^{\cdot}) = 7.1 \pm 1.7$, and $k(CF_3O_2^{\cdot})/k(CCl_3O_2^{\cdot}) = 14.5 \pm 5.1$. These ratios fit a factor of 2.5 for the reactivity effect of a F substituent as compared with that of a Cl substituent ($2.5^2 = 6.25$; $2.5^3 = 15.6$).

Comparison of the rate constants of the perfluorinated methyl- and ethylperoxyl radicals shows that $k(CF_3CF_2O_2^{\cdot})/k(CF_3O_2^{\cdot}) = 1.27 \pm 0.66$. Because of the wide scatter, i.e. the ratio extends from 0.59 to 2.27, it cannot be concluded whether the CF_3 group has a higher or lower activating effect when compared with the F atom at the α-position. It is clear, however, that it

Table 3 Experimental and calculated rate constants for reactions of $CF_2ClO_2^{\cdot}$ and $CF_2BrO_2^{\cdot}$

	$k(CF_2ClO_2^{\cdot})$ ($L\,mol^{-1}\,s^{-1}$)		$k(CF_2BrO_2^{\cdot})$ ($L\,mol^{-1}\,s^{-1}$)	
Alkene	Measured	Calculated	Measured	Calculated
Cyclohexene	4.5×10^5	3.8×10^5	4.9×10^5	3.7×10^5
2-Methy-2-butene	9.4×10^6	7.5×10^6	7.9×10^6	
1-Methylcyclohexene	9.1×10^6	7.2×10^6	7.7×10^6	
1,2-Dimethylcyclohexene	3.2×10^7	2.7×10^7	2.7×10^7	2.6×10^7
2,3-Dimethyl-2-butene	8.3×10^7	8.8×10^7	7.7×10^7	7.9×10^7

has a stronger effect than a Cl atom at the same position, $k(CF_3CF_2O_2^\cdot)$ is always larger than $k(CF_2ClO_2^\cdot)$ (Table 2).

The rate constants for the reactions of $CCl_3O_2^\cdot$ with cyclohexene (c-Hx) and 2,3-dimethyl-2-butene (DMB) have been measured in several solvents (Table 4). For all of the solvents studied, DMB is considerably more reactive than c-Hx, indicating that the main pathway of reaction is the addition to the double bond. It is interesting to note that the rate constants for DMB in the different solvents vary by about an order of magnitude, whereas the values for c-Hx exhibit smaller variations with solvent, i.e. less than factor of three. This is despite the fact that the rate constants for c-Hx are generally lower by two orders of mganitude than those of DMB and are thus expected to have a higher selectivity. A possible explanation for this behaviour is that while DMB reacts predominantly by addition, owing to its activated double bond, c-Hx reacts by both addition and hydrogen abstraction and the latter process exhibits a very weak solvent effect. It should be noted that $^\cdot CCl_3$ radicals also react with c-Hx via addition and hydrogen abstraction with comparable rates [34].

Table 4 Rate constants ($L\,mol^{-1}\,s^{-1}$) for reactions of $CCl_3O_2^\cdot$ radicals with 2,3-dimethyl-2-butene (DMB) and cyclohexene (c-Hx) in various solvents

Solvent	DMB	c-Hx
MeCN	2.3×10^7	8.4×10^4
MeOH	1.4×10^7	9.5×10^4
Me_2CO	1.3×10^7	4.9×10^4
$HCONMe_2$	1.2×10^7	8.0×10^4
HCONHMe	$\approx 1.2 \times 10^7$	6.6×10^4
t-BuOH	7.7×10^6	7.6×10^4
Dioxane	6.9×10^6	9.7×10^4
CCl_4/dioxane	4.3×10^6	
2-PrOH	4.1×10^6	7.9×10^4
Me_2SO	3.3×10^6	1.3×10^5
$(CH_2OH)_2$	$\leqslant 1 \times 10^5$	
$(C_2H_5)_3N$	$\leqslant 3 \times 10^4$	$\leqslant 3 \times 10^4$

Even in the case of DMB, where addition is the predominant reaction, very poor correlation was found between the rate constants and most solvent parameters available in the literature. A somewhat improved correlation was found with the solvent viscosity, but since all the rate constants are much lower than the diffusion-controlled limits, this correlation cannot be due to the rate of collision (although the viscosity might affect the rate of decomposition of the radical–alkene adduct).

The above study could not include water as a solvent due to the insolubility of the alkenes in water. To study the addition reaction in aqueous solutions, rate constants were measured for three unsaturated alcohols, and the results were compared to those obtained in methanol (Table 5). All three alcohols react more rapidly in aqueous solutions than in methanol, but no correlation with solvent properties was attempted since only two solvents were compared.

Table 5 Rate constants ($L\,mol^{-1}\,s^{-1}$) for reactions of $CCl_3O_2^{\cdot}$ radicals with unsaturated alcohols in water and methanol

Alcohol	H_2O/2-PrOH/CCl_4 (90/10/0.1)	MeOH/CCl_4 (9/1)
2-Buten-1-ol	9.0×10^4	2.6×10^4
3-Methyl-2-buten-1-ol	3.8×10^5	5.0×10^4
3-Methyl-2-buten-1-ol	6.8×10^5	1.6×10^5

7 ELECTRON-TRANSFER REACTIONS

Peroxyl radicals oxidize various organic compounds, such as phenothiazines, porphyrins, or sulfides, to the corresponding radical cations. They also oxidize various anions, such as ascorbate, phenolate, and inorganic anions, to the corresponding radicals. These oxidations take place clearly by an electron-transfer process. Certain organic compounds that can be oxidized may do so via electron transfer or hydrogen abstraction, e.g. phenols. For such cases, both mechanisms may be viable and their relative importance will depend on various factors which will be discussed below, such as the redox potentials of the radical and the other reactant, and the nature of the solvent.

It has been noticed earlier [13] that the rate constants for oxidation reactions by peroxyl radicals are increased with increasing number of chlorine atoms on the radical, e.g. in the series $CH_3O_2^{\cdot}$, $CH_2ClO_2^{\cdot}$, $CHCl_2O_2^{\cdot}$ and $CCl_3O_2^{\cdot}$. The ratios of the rate constants for these radicals reacting with different compounds (ascorbate, promethazine, phenolate, tyrosine, etc) varied considerably, but the general trend was $CH_3O_2^{\cdot} \ll CH_2ClO_2^{\cdot} < CHCl_2O_2^{\cdot} < CCl_3O_2^{\cdot}$. The increase in the rate constant has been rationalized by the electron-withdrawing effect of these substituents, which decreases the electron density at the radical site and thus increases its electron affinity and oxidation potential. More recent studies [35] confirmed the order of reactivity of $CH_2ClO_2^{\cdot} < CHCl_2O_2^{\cdot} < CCl_3O_2^{\cdot}$ for the reactions of these radicals with tyrosine and with several purine and pyrimidine bases. The absolute rate constants reported for tyrosine were slightly different from the earlier values, mainly due to solvent effects. Other studies also confirmed this order of reactivity for various other compounds, such as aniline, methoxyphenol, and porphyrins [8, 36].

A quantitative correlation was demonstrated [37] for the reactions of chlorpromazine (ClPz) with several halogenated peroxyl radicals:

$$RO_2^{\bullet} + ClPz \longrightarrow RO_2^{-} + ClPz^{+\bullet} \tag{19}$$

using the rate constants summarized in Table 6 and the polar substituent constants σ* [33]. A good linear correlation was found between log *k* and σ* (Figure 1), with a slope of $\rho^* = 1.7$. The correlation confirms the electrophilic nature of the peroxyl radicals and is in agreement with the electron-transfer mechanism for reaction (19).

A broader correlation of the rate constants with substituent constant

Table 6 Rate constants ($L\,mol^{-1}\,s^{-1}$) for the oxidation of chlorpromazine by peroxyl radicals in aqueous alcohol solutions and the sum of the polar substituent constants of the substituents on the peroxyl radical

Radical	*k* in 2-PrOH/H_2O/RX (6/3/1)	*k* in 2-PrOH/H_2O/RX (1/8/0.01)	σ*
$CH_2ClO_2^{\bullet}$	1.5×10^5	2.5×10^7	1.05
$CHCl_2O_2^{\bullet}$	8.3×10^6	3.6×10^8	1.94
$CCl_3O_2^{\bullet}$	7.2×10^7	1.0×10^9	2.65
$CH_3CCl_2O_2^{\bullet}$	2.3×10^6		1.65
$CF_3CHClO_2^{\bullet}$	3.5×10^6		1.70
$CCl_3CCl_2O_2^{\bullet}$	8.6×10^7		
$CFCl_2O_2^{\bullet}$	1.2×10^8		2.70

was carried out for ascorbate and TMPD (*N,N,N′,N′*-tetramethyl-*p*-phenylenediamine) in aqueous solutions (Figure 1) [38]:

$$RO_2^{\bullet} + Asc^{-} \longrightarrow RO_2^{-} + Asc^{\bullet} \tag{20}$$

$$RO_2^{\bullet} + C_6H_4[N(CH_3)_2]_2 \longrightarrow RO_2^{-} + C_6H_4[N(CH_3)_2]_2^{+\bullet} \tag{21}$$

In these cases, plots of log *k* vs. the polar substituent constants σ* did not yield a good straight line for all of the radicals examined, but rather appeared to fit two sets of radicals with two different lines. The correlation for TMPD indicated separate lines for the methylperoxyl radicals bearing electron-withdrawing groups and for those bearing electron-donating groups, with the slopes differing by nearly an order of magnitude ($\rho^* = 5.6$ for $RO_2^{\bullet}$ with $\sigma_R^* < 0$ and $\rho^* = 0.64$ for $\sigma_R^* > 0$). The radicals bearing electron-donating groups reacted too slowly with ascorbate to permit accurate determination of their rate constants. Among the radicals bearing electron-withdrawing groups the rate constants generally increased with the value of σ* but the linear fit was considerably scattered and it was proposed that the radicals bearing

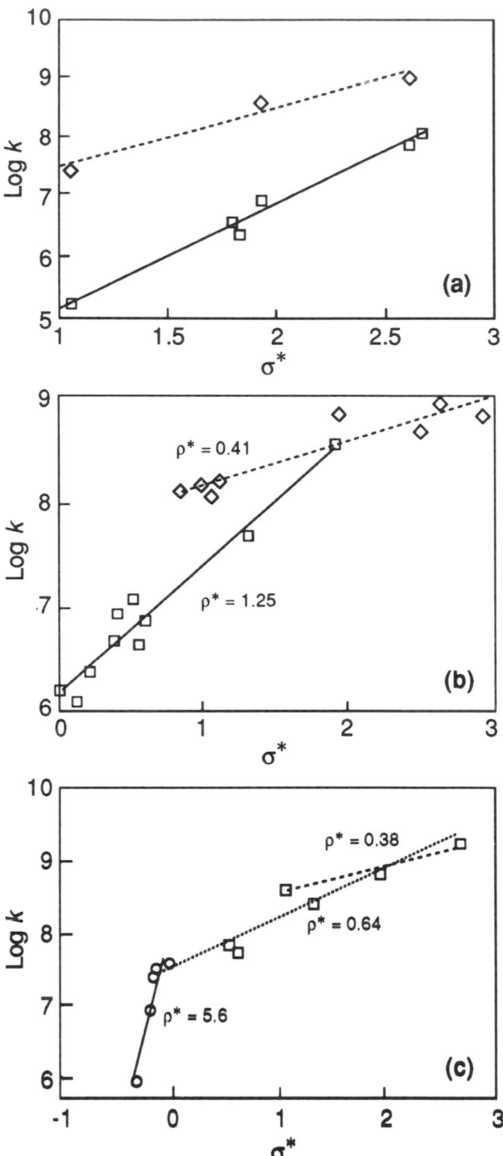

Figure 1 Correlation of log *k* for oxidation of chlorpromazine (a, data from Table 6), ascorbate (b, data from Table 7), and TMPD (c, data from Table 7) by substituted methylperoxyl radicals ($RO_2^{\cdot}$) with the polar substituent constants σ^* for *R* [37, 38].

halogen substituents may fit a different line to those bearing other electron-withdrawing substituents, with the slopes differing by about a factor of three. Different slopes for substituents with positive and negative values of σ^* have been reported before, but deviation of halogen substituents from other electron-withdrawing substituents has not been reported and it is not clear what other effects may play a role in these reactions to result in such behaviour. It should be pointed out, however, that the various reactions studied were monitored in somewhat different media. Although the solvent used was mainly water, varying concentrations of different organic additives were present in the various solutions and these may affect the rate constants. A detailed examination of solvent effects will be given below.

According to the σ* values for the halogens, the order of reactivity of the trihalogenated methylperoxyl radicals should be $CF_3O_2^{\cdot} > CCl_3O_2^{\cdot} > CBr_3O_2^{\cdot}$. Nevertheless, several reports indicated a different order, i.e. $CF_3O_2^{\cdot} > CBr_3O_2^{\cdot} > CCl_3O_2^{\cdot}$. This was found for the oxidation of ascorbate, urate, chlorpromazine, and several phenols (see Table 8, below) [15] as well as for Fe(III) deuteroporphyrin [39]. It should be pointed out that the three radicals were not compared in identical solvent mixtures and, in general, the reactions of $CBr_3O_2^{\cdot}$ were studied in aqueous solutions that contained more alcohol than in the other cases; however, this would tend to lower rather than increase the rate constant (see below). Moreover, the rate constants for the reactions of monohalogenated methylperoxyl radicals with ascorbate in the same solvent mixture (5% 2-PrOH in water) (Table 7) follow the order

Table 7 Rate constants for the oxidation of ascorbate and TMPD by peroxyl radicals in aqueous solutions and the sum of the polar substituent constants for the substituents on the peroxyl radical

Radical	$k_{ascorbate}$ ($L\,mol^{-1}\,s^{-1}$)	k_{TMPD} ($L\,mol^{-1}\,s^{-1}$)	σ*
$(CH_3)_3CO_2^{\cdot}$		1.1×10^6	−0.30
$(CH_3)_2CHO_2^{\cdot}$		9.2×10^6	−0.19
$CH_3CH_2CH_2CH_2O_2^{\cdot}$		2.9×10^7	−0.13
$CH_3CH_2O_2^{\cdot}$		3.3×10^7	−0.10
$CH_3O_2^{\cdot}$	1.7×10^6	4.3×10^7	0.00
$C_6H_5CH_2O_2^{\cdot}$	2.5×10^6		0.22
$ClCH_2CH_2O_2^{\cdot}$	5.0×10^6		0.39
$(C_6H_5)_2CHO_2^{\cdot}$	9×10^6		0.41
$HO_2^{\cdot}$	1.2×10^7		0.49
$HOCH_2O_2^{\cdot}$	4.7×10^6	7.2×10^7	≈0.53
$CH_3COCH_2O_2^{\cdot}$	7.5×10^6	6.6×10^7	0.60
$ICH_2O_2^{\cdot}$	1.3×10^8		0.85
$BrCH_2O_2^{\cdot}$	1.5×10^8		1.00
$ClCH_2O_2^{\cdot}$	1.2×10^8	4.2×10^8	1.05
$FCH_2O_2^{\cdot}$	1.7×10^8		1.10
$NCCH_2O_2^{\cdot}$	5.0×10^7	2.9×10^8	1.30
$(CH_3)_3N^+CH_2O_2^{\cdot}$	4.0×10^8		1.90
$Cl_2CHO_2^{\cdot}$	7.0×10^8	7.4×10^8	1.94
$Br_3CO_2^{\cdot}$	5.0×10^8		2.50
$Cl_3CO_2^{\cdot}$	9.1×10^8	1.7×10^9	2.65
$F_3CO_2^{\cdot}$	6.8×10^8		≈2.9

Table 8 Rate constants for the oxidation of several comounds by peroxyl radicals in aqueous/organic solutions[a] and one-electron oxidation potentials of the compounds

Reactant	pH	Water (vol%)	k (L mol^{-1} s^{-1})				E^b (V)
			$CH_3O_2^{\cdot}$	$CF_3O_2^{\cdot}$	$CCl_3O_2^{\cdot}$	$CBr_3O_2^{\cdot}$	
Chlorpromazine	5	90		1.2×10^9	1.2×10^9		0.78
		60			5.7×10^8	7.7×10^8	
5-Hydroxytryptophan	13	90	7×10^6				0.21
		60			6.0×10^8		
Urate ion	13	90	8×10^6	1.0×10^9	1.5×10^9		0.26
		60		2.9×10^8 (50)	2.7×10^8	4.1×10^8	
	7	60			2.5×10^8	2.7×10^8	
Ascorbate ion	7	90	2×10^6	6.8×10^8	5.8×10^8	5.0×10^8 (80)	0.30
		50		1.9×10^8	1.1×10^8	2.1×10^8 (60)	
Xanthine	13	90	3×10^5	1.0×10^9			0.59
		60			1.1×10^8	1.7×10^8	
Trolox C	4	90	$<1 \times 10^5$	5.0×10^8	$\approx 5 \times 10^8$		0.48
		75		7.0×10^8	4.5×10^8	7.1×10^8	
		50		7.4×10^8	3.6×10^8	4.6×10^8	
Hydroquinone	7	90	$<1 \times 10^6$	7.9×10^7	4.0×10^7		0.46
		60			1.0×10^7	1.8×10^7	
4-Methoxyphenol	7	90		5.2×10^7			0.72
		60			3.4×10^6	9.5×10^6	
Hypoxanthine	13	60			1.5×10^7	2.7×10^7	0.74
Phenol	7	90		2×10^6	$<1 \times 10^5$		0.97

[a] Figures in parentheses represent water vol % when different from that listed in the third column.
[b] vs. (NHE) normal hydrogen electrode.

$FCH_2O_2^{\cdot} > BrCH_2O_2^{\cdot} > ICH_2O_2^{\cdot} > ClCH_2O_2^{\cdot}$, although the differences between the rate constants are not large. The reason for this unexpected order is unclear. It may be speculated that the larger halogen atoms, i.e. Br and I, interact with the terminal O atom through space to create a deeper sink for the electron, thus increasing the driving force for the oxidation reaction beyond the increase expected on the basis of the polar substituent constant σ^*.

The results presented in Table 8 also permit comparison of the reactivities of various compounds with the same radical. For these electron-transfer reactions, correlation of the rate constant with the driving force, or reduction potential, according to the Marcus theory [40], may be expected for similar molecules. It can be seen from Table 8 that the rate constants for the oxidation of phenols and related hydroxy heterocyclic compounds, in general, decrease when the oxidation potential of the compound increases. Deviations from this trend due to differences in the self-exchange rate constants of the various compound/radical pairs are particularly apparent for the case of chlorpromazine, where the rate constant is very high despite the relatively high oxidation potential. Increased reactivity with decreasing oxidation potential was also apparent for the reactions of $CHCl_2O_2^{\cdot}$ with aniline, dimethylaniline, and TMPD [36]. The reactivities of these anilines were higher than those of phenols with similar oxidation potentials, due to the lower self-exchange rates for phenols (related to the need for proton loss). These anecdotal examples call for a quantitative correlation of the rate constants and redox potentials according to the Marcus theory. Such correlation, however, requires the knowledge of the reduction potentials of the peroxyl radicals, for which only estimated values are available at present [41]. Significant correlation cannot be obtained by using estimated reduction potentials. Furthermore, such a correlation should be valid for truely outer-sphere electron-transfer processes but may be only approximately correct for some other types of reactions. As discussed below, electron-transfer reactions of peroxyl radicals were suggested to involve intimate interaction or complexation of the reactants and the solvent and thus do not qualify as truely outer-sphere reactions.

8 SOLVENT EFFECTS ON OXIDATION KINETICS

As mentioned above, the rate constants for the reactions of peroxyl radicals in aqueous solutions containing organic additives, reported by various groups, often differed considerably, beyond the acceptable margin of uncertainty. It appeared that most of these differences could be ascribed to changes in the nature and fraction of the organic co-solvent that was often added.

In an early study of the reactions of $CF_3CHClO_2^{\cdot}$ (the peroxyl radical derived from the anesthetic halothane, $CF_3CHClBr$) it was shown that the rate constant for oxidation of ABTS by this radical in aqueous solutions decreased as the concentration of t-BuOH was increased [42]. In another study of the oxidation of iron(III) porphyrin by $CCl_3O_2^{\cdot}$ radicals, the rate constant was 20 times lower in neat 2-PrOH than in 50% aqueous 2-PrOH [8]. The rate constant for the oxidation of chlorpromazine by peroxyl radicals was also found to decrease considerably when the fraction of water in the solvent decreased [36]. For example, for $CCl_3O_2^{\cdot}$ in various H_2O/2-PrOH/CCl_4 mixtures the rate constant varied by a factor of 50; for $CH_2ClO_2^{\cdot}$ a similar factor was found. These changes in rate constants were ascribed to changes in the solvent dielectric constant, as this parameter is expected to affect the electron-transfer rates. Furthermore, it was pointed out that the ratio of the reactivities of the three halogenated alkylperoxyl radicals, $CCl_3O_2^{\cdot}/CHCl_2O_2^{\cdot}/CH_2ClO_2^{\cdot}$, changed with solvent, from 1/0.67/0.07 in 89% water to 1/0.11/0.002 in 30% water.

In order to study solvent effects in a more systematic fashion, the effects of methanol, 2-propanol, and dioxane, and their fractions in the aqueous solution, on the rate constants for several peroxyl radical reactions have been examined [43]. The rate constants were always lower in the neat organic solvent than in water, by about an order of magnitude. However, the rate constant was not always the highest in nearly pure water. The rate constants for the oxidation of some compounds by $CF_3O_2^{\cdot}$, $CCl_3O_2^{\cdot}$, and $CBr_3O_2^{\cdot}$ in aqueous methanol exhibited a maximum at ca. 25% MeOH, were slightly lower in 5% MeOH and much lower in 100% MeOH (Figure 2). This

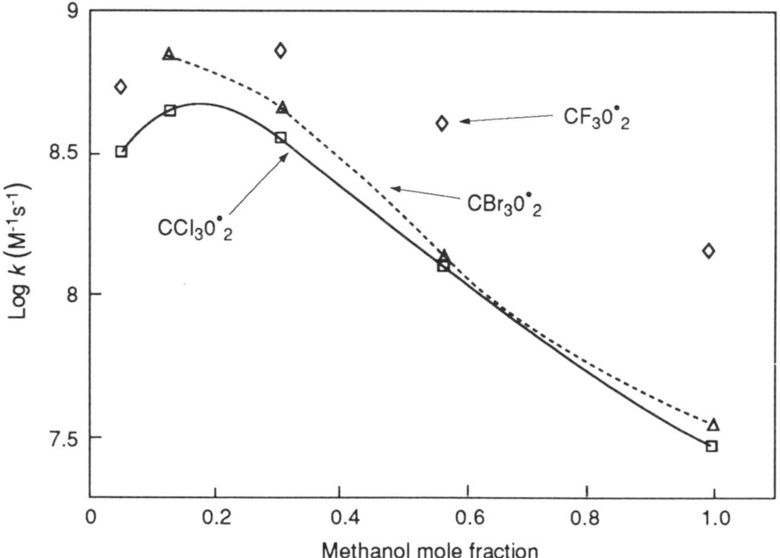

Figure 2 Effect of the methanol content in water on the rate constant for oxidation of Trolox (a vitamin E analogue) by peroxyl radicals: (◇) $CF_3O_2^{\cdot}$; (□) $CCl_3O_2^{\cdot}$; (△) $CBr_3O_2^{\cdot}$ [43].

behaviour has been ascribed to the structure-making properties of methanol in water [44]. The results obtained in these various mixtures, however, did not permit quantitative correlation with the solvent properties.

To correlate the rate constant with solvent parameters it is necessary to carry out the measurements in a wide variety of solvents, preferably not solvent mixtures. This was essentially achieved for reactions of the $CCl_3O_2^{\bullet}$ radical, which was produced by the pulse radiolysis of many solvents containing small amounts of CCl_4 as the source for the peroxyl radicals. The rate constants for reactions of $CCl_3O_2^{\bullet}$ with Trolox (6-hydroxy-2,5,7,8-tetramethylchroman-2-carboxylic acid, a vitamin E analogue) were determined in water and in 12 organic solvents [43]. A good correlation was found between the rate constants and a combination of the solvent polarity and basicity, except for the results for water and CCl_4, which showed considerable deviations. The effect of solvent polarity was as expected for an electron-transfer reaction, but the effect of basicity may be specific to Trolox and will be discussed below. The high rate constant found in water was thought to indicate the participation of the solvent in the reaction as a proton donor to the incipient hydroperoxide anion to form the neutral hydroperoxide, which is more stable under the conditions used. Further evidence for this intimate involvement of the solvent was obtained from kinetic-isotope-effect studies [43]. The rate constants for the oxidation of several compounds by peroxyl radicals were measured in H_2O and D_2O. In all cases, $k(H_2O)/k(D_2O)$ was approximately 2. Since the reactions examined in these cases were purely electron transfer in nature, with no possibility for hydrogen-abstraction, the kinetic isotope effect that was found was interpreted to indicate that protons from the solvent are involved in the electron-transfer process. It was suggested that the solvent forms hydrogen bonds with the peroxyl radical and donates a proton to the incipient hydroperoxide anion in the transition state, i.e. that electron transfer and proton transfer take place in concert:

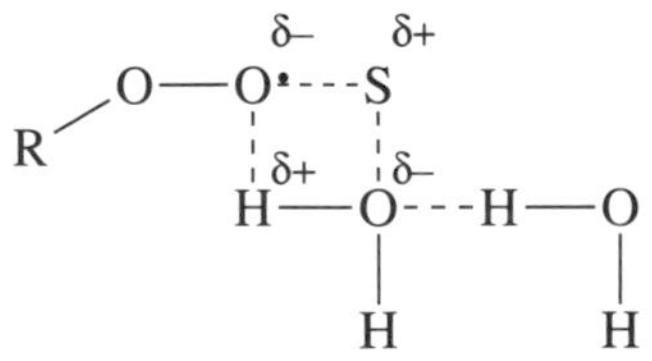

```
           δ–    δ+
      O—O•---S
R ⁄        ¦     ¦
          ¦ δ+  ¦ δ–
          H—O---H—O
                |       |
                H       H
```

A transition state of this type should have a negative activation entropy, since immobilization of water molecules in the transition state involves a profound reorganization of the solvent shell. Measurement of the rate constants for reactions of $CH_3O_2^{\bullet}$ and the dioxanylperoxyl radical with TMPD as a function of temperature showed that the entropy of activation was indeed negative [43].

More recently, the rate constants for the reaction of $CCl_3O_2^{\cdot}$ with chlorpromazine were measured in 14 different solvents covering a wide range of solvent polarity and basicity, while the data for Trolox were extended by adding two solvents of extreme polarity or basicity [45]. The rate constants for Trolox and chlorpromazine (Table 9) range from about 1×10^7 to 1×10^9 L mol^{-1} s^{-1} and are usually higher for ClPz than for Trolox, but there was almost no correlation between the two sets of rate constants. The data in Table 9 were correlated with a number of different solvent parameters and found to correlate with the Hildebrand solubility parameter (cohesive energy density) of the solvent better than with any other single solvent parameter.

Table 9 Rate constants for reactions of $CCl_3O_2^{\cdot}$ radicals with trolox and chlorpromazine in various solvents, cohesive pressure (δ) for the neat solvents, and pK_a values for the solvents when dissolved in water or in acetonitrile

Solvent	k (Trolox) (L mol^{-1} s^{-1})	k (ClPz) (L mol^{-1} s^{-1})	δ (MPa)	pK_a	
				H_2O	CH_3CN
H_2O	5.8×10^8	1.1×10^9	47.9	2.3	−1.7
$HCONH_2$	2.7×10^8	1.0×10^8	39.3	6.1	−1.5
$HCONHCH_3$	5.4×10^7	1.7×10^8	32.9	6.1	−1.3
$(CH_2OH)_2$	4.6×10^7	4.9×10^7	29.9	1.5	−2
CH_3OH	3.1×10^7	8.6×10^7	29.6	2.4	−2
$HCON(CH_3)_2$	2.1×10^7	3.8×10^7	24.8	6.1	−1.1
$(CH_3)_2SO$		8.9×10^7	24.5	5.8	−1.5
CH_3CN	1.2×10^7	5.8×10^7	24.3		−10
$(CH_3)_2CHOH$	2.1×10^7	3.1×10^7	23.5	2.8	−2
Pyridine	8.0×10^7	1.8×10^7	21.9	12.3	5.2
$(CH_3)_3COH$	2.1×10^7	1.5×10^7	21.7	3.4	−2
p-Dioxane	1.5×10^7	1.5×10^7	20.5	0.6	−2.4
$(CH_3)_2CO$	9.2×10^6	1.4×10^7	20.2	−0.1	−3
CCl_4	4.6×10^7		17.6		
CCl_4–dioxane		1.1×10^7			
$(C_2H_5)_3N$	9.5×10^7		15.1	18.7	11.0
$(C_2H_5)_2O$	1.4×10^7		15.1	0.1	−2.4

For the reaction of $CCl_3O_2^{\cdot}$ with ClPz there was a poor correlation of log k with the dielectric constant ϵ ($R^2 = 0.47$), a somewhat improved correlation with Reichard's solvent polarity parameter [46] $E_T(30)$ ($R^2 = 0.75$), and the best fit with Hildebrand's solubility parameter δ ($R^2 = 0.83$) or δ^2 ($R^2 = 0.79$) (Figure 3). Other single-parameter correlations gave poor fits. Combining δ or δ^2 with any other solvent parameter showed only a marginal increase in the values of R^2. There was no improvement in the fit when the solvent basicity was included in the correlation.

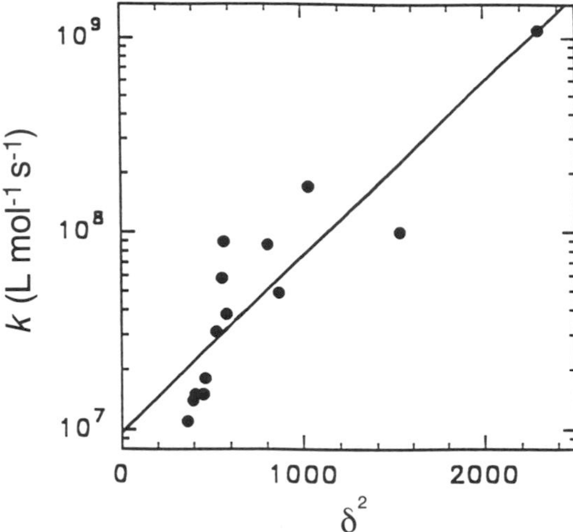

Figure 3 Correlation of the rate constant for oxidation of chlorpromazine by $CCl_3O_2^{\bullet}$ radicals in various solvents with Hildebrand's solubility parameter δ^2 (cohesive energy density) (data shown in Table 9) [45].

The solubility parameter, δ [47], is defined as the square root of the cohesive energy density, i.e. the energy required for breaking the bonds between the molecules in the solvent. It is derived from the latent heat of vaporization for the solvent: $\delta = [(\Delta H_v - RT)/V_M]^{1/2}$, where ΔH_v is the molar heat of vaporization at zero pressure and V_M is the molar volume. The parameter δ^2, which has the units of pressure, has been called [48] the cavity term or the cavity effect, and was suggested to be a measure of the work necessary to separate the solvent molecules and create a suitably sized cavity for the solute. The requirements for an increase or decrease in the space available while going from reactants to a transition state or an intermediate for a reaction should be essentially independent of solvent. The relationship of the rate constant to δ could be an indication of this requirement, with a positive coefficient indicating that less space is required for the transition state or the intermediate, while a negative coefficient indicates that these occupy more space. High values of δ result from strong solvent–solvent interactions, typically in the form of hydrogen bonding and dipole interactions. This ability of the solvent molecules to form intermolecular bonds indicates the possibility of solvent–solute interactions. Therefore, an alternative explanation of the positive effect of δ on electron-transfer reactions is that the transition state will be usually more polar than the reactants and hence will be more solvated in high-δ solvents. This will be especially true for oxygen-centred radicals due to hydrogen bonding in the transition state.

The correlation of the rate constants for Trolox with δ was considerably poorer than that for ClPz (R^2 was only 0.60). The plot shows, however, that the main points deviating from the line are those for the basic solvents. The correlation was improved considerably by taking into account the basicity of the solvent in the form of $\log k = C_0 + C_1\delta_H^2 + C_2 pK_a$. The coefficients of this correlation were: $C_0 = 6.81 \pm 0.14$, $C_1 = (7.9 \pm 0.7) \times 10^{-4}$, $C_2 = 0.048 \pm 0.007$, and $R^2 = 0.94$ (Figure 4) when the pK_a values in acetonitrile

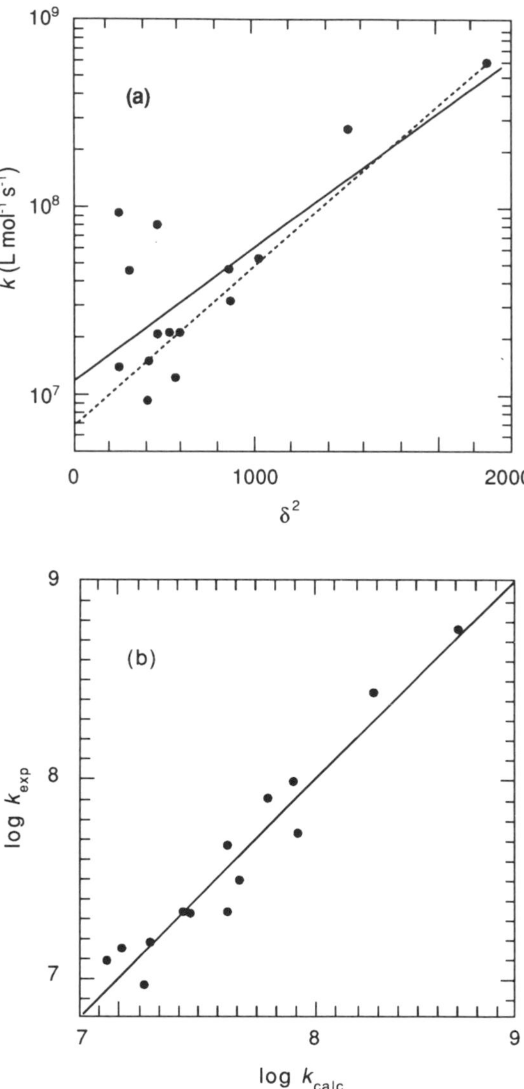

Figure 4 Correlation of the rate constant for oxidation of Trolox by $CCl_3O_2^{\cdot}$ radicals in various solvents with the solvent parameters: (a) δ^2 only, $R^2 = 0.60$; (b) according to the equation $\log k = C_0 + C_1\delta^2 + C_2 pK_a$, where $C_0 = 6.81$, $C_1 = 7.9 \times 10^{-4}$, $C_2 = 0.048$, and the values of pK_a in acetonitrile are taken from Table 9. $R^2 = 0.94$.

[49] are used; $C_0 = 7.04 \pm 0.14$, $C_1 = (8.3 \pm 0.7) \times 10^{-4}$, $C_2 = 0.064 \pm 0.009$, and $R^2 = 0.94$ when the pK_a values in water [50] are taken. These correlations did not inlcude CCl_4 as a solvent since its basicity is unknown. To fit CCl_4 to the same line, it would have to have a pK_a value in acetonitrile of 11.9, close to that of pyridine, which is very unlikely. A possible reason that the reaction in CCl_4 has a higher rate constant than expected is solvent polarizability, which is higher for aromatics and chlorinated aliphatics than for non-chlorinated aliphatic compounds [51]. Another possibility is that the reaction in the low-polarity solvent CCl_4 is via hydrogen abstraction rather than electron transfer [52].

The fact that the rate constant for the reaction of $CCl_3O_2^{\bullet}$ with Trolox also depends on solvent basicity indicates that the solvent participates in the removal of a proton. This effect may occur via two possible mechanisms. The first process is the de-protonation of the phenolic moiety, ArOH, by a solvent molecule, S, leading to the formation of the phenolate anion, which is much more reactive than the neutral phenol in electron transfer reactions [53]:

$$\mathrm{ArOH + S \longrightarrow ArO^- + SH^+} \tag{22}$$

This process may occur to some extent prior to the reaction with the peroxyl radical, but mainly in strongly basic solvents such as triethylamine (TEA). The second mechanism is by removing a proton from ArOH in the transition state, in a process concerted with the electron transfer, to form the neutral radical $ArO^{\bullet}$:

$$\mathrm{CCl_3O_2^{\bullet} + ArOH + S \longrightarrow CCl_3O_2^- + ArO^{\bullet} + SH^+} \tag{23}$$

Subsequently, or concurrently, a proton is transferred from a solvent molecule or from SH^+ to the incipient $CCl_3O_2^-$ to form CCl_3O_2H:

$$\mathrm{CCl_3O_2^- + SH^+ \longrightarrow CCl_3O_2H + S} \tag{24}$$

The sum of the above reactions appears as a hydrogen-atom transfer:

$$\mathrm{CCl_3O_2^{\bullet} + ArOH + S \longrightarrow CCl_3O_2H + ArO^{\bullet} + S} \tag{25}$$

although the mechanism involves electron transfer and the solvent is deeply involved in the process.

The rate constants for oxidation of hematoporphyrin by $CCl_3O_2^{\bullet}$ in different solvents (Table 10) [54] exhibited a similar behaviour to that of ClPz. The best correlation was obtained with the cohesive energy density δ and not with the dielectric constant [$\log k$ vs. $1/\epsilon$ or $(\epsilon - 1)/(2\epsilon + 1)$], despite the fact that the reaction involves electron transfer and produces formally charged species. This finding indicated again that the critical step in this reaction involves changes in the solvent cavity required for the reaction to proceed, whereas the electron-transfer step within the activated complex in

Table 10 Rate constants for reactions of $CCl_3O_2^{\bullet}$ radicals with hematoporphyrin in various solvents and cohesive pressures of the neat solvents

Solvent[a]	k ($L\,mol^{-1}\,s^{-1}$)	δ (MPa)
Water/2-PrOH (9/1) pH 7 (0.1%)	2.2×10^8	47.9
Formamide (1%)	1.5×10^8	39.3
N-methylacetamide (6%)	5.0×10^7	29.9
MeOH (10%)	6.2×10^7	29.6
Tetramethylene sulfone (5%)	7.9×10^7	27.4
Propylene carbonate (5%)	6.2×10^7	27.2
N,N-dimethylformamide (10%)	5.3×10^7	24.8
Dimethyl sulfoxide (5%)	1.0×10^8	24.5
1-Methyl-2-pyrrolidinone (10%)	4.4×10^7	23.1
N,N-dimethylacetamide (10%)	3.0×10^7	22.1
Pyridine (10%)	4.8×10^7	21.9
p-Dioxane (10%)	4.8×10^7	20.5
Acetone (10%)	3.5×10^7	20.2
Tetrahydrofuran (10%)	3.4×10^7	18.6

[a] Figures in parentheses represent % CCl_4 in solvent system.

probably very fast. Since the solvent dielectric constant is one of the factors that determine the cohesive energy, its effect on reactant polarization and electron transfer may be incorporated into the overall effect of δ. However, other factors that contribute to the cohesive energy, such as hydrogen bonding and van der Waals forces, appear to have a more dominant effect on the rate constant. That the rate constant for the oxidation reaction increases with increasing solvent cohesion suggests that the reactants form an activated complex whose volume is less than the sum of those of the reactants, i.e. that the reaction has a negative volume of activation. This, however, has not been verified experimentally and would be an illuminating addition to this topic.

The above discussion dealt with the solvent effect due to the bulk properties of the solvent. Effects due to specific interactions of the solvent were also reported and these provided evidence for stronger bonding between the peroxyl radical and the porphyrin. The rate constant for the oxidation of zinc porphyrin by $CCl_3O_2^{\bullet}$ in different solvents [55] varied between 3×10^7 and $3 \times 10^9\,L\,mol^{-1}\,s^{-1}$ but did not correlate with solvent polarity or cohesive energy. Furthermore, similar rate constants were obtained with solvent/CCl_4 ratios of 9/1 or 1/9. With pyridine, even a 1% addition decreased the rate constant by a factor of 50. These results indicated that the effect of the solvent on the rate constant was not due to the bulk properties of the solvent, but rather to complexation of the solvent as an axial ligand on the Zn centre of the polyphyrin. This was further confirmed by the good correlation found between the rate constant and the energies of the absorption peaks of the porphyrin in each solvent, indicating that k

decreases when the solvent molecule was more strongly bound as an axial ligand. This dependence suggests that the reaction proceeds by an inner-sphere mechanism:

$$S + ZnP \rightleftharpoons S\text{—}ZnP \quad (26)$$

$$CCl_3O_2^{\cdot} + ZnP \rightleftharpoons CCl_3O_2\text{—}ZnP \quad (27)$$

$$CCl_3O_2^{\cdot} + S\text{-}ZnP \rightleftharpoons CCl_3O_2\text{—}Zn(S)P \rightleftharpoons CCl_3O_2\text{—}ZnP + S \quad (28)$$

$$CCl_3O_2\text{—}ZnP \longrightarrow CCl_3O_2^- + ZnP^{+\cdot} \quad (29)$$

$$S + ZnP^{+\cdot} \rightleftharpoons S\text{—}ZnP^{+\cdot} \quad (30)$$

A kinetic study of the reaction of $(CH_3)_2C(OH)O_2^{\cdot}$ with Fe^{III} porphyrins indicated that the mechanism involves an intermediate complex as above, and that this complex decomposes to the final products by acid-dependent and acid-independent pathways [56]:

$$RO_2^{\cdot} + Fe^{III}P \rightleftharpoons [RO_2Fe^{III}P]^{\cdot} \quad (31)$$

$$[RO_2Fe^{III}P]^{\cdot} \longrightarrow RO_2^- + Fe^{III}P^{+\cdot} \quad (32)$$

$$[RO_2Fe^{III}P]^{\cdot} + H^+ \longrightarrow RO_2H + Fe^{III}P^{+\cdot} \quad (33)$$

This mechanism may be operative in many reactions of peroxyl radicals with metalloporphyrins. The lifetime of the intermediate complex, however, would depend on the metal and the radical; it is probably much shorter with Zn than with Fe, due to the difference in the strength of the ligand bonding, and presumably shorter for $CCl_3O_2^{\cdot}$ than for $(CH_3)_2C(OH)O_2^{\cdot}$ due to stronger polarization and thus faster cleavage of the complex in the case of the $CCl_3O_2^{\cdot}$ radical.

In the above studies, complexation of the peroxyl radical to the metal was deduced from kinetic evidence but no spectral observation of these short-lived intermediates was reported. In the reaction of peroxyl radicals with aqueous ferrous ions, however, the intermediate complex has been observed by kinetic spectrophotometry on the μs time-scale and its subsequent decay kinetics were measured as a function of the conditions [57, 58]. The results suggested formation of a peroxyl-iron complex with a rate constant near $1 \times 10^6\,L\,mol^{-1}\,s^{-1}$ for a number of peroxyl radicals, similar to the rate of exchange of a water molecule from the coordination sphere of the iron:

$$RO_2^{\cdot} + Fe_{aq}^{2+} \longrightarrow RO_2^-Fe^{3+} \quad (34)$$

Subsequently, this complex decays on a slower time-scale via several routes:

$$RO_2^-Fe^{3+} + H_2O \longrightarrow RO_2H + OH^- + Fe_{aq}^{3+} \quad (35)$$

$$RO_2^{-}Fe^{3+} + H^+ \longrightarrow RO_2H + Fe_{aq}^{3+} \quad (36)$$

$$RO_2^{-}Fe^{3+} + Fe_{aq}^{2+} + H^+ \longrightarrow RO^{\bullet} + OH^- + 2Fe_{aq}^{3+} \quad (37)$$

The rate constants for $RO_2^{\bullet} = (CH_3)_2C(OH)O_2^{\bullet}$ are $k_{35} = 1 \times 10^2\ s^{-1}$, $k_{36} = 3 \times 10^4\ mol^{-1}\,s^{-1}$, and $k_{37} = 5 \times 10^3\ L\,mol^{-1}\,s^{-1}$.

9 TEMPERATURE EFFECTS

As mentioned above, the rate constants for the reactions of $CH_3O_2^{\bullet}$ and the peroxyl radical derived from dioxane with TMPD and for the reaction of the latter radical with ABTS in aqueous solutions were measured as a function of temperature and showed slightly negative activation entropies. Even more negative values were reported for the reactions of the cyclohexylperoxyl radical with tocopherols in cyclohexane [59]. These negative values were invoked to support an electron-transfer mechanism for these reactions which involves a strongly hydrogen-bonded transition state. The activation energies measured were between 14 and 27 kJ mol^{-1}.

A wider study of the temperature effect on the rate constants was carried out on the reactions of $CCl_3O_2^{\bullet}$, $CHCl_3O_2^{\bullet}$, $CH_2ClO_2^{\bullet}$, and $CH_3O_2^{\bullet}$ with chlorpromazine, Trolox, ascorbate, and urate in aqueous solutions [60]. The temperature dependence of the reactions of $CCl_3O_2^{\bullet}$ was also studied over a range of 2-PrOH/H_2O mixtures. The Arrhenius parameters and room temperatuer rate constants (k_{298}) are summarized in Table 11. It is found that a decrease in the rate constant, either due to an increase in the fraction of the organic solvent or to a decrease in the number of chlorine atoms in the radical, is caused by lower pre-exponential factors (A) and not by higher activation energies. Rather, the activation energy decreased in most cases where the rate constant decreased, but the decrease in the A-factor more than compensates for the lower E_a, thus resulting in a lower k_{298}. In many cases of low rate constants, the activation energy is considerably less than the activation energy for diffusion (17.5 kJ mol^{-1}). However, since the observed rate constant is much below the diffusion limit, the major contribution to the former stems from the rate of the chemical reaction (i.e. $k_{obs} = (k_{reac}k_{diff})/(k_{reac} + k_{diff})$ and when $k_{diff} \gg k_{reac}$, then $k_{obs} = k_{reac}$).

The A-factor and the activation energy vary mostly as a linear function in such a manner as to result in opposite effects on the measured rate constant, i.e. as the activation energy increases, leading to a *lower* rate constant, the A-factor also increases, and leads to a *higher* rate constant. This dependence is clearly seen in Figure 5. This phenomenon has been described before as the isokinetic relationship [61]:

$$E_a = R\beta \ln A + e_0 \quad (38)$$

where R is the gas constant, β is the isokinetic temperature, and e_0 is the

Table 11 Arrhenius parameters and room-temperature rate constants for reactions of peroxyl radicals

Compound	Radical	Solvents	Volume ratios	E_a (kJ mol^{-1})	log A	k_{298} (L mol^{-1} s^{-1})
ClPz	$CCl_3O_2^{\cdot}$	H_2O/2-PrOH/CCl_4	95/5/0.05	≈ 28	14.1	1.4×10^9
			90/10/0.06	23.7 ± 1.0	13.2	1.2×10^9
			80/20/0.01	27.1 ± 0.8	13.8	1.2×10^9
			69/29/2	11.1 ± 5.0	11.0	1.1×10^9
			49/49/2	9.4 ± 1.0	10.1	2.9×10^8
			9/89/2	8.7 ± 1.0	9.4	6.9×10^7
			0/50/50	5.6 ± 2.1	8.3	2.0×10^7
		Dioxane/2-PrOH/CCl_4	40/20/40	1.1 ± 1.0	7.2	1.1×10^7
	$CHCl_2O_2^{\cdot}$	H_2O/2-PrOH/$CHCl_3$	90/10/0.1	14.0 ± 1.1	11.2	5.6×10^8
			9/81/10	11.5 ± 1.1	8.6	3.4×10^6
	$CH_2ClO_2^{\cdot}$	H_2O/2-PrOH/CH_2Cl_2	90/10/0.5	6.1 ± 1.2	8.6	3.6×10^7
			66/33/1	8.8 ± 0.2	7.9	2.3×10^6
Trolox	$CCl_3O_2^{\cdot}$	H_2O/2-PrOH/CCl_4	90/10/0.1	16.6 ± 1.0	11.7	6.9×10^8
			49/49/2	5.8 ± 1.3	9.1	1.2×10^8
			0/96/4	11.7 ± 1.3	9.3	1.9×10^7
	$CHCl_2O_2^{\cdot}$	H_2O/2-PrOH/$CHCl_3$	90/10/0.1	17.2 ± 2.8	11.5	2.7×10^8
	$CH_2ClO_2^{\cdot}$	H_2O/2-PrOH/CH_2Cl_2	90/10/0.5	6.4 ± 5.8	9.1	9.5×10^7
Ascorbate	$CCl_3O_2^{\cdot}$	H_2O/2-PrOH/CCl_4	90/10/0.1	10.5 ± 1.4	10.4	3.9×10^8
	$CH_3O_2^{\cdot}$	H_2O/DMSO	90/10	15.2 ± 0.7	8.9	1.9×10^6
Urate	$CH_3O_2^{\cdot}$	H_2O/DMSO	90/10	9.4 ± 0.9	8.5	7.0×10^6

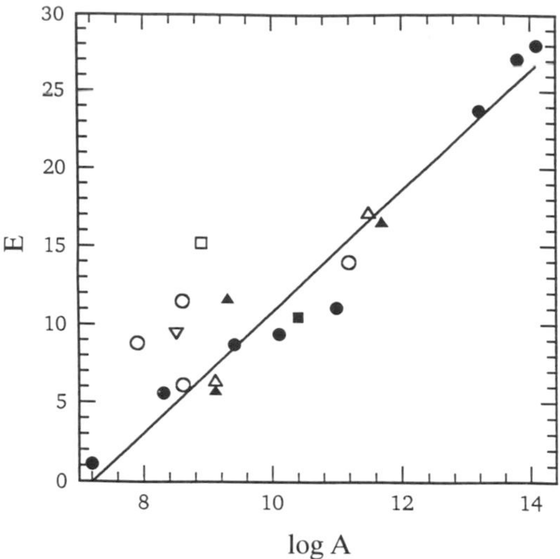

Figure 5 Isokinetic relationship between the activation energy (E, in kJ mol^{-1}) and the pre-exponential factor ($\log A$, where A is in L $mol^{-1} s^{-1}$) for the reactions of peroxyl radicals with organic compounds. The line was fitted to the data for $CCl_3O_2^{\bullet}$ only: (●) chlorpromazine; (▲) Trolox; (■) ascorbate. The data for the other radicals, $CHCl_2O_2^{\bullet}$, $CH_2ClO_2^{\bullet}$, are also shown for comparison: (○) chlorpromazine; (△) ascorbate; (▽) urate ([60], data shown in Table 11).

intercept of the line. The line for the results for $CCl_3O_2^{\bullet}$ alone shows a good fit ($R^2 = 0.97$) and corresponds to $\beta = 202$ K.

This behaviour has been explained in terms of a reversible formation of a transient intermediate. For peroxyl radicals an equilibrium reaction was assumed:

$$ROO^{\bullet} + S \rightleftharpoons ROO\text{-}S^{\bullet} \tag{39}$$

which is followed by the electron-transfer reaction, probably assisted by proton donation from water:

$$ROO\text{-}S^{\bullet} + H_2O \longrightarrow ROOH + S^{+} + OH^{-} \tag{40}$$

If a steady state in ROO—$S^{\bullet}$ is assumed, then the observed rate constant will be given by:

$$k_{obs} = \frac{k_{39}k_{40}}{k_{-39} + k_{40}} \tag{41}$$

or, as a function of temperature:

$$k_{obs} = \frac{A_{39}A_{40} \exp[-(E_{39} + E_{40})/RT]}{A_{-39} \exp(-E_{-39}/RT) + A_{40} \exp(-E_{40}/RT)} \tag{42}$$

If $k_{40} \gg k_{-39}$, then a "normal" Arrhenius expression will result:

$$k_{obs} = A_{39} \exp(-E_{39}/RT) \quad (43)$$

If, on the other hand, $k_{40} \ll k_{-39}$, then

$$k_{obs} = \frac{A_{39}A_{40} \exp[-(E_{39} + E_{40})/RT]}{A_{-39} \exp(-E_{39}/RT)} = \frac{A_{39}A_{40}}{A_{-39}} = \exp[-(E_{39} + E_{40} - E_{-39})/RT] \quad (44)$$

Although the rate constants for the formation and decomposition of the complex (reaction (39)) are unknown, the rate constant for reaction (40) is probably related to the driving force for the reaction and to the ability of the solvent to promote the electron transfer. In any given solvent, the driving force is expected to increase with the degree of halogen substitution on the peroxyl radical. For a given reaction, the rate constant was related to both the dielectric constant of the solvent and to the proton-donating ability of the solvent. The electron-transfer step is expected to have little temperature dependence, possibly even a negative value due to the negative temperature dependence for the dielectric constant, and the variation in its rate constant is probably due to a variation in the pre-exponential factor [62]. If this is correct, then the observed activation energy varies from E_{39} to $(E_{39} - E_{-39})$ as k_{40} becomes slower relative to k_{-39}. The observed pre-exponential factor also ranges from A_{39} to $A_{39}A_{40}/A_{-39}$. Under these conditions, where $k_{40} \ll k_{-39}$, A_{40} is probably much less than A_{-39}. Therefore, the observed pre-exponential factor also decreases. The proposed mechanism for these electron-transfer reactions, then, leads to a kinetic expression which is in basic agreement with the results.

10 MULTI-ELECTRON TRANSFER REACTIONS

Although most oxidation reactions by peroxyl radicals are via one-electron-transfer mechanisms, several cases have been suggested to involve the transfer of more than one electron in a single reaction step. This was particularly observed for the oxidation of I^- and of organic sulfides.

Oxidation of dimethyl sulfide by $CCl_3O_2^{\cdot}$ and $CHCl_2O_2^{\cdot}$ was found [63] to lead to formation of the dimer radical cation, presumably by a mechanism similar to that of oxidation of this sulfide by OH and other radicals [64]:

$$CCl_3O_2^{\cdot} + Me_2S \longrightarrow Me_2S^{+\cdot} + CCl_3O_2^- \quad (45)$$

$$Me_2S^{+\cdot} + Me_2S \rightleftharpoons (Me_2S \therefore SMe_2)^+ \quad (46)$$

The yield of the radical cation, however, was only 75% that of the yield of $CCl_3O_2^{\cdot}$ and only 50% that of $CHCl_2O_2^{\cdot}$. From pulse conductometric measurements it was suggested that peroxyl radicals add to the sulfide to form an intermediate which may undergo two parallel reactions, one leading to the radical cation and the other to the sulfoxide:

$$CCl_3O_2^{\cdot} + Me_2S \longrightarrow (CCl_2O\text{–}O\text{–}SMe_2)^{\cdot} \qquad (47)$$

$$(CCl_3O\text{—}O\text{—}SMe_2)^{\cdot} + Me_2S \longrightarrow CCl_3OO^{-} + (Me_2S \therefore SMe_2)^{+} \qquad (48)$$

$$(CCl_3O\text{—}O\text{—}SMe_2)^{\cdot} \longrightarrow CCl_3O^{\cdot} + OSMe_2 \qquad (49)$$

The relative contributions of the two paths depend on the electron affinity of the peroxyl radical. The intermediate adduct will be slightly more polar with $CCl_3O_2^{\cdot}$ than with $CHCl_2O_2^{\cdot}$, and thus will produce the radical cation more readily. Weaker oxidizing peroxyl radicals will favour decomposition of the adduct to the sulfoxide and alkoxyl radical. The chlorinated alkoxyl radicals produced will release Cl atoms rapidly. Formation of the sulfoxide was confirmed by product analysis. From experiments with $R^{18}O^{18}O^{\cdot}$, however, it was concluded that the oxygen in the sulfoxide does not originate in the peroxyl radical. Therefore, it was suggested that the intermediate adduct must be hydrated before it decomposes so that the sulfoxide O-atom comes from the water:

$$(CCl_3O-O-SMe_2)^{\cdot} + H_2O \longrightarrow CCl_3O-\underset{\displaystyle H}{\underset{|}{O}}\cdots\overset{\displaystyle OH}{\overset{|}{S}}Me_2 \qquad (50)$$

The overall process constitutes a transfer of an oxygen atom or two-electron process in a single step.

By comparison with the sulfides, the reactions of $CCl_3O_2^{\cdot}$ with a selenide [65] and with tellurides [66] were suggested to proceed mainly via one-electron transfer. On the other hand, the reactions of tellurides with a weaker oxidant, i.e. the $(CH_3)_2C(OH)CH_2O_2^{\cdot}$ radical, took place by oxygen-atom transfer. In the latter case, the intermediate adduct was suggested to be very short-lived and the reaction is viewed as a one-step O-atom transfer with a polar transition state. Reactions of non-halogenated alkylperoxyl radicals with phosphines and phosphites take place via adduct formation, with the final products being partly those produced by O-atom transfer to the phosphorus [67]. These are discussed in more detail in other chapters of this book. The reaction of $CCl_3O_2^{\cdot}$ with indole was shown to form some adducts which decompose to products by ring opening [68]; the mechanism is not strictly an O-atom transfer, although the O—O bond of the peroxyl radical is ruptured in the process.

The reaction of $CCl_3O_2^{\cdot}$ with I^- also was shown to involve multi-electron transfer [69]. Pulse conductometric and spectrophotometric measurements indicated that the reaction leads to simultaneous formation of both I_3^- and I_2^-. This was interpreted in terms of a concerted two-electron transfer to form I_2 and an alkoxyl radical; the latter then oxidizes another I^- to give an I atom, and finally the I_2 and the I atom each bind another I^- ion to give the more stable complexes that are observed:

$$CCl_3O_2^{\cdot} + I^- + H_2O \rightleftharpoons [CCl_3O\text{—}O(H)\text{–}I]^{\cdot} + OH^- \quad (51)$$

$$[CCl_3O\text{—}O(H)\text{—}I]^{\cdot} + I^{\cdot} \longrightarrow CCl_3O^{\cdot} + OH^- + I_2 \quad (52)$$

$$CCl_3O^{\cdot} + I^- \longrightarrow CCl_3O^- + I^{\cdot} \quad (53)$$

$$I^{\cdot} + I^- \rightleftharpoons I_2^- \quad (54)$$

$$I_2 + I^- \rightleftharpoons I_3^- \quad (55)$$

The rate constants for the formation of I_2^- and I_3^- were identical, with values of $2.0 \times 10^8\,L\,mol^{-1}\,s^{-1}$ at pII 3 and $4.0 \times 10^7\,L\,mol^{-1}\,s^{-1}$ at $pH > 6$. The pH dependence showed a sigmoidal curve corresponding to a pK_a value of about 4. This was not interpreted as a true acid–base equilibrium, but rather as an effect of protons on the above reactions. Reaction (51) will be accelerated if H_3O^+ participates instead of H_2O, and reaction (52) may be enhanced as well.

11 ARYLPEROXYL RADICALS

All of the aliphatic peroxyl radicals discussed above have absorption spectra in the UV region, generally with λ_{max} near 250 nm and little absorbance above 300 nm. Therefore, their reactions could not be followed by utilizing these absorptions if other compounds were present that absorb in the UV region. In contrast, arylperoxyl radicals exhibit absorption peaks in the visible range which permit measurement of their reaction rates by following the decay of these absorptions.

Phenylperoxyl radicals were produced by flash photolysis of iodobenzoate ions [70] and by pulse radiolysis of halogenated aromatic compounsd [71, 72]. In the former method, photochemical homolysis of the C—I bond produces an I atom and a phenyl radical. In the latter method, reductive dehalogenation produces a halide ion and a phenyl radical. The phenyl radicals react very rapidly with O_2 ($k \approx 2 \times 10^9\,L\,mol^{-1}\,s^{-1}$) to form phenylperoxyl radicals.

The 4-carboxyphenylperoxyl radical was found [72] to react with 4-methoxyphenolate ion by electron transfer to form the 4-methoxyphenoxyl radical, having a rate constant of $2.7 \times 10^8\,L\,mol^{-1}\,s^{-1}$. The reaction with the neutral 4-methoxyphenol at pH 7, however, was too slow to be measured, with an upper limit of $k \leqslant 1 \times 10^7\,\mathrm{L\,mol^{-1}\,s^{-1}}$ being estimated. For the unsubstituted phenolate ion (pH 11.2) the rate constant was also too slow to measure, i.e. $k \leqslant 2 \times 10^7\,\mathrm{L\,mol^{-1}\,s^{-1}}$. The 4-carboxyphenylperoxyl radical oxidized several other compounds moderately rapidly, e.g. ascorbate at pH 7, $k = 8.7 \times 10^7\,\mathrm{L\,mol^{-1}\,s^{-1}}$, Trolox at pH 7, $k = 1.0 \times 10^8\,\mathrm{L\,mol^{-1}\,s^{-1}}$, and at pH 11.5, $k = 2.6 \times 10^8\,\mathrm{L\,mol^{-1}\,s^{-1}}$, and ABTS, $k = 9 \times 10^8\,\mathrm{L\,mol^{-1}\,s^{-1}}$.

The rate constant found [72] for ascorbate indicates that phenylperoxyl radicals are more reactive than most alkylperoxyl and substituted alkylperoxyl radicals, except for those containing α-halogens. The rate constants for ClPz and ABTS are even higher than those of $CH_2ClO_2^{\bullet}$ but lower than those of $CCl_3O_2^{\bullet}$ with the same compounds. The high reactivity of phenylperoxyl radicals is clearly due to the electron-withdrawing property of the phenyl group which decreases the electron density at the oxyl site and thus increases its electron affinity and reduction potential.

The reactivity of 4-substituted phenylperoxyl radicals is affected by the electron-withdrawing properties of the 4-substituent group. For example [72], for the phenylperoxyl radicals with H, CH_3, OH, Cl, and CN at the 4-position, the rate constant for oxidation of ABTS varied between 3.4×10^8 and $2.5 \times 10^9\,\mathrm{L\,mol^{-1}\,s^{-1}}$. A plot of $\log k$ vs. the Hammett substituent constant, σ_p gave a straight line with a slope of $\rho = 0.88$.

The absorption spectra and the reactivities of arylperoxyl radicals were also studied for species in which the aryl group is larger than phenyl, i.e. biphenyl, 2-naphthyl, 1-naphthyl, and 9-phenanthryl [73]. Because of the low solubilities of the aryl halide precursors in water, these radicals were studied in MeOH. Higher alcohols were less suitable, owing to their higher reactivity toward aryl radicals which competes with the formation of the arylperoxyl radical. The absorption peak of the arylperoxyl is red-shifted for the higher aryl groups (Table 12) and reaches 700 nm for the phenanthrylperoxyl. In parallel with this red shift, the reactivity of the radicals toward several compounds increases (Table 12). Figure 6 shows that a good linear

Table 12 Absorption maxima of arylperoxyl radicals in neat methanol and rate constants for their reactions with several reductants in 60% MeoH or 30% t-BuOH

Radical	λ_{max} (nm)	E (eV)	k (ABTS) ($\mathrm{L\,mol^{-1}\,s^{-1}}$)		k (Trolox) ($\mathrm{L\,mol^{-1}\,s^{-1}}$		k (ClPz) in t-BuOH/H_2O ($\mathrm{L\,mol^{-1}\,s^{-1}}$)
			MeOH/H_2O	t-BuOH/H_2O	MeOH/H_2O	t-BuOH/H_2O	
$C_6H_5O_2^{\bullet}$	470	2.64	3.8×10^7	5.1×10^8	2.7×10^7	1.0×10^7	4.2×10^6
4-C_6H_5-$C_6H_4O_2^{\bullet}$	550	2.25	6.6×10^8	5.9×10^8			
2-$C_{10}H_7O_2^{\bullet}$	575	2.16	6.0×10^8	7.2×10^8	1.2×10^8	4.1×10^8	1.7×10^7
1-$C_{10}H_7O_2^{\bullet}$	650	1.91	1.6×10^9	1.2×10^9	2.8×10^8	5.6×10^8	1.2×10^8
9-$C_{14}H_9O_2^{\bullet}$	700	1.77	2.1×10^9	1.2×10^9			

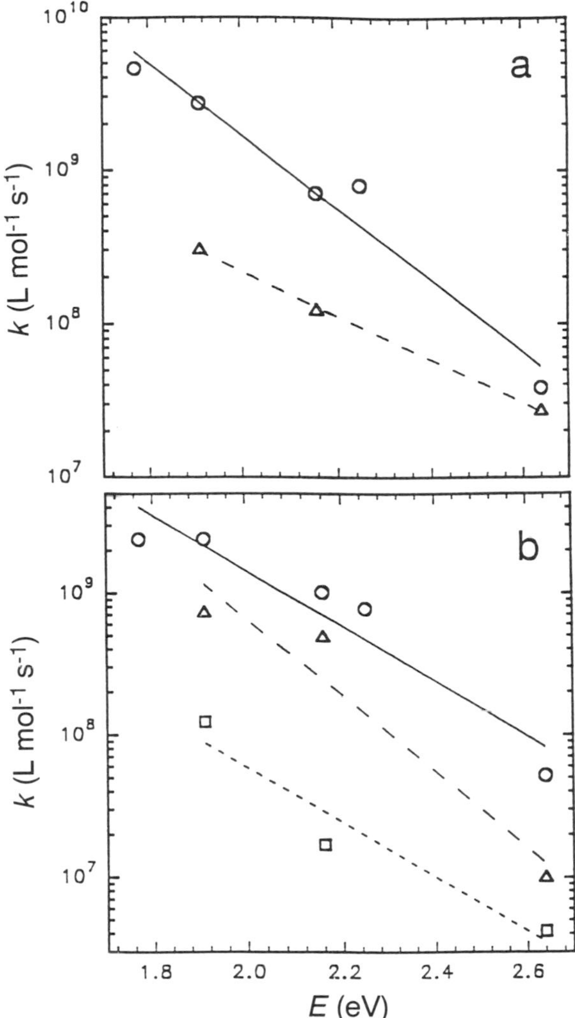

Figure 6 Correlation of the rate constants for oxidation by arylperoxyl radicals with the energies of the absorption maxima for these radicals: (a) (○) ABTS and (△) Trolox in 60% MeOH; (b) (○) ABTS, (△) Trolox and (□) chlorpromazine in 30% t-BuOH (data shown in Table 12) [73].

correlation is obtained between $\log k_{act}$ (corrected for diffusion according to $1/k_{act} = 1/k_{obs} - 1/k_{diff}$) and the energy of the absorption peak ($E = hc/\lambda$). This correlation suggests that the driving force for the electron transfer is inversely related to the energy difference between the singly occupied molecular orbital (SOMO) of the radical and the penultimate occupied molecular orbital which is responsible for the visible-range absorption (the highest wavelength absorption peak of phenylperoxyl was interpreted as an excitation of a ring π-electron into the singly occupied π-orbital on the oxygen) [74]. Possibly, as the π-system becomes more extended all the energy levels are compressed and this reduces both the energy difference between the SOMO and the penultimate occupied orbital (optical absorption energy) as well as the energy difference between the SOMO and the continuum level (ionization potential). A decrease in ionization potential correlates with an increase in electron affinity and, assuming similar effects of solvation in the whole series, the reduction potential will increase. In other words, there is a strong interaction between the aromatic π-system and the unpaired electron on the terminal oxygen which affects both the absorption spectrum and the reactivity.

Phenylperoxyl radicals containing one or more halogen atoms on the ring were studied in various H_2O/MeOH mixtures [75]. These measurements were restricted by the limited solubility of the precursors. The rate constants for reaction with ABTS (Table 13) indicate that all of the halogenated radicals are more reactive than the unsubstituted phenylperoxyl radical. Among the halogenated species, the order of reactivity is Br > Cl > F, and for each halogen the order of reactivity of the different isomers is $o > m > p$. The halogens exert two effects on the phenylperoxyl radical: an inductive effect, which lowers the electron density on the oxyl group and thus makes the peroxyl radical a stronger oxidant, and a mesomeric effect which acts in the opposite direction. The observation that all halogen-substituted phenylperoxyl radicals are more reactive than the unsubstituted species

Table 13 Rate constants for reactions of halogenated phenylperoxyl radicals with ABTS in aqueous alcohol solutions

Substituents	k (L mol^{-1} s^{-1})			
	0.5% 2-PrOH	33% MeOH	60% MeOH	100% MeOH
H	6.6×10^8	2.9×10^8	4.8×10^7	
4-F	1.0×10^9	2.0×10^8	6.8×10^7	
3-F	1.7×10^9	7.4×10^8	1.8×10^8	
2-F	2.1×10^9	8.1×10^8	2.3×10^8	5.2×10^7
4-Cl	1.1×10^9	6.7×10^8	2.0×10^8	
3-Cl	2.7×10^9	1.1×10^9	3.2×10^8	3.3×10^7
2-Cl	2.3×10^9	1.2×10^9	4.9×10^8	5.5×10^7
4-Br		9.9×10^8	2.8×10^8	
3-Br		1.4×10^9	3.8×10^8	3.2×10^7
2-Br		1.3×10^9	4.7×10^8	4.1×10^7
3,5-F_2		9.3×10^8	3.7×10^8	
3,4-F_2		6.2×10^8	1.8×10^8	
3,4,5-F_3		8.5×10^8	3.5×10^8	
3,5-Cl_2		2.1×10^9	8.8×10^8	1.2×10^8
2,4,5-Cl_3			8.3×10^8	3.8×10^8

indicates that the inductive effect predominates. The order of reactivity outlined above is in line with the Hammett σ or the Brown σ^+ parameters, which are in the order Br > Cl > F and $m > p$ (no general constants are accepted for the *ortho*-isomer). Furthermore, substituent constants for the di- and tri-substituted species are unknown. Therefore, correlation of all of the rate constants given in Table 13 with substituent constants is not feasible. Instead, the pK_a values for halogenated phenols were utilized as a measure of the effect of the halogens on the electron density at the oxygen. The plot (Figure 7) shows a monotonic increase in $\log k$ as the pK_a decreases, but not a strict linear correlation. The curve bends toward a plateau level at $\log k > 9$.

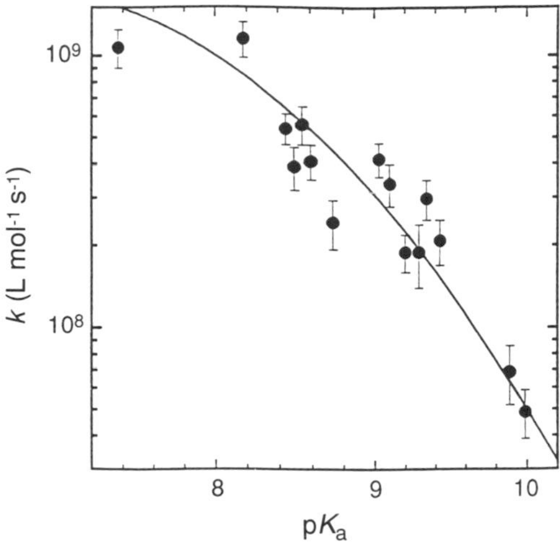

Figure 7 Correlation of the rate constant for oxidation of ABTS by various halogenated phenylperoxyl radicals with the pK_a values of similarly substituted phenols (data shown in Table 13) [75].

If the pK_a scale represents a measure of the reduction potential of the peroxyl radical, and assuming that the oxidation of ABTS by these peroxyl radicals is via an outer-sphere electron-transfer mechanism, then the Marcus theory [40] predicts that the dependence of $\log k$ on pK_a should be a parabola. By using the Marcus equations, the reduction potential for the phenylperoxyl radical was estimated to be 0.7 V vs. the normal hydrogen electrode (NHE), i.e. similar to the values estimated for alkylperoxyl but lower than the reduction potential for the phenoxyl radical [76]. This is in line with the rate constants for these two radicals with ABTS and with the weakening of the electron-withdrawing effect of the phenyl group with the distance from the oxygen.

Comparison of the reactivities of 4-methoxy substituted phenylperoxyl and phenoxyl radicals [77] showed the opposite trend from that seen with the unsubstituted species. Oxidation of ascorbate, ABTS, and sulfite ions took place with higher rate constants in the case of 4-$CH_3OC_6H_4OO^{\bullet}$ than in the case of 4-$CH_3OC_6H_4O^{\bullet}$. Furthermore, 4-$CH_3OC_6H_4OO^{\bullet}$ reacts with 4-$CH_3OC_6H_4OH$ to yield 4-$CH_3OC_6H_4OOH$ and 4-$CH_3OC_6H_4O^{\bullet}$. This difference in behaviour between the 4-methoxy and the unsubstituted radical was rationalized by the electron-donating effect of the 4-methoxy group, which increases the electron density on the phenoxyl oxygen to a larger extent than on the more distant terminal peroxylic oxygen.

Other studies on phenylperoxyl radicals concentrated on the mechanism of decay of these species and discovered intramolecular electron-transfer processes within certain substituted radicals [77]. In principle, phenylperoxyl radicals may decay via mechanisms similar to those for aliphatic peroxyl radicals [3], i.e. formation of an unstable tetroxide and subsequent elimination of O_2. This may lead to formation of two identical phenoxyl radicals:

$$2C_6H_5OO^{\bullet} \longrightarrow C_6H_5OOOOC_6H_5 \longrightarrow 2C_6H_5O^{\bullet} + O_2 \quad (56)$$

or their combination products (mainly hydroxylated biphenyl) [78]:

$$2C_6H_5OO^{\bullet} \longrightarrow C_6H_5OOOOC_6H_5 \longrightarrow (C_6H_4)_2(OH)_2 + O_2 \quad (57)$$

The decay was studied for three substituted phenylperoxyl radicals. The 4-$CH_3OC_6H_4OO^{\bullet}$ radical was found to decay with a second-order rate law and to yield the expected 4-$CH_3OC_6H_4O^{\bullet}$ radical, but only partially, due to competing processes. The 4-OH and the 4-$N(CH_3)_2$ derivatives, however, exhibited more complex behaviour, which was interpreted as intramolecular electron transfer.

The 4-hydroxyphenylperoxyl radical, absorbing at ca. 530 nm, underwent a rapid decay that was accompanied by absorption buildup at ca. 430 nm [77]. The decay followed a first-order rate law and the rate constant was nearly independent of the radical concentration and the 4-bromophenol precursor concentration, but was linearly dependent on buffer concentration at constant pH. These results suggested that the 4-hydroxyphenylperoxyl radical undergoes a base-catalyzed intramolecular electron transfer to yield the 4-hydroperoxyphenoxyl radical:

$$HOC_6H_4OO^{\bullet} + HPO_4^{2-} \longrightarrow {}^{-}OC_6H_4OO^{\bullet} + H_2PO_4^{-} \quad (58)$$

$${}^{-}OC_6H_4OO^{\bullet} \rightleftharpoons {}^{\bullet}OC_6H_4OO^{-} \quad (59)$$

$${}^{\bullet}OC_6H_4OO^{-} + H_2PO_4^{-} \longrightarrow {}^{\bullet}OC_6H_4OOH + HPO_4^{2-} \quad (60)$$

The pK_a of $HOC_6H_4OO^{\bullet}$ was expected to be lower than that of phenol (10.0), due to the electron-withdrawing effect of the dioxyl group, and by comparison with the pK_a values of other 4-substituted phenols [79], it was estimated to lie between 8.5 and 9.5. The pK_a of $^{\bullet}OC_6H_4OOH$ may be slightly lower than that observed for aliphatic hydroperoxides, which is near 12 [80], due to the electron-withdrawing effect of the phenyl group. The reaction of the peroxyl radical with the hydrogenphosphate ion results in partial ionization of the phenolic OH group. Subsequently, an intramolecular electron transfer occurs between the phenolic and the peroxyl moieties, resulting in partial formation of the phenoxyl radical. Protonation of the hydroperoxide (reaction (60)), which has a higher pK_a value than the peroxyl radical, will pull reaction (59) to the right. From the linear dependence of the observed rate constant on phosphate concentration, a second-order rate constant of $4.0 \times 10^8\,L\,mol^{-1}\,s^{-1}$ was determined for reaction (58). Because of the rapid intramolecular process in the 4-hydroxyphenylperoxyl radical, it was not possible to examine the second-order decay of this radical to find out whether any phenoxyl radical is produced by such a process, nor was it possible to study the oxidation of other compounds by this peroxyl radical.

The 4-(*N*,*N*-dimethylamino)phenylperoxyl radical exhibited λ_{max} at 600 nm in neutral solutions, but at 500 nm at pH 3 [77] due to the acid–base equilibrium:

$$(CH_3)_2NC_6H_4OO^{\bullet} + H^+ \longrightarrow (CH_3)_2HN^+C_6H_4OO^{\bullet} \qquad (61)$$

Decay of the peroxyl radical results in the formation of transient species absorbing at 500 nm, which may be ascribed to several possible products:
(a) Dimethylaminophenoxyl radicals produced by second-order decay of the peroxyl radicals:

$$2(CH_3)_2NC_6H_4OO^{\bullet} \longrightarrow 2(CH_3)_2NC_6H_4O^{\bullet} + O_2 \qquad (62)$$

(b) The radical cation $HOOC_6H_4N(CH_3)^{+}{}_2^{\bullet}$ formed by intramolecular electron transfer:

$$(CH_3)_2NC_6H_4OO^{\bullet} \rightleftharpoons (CH_3)_2N^+C_6H_4OO^- \qquad (63)$$

$$(CH_3)_2N^+C_6H_4OO^- + H_2PO_4^- \longrightarrow (CH_3)_2N^{+\bullet}C_6H_4OOH + HPO_4^{2-} \qquad (64)$$

(c) The radical cation $BrC_6H_4N(CH_3)^{+}{}_2^{\bullet}$ formed by oxidation of the parent compound by the peroxyl radical:

$$(CH_3)_2NC_6H_4OO^{\bullet} + (CH_3)_2NC_6H_4Br \longrightarrow (CH_3)_2NC_6H_4OO^- + (CH_3)_2N^{+\bullet}C_6H_4Br \qquad (65)$$

The decay of the peroxyl radical obeyed neither a pure first-order-rate law nor a second-order one. Individual traces fitted first-order better than second-order, but the rate constant was dependent on the initial radical concentration, indicating a considerable second-order component. A plot of $1/t_{1/2}$ vs. initial radical concentration (calculated from the absorbance by using

$\epsilon = 1200\,L\,mol^{-1}\,cm^{-1}$) [81] gave a straight line which yielded $k = 3.5 \times 10^4\,s^{-1}$ for the first-order contribution and $k = 3 \times 10^9\,L\,mol^{-1}\,s^{-1}$ for the second-order contribution [77]. The observed rate constant was independent of the concentration of the starting material, ruling out mechanism (c). Therefore, the first-order component of the decay indicated that the intramolecular process (b) has a pronounced contribution and is accompanied by second-order radical–radical reactions involving all of the radicals present in this system, i.e. mechanism (a) and mixed radical decay.

At pH 3, where the peroxyl radical is protonated, the first-order component of the decay kinetics became insignificant and the process fitted second-order kinetics quite well. The peroxyl radical absorption at 500 nm decayed over 50–100 μs to yield a species which also absorbed at 500 nm and which decaysed over several ms. The two decay processes were well resolved. The faster process was a second-order decay of the peroxyl radicals to form a product absorbing at 500 nm, which was ascribed to the phenoxyl radical produced by the reaction:

$$2(CH_3)_2HN^+C_6H_4OO^\bullet \longrightarrow 2(CH_3)_2HN^+C_6H_4O^\bullet + O_2 \tag{66}$$

From the yield of phenoxyl radicals, however, it was concluded that this reaction accounts for only 10–15% of the radical decay. The rest of the radicals decay either by reactions with other radicals present in solution (e.g. peroxyl radicals derived from the alcohol), or by producing directly stable products that are not likely to absorb at 500 nm (biphenyl derivatives such as those formed by the coupling of phenoxyl radicals).

$$2(CH_3)_2NC_6H_4OO^\bullet \longrightarrow [(CH_3)_2NC_6H_3OH]_2 + O_2 \tag{67}$$

12 PYRIDYLPEROXYL RADICALS

Pyridylperoxyl radicals have been produced by radiolytic reduction of halogenated pyridines in the presence of oxygen [82]. Solvated electrons reduce halogenated pyridines (e.g. BrPy) very rapidly to form the radical anions:

$$BrPy + e_{aq}^- \longrightarrow (BrPy)^{-\bullet} \tag{68}$$

which may undergo rapid dehalogenation to yield the pyridyl radical:

$$(BrPy)^{-\bullet} \longrightarrow Br^- + Py^\bullet \tag{69}$$

and this subsequently reacts with O_2:

$$Py^\bullet + O_2 \longrightarrow PyO_2^\bullet \tag{70}$$

Radical anions of pyridines, however, may undergo rapid protonation on the nitrogen to produce pyridinyl radicals [83], in competition with the dehalogenation process:

$$(BrPy)^{-\cdot} + H_2O \longrightarrow BrPyH^{\cdot} + OH^- \qquad (71)$$

Pulse radiolysis experiments with 2-chloro- and 2-bromopyridine in aqueous alcohol solutions under air led to the formation of the 2-pyridylperoxyl radical with a visible absorption band ($\lambda_{max} = 440\,nm$). Experiments with 3-chloro-, 3-bromo-, and 4-chloropyridine, however, did not show any significant absorptions in the 400–600 nm region, which may be ascribed to $PyO_2^{\cdot}$ radicals, owing to rapid protonation of their radical anions which competes with the dehalogenation reaction. Thus protonation of the radical anion is more rapid with the 3- and 4-halopyridines than with the 2-halopyridine. The pK_a values for 2-substituted pyridines are considerably lower than those of the 3- and 4-substituted analogues, due to a steric effect of the substituent at position 2 which restricts the rate of protonation on the nitrogen. As a result of this steric effect, the electron adduct of 2-halopyridine undergoes dehalogenation before protonation, whereas those of the 3- and 4-analogues protonate rather than dehalogenate. Since protonation of pyridine radical anions takes place even at pH 14, an attempt was made to obtain the 3- and 4-pyridyl radicals by irradiation in a less protic medium. In neat MeOH the peroxyl radicals were observed: 3-pyridylperoxyl exhibited λ_{max} at ca. 450 nm but 4-pyridylperoxyl exhibited only a broad absorption in the same range with no clear maximum.

Evidence for the dehalogenation vs. protonation processes was obtained from the yield of halide ions in the γ-radiolysis of halopyridine solutions. The yield of Cl^- from 2-chloropyridine and the yield of Br^- from 2-bromopyridine in deoxygenated aqueous methanol solutions were both higher than the yield of solvated electrons, indicating not only dehalogenation by e_{aq}^- (reactions (68) and (69)), but also by the alcohol radical:

$$\dot{C}H_2OH + BrPy \longrightarrow CH_2O + H^+ + (BrPy)^{-\cdot} \qquad (72)$$

The latter reaction is too slow to be important under the pulse radiolysis conditions, but may have a significant contribution under the low dose rates of the γ-radiolysis experiments. The radiolytic yield of Br^- was ca. 1.4 $\mu mol\,J^{-1}$, i.e. more than twice the total yield of radicals, indicating debromination by a chain process. The chain propagation step must involve reaction of the pyridyl radical with the alcohol:

$$Py^{\cdot} + CH_3OH \longrightarrow PyH + \dot{C}H_2OH \qquad (73)$$

Because of this chain process, it was not possible to determine the primary yield of pyridyl radicals from the reaction of e_{aq}^- with the various bromopyridines. With chloropyridines, however, the chain was relatively short and the yield of Cl^- was ca. 0.6 μmol J^{-1} with 2-chloropyridine but only ca. 0.1 μmol J^{-1} with 3-chloropyridine. This low yield is in accord with the abscnce of a peroxyl radical absorption in the pulse radiolysis experiments.

The rate constants for the oxidation of Trolox, chlorpromazine, and ABTS by $2\text{-}PyO_2^{\cdot}$ (Table 14) were found to be strongly dependent on the solvent composition, varying in each case by about two orders of magnitude. The rate constants for the reactions of $3\text{-}PyO_2^{\cdot}$ and $4\text{-}PyO_2^{\cdot}$ were measured only with ABTS in neat MeOH, giving values of 2.1×10^7 and 1.4×10^8 L mol^{-1} s^{-1}, respectively, indicating an order of reactivity of $3\text{-}PyO_2^{\cdot} < 2\text{-}PyO_2^{\cdot} < 4\text{-}PyO_2^{\cdot}$.

Table 14 Rate constants for reactions of 2-pyridylperoxyl radicals in aqueous/organic solvent mixtures

Organic cosolvent	Content (vol%)	k (L mol^{-1} s^{-1})		
		ABTS	Trolox	ClPz
t-BuOH	5	2.1×10^9	2.9×10^8	8.3×10^8
	30	3.8×10^8	1.1×10^8	4.7×10^7
	67	6.7×10^7	1.1×10^7	2.8×10^6
Dioxane	5		1.9×10^8	5.6×10^8
	50	8.6×10^7	4.8×10^6	1.2×10^7
Ethylene glycol	50	6.1×10^8	3.9×10^7	1.2×10^8
	100			5.2×10^6
DMSO	50	1.5×10^8	6.7×10^6	2.1×10^7
Acetonitrile	84	3.3×10^7		
	100			5.5×10^6
MeOH	2, 10	1.9×10^9		
	30	1.1×10^9		
	60	4.9×10^8		
	80	8.4×10^7		
	95	4.3×10^7		
	100	2.8×10^7		

The reactions of $2\text{-}PyO_2^{\cdot}$ take place with rate constants that are about an order of magnitude higher than those of the phenylperoxyl radical. This is due to the electron-withdrawing properties of the nitrogen in pyridine [84], making $PyO_2^{\cdot}$ a stronger oxidant than $PhO_2^{\cdot}$. The higher arylperoxyl radicals, however, are more reactive than 2-pyridylperoxyl.

The rate constants for the oxidation of ClPz by $CCl_3O_2^{\cdot}$ in different solvents have been found to give a good correlation between log k and the solvent cohesive pressure, δ^2. A similar correlation was found for Trolox when the results in highly basic solvents are omitted. These correlations have been rationalized on the basis of a relationship between the energy required to separate solvent molecules in a vaporization process (related to δ^2) [47], which is a measure of intermolecular forces, and the energy driving the reaction due to the relative solvation of the reactants and the transition state of the reaction. The results for 2-pyridylperoxyl can be correlated in the same fashion and indicate similar solvation properties as in the reaction of $CCl_3O_2^{\bullet}$. Since the rate constants for 2-pyridylperoxyl were measured in solvent mixtures rather than in neat solvents, the value of δ for the mixture was calculated from the values of the neat solvents [47] and their volume fraction in the mixture. Figure 8 shows the correlations for 2-pyridylperoxyl, and compares them with the results for $CCl_3O_2^{\bullet}$ obtained previously in almost neat solvents, as well as in various solvent mixtures. The previous correlations for $CCl_3O_2^{\bullet}$ were carried out with almost neat solvents only, but the correlations of Figure 8 include the results for mixed solvents as well, and good fits are found for the full set of results (correlation coefficients of $R^2 = 0.89$ for both ClPz and Trolox). The correlation for the pyridylperoxyl radical is of similar quality ($R^2 = 0.92$ for ClPz, 0.81 for Trolox, and 0.94 for ABTS). In all cases, $CCl_3O_2^{\bullet}$ is more reactive than 2-pyridylperoxyl and its correlation line is less steep. The correlations for 2-pyridylperoxyl with the three reductants give nearly parallel lines. The results indicate that the dependence of the rate constant on the solvent cohesive pressure probably holds for the oxidation reactions of many peroxyl radicals and for many solvent mixtures. Comparison of the experimental rate constants with those calculated from the correlation equations ($\log k = A + B\delta^2$) shows that one can predict the rate constant from δ^2 to better than a factor of three and this may have some utility in predicting the rates of reaction for peroxyl radicals in more complex media from measurements in several solvents. The lines in Figure 8 also indicate that comparison of the reactivities of different radicals with the same compound may be less illuminating in water, where the rate constants may approach the diffusion-controlled limit, than in aqueous/organic mixtures, where the rate constants are lower and differences between them are more pronounced.

13 VINYLPEROXYL RADICALS

The formation and reactions of vinylperoxyl radicals have been studied by two groups [71, 85]. Various vinylperoxyl radicals were produced from the corresponding chlorinated alkenes by the general method described above. A specific method, however, was also utilized to produce vinylperoxyl radicals, i.e. by producing the vinyl radical from an alkyne. This is achieved by the

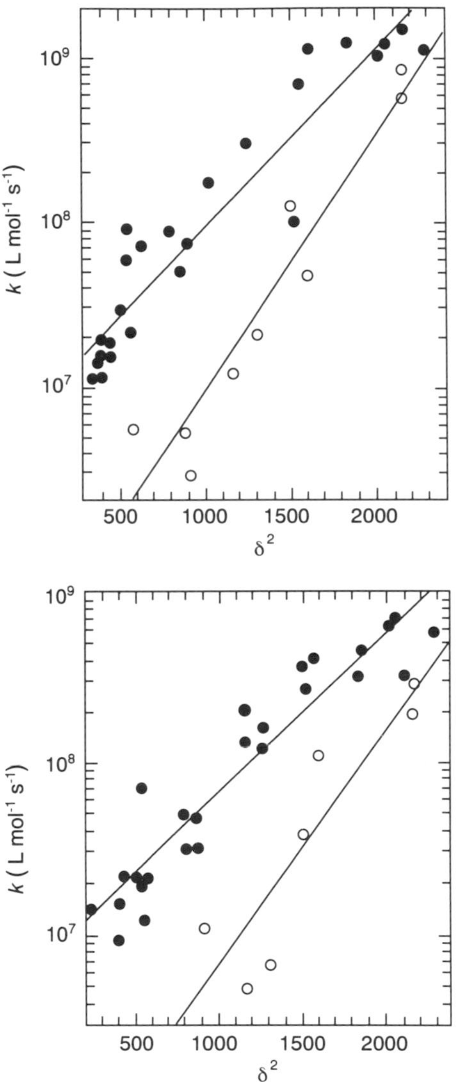

Figure 8 Correlation of the rate constants for oxidation of chlorpromazine (top) and Trolox (bottom) by the 2-pyridylperoxyl radical (○) and $CCl_3O_2^{\cdot}$ radical (●) with the cohesive pressure of the solvent mixture. (For the reaction of $CCl_3O_2^{\cdot}$ with Trolox, the results for triethylamine, pyridine, and CCl_4 were not included due to the effects discussed in the text.) The data are given in Tables 9 and 14 and in Refs 45 and 82.

addition of a hydrogen atom to the triple bond ($k = 2.2 \times 10^9\,\mathrm{L\,mol^{-1}\,s^{-1}}$ for acetylene) [86]:

$$CH{\equiv}CH + H^{\bullet} \longrightarrow CH_2{=}CH^{\bullet} \tag{74}$$

or by addition of e_{aq}^- to the triple bond, followed by protonation [86, 87]:

$$CH{\equiv}CH + e_{aq}^- \longrightarrow (CH{\equiv}CH)^{-\bullet} \tag{75}$$

$$(CH{\equiv}CH)^{-\bullet} + H_2O \longrightarrow CH_2{=}CH^{\bullet} + OH^- \tag{76}$$

Thus, vinyl, carboxyvinyl, and dicarboxyvinyl radicals were produced in irradiated aqueous solutions from acetylene, acetylenecarboxylic, and acetylenedicarboxylic acid, respectively. All of these vinyl radicals react rapidly with oxygen to produce the corresponding vinylperoxyl radicals. Vinyl radicals, however, can also add to the alkyne and this reaction competes with the formation of the peroxyl radical at high alkyne concentrations.

Vinylperoxyl radicals exhibit absorption spectra and reactivities which are similar to those of arylperoxyl rather than alkylperoxyl. This is due to the interaction between the double bond π-electrons and the oxyl unpaired electrons. The absorption peak of vinylperoxyl is at 440 nm and is shifted to higher wavelengths by various substituents (ca. 480 nm for monochloro and monocarboxy, ca. 520 nm for dicarboxy, 580 nm for trichloro, and 540 and 690 nm for phenyl).

Vinylperoxyl radicals oxidize organic reductants such as ABTS, chlorpromazine, ascorbate, and Trolox, with rate constants in the range 10^5–$10^9\,\mathrm{L\,mol^{-1}\,s^{-1}}$, depending on the radical and the reductant. The rate constants are higher than those found for oxidation of the same compounds by alkylperoxyl radicals. This increase results from an electron-withdrawing effect by the vinyl group, which lowers the electron density on the oxyl site and thus increases its tendency to oxidize the substrates. Similar effects are exerted by other electron-withdrawing substituents, such as halogens, carbonyl, cyano, or phenyl groups. A correlation has been demonstrated previously between $\log k$ for the oxidation of ascorbate by substituted alkylperoxyl radicals and the Taft σ^* substituent constant, which is a measure of the electron-withdrawing effect of the substituent. By comparison with this correlation, however, the rate constants for vinylperoxyl and phenylperoxyl are higher than expected on the basis of their σ^* values, suggesting that the effect of the vinyl and phenyl groups is to withdraw electrons from the oxyl site by an inductive effect, as well as through mesomeric structures with decreased electron densities on the oxygen. Such an interaction between the unpaired electron on the oxygen and the π-system of the phenyl or vinyl group is also the cause of the absorption in the visible range [74].

The rate constants for oxidation by the carboxyvinylperoxyl and phenylvinylperoxyl radicals are higher than those of the unsubstituted vinylperoxyl radicals. This increase in reactivity is clearly due to the electron-withdrawing properties of these substituents. Furthermore, there is an effect of pH on the rate constant. The rate of oxidation of the ascorbate anion by the vinylperoxyl radical in neutral solutions is much higher than that of ascorbic acid at low pH, as has been found for many other radical reactions. However, the finding that the rate constants for ClPz, ABTS, and Trlox are lower at pH 7 than at pH 1 is clearly due to changes in the ionic state of the radical and not of the reductants. This indicates that the rate-enhancing effect of a carboxylic acid group is stronger than that of the carboxylate anion, as expected from their inductive effects [88].

These trends of the substituent effect on the reactivity hold for all of the reductants studied, and the order of reactivity of the different reductants is the same (ClPz > ABTS > Trolox) for all of the radical studied. The finding that the rate constant for oxidation of ClPz is always higher than that of Trolox, despite the fact that the one-electron oxidation potential of Trolox (0.48 V vs. NHE at pH 7) is lower than that of ClPz (ca. 0.8 V) [89], stresses the difference between these reactions. The reaction with ClPz is a simple electron-transfer process, whereas that with Trolox requires deprotonation (before, or concerted with the electron transfer) and thus is relatively slower. The finding that the vinylperoxyl radical oxidizes ClPz with a measurable rate constant, but does not oxidize ascorbic acid on the pulse radiolysis time-scale suggests that the oxidation potential of this radical is close to that of ascorbic acid (ca. 1.0 V vs. NHE), i.e. clearly higher than that of the methylperoxyl radical, which was estimated to be ca. 0.6–0.7 V.

REFERENCES

1. (a) C. von Sonntag and H.-P. Schuchmann, *Angew. Chem. Int. Ed. Engl.* **30,** 1229 (1991); (b) C. von Sonntag, *The Chemical Basis of Radiation Biology*, Taylor and Francis, London (1987).
2. P. Neta, R.E. Huie and A.B. Ross, *J. Phys. Chem. Ref. Data* **19,** 413 (1990). Included and updated in the NDRL/NIST Solution Kinetics Database, *NIST Standard Reference Database 40*, National Institute of Standards and Technology, Gaithersburg (1994).
3. (a) K.U. Ingold, *Acc. Chem. Res.* **2,** 1 (1969); (b) K.U. Ingold, in *Free Radicals*, edited by J.K. Kochi, John Wiley & Sons, New York (1973), p. 37; (c) J.A. Howard, *Adv. Free Radical Chem.* **4,** 49 (1972).
4. J. A. Howard and J. C. Scaiano, in *Landolt-Börnstein Numerical Data and Functional Relationships in Science and Technology*, New Series, Group II, *Atomic and Molecular Physics*, edited by K.-H. Hellwege and O. Madelung, Springer-Verlag (Berlin) (1984), Vol. 13, edited by H. Fischer, Part D, p. 205.
5. (a) T.F. Slater, in *Biochemical Mechanisms of Liver Injury*, edited by T.F. Slater, Academic Press, London (1978) p. 1; (b) R.O. Recknagel and E.A. Glende, Jr, *CRC Crit. Rev. Toxicol.* **2,** 263 (1973); (c) D. Brault, *Environ. Health Perspect.* **64,** 53 (1985).

6. G.V. Buxton, C.L. Greenstock, W.P. Helman and A.B. Ross, *J. Phys. Chem. Ref. Data* **17,** 513 (1988). Included and updated in the NDRL/NIST Solution Kinetics Database, *NIST Standard Reference Database 40*, National Institute of Standards and Technology, Gaithersburg (1994).
7. D. Veltwisch, E. Janata and K.-D. Asmus, *J. Chem. Soc. Perkin Trans. 2* 146 (1980).
8. D. Brault and P. Neta, *J. Phys. Chem.* **88,** 2857 (1984).
9. Y. Illan, J. Rabani and A. Henglein, *J. Phys. Chem.* **80,** 1558 (1976).
10. E. Bothe, G. Behrens and D. Schulte-Frohlinde, *Z. Naturforsch.* **32B,** 886 (1977).
11. G.E. Adams and R.L. Willson, *Trans. Faraday Soc.* **65,** 2981 (1969).
12. D.E. Cabelli and B.H.J. Bielski, *J. Phys. Chem.* **87,** 1809 (1983).
13. (a) J.E. Packer, R.L. Willson, D. Bahnemann and K.-D. Asmus, *J. Chem. Soc. Perkin Trans. 2* 296 (1980); (b) J.E. Packer, R.L. Willson, D. Bahnemann and K.-D. Asmus, *J. Chem. Soc. Perkin Trans. 2* 1133 (1983).
14. R.E. Huie and P. Neta, *Int. J. Chem. Kinet.* **18,** 1185 (1986).
15. R.E. Huie, D. Brault and P. Neta, *Chem.-Biol. Interact.* **62,** 227 (1987).
16. L.J. Mittal and J.P. Mittal, *Radiat. Phys. Chem.* **28,** 363 (1986).
17. S. Padmaja and R.E. Huie, *Biochem. Biophys. Res. Commun.* **195,** 539 (1993).
18. R.E. Huie and S. Padmaja, *Free Radical Res. Commun.* **18,** 195 (1993).
19. T.J. Wallington, P. Dagaut and M.J. Kurylo, *Chem. Rev.* **92,** 667 (1992).
20. (a) D.A. Wink, J.A. Cook, M.C. Krishna, I. Hanbauer, W. DeGraff, J. Gamson and J.B. Mitchell, *Arch. Biochem. Biophys.* **319,** 402 (1995); (b) S.P.A. Goss, N. Hogg and B. Kalyanaraman, *Chem. Res. Toxicol.* **8,** 800 (1995).
21. (a) N.A. Porter, *Acc. Chem. Res.* **19,** 262 (1986); (b) N.A. Porter, K.A. Mills and R.L. Carter, *J. Am. Chem. Soc.* **116,** 6690 (1994).
22. B.A. Wagner, G.R. Buettner and C.P. Burns, *Biochemistry* **33,** 4449 (1994).
23. W. Tsang, in *Energetics of Free Radicals*, edited by A. Greenberg and J. Liebman, Chapman and Hall, New York (1996), p. 22.
24. E.T. Denisov and T.G. Denisova, *Kinet. Catal.* (*Engl. Transl.*) **34,** 173 (1993).
25. S. Mosseri, Z.B. Alfassi and P. Neta, *Int. J. Chem. Kinet.* **19,** 309 (1987).
26. L.G. Forni, J.E. Packer, T.F. Slater and R.L. Willson, *Chem.-Biol. Interact.* **45,** 171 (1983).
27. D. Brault, P. Neta and L.K. Patterson, *Chem.-Biol. Interact.* **54,** 289 (1985).
28. P. Koelewign, *Recl. Trav. Chim. Pays-Bas* **91,** 1275 (1972).
29. G.S. Nahor and P. Neta, *Int. J. Chem. Kinet.* **23,** 941 (1991).
30. Z.B. Alfassi, R.E. Huie and P. Neta, *J. Phys. Chem.* **97,** 6835 (1993).
31. L.C.T. Shoute, Z.B. Alfassi, P. Neta and R.E. Huie, *J. Phys. Chem.* **98,** 5701 (1994).
32. J. March, *Advanced Organic Chemistry*, 3rd Ed., John Wiley & Sons, New York (1985), Ch. 14, p. 608.
33. (a) R.W. Taft, Jr, in *Steric Effects in Organic Chemistry*, edited by M.S. Newman, John Wiley & Sons, New York (1956), p. 619; (b) K.B. Weiberg, *Physical Organic Chemistry*, John Wiley & Sons, New York (1964), p. 415.
34. E.S. Huyser, *J. Org. Chem.* **26,** 3261 (1961).
35. S.K. Kapoor and C. Gopinathan, *Int. J. Chem. Kinet.* **24,** 1035 (1992).
36. Z.B. Alfassi, S. Mosseri and P. Neta, *J. Phys. Chem.* **93,** 1380 (1989).
37. Z.B. Alfassi, S. Mosseri and P. Neta, *J. Phys. Chem.* **91,** 3383 (1987).
38. P. Neta, R.E. Huie, S. Mosseri, L.V. Shastri, J.P. Mittal, P. Maruthamuthu and S. Steenken, *J. Phys. Chem.* **93,** 4099 (1989).
39. D. Brault and P. Neta, *J. Phys. Chem.* **91,** 4156 (1987).
40. (a) R.A. Marcus, *J. Chem. Phys.* **24,** 966 (1956); (b) R.A. Marcus, *J. Chem. Phys.* **26,** 872 (1957); (c) R.A. Marcus, *Annu. Rev. Phys. Chem.* **15,** 155 (1964); (d) R.A.

Marcus and N. Sutin, *Biochim. Biophys. Acta* **811,** 265 (1985).
41. (a) G. Merenyi, J. Lind and L. Engman, *J. Chem. Soc. Perkin Trans. 2* 2551 (1994); (b) M. Jonsson, *J. Phys. Chem.* **100,** 6814 (1996).
42. J. Mönig, K.-D. Asmus, M. Schaeffer, T. Slater and R.L. Willson, *J. Chem. Soc.* Perkin Trans. 2 1133 (1983).
43. P. Neta, R.E. Huie, P. Maruthamuthu and S. Steenken, *J. Phys. Chem.* **93,** 7654 (1989).
44. See, e.g. (a) F. Franks and D.J.G. Ives, *Q. Rev. Chem. Soc.* 20 (1996); (b) E.M. Arnett, in *Physico-Chemical Processes in Mixed Aqueous Solvents*, edited by F. Franks, Elsevier, New York (1967), p. 105; (c) Y. Maham and G.R. Freeman, *Can. J. Chem.* **66,** 1706 (1988).
45. Z.B. Alfassi, R.E. Huie and P. Neta, *J. Phys. Chem.* **97,** 7253 (1993).
46. (a) K. Dimroth, C. Reichardt, T. Siepmann and F. Bohlmann, *Ann. Chem.* **661,** 1 (1963); (b) C. Reichardt, *Solvents and Solvent Effects in Organic Chemistry*, VCH, Weinheim (1988), p. 365.
47. (a) J.H. Hildebrand, J.M. Prausnitz and R.L. Scott, *Regular and Related Solutions*, Van Nostrand Reinhold, New York (1970); (b) A.F.M. Barton, *CRC Handbook of Solubility Parameters and Other Cohesion Parameters*, CRC Press, Boca Raton (1983).
48. M.H. Abraham, R.M. Doherty, M.J. Kamlet, J.M. Harris and R.W. Taft, *J. Chem. Soc. Perkin Trans. 2* 913 (1987).
49. K. Izutsu, *Acid–Base Dissociation Constants in Dipolar Aprotic Solvents*, Blackwell Scientific, Oxford (1990).
50. (a) D.D. Perrin, *Dissociation Constants of Organic Bases in Aqueous Solution*, Butterworths, London (1965) and Supplement (1972); (b) E.M. Arnett and G. Scorrano, *Adv. Phys. Org. Chem.* **13,** 83 (1976); (c) G. Perdoncin and G. Scorrano, *J. Am. Chem. Soc.* **99,** 6983 (1977); (d) R.A. Cox, L.M. Druet, A.E. Klausner, T.A. Modro, P. Wan and K. Yates, *Can. J. Chem.* **59,** 1568 (1981); (e) A. Bagno, V. Lucchini and G. Scorrano, *Bull. Soc. Chim. France* 563 (1987); (f) A. Bagno, G. Scorrano and R.A. More O'Ferrall, *Rev. Chem. Intermed.* **7,** 313 (1987); (g) A. Bagno and G. Scorrano, *J. Am. Chem. Soc.* **110,** 4577 (1988).
51. (a) M.J. Kamlet, J.L.M. Abboud, M.H. Abraham and R.W. Taft, *J. Org. Chem.* **48,** 2877 (1983); (b) M. Chastrette, M. Rajzmann, M. Chanon and K.F. Purcell, *J. Am. Chem. Soc.* **107,** 1 (1985).
52. L. Valgimigli, K.U. Ingold and J. Lusztyk, *J. Am. Chem. Soc.* **118,** 3545 (1996).
53. S. Steenken and P. Neta, *J. Phys. Chem.* **86,** 3661 (1982).
54. S. Marguet, P. Hapiot and P. Neta, *J. Phys. Chem.* **98,** 7136 (1994).
55. Z.B. Alfassi, A. Harriman, S. Mosseri and P. Neta, *Int. J. Chem. Kinet.* **18,** 1315 (1986).
56. D. Brault and P. Neta, *Chem. Phys. Lett.* **121,** 28 (1985).
57. (a) G.G. Jayson, J.P. Keene, D.A. Stirling and A.J. Swallow, *Trans. Faraday Soc.* **65,** 2453 (1969); (b) G.G. Jayson, B.J. Parsons and A.J. Swallow, *J. Chem. Soc. Faraday Trans. 1* **69,** 236 (1973); (c) J. Butler, G.G. Jayson and A.J. Swallow, *J. Chem. Soc. Faraday Trans. 1* **70,** 1394 (1974).
58. G.I. Khaikin, Z.B. Alfassi, R.E. Huie and P. Neta, *J. Phys. Chem.* **100,** 7072 (1996).
59. S.V. Jovanovic, I. Jankovic and L. Josimovic, *J. Am. Chem. Soc.* **114,** 9018 (1992).
60. Z.B. Alfassi, R.E. Huie, M. Kumar and P. Neta, *J. Phys. Chem.* **96,** 767 (1992).
61. (a) E. Exner, *Progr. Phys. Org. Chem.* **10,** 411 (1973); (b) W. Linert and R.F. Jameson, *Chem. Soc. Rev.* **18,** 477 (1989).
62. In a recent study of the temperature and solvent dependence of intramolecular electron transfer, the activation energies were found to be small and to fall in a

narrow range, while the pre-exponential factors increased with the dielectric constant fo the medium: H. Heitele, F. Pöllinger, S. Weeren and M.E. Michel-Beyerle, *Chem. Phys. Lett.* **168,** 598 (1990).
63. C. Schöneich, A. Aced and K.-D. Asmus, *J. Am. Chem. Soc.* **113,** 375 (1991).
64. (a) M. Bonifacic, H. Möckel, D. Bahnemann and K.-D. Asmus, *J. Chem. Soc. Perkin Trans. 2* 675 (1975); (b) K.-D. Asmus, D. Bahnemann, M. Bonifacic and H.A. Gillis, *Disc. Faraday Soc.* **63,** 1748 (1977); (c) M. Göbl, M. Bonifacic and K.-D. Asmus, *J. Am. Chem. Soc.* **106,** 5984 (1984).
65. C. Schöneich, V. Narayanaswami, K.-D. Asmus and H. Sies, *Arch. Biochem. Biophys.* **282,** 18 (1990).
66. L. Engman, J. Persson, G. Merenyi and J. Lind, *Organometallics* **14,** 3641 (1995).
67. E. Furimski and J.A. Howard, *J. Am. Chem. Soc.* **95,** 369 (1973).
68. X. Shen, J. Lind, T.E. Eriksen and G. Merenyi, *J. Chem. Soc. Perkin Trans. 2* 555 (1989).
69. M. Bonifacic, C. Schöneich and K.-D. Asmus, *J. Chem. Soc. Chem. Commun.* 1117 (1991).
70. P.M. Sommeling, P. Mulder, R. Louw, D.V. Avila, J. Lusztyk and K.U. Ingold, *J. Phys. Chem.* **97,** 8361 (1993).
71. R. Mertens and C. von Sonntag, *Angew. Chem. Int. Ed. Engl.* **33,** 1262 (1994).
72. Z.B. Alfsasi, S. Marguet and P. Neta, *J. Phys. Chem.* **98,** 8019 (1994).
73. Z.B. Alfassi, G.I. Khaikin and P. Neta, *J. Phys. Chem.* **99,** 265 (1995).
74. M. Krauss and R. Osman, *J. Phys. Chem.* **99,** 11387 (1995).
75. G.I. Khaikin, Z.B. Alfassi and P. Neta, *J. Phys. Chem.* **99,** 11447 (1995).
76. J. Lind, X. Shen, T.E. Eriksen and G. Merenyi, *J. Am. Chem. Soc.* **112,** 479 (1990).
77. G.I. Khaikin, Z.B. Alfassi and P. Neta, *J. Phys. Chem.* **99,** 16722 (1995).
78. M. Ye and R.H. Schuler, *J. Phys. Chem.* **93,** 1898 (1989).
79. J.A. Dean (Ed.), *Lange's Handbook of Chemistry*, 13th Ed., McGraw-Hill, New York (1985), Table **5**–8, p. 5.
80. W.P. Jencks and J. Regenstein, in *Handbook of Biochemistry and Molecular Biology, Physical and Chemical Data*, edited by G.D. Fasman, Vol. 1, 3rd Ed., CRC Press, Cleveland (1976), Section D, pp. 314 and 316.
81. X. Fang, R. Mertens and C. von Sonntag, *J. Chem. Soc. Perkin Trans. 2* 1033 (1995).
82. Z.B. Alfassi, G.I. Khaikin and P. Neta, *J. Phys. Chem.* **99,** 4544 (1995).
83. (a) P. Neta, *Radiat. Res.* **52,** 471 (1972); (b) R.W. Fessenden and P. Neta, *Chem. Phys. Lett.* **18,** 14 (1973); (c) P. Neta and L.K. Patterson, *J. Phys. Chem.* **78,** 2211 (1974).
84. C. Hansch, A. Leo and R.W. Taft, *Chem. Rev.* **91,** 165 (1991).
85. G.I. Khaikin and P. Neta, *J. Phys. Chem.* **99,** 4549 (1995).
86. R.F. Anderson and D. Schulte-Frohlinde, *J. Phys. Chem.* **82,** 22 (1978).
87. P. Neta and R.W. Fessenden, *J. Phys. Chem.* **76,** 1957 (1972).
88. O. Exner, in *Correlation Analysis in Chemistry: Recent Advances*, edited by N.B. Chapman and J. Shorter, Plenum, New York (1978), p. 439.
89. P. Wardman, *J. Phys. Chem. Ref. Data* **18,** 1637 (1989).

10 Reactions of Organic Peroxyl Radicals in Organic Solvents

J.A. HOWARD
National Research Council, Ottawa, Canada

1 INTRODUCTION

Peroxyls, $RO_2^{\cdot}$, are the transient species involved in the rate-controlling propagation and termination reactions for the free-radical chain autoxidation of most organic materials. They are generated in organic solvents by the reaction of alkyls, $R^{\cdot}$, with dissolved oxygen, by the abstraction of the hydroperoxidic hydrogen atom from an alkyl hydroperoxide ROOH, by a free radical $X^{\cdot}$, and by transfer of an electron from an alkyl hydroperoxide to a metal ion or metal complex. They take part in a wide variety of reactions in solution, including hydrogen-atom abstraction, addition to a double bond, intramolecular abstraction and addition, homolytic displacement, oxygen-atom transfer, electron transfer, unimolecular decomposition, self-reaction and reactions with other free radicals. These reactions in organic solvents are the subject of this present chapter. Several books and review articles have appeared over the last 25 years that deal with various aspects of the chemistry of peroxyls [1–21], and original work that is not cited here is referenced in these articles.

2 METHODS OF STUDYING THE KINETICS OF REACTIONS OF PEROXYLS IN ORGANIC SOLVENTS

2.1 OXYGEN ABSORPTION STUDIES

Autoxidation of many organic materials at oxygen pressures above ca. 100 torr is a chain reaction that is described by the following elementary reactions:

Initiation

$$\text{Production of free radicals at a rate } R_i \tag{1}$$

Propagation

$$R^{\cdot} + O_2 \longrightarrow RO_2^{\cdot} \tag{2}$$

$$RO_2^{\cdot} + RH \longrightarrow ROOH + R^{\cdot} \tag{3}$$

Peroxyl Radicals. Edited by Z.B. Alfassi

Termination

$$2RO_2^{\bullet} \longrightarrow \text{non-radical products} \quad (4)$$

The overall rate of autoxidation at chain lengths above ca. 10 is given by the kinetic expression:

$$\frac{-d[O_2]}{dt} = \frac{k_p[RH]R_i^{1/2}}{(2k_t)^{1/2}} \quad (5)$$

where $k_p/(2k_t)^{1/2}$ is the oxidizability of the substrate, RH, k_p is the *overall* homo-propagation rate constant, i.e. the rate constant for reaction (3), and $2k_t$ is the *overall* termination rate constant, i.e. the rate constant for reaction (4). In this scheme the rate-controlling propagation reaction is shown as a hydrogen-atom-transfer reaction but it can be one of the other propagation reactions of alkylperoxyls.

Inhibition of autoxidation by addition of an antioxidant, AH, is an important method of controlling autoxidation. In the simplest case, reaction (4) is replaced by the termination reactions:

$$RO_2^{\bullet} + AH \longrightarrow ROOH + A^{\bullet} \quad (6)$$

$$RO_2^{\bullet} + A^{\bullet} \longrightarrow ROOA \quad (7)$$

with each molecule of AH scavenging two peroxyls. The rate of inhibited autoxidation is given by the rate expression:

$$\left(\frac{-d[O_2]}{dt}\right)_{inh} = \frac{k_p[RH]R_i}{nk_{inh}[AH]_0} \quad (8)$$

where k_{inh} is the rate constant for reaction (6) and n is the number of peroxyls scavenged by each molecule of inhibitor.

Values of $k_p/(2k_t)^{1/2}$ and k_p/nk_{inh} are obtained by measuring the rates of oxygen absorption at a known R_i and substrate concentration. Estimates of these ratios gives qualitative information about the reactivity of peroxyls in solution.

The values of n can be determined from equation (9) by measuring the time (τ) taken to reach the uninhibited rate in the presence of a known concentration of AH:

$$n = \frac{R_i\tau}{[AH]} \quad (9)$$

Isopropylbenzene is an ideal substrate for such studies because it has low values of k_p and $2k_t$ and oxidizes reasonably quickly. There is, therefore, a sudden increase in $d[O_2]/dt$ when [AH] falls to zero. This technique can be used to measure R_i by using an inhibitor with a known value of n [22]. It has also been used by Mahoney *et al.* [23] to determine the concentration of inhibitors remaining in used automobile engine lubricating oils and by Ingold and co-workers [24] to estimate the concentration of natural inhibitors in human blood plasma.

2.2 THE ROTATING SECTOR METHOD

The rotating sector method [25] has been used to evaluate the absolute propagation and termination rate constants for autoxidation of many reactive organic substrates. In this method the average lifetime, τ', of the reaction chains for a photochemically initiated autoxidation is measured. This is achieved by rotating a sectored disc in the light beam and measuring $d[O_2]dt$ at different disc speeds. $2k_t$ is related to τ' and R_i by equation (10):

$$\tau' = 1/(2k_tR_i)^{1/2} \tag{10}$$

The rate constant k_p is then calculated from $k_p/(2k_t)^{1/2}$ and the known value of $2k_t$.

This method was first used successfully to study the autoxidation of tetralin [26] and has subsequently been used to study autoxidation of a wide variety of organic compounds in both the absence and presence of added alkyl hydroperoxide [6, 14]. Extensive compilations of the absolute values of k_p and $2k_t$ determined by this method have been made [6, 10, 15, 19].

The most important conclusion that has been drawn from rotating sector studies of autoxidation is that the oxidizability of an organic material depends *mainly* on three factors, (i) the magnitude of $2k_t$, (ii) the structure of the peroxyl, and (iii) the stabilization energy of the radical formed in the rate-controlling propagation reaction.

2.3 KINETIC ELECTRON SPIN RESONANCE (KESR) SPECTROSCOPY

Peroxyls have been detected by ESR spectroscopy in many systems in which liquid organic materials are being oxidized by O_2 and hydroperoxides are being decomposed homolytically. Tertiary alkylperoxyls are easily recognized by this technique because they have a broad singlet spectrum centered at $g \approx 2.0015$ [27]. They are persistant at ambient temperatures and below and are excellent candidates for studying peroxyl-molecule reactions. Primary and secondary alkylperoxyls are less persistant and are not as useful for this kind of study. They do, however, have small α-hydrogen hyperfine interactions and have the characteristic g-factor which makes them easy to recognize in autoxidizing systems [27].

Kinetic ESR has been used to *directly* determine termination rate constants for hydrocarbon autoxidation. Thus Lebedev, Tsepalov and Shlyapintokh [28] measured the steady-state concentrations of 1,1-dimethylbenzylperoxyl, $[RO_2^{\cdot}]_{ss}$, during autoxidation of isopropyl benzene, at known rates of chain initiation, and found that they are in good agreement with values calculated from:

$$[RO_2^{\cdot}]_{ss} = R_i/(2k_t)^{1/2} \tag{11}$$

utilizing the value of $2k_t$ determined by Melville and Richards using the rotating sector method [29], and a corrected value of R_i.

Termination rate constants and rate constants for self-reaction have been determined for photochemically generated alkylperoxyls by following the second-order radical decay after the initiating light was cut off, using the kinetic expression:

$$-\frac{d[RO_2^{\cdot}]}{dt} = 2k_t\,[RO_2^{\cdot}]^2 \tag{12}$$

KESR has also been used to measure rate constants for tert-peroxyl-molecule reactions with reactants such as aralkanes, hindered and non-hindered phenols, aromatic amines, trivalent phosphorus compounds, alkyl hydroperoxides, organotin and lead compounds, and metal complexes [15]. In this method, the radical decay is monitored in the presence of a large excess of the reactant:

$$-\frac{d[RO_2^{\cdot}]}{dt} = k_{\psi}[RO_2^{\cdot}] \tag{13}$$

where k_{ψ} is the pseudo-first-order rate constant k_p' [RH].

The sensitivity of the KESR method for studying peroxyls falls off rapidly with increasing temperature because of severe line-broadening. Among other factors, line-broadening is inversely proportional to the effective radius of the radical and is thus greater for the smaller peroxyls [30]. For most peroxyls the upper temperature limit at which valid kinetic measurements can be made is determined by the line-broadening rather than by the rate of reaction. For example, the upper limits for $HOCH_2O_2^{\cdot}$ and $(CH_3)_2CHO_2^{\cdot}$ are ca. 235 and 293 K, respectively, whereas that for 2-methylhexa-2-ylperoxyl is ca. 373 K.

The stoichiometry of the reaction between tert-peroxyls and 'reactive' reactants, e.g. phenols and aromatic amines, can be determined by using KESR. In this method, a known steady-state concentration of tert-peroxyls is generated by photolysis of a photochemical source of $R^{\cdot}$ in an O_2-saturated solvent at 178 K or below in the cavity of the spectrometer. A known quantity of substrate, which is less than the anticipated stoichiometeric amount, is then added slowly to the solution. The decrease in intensity of the tert-peroxyl species gives the stoichiometric factor n [31].

2.4 KINETIC ULTRAVIOLET ABSORPTION SPECTROSCOPY

Overall rate constants for self-reaction of $(CH_3)_2CHO_2^{\cdot}$ in cyclohexane, decane and dodecane have been measured over the temperature range 293–373 K by using ultraviolet absorption spectroscopy to monitor the concentration of $(CH_3)_2CHO_2^{\cdot}$ [32].

2.5 OTHER NON-STATIONARY-STATE METHODS

Several other non-stationary-state methods have been developed to estimate $2k_t$ for autoxidation [8]. These include the pre- and after-effect method in which the time taken to reach $(d[O_2]/dt)_{ss}$, after the light for a photochemically initiated autoxidation is switched on, is determined. Alternatively, the time taken for the steady-state rate to fall to zero, after the light is switched off, is estimated. The thermocouple method is a variation of this method where the overall rate is followed during the non-stationary-state regime by measuring the adiabatic evolution of heat. If autoxidation is initiated thermally, an inhibitor can be added to disturb the steady state, or the intensity of the chemiluminescence generated in termination can be monitored upon the sudden addition of oxygen..

2.6 THE HYDROPEROXIDE METHOD

The hydroperoxide (or hydroperoxide-loaded method) has proved to be an extremely useful method for studying reactions of alkylperoxyls in solution [14]. It is based on the discovery by Thomas and Tolman [33] that the rate of autoxidation of isopropylbenzene is increased by ca. 60% by the addition of isopropylbenzene hydroperoxide. This effect is not due to an increase in R_i, as one might expect, but to an increase in $[RO_2^{\cdot}]_{ss}$ caused by a decrease in $2k_t$.

Autoxidation of a substance, in the presences of enough added alkyl hydroperoxide, R′OOH, to ensure that only $R'O_2^{\cdot}$ radicals derived from R′OOH are involved in rate-controlling propagation and termination reactions, is described by reactions (1) and (2), followed by the chain-transfer reaction (14), and the cross-propagation reaction (15):

Chain transfer

$$RO_2^{\cdot} + R'OOH \longrightarrow ROOH + R'O_2^{\cdot} \qquad (14)$$

Cross-propagation

$$R'O_2^{\cdot} + RH \longrightarrow R'OOH + R^{\cdot} \qquad (15)$$

The chain is terminated by the self-reaction of $R'O_2^{\cdot}$:

$$2R'O_2^{\cdot} \longrightarrow \text{non-radical products} \qquad (16)$$

Under these conditions the rate of autoxidation is given by:

$$\frac{-\mathrm{d}[\mathrm{O_2}]}{\mathrm{d}t} = \frac{k_p'[\mathrm{RH}]R_i^{1/2}}{(2k_t')^{1/2}} \tag{17}$$

where k_p' is the rate constant for the reaction of the peroxyl derived from R′OOH, i.e. $R'O_2^{\cdot}$, with RH (reaction 15), and $2k'_t$ is the rate constant for self-reaction of $R'O_2^{\cdot}$.

This method is particularly useful if the hydroperoxide is tert-butyl hydroperoxide, because $(CH_3)_3CO_2^{\cdot}$ has the smallest rate constant for self-reaction. Addition of tert-butyl hydroperoxide to an autoxidation, therefore, increases the chain length and enables unreactive substrates to be studied at long chain lengths. Measurement of the termination rate constant by the rotating sector method gives $2k_t'$ for $(CH_3)_3CO_2^{\cdot}$, a rate constant that cannot be obtained from the autoxidation of isobutane. Similarly $2k_t'$ for other alkylperoxyls, e.g. n-butyl and cyclohexyl, can be determined. Measurement of $k_p'/(2k_t')^{1/2}$ in conjunction with $2k_t'$ enables the relative reactivities of a series of substrates to the same peroxyl to be determined.

Competitive co-oxidations of one or more substrates in the presence of isopropylbenzene and tert-butyl hydroperoxide have been carried out over a temperature range [14]. The yields of hydroperoxides from the subtrate, relative to isopropylbenzene hydroperoxide, give k_p', A_p' and E_p' for the substrates relative to the values of these parameters for isopropylbenzene. This adaptation of the hydroperoxide method has given valuable information about the regioselectivity of $(CH_3)_3CO_2^{\cdot}$.

2.7 CO-OXIDATION

A large amount of information on the reactivity of peroxyls to organic substrates has been obtained from competitive studies of binary mixtures of substrates [20]. There are four propagation reactions for autoxidation of a mixture of two substrates AH and BH, namely the reactions of $AO_2^{\cdot}$ with AH and BH and the reactions of $BO_2^{\cdot}$ with AH and BH. The relative reactivity of AH and BH to a peroxyl is readily determined from the relative decrease in the concentrations of AH and BH, or, at long chain lengths, to the relative yields of the hydroperoxides, AOOH and BOOH. This analysis can be carried out because the selectivity of a peroxyl is independent of its structure [34].

Russell, in 1955 [35], reported that autoxidation of isopropylbenzene is inhibited by low concentrations of the more readily autoxidizable tetralin. He proposed that relatively high concentrations of tetralylperoxyls are formed in a solution containing mostly isopropylbenzene. These radicals terminate much more rapidly than 1,1-dimethylbenzylperoxyls and the $\mathrm{d}[O_2]/\mathrm{d}t$ of a mixture is slower than the $\mathrm{d}[O_2]/\mathrm{d}t$ of pure cumene because of a lower overall

$[RO_2^{\bullet}]_{ss}$. This is a general phenomenon for autoxidation of a substrate with relatively small values of k_p and $2k_t$ in the presence of low concentrations of a compound with relatively large values of k_p and $2k_t$.

The various reactions of alkylperoxyls in organic solvents are now discussed below in more detail.

3 PEROXYL-MOLECULE REACTIONS

3.1 HYDROGEN-ATOM ABSTRACTION

3.1.1 Alkanes

The rate constants for the reaction of $(CH_3)_3CO_2^{\bullet}$ with a number of acyclic and cyclic alkanes have been determined by the hydroperoxide method [14]. From the data presented in Tables 1 and 2 we see that the rate constants for abstraction of primary aliphatic hydrogens are ca. 10^{-5} $M^{-1}s^{-1}$ per active hydrogen and are among the smallest that have been measured for free-radical reactions in solution. Secondary aliphatic hydrogens are one to two orders of magnitude more reactive and tertiary aliphatic hydrogens are two

Table 1 Rate constants for abstraction of a primary and a secondary hydrogen atom from alkanes by $(CH_3)_3CO_2^{\bullet}$ at 303 K[a]

Alkane[b]	$10^5 k_p'$ ($M^{-1}s^{-1}$)
$(CH_3)_3CCH_2C\mathbf{H}_3$	2.0[c]
$(C\mathbf{H}_3)_3CCH_2CH_3$	1.0[c]
$(CH_3)_3CC\mathbf{H}_2CH_3$	70.0[c]
$CH_3CH_2CH(CH_3)C\mathbf{H}_2CH_3$	20
$CH_3CH_2C\mathbf{H}_2CH(CH_3)_2$	10
$CH_3C\mathbf{H}_2CH_2CH(CH_3)_2$	33
$(CH_3)_2CH_2C\mathbf{H}_2CH_3$	15
$\overline{CH_2(CH_2)_4CH_2}$ (cyclic)	26
$CH_3CH_2CH_2CH_2C\mathbf{H}_2CH_3$	55
$CH_3CH_2CH_2C\mathbf{H}_2CH_2CH_3$	35
$\overline{CH_2(CH_2)_3CH_2}$ (cyclic)	87
$CD_3CD_2CD_2(CD_3)C\mathbf{D}_2CD_3$	0.8

[a] Note that these rate constants and all subsequent values of k_p and k_p' (unless stated otherwise) are per active hydrogen atom.
[b] Reactive hydrogen and deuterium atoms are shown in bold type.
[c] Measured at 323 K.

Table 2 Rate constants for abstraction of a tertiary hydrogen atom from alkanes by $(CH_3)_3CO_2^{\cdot}$ at 303 K

Alkane[a]	$10^5k_p'$ ($M^{-1}s^{-1}$)
$(CH_3)_3CCH_2\mathbf{CH}(CH_3)_2$	150
$(CH_3)_2CHCH_2\mathbf{CH}(CH_3)_2$	250
$CH_3CH_2\mathbf{CH}(CH_3)C(CH_3)_3$	320
$CH_3CH_2\mathbf{CH}(CH_3)CH(CH_3)_2$	900
$CH_3CH_2CH_2\mathbf{CH}(CH_3)_2$	700
$CH_3CH_2\mathbf{CH}(CH_3)_2$	700
$CH_3CH_2\mathbf{CH}(CH_3)CH_2CH_3$	700
$CH_3\mathbf{CH}(CH_3)C(CH_3)_3$	130
$CH_3CH_2CH(CH_3)\mathbf{CH}(CH_3)_2$	550
$CD_3CD_2\mathbf{CD}(CD_3)CD_2CD_3$	25

[a]Reactive hydrogen and deuterium atoms are shown in bold type.

to three orders of magnitude more reactive than 1° aliphatic hydrogens. This means that $(CH_3)_3CO_2^{\cdot}$ is more selective than $(CH_3)_3CO^{\cdot}$, but less selective than $Br^{\cdot}$, towards aliphatic hydrogens [36].

Within each type of hydrogen, the 1° hydrogens of the methyl group of 2,2-dimethylbutane are almost twice as reactive as the 1° hydrogens of the tert-butyl group of this alkane. The reactivity of 2° aliphatic hydrogens differ by almost an order of magnitude, with the C–3 hydrogens of 2-methylpentane being the least reactive and the cyclopentane hydrogens the most reactive. For n-hexane the hydrogens attached to the penultimate carbon are slightly more reactive than those attached to C–3. The cyclohexane hydrogens are less reactive than the two types of 2° hydrogen of n-hexane.

Tertiary-hydrogen-atom reactivity depends on the size of the alkyl group on the adjacent carbon and the reactivity decreases with an increase in size. The reactivity of this type of hydrogen is less sensitive to the bulk of an alkyl group attached to the carbon bearing the reactive hydrogen, partly because the relief of steric strain counterbalances the effect of the size of the substituent.

There is a primary kinetic isotope effect of 25–28 when the 2° and 3° hydrogens of 3-methylpentane are replaced by deuterium. These isotope effects are well above the theoretical limit of 9.5 for C—H bond cleavage, suggesting that factors other than the complete loss of zero-point energies of stretching and bending vibrations in the transition state influence the magnitude of $k_p'(H)/k_p'(D)$. The rate constants for methylcyclohexane and methylcyclohexane-d_{14} measured over a temperature range give $A_p'(D) = (3–5)A_p'(H)$ and $E_p'(D) = E_p'(H) + (40.5–55)$ kJ mol^{-1}, which suggests an entropy of activation difference of ca. 12.5 J $deg^{-1}mol^{-1}$ in favour of the deuterated compound and an enthalpy of activation energy difference of ca. 6.3 kJ mol^{-1} above the maximum. An interesting example of regioselectivity occurs in the reaction of peroxyls with dimethylcyclohexanes, where it has been shown that the equatorial 3° hydrogens are approximately four times more reactive than the axial 3° hydrogens [37].

The Arrhenius parameters for the reactions of alkanes with $(CH_3)_3CO_2^{\bullet}$ are given in Table 3. The differences in reactivity are the result of differences in both the entropy and enthalpy of activation. More specifically, the high *A*-factor for cyclohexane is consistent with an increase in entropy upon conversion of a tight cyclohexane ring to a looser cyclohexyl radical, while the high activation energy for this compound arises because an unstrained sp^3 hybridized carbon is converted to a strained sp^2 carbon. The *A*-factor for cyclopentane is similar to the *A*-factor for an acyclic aliphatic 2° hydrogen, which means that the enhanced reactivity of this substrate is entirely due to a low activation energy. The latter reflects the relief of strain upon removal of a hydrogen atom from cyclopentane.

Table 3 Arrhenius parameters for abstraction of a hydrogen atom from alkanes by $(CH_3)_3CO_2^{\bullet}$

Alkane	A[a]	$\log A_p'$[b]	E_p' (kJ mol^{-1})
3-Methylpentane	3°	9.5	68.2
	2°	8.9	73.3
Cyclohexane	2°	10.4	81.6
Cyclopentane	2°	8.8	69.9
2,2,4-Trimethylpentane	3°	8.3	64.9
2,2,3-Trimethylpentane	3°	7.5	55.3

[a] Type of hydrogen.
[b] A_p' in units of $M^{-1}s^{-1}$.

3.1.2 Aralkanes

The rate constants k_p and k_p' for some aralkanes are listed in Table 4. As expected, the value of k_p for toluene with reactive 1° hydrogens is smaller than k_p for ethylbenzene with reactive 2° hydrogens, but k_p for ethylbenzene is larger than k_p for cumene with a reactive 3° hydrogen. This reactivity order is different to that for other radicals, where 1° hydrogens are less reactive than 2° hydrogens, which in turn are less reactive than 3° hydrogens. The rate constants for the reactions of toluene, ethylbenzene, and cumene with the

Table 4 Rate constants for abstraction of a hydrogen atom from aralkanes by the peroxyl derived from the substate (k_p) and by $(CH_3)_3CO_2^{\cdot}$ (k_p') at 303 K[a]

Aralkane	k_p ($M^{-1}s^{-1}$)	k_p' ($M^{-1}s^{-1}$)	k_p/k_p'
Toluene	0.08	0.012	6.7
Ethylbenzene	0.65	0.1	6.5
Diphenylmethane	1.05	0.25	4.2
Cumene	0.18	0.16	1.12
Tetralin	1.6	0.5	3.2
9,10-Dihydroanthracene	82.5	17.5	4.7

[a]Values are given per active hydrogen.

same peroxyl, $(CH_3)_3CO_2^{\cdot}$, do, however, follow the normal reactivity pattern, with relative values of 1 / 8.3 / 13.3. The last column in Table 4 gives values of the ratio k_p/k_p', which is a measure of the relative reactivities of peroxyls, and in all cases primary and secondary peroxyls are more reactive than tert-peroxyls. This means that the anomolous order of the relative values of k_p is the result of a difference in peroxyl reactivity.

The $(CH_3)_3CO_2^{\cdot}$ radical abstracts a 2° hydrogen from the cyclic aralkane, tetralin, five times more readily than it does from the acyclic aralkane, ethylbenzene. This enhanced reactivity of cyclic substrates is quite general in peroxyl chemistry. A possible explaination is that the 2° hydrogens in ethylbenzene have low reactivity because of steric inhibition to resonance in the incipient 1-phenylethyl, which arises from an interaction between the CH_3 group and the *ortho*-hydrogens in the ring. Alternatively, the enhanced reactivity of tetralin may be associated with the relief of ring strain upon removal of an α–C—H hydrogen atom. Diphenylmethane and 9,10-dihydroanthracene, aralkanes with 2° hydrogens activated by two phenyl rings, also show the same pattern of reactivity, although diphenylmethane is only twice as reactive as ethylbenzene, while 9,10-dihydroanthracene is ca. 35 times as reactive as tetralin. In general, rate enhancement by a second phenyl ring is not as great as that of the first, because the two phenyl rings are not coplanar in the incipient radical.

There is a substantial deuterium isotope effect for abstraction of a benzylic hydrogen by tert-peroxyls, e.g. $k_p'(H)/k_p'(D) = 8.0$ and 20 for the reactions of 1,1-dimethylbenzylperoxyl and $(CH_3)_3CO_2^{\cdot}$, with isopropylbenzene respectively. These values suggest a symmetrical transition state for the hydrogen transfer. The value of 8 is close to the maximum for loss of the stretching vibration, while the value of 20 suggests loss of stretching and one or more of the bending modes in the reactant.

Table 5 gives the relative reactivities of toluene, ethylbenzene and

Table 5 Relative values of k_p'/H for abstraction of a hydrogen atom from aralkanes by peroxyls at 303 K

Peroxyl	Aralkane		
	$C_6H_5CH_3$	$C_6H_5CH_2CH_3$	$C_6H_5C(CH_3)_2H$
$CH_3(CH_2)_2CH_2O_2^{\cdot}$	1	6	13
$CH_3CH_2CH(CH_3)O_2^{\cdot}$	1	7.5	12
$(CH_3)_3CO_2^{\cdot}$	1	6	13
$C_6H_5C(CH_3)_2O_2^{\cdot}$	1	9	18

isopropylbenzene to four different peroxyls, namely a primary peroxyl, ($CH_3(CH_2)_2CH_2O_2^{\cdot}$), a sec-peroxyl, ($CH_3CH_2CH(CH_3)O_2^{\cdot}$), and two tert-peroxyls, ($C_6H_5C(CH_3)_2O_2^{\cdot}$ and $(CH_3)_3CO_2^{\cdot}$). In all of the cases, the relative reactivities of the aralkanes are constant, which confirms the suggestion that the relative substrate reactivities are independent of the structure of the alkylperoxyl [34]. When comparing the relative reactivities of aralkanes to peroxyls with other radicals, we again find that this class is more selective than $(CH_3)_3CO^{\cdot}$ (1 / 3.2 / 6.8) and less selective than $Br^{\cdot}$ (1 / 17 / 37) [36].

The rate constants for the reactions of *meta*- and *para*-substituted toluenes with $(CH_3)_3CO_2^{\cdot}$ are listed in Table 6 [14]. These rate constants depend on both the electron-donating and electron-withdrawing capacities of the ring substituents. A plot of log ($k_p'(M^{-1}s^{-1})$) against the σ- and σ^+-substituent

Table 6 k_p' per methyl group for reactions of $(CH_3)_3CO_2^{\cdot}$ with ring-substituted toluenes at 303 K

Substituent	k_p' ($M^{-1}s^{-1}$)
4–CH_3O	0.083
4–C_6H_5O	0.066
4–CH_3	0.056
3–CH_3	0.035
None	0.031
4–Cl	0.030
3–CH_3O	0.027
4–$CH_3OC(O)$	0.025
3–Cl	0.0185
4–CN	0.014
4–NO_2	0.0145
3–CN	0.011
3–NO_2	0.0105
4–$CH_3C(O)$	0.008

constants gives a better correlation with σ^+, resulting in log ($k_p'(M^{-1}s^{-1})$) = $0.56\sigma^+ - 1.5$. This correlation suggests that there is a polar contribution to the transition state for the hydrogen-atom-transfer reaction, i.e.

$$X\text{-}C_6H_4\text{-}\overset{\delta+}{CH_2} \cdot H : \overset{\delta-}{OOR}$$

The rate constants k_p and k_p' for some α-substituted toluenes [15] are listed in Table 7. The values of k_p are markedly dependent on the peroxyl and substrate reactivity. The influence of α-substitution on the peroxyl reactivity is apparent from the ratio k_p/k_p', which is as high as ca. 4×10^4 for benzaldehyde and ca. 3×10^2 for benzyl acetate. The peroxyl reactivity increases as the electron-withdrawing capacity of the α-substituent increases

Table 7 k_p/H and k'_p/H[a] for reactions of α-substituted toluenes with peroxyls

Substrate	k_p ($M^{-1}s^{-1}$)	k_p' ($M^{-1}s^{-1}$)	k_p/k_p'
Benzaldehyde	33000	0.85	≈40000
Benzyl ether	7.5	0.3	25
Benzyl alcohol	2.4	0.065	37
Benzyl ketone	0.82	0.045	18
Benzyl cyanide	1.56	0.01	156
Benzyl chloride	1.5	0.008	190
Benzyl acetate	2.3	0.0075	307

[a]For reaction with $(CH_3)_3CO_2^{\cdot}$.

and there is a good correlation between $\log(k_p/k_p')$ and the σ-*meta*-substituent constants. This is associated with the fact that the electron distribution in peroxyls has contributions from two canonical structures:

$$R-\ddot{O}-\dot{O} \longleftrightarrow R-\overset{+}{\dot{O}}-\overset{-}{\ddot{O}}$$

The dipolar structure is favoured by the electron-releasing R groups, and the more electrophilic structure, with less electron delocalization, is favoured by the electron-withdrawing groups. The values of log k_p' give a better correlation with ρ_p^+ than the σ_p and σ_m substituent constants, suggesting that both inductive and resonance effects influence the reactivity of the labile hydrogens of the substrate.

The reactivity of the benzylic hydrogens towards peroxyls is influenced by steric effects, e.g. the tert-H groups of *o*-methylisopropylbenzenes are less reactive than that of isopropylbenzene because of steric hinderance to the resonance stabilization of the incipient *o*-substituted isopropylbenzyls [20]. The influence of steric hinderance on the substrate reactivity is also apparent in the reactivity of the secondary α–C—H bonds which decrease in the order, $C_6H_5CH_2CH_3 > C_6H_5CH_2CH_2CH_3 > C_6H_5CH_2CH_2CH_2CH_3 > C_6H_5CH_2C(CH_3)_2H$ [20].

The major reaction of tert-butylperoxyls with 1- and 2-methylnaphthalene is hydrogen abstraction from the methyl group, with rate constants per active hydrogen, *k*/H, of 0.002 and 0.0016 $M^{-1}s^{-1}$, respectively, at 303 K [38]. These rate constants are slightly larger than the corresponding rate constant for toluene because of a difference in the C—H bond strength. On the other hand, although 2-isopropylnaphthalene is slightly more reactive than isopropylbenzene, 1-isopropylnaphthalene is only about one tenth as reactive as 2-isopropylnaphthalene at 383 K [20].

Relative reactivity data for a large number of activated and unactivated C—H bonds of aralkanes have been collated by Belyakov *et al.* [20]. For example, in isopropylbenzene the activated 3° hydrogen is 2.5×10^3 times more reactive than the unactivated 1° hydrogens, and in n-propylbenzene the activated α-hydrogen is 26 times more reactive than the unactivated β-hydrogens.

3.1.3 Alkenes

Peroxyls can either abstract an allylic hydrogen from an alkene (reaction (18)), or add to the double bond to give a β-alkylperoxyalkyl (reaction (19)). The overall propagation rate constant, k_p (overall), is therefore made up of contributions from k_p (abstraction) and k_p (addition). The relative importance of these two reactions depends on the structure of the alkene. A further complication to alkene autoxidation arises because the β-alkylperoxyalkyl can either add O_2 to give a β-alkylperoxyalkylperoxyl (reaction (20)), or undergo an intramolecular S_H2 reaction to give an epoxide and an alkoxyl (reaction (21)). The importance of reaction (21) increases as the oxygen pressure falls. The major product of the reaction of an allylperoxyl with an alkene is therefore either a β-alkylperoxylalkyl hydroperoxide, polyperoxide, or α-unsaturated hydroperoxide, depending on the structure of the alkene:

$$RO_2^{\bullet} + -CH_2CH{=}CH- \longrightarrow ROOH + -\dot{C}HCH{=}CH- \quad (18)$$

$$RO_2^{\bullet} + -CH_2CH{=}CH- \longrightarrow ROO\overset{|}{C}H\dot{C}HCH_2- \quad (19)$$

$$ROO\overset{|}{C}H\dot{C}HCH_2 + O_2 \longrightarrow ROO\overset{|}{C}HCH(O_2^{\bullet})CH_2- \quad (20)$$

$$-CH_2\dot{C}HCHOOR \longrightarrow RO^{\bullet} + -CH_2CH{-}\overset{|}{C}H \text{ (epoxide, with O bridging)} \quad (21)$$

Two important features of alkene autoxidation are allylic isomerization and allylperoxyl isomerization:

$$\text{CH}_2{=}\text{CH}{-}\text{CH}_2 \longleftrightarrow [\text{CH}_2{\cdots}\text{CH}{\cdots}\text{CH}_2]^{\bullet} \longleftrightarrow \text{CH}_2{-}\text{CH}{=}\text{CH}_2 \tag{22}$$

$$\text{CH}_2{=}\text{CH}{-}\text{CH}_2\text{O}_2^{\bullet} \rightleftharpoons {}^{\bullet}\text{O}_2\text{CH}_2{-}\text{CH}{=}\text{CH}_2 \tag{23}$$

Thus 2,3-dimethylbut-2-ene autoxidizes to 2,3-dimethylbut-1-enyl-2-hydroperoxide (reaction (24)), either because the incipient 1° alkyl isomerizes to the 3° alkyl, (reaction (25)), or because the 1° alkylperoxyl rearranges to the 3° alkylperoxyl (reaction (26)):

$$(CH_3)_2C{=}C(CH_3)_2 + O_2 \longrightarrow (CH_3)_2C(OOH)C(CH_3){=}CH_2 \tag{24}$$

$$(CH_3)_2C{=}C(CH_3){-}\dot{C}H_2 \longleftrightarrow [(CH_3)_2C{\cdots}C(CH_3){\cdots}CH_2]^{\bullet} \longleftrightarrow (CH_3)_2\dot{C}{-}C(CH_3){=}CH_2 \tag{25}$$

$$(CH_3)_2C{=}C(CH_3){-}CH_2O_2^{\bullet} \rightleftharpoons (CH_3)_2C(O_2^{\bullet}){-}C(CH_3){=}CH_2 \tag{26}$$

The values of k_p/H and k_p'/H (for $(CH_3)_3CO_2^{\bullet}$) for some alkenes (Table 8) reveal that the 1° hydrogens of 2,3-dimethylbut-2-ene are about ten times more reactive than the 1° hydrogens of toluene. In addition, the peroxyls derived from 2,3-dimethylbut-2-ene and $(CH_3)_3CO_2^{\bullet}$ have the same reactivity to these hydrogens. This latter observation is consistent with a chain-carrying tert-peroxyl (2,3-dimethylbut-1-enyl-3-peroxyl) for autoxidation of 2,3-dimethylbut-2-ene.

Table 8 k_p/H and k_p'/H[a] for hydrogen atom abstraction from alkenes by peroxyls at 303 K

Alkene	k_p ($M^{-1}s^{-1}$)	k_p' ($M^{-1}s^{-1}$)	k_p/k_p'
2,3-Dimethylbut-2-ene	0.14	0.14	1.0
Oct-1-ene	0.5	0.084	5.9
2,5-Dimethylhex-3-ene	1.2	1.1	1.1
Cyclohexene	1.5	0.75	2.0
Cyclohexa-1,4-diene	370	5.9	63

[a]Reaction with $(CH_3)_3CO_2^{\bullet}$.

The 2° acyclic allylic hydrogen atoms in oct-1-ene are less reactive than the 2° cyclic allylic hydrogen atoms in cyclohexene to $(CH_3)_3CO_2^{\cdot}$, while the large value observed for k_p of oct-1-ene results from the involvement of 1° allylperoxyls in propagation. The 3° hydrogens of 2,5-dimethylhex-3-ene are close to being ten times more reactive than the 3° hydrogen of cumene.

The value of k_p for cyclohexa-1,4-diene is about 100 times larger than k_p for cyclohexene while the difference in reactivity towards $(CH_3)_3CO_2^{\cdot}$ is only approximately a factor of ten. This difference is because the cyclohexadienylperoxyl decomposes at ambient temperatures to benzene and $HO_2^{\cdot}$, and the chain is propagated and terminated by $HO_2^{\cdot}$ rather than $RO_2^{\cdot}$ [39]. From k_p and k_p' it can be concluded that $HO_2^{\cdot}$ is about 100 times more reactive than a tert-alkylperoxyl to H-atom abstraction. This may be an overestimation as it has been suggested that $HO_2^{\cdot}$ has about the same reactivity in H-atom abstraction as a sec-alkylperoxyl [38]:

$$\text{C}_6\text{H}_7^{\cdot} + O_2 \longrightarrow \left[\text{(H, } O_2^{\cdot}\text{)-cyclohexa-2,5-dienyl} \text{ or } \text{(H, } O_2^{\cdot}\text{)-cyclohexa-2,4-dienyl} \right] \longrightarrow \text{C}_6\text{H}_6 + HO_2^{\cdot} \qquad (27)$$

3.1.4 Aldehydes

Acylperoxyls and aroylperoxyls abstract hydrogen atoms extremely fast from aldehydes, e.g. $k_p = 3.3 \times 10^4\ M^{-1}s^{-1}$ for the reaction of benzoylperoxyl with benzaldehyde at 303 K. The reaction of $(CH_3)_3CO_2^{\cdot}$ with benzaldehyde is much slower ($k_p' = 0.085\ M^{-1}s^{-1}$), giving $k_p/k_p' = 4 \times 10^5$. The rapid rate of autoxidation of aldehydes ($D_{C\text{–}H}$ is ca. 365 kJ mol^{-1}) is therefore due principally to an extremely reactive chain-propagating peroxyl.

3.1.5 Ethers, thioethers and aliphatic amines

Some typical values of k_p/H and k_p'/H for these compounds are given in Table 9. The order of substrate reactivity is, aliphatic amine $\gg$ thioether $>$ ether, with 5-membered rings being more reactive than 6-membered rings. With regards to the peroxyl reactivity, α-substituted sec-peroxyls are 10–20 times more reactive than $(CH_3)_3CO_2^{\cdot}$, and 2–5 times more reactive than sec-alkylperoxyls.

The enhanced reactivity of aliphatic amines is attributed to polar effects on the transition state and to the stabilization of α-aminoalkyls [40]. These product radicals are stabilized by conjugation between the unpaired electron and the nitrogen lone pair and as a result the hydrogen abstraction shows a pronounced stereoelectronic effect, with abstraction being most facile when the C—H bond being broken is eclipsed with the axis of the nitrogen lone-pair orbital.

Table 9 Rate constants for reactions of $RO_2^{\cdot}$ and $(CH_3)_3CO_2^{\cdot}$ with various ethers, thioethers and aliphatic amines at 303 K

Compound	k_p ($M^{-1}s^{-1}$)	k_p' ($M^{-1}s^{-1}$)	k_p/k_p'
$(C_2H_5)_3N$	30	1.92	15.6
Pyrrolidine	–	47.5	–
Tetrahydrofuran	1.1	0.085	13
Tetrahydothiophene	6.4	0.1	64
Tetrahydothiophene-α-d_4	0.58	0.015	39
Piperidine	–	2.6	–
Tetrahydropyran	0.14	0.006	23
Tetrahydrothiopyran	1.5	0.02	75
Benzyl phenyl ether	0.75	0.1	7.5
Benzyl phenyl sulfide	9.5	0.16	59

In 1963, Thomas reported that triethylamine and tri-n-butylamine are efficient inhibitors of the liquid-phase autoxidation of isopropylbenzene at 343 K [41]. This unexpected result was attributed to the formation of a charge-transfer complex between the chain-carrying 1,1-dimethylbenzylperoxyl and the amine, followed by rapid reaction of the complex with a second peroxyl to give non-radical products:

$$RO_2^{\cdot} + R_3N \longrightarrow [RO_2^{-} + R_3N^{+}] \quad (28)$$

$$[RO_2^{-} + R_3N^{+}] + RO_2^{\cdot} \longrightarrow \text{non-radical products} \quad (29)$$

The reaction kinetics are not entirely consistent with this mechanism and it has subsequently been shown that the oxidation of isopropylbenzene containing low concentrations of triethylamine behaves as a straightforward co-oxidation in which the 'inhibitor' has much larger propagation (k_p = 270 $M^{-1}s^{-1}$) and termination ($2k_t$ = 1×10^9 $M^{-1}s^{-1}$) rate constants than isopropylbenzene. Furthermore, these rate constants are larger than those expected for α-(diethylamino)ethylperoxyl, $(C_2H_5)_2NCH(CH_3)O_2^{\cdot}$, and it was shown by radical-trapping experiments with 2,6-di-tert-butyl-4-methylphenol that $HO_2^{\cdot}$ is the chain-carrying radical for autoxidation of triethylamine [42].

3.1.6 Phenols

Hydrocarbon autoxidation is inhibited or retarded by the addition of mM concentrations of a phenol. There has been some controversy over the mechanism of this reaction but it is now firmly established that chain-carrying peroxyls abstract the phenolic hydrogen to give an alkyl hydroperoxide and a phenoxyl (reaction (30)), where the latter is less reactive than the peroxyl toward propagation [3]:

$$RO_2^{\cdot} + AOH \longrightarrow ROOH + AO^{\cdot} \quad (30)$$

The kinetics of this reaction can most easily be obtained from the rates of inhibited hydrocarbon autoxidation if the chain-transfer reactions (31) and (32) are not important, i.e. when the rate expression (8) is obeyed:

$$AO^{\cdot} + RH \longrightarrow AOH + R^{\cdot} \quad (31)$$

$$AO^{\cdot} + ROOH \longrightarrow AOH + RO_2^{\cdot} \quad (32)$$

Styrene is often used as the autoxidizable hydrocarbon because it is relatively unreactive to phenoxyls and the principal reaction product is a polyperoxide, which means that the chain-transfer reactions (31) and (32) are not important and the rate expression (8) is obeyed. In addition, k_p is large for styrene and inhibited rates of autoxidation have long chain lengths. The kinetic expression (8) is also obeyed for autoxidation of aralkanes (tetralin and 9,10-dihydroanthracene) if the OH function of the phenol is protected by two bulky *ortho*-groups.

Absolute values of k_{30} have been measured for a large number of phenols [15], and a few selected examples for some hindered phenols are listed in Table 10. It should be remembered that the determination of k_{30} depends on an accurate determination of k_p, which in turn depends on an accurate determination of $2k_t$. This means that values of k_{30} are probably accurate to within a factor of approximately two. A comparison of k_{30} for the reaction of tetralylperoxyl with 2,6-di-t-butyl-4-methylphenol and k_p for tetralin shows that the hindered phenol is ca. 10^4 times more reactive than the aralkane. The replacement of hydrogen by deuterium in the phenolic O—H group increases the inhibited rates of autoxidation of isopropylbenzene and styrene by factors of approximately four and ten respectively. This is consistent with the simple H-atom-transfer mechanism shown in reaction (30).

Table 10 Rate constants for reactions of peroxyls with phenols

Peroxyl	Phenol[a]	T (K)	$10^{-4}k_{30}$ ($M^{-1}s^{-1}$)
1-Tetralyl	BMeP	338	3.6
Poly(peroxystyryl)	BMeP	338	1.78
Poly(peroxystyryl)	BMeP (OD)[b]	338	0.18
9,10-Dihydro-9-anthracyl	BMeP	303	1.10
Poly(peroxystyryl)	BP	338	0.49
Poly(peroxystyryl)	BMeOP	338	7.8

[a] BMeP, 2,6-di-t-butyl-4-methyl phenol; BP, 2,6-di-t-butyl phenol; BMeOP, 2,6-di-t-butyl-4-methoxy phenol
[b] OD, deuterated species.

Electron-attracting and electron-donating *para*-substituents decrease and increase, respectively, the reactivity of phenols, and the relative rate constants correlate with Brown's electrophilic σ^+-substituent constants. For example, the relative rate constants for the reaction of poly(peroxystyryl)peroxyl with ring-substituted phenols at 338 K obey the expression:

$$\log(k^{X}_{30}/k^{O}_{30}) = 0.002 - 1.58\sigma^+$$

where k^{X}_{30} and k^{O}_{30} are the rate constants for the substituted and unsubstituted phenols, respectively. A similar study of the reaction of 1-tetralylperoxyls with 2,6-di-t-butyl-4-substituted phenols gives $\rho^+ = -1.36$.

The transition state for the hydrogen-atom-transfer reaction (30) is represented by the three resonance forms I, II and III:

$$\underset{\text{I}}{XC_6H_4O{:}H^{\cdot}OOR} \longleftrightarrow \underset{\text{II}}{XC_6H_4O^{+\cdot}H{:}^{-}OOR} \longleftrightarrow \underset{\text{III}}{XC_6H_4O^{\cdot}H{:}OOR}$$

where I resembles the reactants, III resembles the products and II is a polar contribution. The correlation of k_{30} with the σ^+-substituent constants means that structure II contributes to the transition state but the magnitude of the ρ-factors are consistent with a moderate degree of charge separation. However, it is possible that these substituent effects are simply due to the influence of the substituent on the thermodynamic stability of the incipient phenoxyl.

Reactions of tert-butylperoxyl with phenols have been studied directly by KESR spectroscopy and the rate expression (13) is usually obeyed. Absolute rate constants have been determined from 173 to 236 K for the reaction with 2,6-di-tert-butyl-4-substituted phenols and pre-exponential factors and activation energies fall within the range 10^4–10^5 $M^{-1}s^{-1}$ and 2–4 kJ mol^{-1}, respectively (Table 11). Extrapolation of these data to ambient temperatures gives rate constants that are in good agreement with those determined indirectly from oxygen absorption studies.

Table 11 Rate constants and Arrhenius parameters for reactions of $(CH_3)_3CO_2^{\cdot}$ with various (4-Substituted) 2,6-di-tert-butyl phenols (BPs)

Phenol	$10^{-4}\,k_{30}$ ($M^{-1}s^{-1}$)	$\log A_{30}$ [a]	E_{30} (kJ mol^{-1})
BMeOP	2.3	4.7(2)	1.7(8)
BMeP	0.5	4.6(2)	3.3(8)
BP	0.34	4.3(2)	4.1(6)
BClP	0.5	4.3(2)	2.1(8)
BCNP	0.1	4.0(3)	4.1(12)

[a] A_{30} in units of $M^{-1}s^{-1}$.

Both A_{30} and E_{30} are low for a hydrogen-atom-transfer reaction and it has been proposed that the initial reaction involves the reversible formation of a peroxyl-hindered phenol complex. Hydrogen-atom transfer occurs within the complex, which then decomposes to give the reaction products:

$$RO_2^{\bullet} + \text{(2,6-R}_2\text{-4-X-phenol, ArOH)} \rightleftharpoons RO_2\cdots H{:}O\text{-Ar} \;\; (\mathbf{1}) \tag{33}$$

$$\mathbf{1} \rightleftharpoons RO_2{:}H\cdots O\text{-Ar} \;\; (\mathbf{2}) \tag{34}$$

$$\mathbf{2} \rightleftharpoons ROOH + {}^{\bullet}O\text{-Ar} \tag{35}$$

where R is $(CH_3)_3C$ and X is the 4-substituent. The overall rate constant is given by:

$$2k_{30} = 2k_{34}k_{33}/k_{-33}$$

The overall activation energy (E_{30}) is lower than for a normal H-atom-transfer reaction because of the negative heat of formation of the hydrogen-bonded complex, and the pre-exponential factor is low by an amount equivalent to the loss in entropy on formation of the complex. Zavitsas and Chatgilialoglu [43] have suggested an alternative explanation for the low activation energies and pre-exponential factors for H-atom transfer between the oxygen atoms. This is based on a low triplet repulsion energy between the oxygen atoms in a tight transition state.

The value of ρ^+ for the reactions of $(CH_3)_3CO_2^{\bullet}$ with 4-substituted phenols is -1.0 and -1.23 at 236 and 173 K, respectively. This temperature effect is consistent with the suggestion that differences in k_{30} depend mainly on differences in E_{30}.

The kinetic behaviour of non-hindered phenols that give reactive phenoxyls in reaction (30) is complex in hydrogen-atom-donating substrates. For example, the inhibited rates of autoxidation of tetralin in the presence of phenol are proportional to $R_i^{1/2}$ and $[AOH]^{-1/2}$. These kinetics prompted Hammond and coworkers [22] to suggest that the mechanism of inhibition by phenols involves the reversible formation of a peroxyl–phenol complex, followed by reaction of the complex with a second peroxyl:

$$RO_2^{\bullet} + C_6H_5OH \rightleftharpoons [C_6H_5O{:}H^{\bullet}OOR] \quad (36)$$

$$[C_6H_5O{:}H^{\bullet}OOR] + RO_2^{\bullet} \longrightarrow \text{non-radical products} \quad (37)$$

Somewhat later, Mahoney [3] showed that $(d[O_2]/dt)_{inh}$ in this system and in the phenol-inhibited autoxidation of 9,10-dihydroanthracene is also proportional to $[RH]^{3/2}$. These kinetics are consistent with a reaction scheme that includes the chain-transfer reaction (31) and the more complex scheme of Hammond and coworkers [22] is not required to explain the reaction kinetics.

Direct studies of the reactions of $(CH_3)_3CO_2^{\bullet}$ with phenol and β-naphthol have shown that equation (13) is obeyed. For phenol, $\log(A_{30}(M^{-1}s^{-1})) = 7.2$ and $E_{30} = 21.8$ kJ mol^{-1}, which gives an extrapolated value of 6×10^3 M^{-1}s^{-1} at 333 K. This is in good agreement with $k_{30} = 5 \times 10^3$ M^{-1}s^{-1} for poly(peroxystyryl)peroxyl at 338 K and 8.2×10^3 M^{-1}s^{-1} for 9-peroxy-9,10-dihydroanthracene at 333 K [15].

The value of k_{30} for β-naphthol in 1 M acetonitrile is four times smaller than it is in isopentane, which is consistent with the recent finding by Wayner *et al.* that the O–H bond strength of phenol is 29 kJ mol^{-1} stronger in acetonitrile than it is in isooctane [44].

3.1.7 Alkyl hydroperoxides

The rate constants for the reactions of peroxyls with alkyl hydroperoxides, i.e. the chain transfer-reaction (14), have been estimated both from hydrocarbon autoxidation and directly by KESR [15]. A kinetic study of the effect of isopropylbenzene hydroperoxide on the autoxidation of tetralin gives $k_{14} = 2.5 \times 10^3$ M^{-1}s^{-1} at 303 K for the reaction of 1-tetralylperoxyl with this hydroperoxide, while a direct study of the reaction of $(CH_3)_3CO_2^{\bullet}$ with 1-tetralin hydroperoxide in isopentane gives $\log(A_{15}(M^{-1}s^{-1})) = 6.0 \pm 0.5$ and $E_{15} = 18.8 \pm 2.1$ kJ mol^{-1}. There is a substantial isotope effect on this reaction when the hydroperoxidic hydrogen is replaced by deuterium, i.e. $k_{15}(H)/k_{15}(D) = 9$ at 294 K, which is consistent with a symmetric transition state.

3.1.8 Alkyl halides

The $(CH_3)_3CO_2^{\bullet}$ radical abstracts a hydrogen atom from alkyl halides, giving the rate constants which are listed in Table 12. The most interesting feature

Table 12 Rate constants for reactions of $(CH_3)_3CO_2^{\cdot}$ with alkyl halides at 303 K

Alkyl halide[a]	$10^3 k_p'$ ($M^{-1}s^{-1}$)
$CH_3CH_2\mathbf{CH}(CH_3)CH_2Br$	9.0
$CH_3\mathbf{CH}(CH_3)CH_2Br$	13.0
$CH_3\mathbf{CH}(CH_3)CH_2CH_2Br$	1.0
$(CH_3)_2CH\mathbf{CH_2}CH_2Br$	0.04
$CH_3CH_2\mathbf{CH}(CH_3)CH_2Cl$	3.5
$(CH_3)_2\mathbf{CH}CH_2CH_2Cl$	3.0

[a]Reactive hydrogen atoms are shown in bold type.

of the data is the enhanced reactivity of a 3° hydrogen which is β to a bromine substituent. Thus the 3° hydrogen of 1-bromo-2-methylbutane is approximately ten times more reactive than the 3° hydrogen of 1-bromo-3-methylbutanc, whcrcas it would be expected to be less reactive on purely electronic grounds. The activating effect of a β-bromine substituent is attributed to anchimeric assistance, whereby the substituent interacts with the developing singly occupied p orbital on the neighbouring carbon atom. This is supported by a decrease in the enthalpy of activation for abstraction of the 3° hydrogen from 1-bromo-2-methylbutane when compared with the 3° hydrogen of 3-methylpentane [45]. The formation of a bridged β-bromoalkyl has received support from the observation that 1-bromo-2-methylbutane undergoes autoxidation with some stereochemical control [46].

3.1.9 Aromatic amines

Kinetic studies of the influence of aromatic amines, such as diphenylamine and *N*-phenylnaphth-1-ylamine, on hydrocarbon autoxidation at ambient temperatures have established that these compounds inhibit the reaction by donating an amino hydrogen to the chain-propagating peroxyl [3, 5]:

$$RO_2^{\cdot} + \rangle NH \longrightarrow ROOH + \rangle N^{\cdot} \qquad (38)$$

Values of k_{38} at 338 K from the inhibited autoxidation of styrene are 2×10^4 for diphenylamine, $5 \times 10^4\ M^{-1}s^{-1}$ for *N*-phenylnaphth-2-ylamine, and $7 \times 10^4\ M^{-1}s^{-1}$ for *N*-phenylnaphth-1-ylamine [15], which makes these compounds either equally or more reactive than 2,6-di-t-butyl-4-methylphenol (BMeP). Replacing the amino hydrogen by deuterium reduces the rate constant by a factor of 3–4, thus confirming the hydrogen-atom-transfer mechanism.

An excellent correlation exists between the values of k_{38} for poly(peroxystyryl)peroxyls and Brown's electrophilic substituent constants (σ^+) for ring-substituted diphenylamines and *N*-methyl anilines with $\rho^+ = -0.89$ and -1.6, respectively. This indicates that there is charge separation in the transition state, i.e. the dipolar structure:

$$X\text{-}C_6H_4\text{-}\overset{\overset{\displaystyle H}{|}\ \delta+}{N} \cdot H : \overset{\delta-}{OOR}$$

makes an important contribution to this state.

Rate constants over the temperature range 176–238 K have been obtained directly for the reactions of tert-peroxyls with naphth-1-ylamine (AN) and *N*-phenylnaphth-1-ylamine (PAN) by KESR [15]. This work yields the Arrhenius parameters $\log(A_{38}'(M^{-1}s^{-1})) = 3.9 \pm 0.5$ and 5.1 ± 0.5 for AN and PAN, respectively, while $E_{38}' = 4.2 \pm 2$ kJ mol^{-1} for both amines. Thus PAN is about ten times more reactive than AN because of a difference in $\Delta S^{\ddagger}$. The chain-transfer reactions:

$$>N^{\bullet} + RH \longrightarrow >NH + R^{\bullet} \tag{39}$$

and

$$>N^{\bullet} + ROOH \longrightarrow >NH + RO_2^{\bullet} \tag{40}$$

do not occur for both these aromatic amines since they react with exactly two tert-peroxyls and added $(CH_3)_3COOH$ has no effect on the rates of peroxyl decay.

This direct method cannot be used to measure the rate constants for the reactions of tert-peroxyls with diphenylamine, *N*-phenylnaphth-2-ylamine and *N,N'*-diphenyl-*p*-phenylenediamine. This is because an aminoxyl is produced by the reaction of the peroxyl with the aminyl formed in reaction (38) (see later), and the central line of the triplet aminoxyl spectrum interferes with the peroxyl spectrum.

The inhibited rate of autoxidation can be proportional to $[AH]^{1/2}$ and $R_i^{1/2}$ and a mechanism analogous to reactions (36) and (37), with the aromatic amine replacing the phenol, has been proposed. These kinetics are, however, the result of the chain-transfer reactions (39) and (40). Inhibition by *N,N,N'*,*N'*-tetramethyl-*p*-phenylenediamine has also been presented as evidence against the H-atom-transfer mechanism. This amine is in fact a pro-oxidant, and it is either the radical cation, $[(CH_3)_2NC_6H_4N(CH_3)_2]^{+}$, or a reaction product that scavenges the peroxyls [5].

The apparent stoichiometric factors, n, for inhibition by secondary aromatic amines can be >12, rather than the more usual 2 found for hindered phenols. This catalytic inhibition process occurs because an alkyl reacts with an aminoxyl, rather than with O_2, to give an *N*-alkoxy-dialkylamine which can decompose to regenerate the amine [47]:

$$\rangle NO^{\bullet} + R^{\bullet} \longrightarrow \rangle NOR \quad (41)$$

$$\rangle NOR \longrightarrow \rangle NH + \text{other products} \quad (42)$$

3.1.10 Hydroxylamines

Compounds such as 2,2,6-6-tetramethyl-4-piperidone hydroxylamine and 4,4′-dimethoxydiphenyl hydroxylamine are efficient inhibitors of styrene autoxidation because they react rapidly with peroxyls:

$$RO_2^{\bullet} + \rangle NOH \longrightarrow ROOH + NO^{\bullet} \quad (43)$$

Direct studies have shown that *N*-t-butylnaphth-2-ylhydroxylamine reacts more rapidly with peroxyls than the corresponding amine because of a smaller activation energy. This is in accord with the lower bond strength of NO—H (291 kJ mol^{-1}) compared with that of N—H (335 kJ mol^{-1}) [48].

3.1.11 Silanes

The silane $C_6H_5Si(CH_3)_2H$ is about 50 times less reactive than $C_6H_5C(CH_3)_2H$ towards $(CH_3)_3CO_2^{\bullet}$ and has a reactivity similar to that of a 3° aliphatic hydrogen [14]. The principal reason for the low activity of this silane is an ineffective stabilization of the silicon-centred radical by the phenyl group. The $(CH_3)_3CO_2^{\bullet}$ radical has a similar reactivity to hexamethylsilane as it does to 1° aliphatic hydrogen atoms, indicating that an α-silicon atom does not activate a C—H bond to peroxyl metathesis. In contrast, $(CH_3CH_2)_3SiH$ has an enhanced reactivity towards peroxyls, suggesting that a β-silicon atom activates a neighbouring C—H bond.

3.1.12 Hydrobromic acid

Hydrocarbon autoxidation is catalyzed by trace amounts of HBr because the latter reacts rapidly with peroxyls by a chain transfer reaction analogous to reaction (14). The slow propagation reaction (3) is replaced by the much faster reaction of $Br^{\cdot}$ with the substrate [12]. This chemistry is made use of commercially in cobalt-catalyzed autoxidations carried out in the presence of a source of HBr.

3.2 INTRAMOLECULAR HYDROGEN ABSTRACTION

An important kinetic pathway available to larger peroxyls is unimolecular isomerization by internal H-atom metathesis, a reaction that gives rise to polyhydroperoxyl products, $R(OOH)_n$, and many of the other diverse products seen in autoxidation [4]. For example, dibenzyl ether undergoes intra- as well as intermolecular hydrogen-atom transfer to give a dihydroperoxide and a monohydroperoxide as the major reaction products:

$$C_6H_5CH_2OCH(O_2^{\cdot})C_6H_5 \longrightarrow C_6H_5\dot{C}HOCH(OOH)C_6H_5 \qquad (44)$$

$$C_6H_5\dot{C}HOCH(OOH)C_6H_5 + O_2 \xrightarrow{RH} C_6H_5CH(OOH)OCH(OOH)C_6H_5 \qquad (45)$$

$$C_6H_5CH_2OCH(O_2^{\cdot})C_6H_5 + C_6H_5CH_2OCH_2C_6H_5 \longrightarrow C_6H_5CH_2OCH(OOH)C_6H_5 + C_6H_5CH_2O\dot{C}HC_6H_5 \qquad (46)$$

The overall propagation rate constant is made up of contributions from k_{44} and k_{46}, and their relative importance depends on the concentration of the substrate.

Liquid-phase autoxidation of hexadecane is one of the most thoroughly studied systems in which α,γ- and α,δ-intramolecular H-atom transfers are important reaction pathways, with α,γ- and α,δ-dihydroperoxides as the major reaction products [49]:

$$-\underset{}{\overset{O_2^{\cdot}}{\overset{|}{C}}}H(CH_2)_mCH_2- \longrightarrow -\overset{OOH}{\overset{|}{C}}H(CH_2)_m\dot{C}H- \longrightarrow -\overset{OOH}{\overset{|}{C}}H(CH_2)_m\overset{O_2^{\cdot}}{\overset{|}{C}}H- \longrightarrow -\overset{OOH}{\overset{|}{C}}H(CH_2)_m\overset{OOH}{\overset{|}{C}}H- \qquad (47)$$

where $m = 1$ and 2 and the last step is an intermolecular reaction. The hydroperoxyalkylperoxyl can also abstract a hydrogen intramolecularly to give trihydroperoxide. Studies at different oxygen pressures show that the concentration of $HOORO_2^{\bullet}$ decreases more than the concentration of $RO_2^{\bullet}$ with decreasing oxygen pressure. This is because the hydroperoxyalkyls undergo intramolecular reactions to produce cyclic ethers:

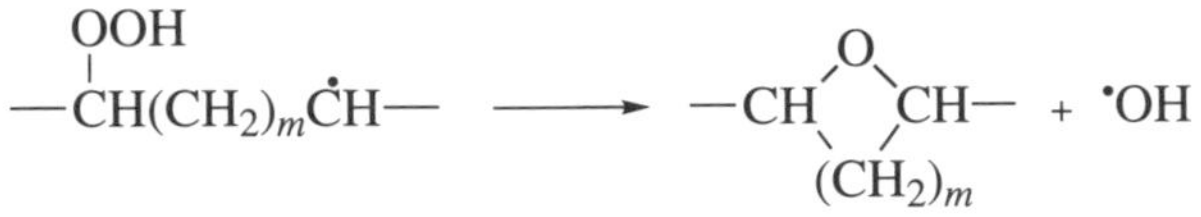

$$\text{—}\overset{\overset{\text{OOH}}{|}}{\text{C}}\text{H(CH}_2)_m\dot{\text{C}}\text{H—} \longrightarrow \text{—CH}\underset{(\text{CH}_2)_m}{\overset{\text{O}}{\frown}}\text{CH— } + {}^{\bullet}\text{OH} \quad (48)$$

The rate constants k_{47}/H for m = 1 and 2 at 433 K are 26 and 18 s^{-1}, respectively. The corresponding Arrhenius parameters are $\log(A_{47}(s^{-1}))$ = 10.1 and 11.3, and E_{47} = 71 and 84 kJ mol^{-1}. The activation energy for α,γ-abstraction which involves a six-membered-ring transition state is about the same as that of intermolecular abstraction:

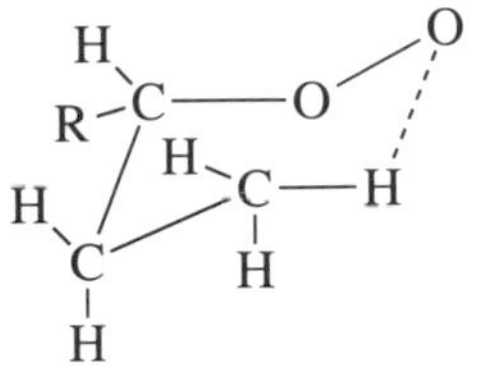

[six-membered-ring transition state structure: R(H)C—O—O···H—C(H)(H), ring closed through CH₂]

The activation energy for α,δ-abstraction involving a seven-membered ring is higher, but the 12.5 kJ mol^{-1} difference is less than the ring-strain difference, which should be about 30 kJ mol^{-1}.

3.3 ADDITION

3.3.1 Alkenes and alkynes

Peroxyls react with most alkenes and alkynes by simultaneous abstraction (reaction (18)), and addition (reaction (19)), to give a complex mixture of products, the yields of which give a rough measure of the relative rates of these two reactions [1]. Ethylene and tert-butylethylene have no allylic hydrogens and peroxyls add exclusively to these alkenes. Isobutylene has 1° allylic hydrogens and undergoes ca. 80% addition. The importance of addition decreases as the allylic hydrogens become more reactive, with addition representing only ca. 10% of the overall propagation process for cyclohexene, 3-methyl-1-butene, hex-1-yne, and hex-3-yne. For most other alkenes the addition–abstraction competition is balanced between 1 / 2 and 2 / 1. When peroxyls add to an alkene they do so at the least substituted side of unsymmetrically substituted alkenes.

Mayo has estimated $k_{19}/(2k_t)^{1/2}$ for a large number of acyclic and cyclic alkenes [1]. He assumed that differences in this ratio represent differences in k_{19} and concluded that the rates of addition correlate with the calculated excitation energies of the alkene. There are few absolute values of k_{19} for alkenes but we have found that the rate constant for the addition of

$(CH_3)_3CO_2^{\cdot}$ to oct-1-ene is ca. 0.001 $M^{-1}s^{-1}$ at 303 K, i.e. approximately two orders of magnitude smaller than the rate constant for abstraction [50].

Aralkenes and terminal alkenes, e.g. styrene and methylmethacrylate, which give highly stabilized β-alkylperoxylalkyls, react with peroxyls exclusively by addition to give radicals that combine with O_2 to give β-alkylperoxyalkylperoxyls. The reaction products are alternating 1 / 1 polyperoxides; styrene, for example, gives a copolymer with a molecular weight of about 3.6 kDa:

$$RO_2^{\cdot} + RCH{=}CH_2 \longrightarrow ROOCH_2\dot{C}HR \quad (49)$$

$$ROOCH_2\dot{C}HR + O_2 \longrightarrow ROOCH_2CH(R)O_2^{\cdot} \quad (50)$$

$$ROOCH_2CH(R)O_2^{\cdot} + RCH{=}CH_2 \longrightarrow ROOCH_2CH(R)OOCH_2\dot{C}HR \quad (51)$$

As the O_2 pressure falls, more styrene is incorporated into the polymer and epoxide formation increases via an intramolecular S_H2 reaction analogous to reaction (21), a reaction that causes the polymer to unzip.

The rate constants k_p and k_p' and the rate constant ratio k_p/k_p' for a variety of vinyl compounds at 303 K are listed in Table 13. From this data it is apparent that α-substituted sec-peroxyls, $RCH(X)O_2^{\cdot}$, are significantly more reactive towards addition than $(CH_3)_3CO_2^{\cdot}$ and the reactivity increases in the order, $—CH(OC(O)CH_3)O_2^{\cdot} > —CH(CN)O_2^{\cdot} > —CH(OC_2H_5)O_2^{\cdot} > —CH(C(O)OCH_3)O_2^{\cdot} \sim —CH(C_6H_5)O_2^{\cdot}$. This reactivity sequence is identical to that observed for α-monosubstituted benzylperoxyls towards H-atom abstraction. The α-substituted tert-peroxyls are more reactive than $(CH_3)_3CO_2^{\cdot}$, but are less reactive than α-substituted sec-peroxyls. Substrate reactivity towards $(CH_3)_3CO_2^{\cdot}$ decreases as the electron-withdrawing capacity of the substituent increases. The influence of α-substitution on substrate reactivity is

Table 13 Rate constants and Arrhenius parameters for addition of peroxyls to vinyl monomers

Vinyl monomer	k_p[a] ($M^{-1}s^{-1}$)	k_p'[b] ($M^{-1}s^{-1}$)	k_p/k_p'	Log A_p'[c]	E'_p (kJ mol^{-1})
Isopropenyl acetate	0.2	0.01	20	10.3	71.6
Methyl methacrylate	1.0	0.08	12.5	–	–
Methyl acrylonitrile	4.5	0.094	50	–	–
cis-1,2-Diphenylethylene	–	0.12	–	–	–
trans-1,2-Diphenylethylene	–	0.44	–	–	–
Styrene	41	1.3	31.5	10.7	62
α-Methylstyrene	10	2.9	3.4	–	–
1,1-Diphenylethylene	–	9.5	–	10.2	54.4

[a] Measured at 303 K.
[b] Addition of $(CH_3)_3CO_2^{\cdot}$.
[c] A_p in units of $M^{-1}s^{-1}$.

the reverse of the influence of α-substitution on peroxyl reactivity.

The values of k_p' depend principally on the stability of the incipient radical, and the empirical equation:

$$\log(k_p'(M^{-1}s^{-1})) = 0.2E_s - 4.3$$

where E_s is an estimate of the β-alkylperoxylalkyl resonance stabilization energy, is obeyed. The $(CH_3)_3CO_2^{\cdot}$ radical is approximately four times more reactive to *trans*-1,2-diphenylethylene than it is to the *cis*-isomer, despite the fact that the latter is thermodynamically less stable. This is because the β-alkylperoxylalkyl from the *cis*-isomer is ca. 12.5 kJ mol^{-1} less stable than the radical from the *trans*-isomer.

The rate constants for addition of $(CH_3)_3CO_2^{\cdot}$ to ring-substituted styrenes (Table 14) depend on the nature of the substituent. A free-energy plot of these rate constants against those found for H-atom abstraction from ring-substituted toluenes by $(CH_3)_3CO_2^{\cdot}$ suggests that incipient radical stabilization and polar effects influence peroxyl addition and abstraction to a similar degree.

Table 14 Rate constants for addition of peroxyls to ring-substituted styrenes at 303 K

Substituent	k_p ($M^{-1}s^{-1}$)	$k_p'^{a}$ ($M^{-1}s^{-1}$)
4–CH_3O	81	2.15
4–CH_3	–	2.15
None	41	1.3
4–Cl	–	1.6
3–Cl	–	1.2
3–NO_2	–	0.73
4–NO_2	–	1.0

[a] Addition of $(CH_3)_3CO_2^{\cdot}$.

3.3.2 Polynuclear aromatics

A minor reaction pathway for the reactions of peroxyls with 1-methyl- and 2-methyl-naphthalenes is addition to the naphthalene ring, with $k_{19} = 7.2 \times 10^{-6}$ M^{-1}s^{-1} compared to 0.13 M^{-1}s^{-1} for abstraction of the active hydrogen. This reaction may appear to be minor but it does have a profound effect on the autoxidation of methyl naphthalenes [38]. Peroxyls add exclusively to the *meso*-positions of anthracene, α-benzoanthracene and tetracene, with rate constants per equivalent position of 31, 4.3 and 5000 M^{-1}s^{-1}, respectively [15].

3.4. INTRAMOLECULAR ADDITION

An important reaction pathway for long unsaturated peroxyls is intramolecular addition or cyclization to give an alkyl containing a cyclic peroxide. The alkyl produced by cyclization will react rapidly with O_2 to give a substituted peroxyl, which can then abstract a hydrogen atom internally or externally, or add to a double bond internally or externally. Porter has formulated a number of rules that govern peroxyl cyclization [17]. These are as follows: (i) 5-*exo* and 6-*exo* cyclization products are formed to the exclusion of 6-*endo* and 7-*endo* products, (ii) 5-*exo* and 6-*exo* cyclizations are stereoselective with a preference for *cis*-substitution, and (iii) peroxyls generated from 5-*exo* and 6-*exo* cyclizations have a multitude of reaction pathways available for further reaction. These include S_H2 attack of the carbon radical on the peroxide bond, reaction of the carbon radical with O_2 which leads to monocyclic or polycyclic peroxides, or a second cyclization to give a bicyclic *endo*-peroxide radical. A rate constant of ca. 8×10^2 s^{-1} has been estimated for 5-*exo*-peroxyl ring closure, with 6-*exo*-ring closure being approximately two orders of magnitude slower.

Courtneidge has used the hydroperoxide method to prepare polyfunctional peroxides from polyunsaturated alkenes, e.g. the autoxidation of *cis*-2,6-dimethyl-2,6-octadiene in the presence of $(CH_3)_3COOH$ provides good yields of the hydroperoxy-1,2-dioxane **3** and the hydroperoxy dioxalane **4** [51]:

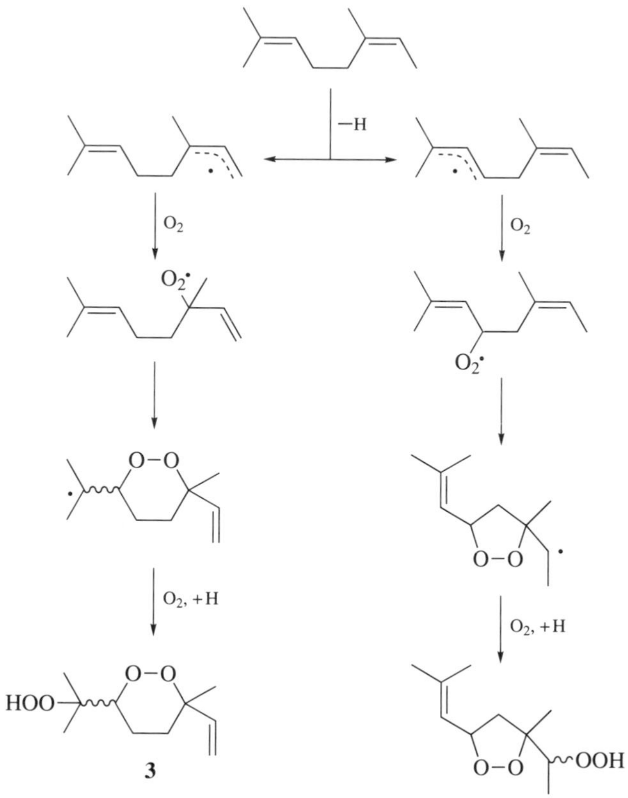

3

4

3.5 S_H2 REACTIONS

Alkyl derivatives of many organometallic compounds, MR_n (M = Li, Mg, Zn, Cd, B, Al, or T1), react with O_2 to give the corresponding alkylperoxyorganometallic compounds by displacement of the alkyl by the alkylperoxyl [52]:

$$RO_2^{\cdot} + MR_n' \longrightarrow ROOMR_{n-1}' + R' \qquad (50)$$

These autoxidations are usually fast and difficult to study kinetically but powerful peroxyl scavengers do inhibit the reaction. The radical nature of the reaction for boron was substantiated by the finding that optically active 1-phenylethyldihydroxyborane autoxidizes to racemic 1-phenylethyl hydroperoxide and both the *exo*- and *endo*-norborn-2-yl boranes react to give a mixture of 76% *exo*- and 24% *endo*-norbornylperoxyboranes.

Ingold and coworkers [53, 54] showed that the rate equation (5) is obeyed for a series of organoboranes and estimated the rate constants for reaction (50) where M = B. For example, k_{50} (per carbon–boron bond) = 2×10^6 $M^{-1}s^{-1}$ for tri-n-butylborane and 1.6×10^4 $M^{-1}s^{-1}$ for sec-butylboronic anhydride. These rate constants are large because a weak M—C bond is broken and a strong M—O bond is formed [6, 15].

Evidence for the displacement of an alkyl from the Mg of Grignard reagents by an alkylperoxyl during autoxidation is provided by a product study of the autoxidation of 5-hexenylmagnesium bromide. With excess O_2 at low temperatures, 5-hexenylperoxymagnesium bromide is formed, but if the O_2 pressure is reduced, or if the temperature is raised, a substantial amount of cyclopentylmethylperoxymagnesium bromide is formed by cyclization of the intermediate 5-hexenyl [52].

3.6 OXYGEN ATOM-TRANSFER

Trialkyl phosphites and trialkylphosphines react rapidly with atmospheric O_2 to give pentavalent oxides by a free-radical chain process. Peroxyls are intermediates in these reactions and are produced and react by the following series of reactions:

$$RO_2^{\cdot} + P(OR)_3 \longrightarrow [ROOP(OR)_3] \longrightarrow RO^{\cdot} + OP(OR)_3 \qquad (51)$$

$$RO^{\cdot} + P(OR)_3 \longrightarrow [ROP(OR)_3] \xrightarrow{O_2} RO_2^{\cdot} + OP(OR)_3 \qquad (52)$$

Peroxyls add to the trivalent phosphorus compound to give a phosphoranyl, which decomposes by O—O bond cleavage. Absolute rate constants and activation parameters for the reactions of $(CH_3)_3CO_2^{\cdot}$ with several trivalent phosphorus compounds have been obtained by KESR [15], and some of these data are listed in Table 15. It has been concluded that most trialkyl

Table 15 Rate constants and activation parameters for reactions of $(CH_3)_3CO_2^{\cdot}$ with some trivalent phosphorus compounds

Substrate	$k_{51}{}^{a}$ ($M^{-1}s^{-1}$)	$\log A_{51}{}^{b}$	E_{51} (kJ mol^{-1})
$P(OCH_3)_3$	11	5.7	24
$P(OC_6H_5)_3$	26	7.3	20
$P(C_6H_5)_3$	200	6.0	12.5
$HP(C_6H_5)_2$	180	5.0	10.5
$CH_3P(C_6H_5)_2$	900	5.1	6.3
$CH_3OP(C_6H_5)_2$	120	3.7	5.4
$ClP(C_6H_5)_2$	100	3.7	5.4

[a] Measured at 178 K.
[b] A_{51} in units of $M^{-1}s^{-1}$.

phosphites have approximately the same reactivity to $(CH_3)_3CO_2^{\cdot}$, although steric inhibition to addition has been invoked to explain the low reactivity of tri-tert-butyl phosphite. Triphenyl phosphite can inhibit autoxidation because a phenoxyl is formed in reaction (51). The $(CH_3)_3CO_2^{\cdot}$ radical reacts rapidly with triphenylphosphines because of low activation energies. The reactivities of α-substituted diphenylphosphines increase in the order, $ClP(C_6H_5)_2 < CH_3OP(C_6H_5)_2 < P(C_6H_5)_3 < CH_3P(C_6H_5)_2$, i.e. the reactivity increases as the inductive electron-donating capacity of the α-substituent increases.

3.7 METAL COMPLEXES

Although complexes of transition metals such as cobalt, manganese and iron are usually thought of as initiators of autoxidation, high concentrations inhibit autoxidation by electron transfer from the complex to the peroxyl to give the alkylperoxyl anion [2], for example:

$$RO_2^{\cdot} + Co^{2+} \longrightarrow RO_2^{-} + Co^{3+} \tag{53}$$

$$RO_2^{\cdot} + Mn^{2+} \longrightarrow RO_2^{-} + Mn^{3+} \tag{54}$$

In non-polar media this reaction may not be an outer-sphere electron transfer, but is probably best represented by the following addition reaction:

$$RO_2^{\cdot} + ML_X \longrightarrow ROOML_X \tag{55}$$

followed by electron transfer within the complex $ROOML_X$.

Inhibition by cupric salts (acetate, heptanoate, or stearate) has been ascribed to complex formation between a chain-carrying peroxyl and the cupric salt, with the complexed radical being less reactive to propagation and more reactive to termination:

Propagation

$$(RO_2^{\cdot}\ldots Cu^{2+}) + RH \longrightarrow ROOH + R^{\cdot} + Cu^{2+} \quad (56)$$

Termination

$$2(RO_2^{\cdot}\ldots Cu^{2+}) \longrightarrow \text{non-radical products} + Cu^{2+} \quad (57)$$

$$(RO_2^{\cdot}\ldots Cu^{2+}) + RO_2^{\cdot} \longrightarrow \text{non-radical products} + Cu^{+} \quad (58)$$

KESR has been used to demonstrate that $(CH_3)_3CO_2^{\cdot}$ reacts rapidly with cobalt acetylacetonate in the temperature range 183–213 K by a bimolecular reaction with $\log(nk(M^{-1}s^{-1})) = 8.9 \pm 0.7 - (19.7 \pm 6.3)/\theta$, where $\theta = 2.303RT$ kJ mol^{-1} [55]. It has been proposed that $(CH_3)_3CO_2^{\cdot}$ forms a long-lived complex-bonded peroxyl with Co(acac)$_2$ [56], but there is no evidence for such a complex in the direct KESR studies of the reaction of $(CH_3)_3CO_2^{\cdot}$ with Co(acac)$_2$ [55]. Ingold has suggested that the observation of long-lived $(CH_3)_3CO_2^{\cdot}$ in the $(CH_3)_3COOH$–Co(acac)$_2$ system is due to continuous generation of the radical from excess hydroperoxide by the Co complex [57].

The $(CH_3)_3CO_2^{\cdot}$ radical reacts with VO(acac)$_2$, with rate constants that fit the equation $\log(k_{59}(M^{-1}s^{-1})) = 6.0 \pm 0.2 - (6 \pm 1)/\theta$, by the following reaction sequence [58]:

$$RO_2^{\cdot} + VO(acac)_2 \longrightarrow ROOVO(acac)_2 \quad (59)$$

$$ROOVO(acac)_2 \longrightarrow \text{reaction products} \quad (60)$$

Initial kinetic studies of hydrocarbon autoxidation in the presence of zinc dialkyldithiophosphates (Zn[(RO)$_2$PS$_2$]$_2$) have demonstrated that these metal complex antioxidants react with alkylperoxyls and that the kinetic expression (8) is obeyed [13]. On the assumption that each molecule of Zn[(RO)$_2$PS$_2$]$_2$ scavenges two peroxyls, $\langle k_{inh}\rangle \approx 4.2 \times 10^3$ M^{-1}s^{-1} at 323 K, which makes them approximately an order of magnitude less reactive than BMeP. More potent peroxyl scavengers are produced in these systems because $d[O_2]/dt$ decreases dramatically during the induction period. The zinc salts of dialkyldithiocarbamic acid (Zn[R$_2$NCS$_2$]$_2$) and xanthic acid also scavenge peroxyls. It has been shown that $(d[O_2]/dt)_{inh}$ is proportional to $[RH]^{3/2}$ for the Zn[RO)$_2$PS$_2$]$_2$-inhibited autoxidation of styrene [13]. This has been taken as evidence that poly(peroxystyryl)peroxyl displaces $(RO)_2PS_2^{\cdot}$ from the metal complex, with the sulfur-centred radical propagating the chain less efficiently than a peroxyl:

$$RO_2^{\cdot} + Zn[(RO)_2PS_2]_2 \longrightarrow ROOZnS_2P(OR)_2 + (RO)_2PS_2^{\cdot} \quad (61)$$

The nickel complexes $Ni[(RO)_2PS_2]_2$ and $Ni[R_2NCS_2]_2$ are able to scavenge two alkylperoxyls, and values of k_{inh} fall within the range (3–30) $\times 10^3\ M^{-1}s^{-1}$ at 323 K, with $Ni[R_2NCS_2]_2$ being slightly more reactive than $Ni[(RO)_2PS_2]_2$. The radical-derived products suggest that reaction involves an oxygen-atom-transfer reaction from the peroxyl to the Ni(*II*) complex [13]. Direct studies using KESR have confirmed that $(CH_3)_3CO_2^{\bullet}$ reacts rapidly with $Zn[(RO)_2PS_2]_2$, $Zn[R_2NCS_2]_2$, $Ni[(RO)_2PS_2]_2$, and $Ni[R_2NCS_2]_2$ [13].

$Cu[(RO)_2PS_2]_2$ and $Cu[R_2NCS_2]_2$ are extremely efficient peroxyl-scavenging antioxidants for the autoxidation of styrene and 9,10-dihydroanthracene, with $k_{inh} > 1 \times 10^6\ M^{-1}s^{-1}$ [13]. This makes them as efficient, or even more efficient than the most reactive phenolic antioxidants. Values of n for these complexes are 3–5 and the peroxyl-scavenging ability lasts longer than for most peroxyl-scavenging antioxidants. An important feature of the cupric complexes is their paramagnetism, e.g. $Cu[Et_2NCS_2]_2$ has the magnetic parameters $\langle a_{63}\rangle$ = 246.6 MHz and $\langle g\rangle$ = 2.0458. Consequently, these complexes can be monitored during their reaction with peroxyls by ESR spectroscopy. By using this technique, it has been shown that $Cu[R_2NCS_2]_2$ reacts with one $(CH_3)_3CO_2^{\bullet}$ to give $Cu[R_2NCS_2][R_2NCSS(O)]$, **5** ($\langle a_{63}\rangle$ = 251.4 MHz, $\langle g\rangle$ - 2.0671). This complex reaches a maximum concentration equal to ca. 50% of the $Cu[Et_2NCS_2]_2$ when ca. 90% of the latter has been destroyed. The concentration of **5** then slowly decreases just after the complete disappearance of $Cu[R_2NCS_2]_2$. Two other transient Cu^{2+} complexes $Cu[R_2NCSS(O)]_2$ (**6**) and $Cu[R_2NCSSO_2][R_2NCSS(O)]$ (**7**) then build up and decay [13]:

$$RO_2^{\bullet} + R_2NC(S)_2Cu(S)_2CNR_2 \longrightarrow RO^{\bullet} + R_2NC(S)_2Cu(S)(S{=}O)CNR_2 \quad \mathbf{5} \qquad (62)$$

$$RO_2^{\bullet} + R_2NC(S)_2Cu(S)(S{=}O)CNR_2 \longrightarrow RO^{\bullet} + R_2NC(S)(S{=}O)Cu(S)(S{=}O)CNR_2 \quad \mathbf{6} \qquad (63)$$

$$RO_2^{\bullet} + R_2NC(S)(S{=}O)Cu(S)(S{=}O)CNR_2 \longrightarrow RO^{\bullet} + R_2NC(S)(S{=}O)Cu(S)(SO_2)CNR_2 \quad \mathbf{7} \qquad (64)$$

3.8 SPIN TRAPS

There is general agreement that peroxyls are trapped by nitrones to give peroxyl-substituted aminoxyls [59–61]. For example, 2-cyano-2-propylperoxyl and $(CH_3)_3CO_2^{\cdot}$ add to α-phenyl *N*-tert-butylnitrone (PBN) to give adducts with a_N = 36 MHz, a^{β}_H = 3.45 MHz, and a_{13} = 14.1 MHz at 205 K, and a_N = 37.7 MHz, a^{β}_H = 2.67 MHz, a_{17O} = 8.1 MHz and g = 2.0064 at 193 K, respectively:

$$RO_2^{\cdot} + C_6H_5CH{=}\overset{\overset{O}{\uparrow}}{N}C(CH_3)_3 \longrightarrow C_6H_5\underset{\underset{OOR}{|}}{C}H{-}\overset{\overset{O^{\cdot}}{|}}{N}C(CH_3)_3 \quad (65)$$

It has been suggested that these aminoxyls are not stable above ca. 230 K and decompose initially by O—O bond scission to give alkoxyls which add to the spin trap to give the more stable alkoxyl–PBN adduct. Niki *et al.* [61] have, however, successfully added $(CH_3)_3CO_2^{\cdot}$ and 1-tetralylperoxyl to PBN and methyl-*N*-durylnitrone at room temperature and concluded that the adducts are light sensitive.

Aminoxyls produced by the addition of peroxyls to nitroso compounds are also unstable and even trifluoronitrosomethane does not give the spin adduct, $ROON(O^{\cdot})CF_3$. Instead, the alkoxyl adduct is formed, even at low temperatures (173 K) [62].

3.9 BIOLOGICALLY IMPORTANT MOLECULES

3.9.1 Tocopherols

Living organisms are protected from autoxidation by a number of natural antioxidants. Vitamin E, which is made up of one or more of the four structurally related tocopherols:

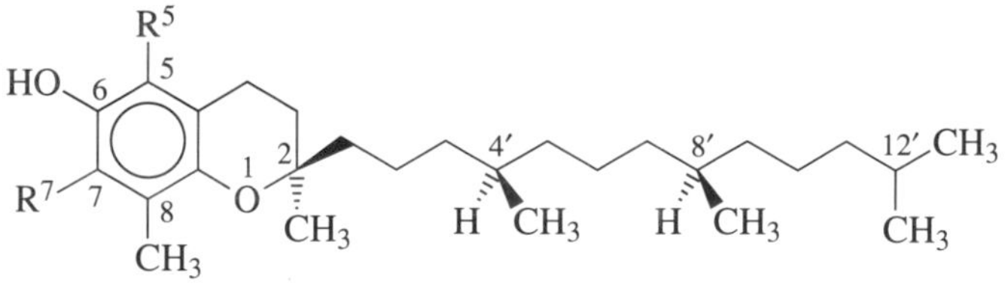

[Structure of tocopherol: chroman ring with HO at 6, R^5 at 5, R^7 at 7, CH_3 at 8, O at 1, 2-CH_3, phytyl side chain with H/CH_3 at 4′ and 8′, CH_3 at 12′ and terminal CH_3]

α-T: $R^5 = R^7 = CH_3$
β-T: $R^5 = CH_3$; $R^7 = H$
γ-T: $R^5 = H$; $R^7 = CH_3$
δ-T: $R^5 = R^7 = H$

is the most important lipid-soluble natural antioxidant [18] and reacts with peroxyls by a mechanism identical to that found for hindered phenols.

Table 16 Rate constants for reactions of poly(peroxystyryl)peroxyl with the tocopherols at 303 K

Tocopherol	$10^{-6}k_{30}$ ($M^{-1}s^{-1}$)
α–T	3.2
β–T	1.3
γ–T	1.4
δ–T	0.44

Absolute values of k_{30} for the reactions of poly(peroxystyryl)peroxyl with the tocopherols have been measured in organic media and are listed in Table 16 [15].

Tocopherol reactivity increases in the order, α-T > β–T ≈ α–T > δ–T, which parallels their relative biological activity. The α–T form is approximately two orders of magnitude more reactive than the commercial antioxidant BMeP, and approximately ten times more reactive than the structurally analogous 4-methoxy-2,3,5,6-tetramethylphenol. This enhanced reactivity is related to the fact that the oxygen in the 1-position of α-topheroxyl is in a locked conformation, whereby its p-type long-pair orbital has maximum overlap with the p orbitals of the benzene ring, thus giving a highly resonance-stabilized phenoxyl. The phytyl chain has little effect on the reactivity as 6-hydroxy-2,2,5,7,8-pentamethyl chroman has about the same reactivity as α–T.

3.9.2 Ascorbates

Ascorbic acid (vitamin C) and its derivatives are important biological antioxidants that can either destroy peroxyls or regenerate tocopherols from their tocopheroxyls. Barclay, Dakin and Zahalka [63] have measured the rate constant for the reaction of poly(peroxystyryl)peroxyl with ascorbyl palmitate in styrene at 303 K (reaction (66)), and obtained $k_{66} = 1.1 \times 10^6$ $M^{-1}s^{-1}$, showing that ascorbyl palmitate has about the same reactivity as α-tocopherol:

$$RO_2^{\bullet} + \text{[ascorbyl palmitate: HO, OH, OH, R']} \longrightarrow ROOH + \text{[HO, O}^{\bullet}\text{, OH, R']} \quad (66)$$

where R′ is palmitate.

3.9.3 Polyenes

Carotenoids and retinoids are polyunsaturated compounds that react rapidly with peroxyls to give highly resonance-stabilized radicals. In the case of β-carotene (β–C) a polyperoxide and 5,6-epoxy-β-carotene are the major reaction products, with the latter product suggesting that addition to the 5-position is an important propagation reaction. The structure of the propagating peroxyl has, however, not been determined [64]:

$$\xrightarrow{RO_2^{\bullet}} \quad \textbf{8} \text{ (–OOR)} \qquad (67)$$

At low O_2 concentrations, **8** may scavenge a second peroxyl which may be the reason why β–C functions as an antioxidant. Alternatively, k_p and $2k_t$ for β–C may be so large that small quantities can inhibit autoxidation by a mechanism similar to that observed for trialkylamines. The best estimates give a value of $k_p'((CH_3)_3CO_2^{\bullet})$ of ca. 3×10^3 $M^{-1}s^{-1}$ at 303 K, but there are no estimates of $2k_t$. Peroxyls react rapidly with all-*trans*-retinoic acid, but in this case the 5,6-epoxide is the major product, with little or no oligomeric polyperoxide. The host of reaction products given by carotenoids and retinoids must be associated with the number of different peroxyls that can be generated in propagation and the ease of oxidation of the initial reaction products.

3.9.4 Polyunsaturated fatty acids

These compounds are building blocks for the phospholipids and glycolipids that are the important components of biological membranes. They also serve as hormones, intracellular messengers and fuel hormones. The reactions of peroxyls with polyunsaturated fatty acids (PUFAs) are fast and disrupt the important structural and protective function of membranes. Consequently, they have to be protected from oxidative degradation by the presence of vitamin E.

Peroxyls initially abstract a hydrogen from methyl linoleate (k_p/H = 31 $M^{-1}s^{-1}$ in chlorobenzene at 303 K):

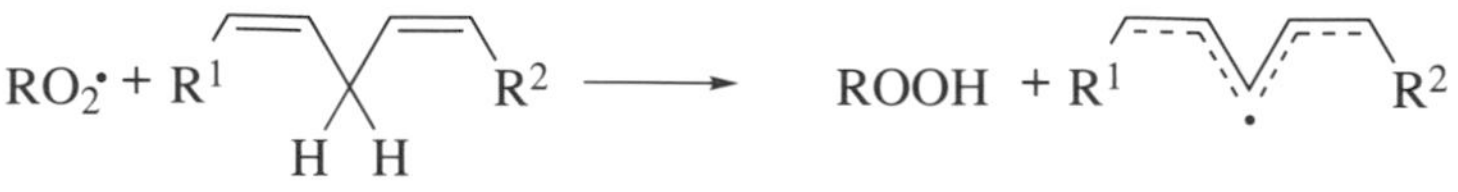

$$RO_2^{\bullet} + R^1\text{–CH=CH–CH}_2\text{–CH=CH–}R^2 \longrightarrow ROOH + R^1\text{–CH=CH–}\dot{C}H\text{–CH=CH–}R^2 \quad (68)$$

where R^1 is $CH_3(CH_2)_4$ and R^2 is $(CH_2)_7COOH$.

The incipient radical isomerizes and 9- and 13-hydroperoxy-substituted 9-*cis*-/11-*trans*-, 10-*trans*-/12-*cis*-, 9-*trans*-/11-*trans*-, and 10-*trans*-/12-*trans*-octadecadienoates are the major reaction products, with the *cis*-, *trans*- to *trans*-, *trans*- ratios related to the hydrogen-atom-donating ability of the medium:

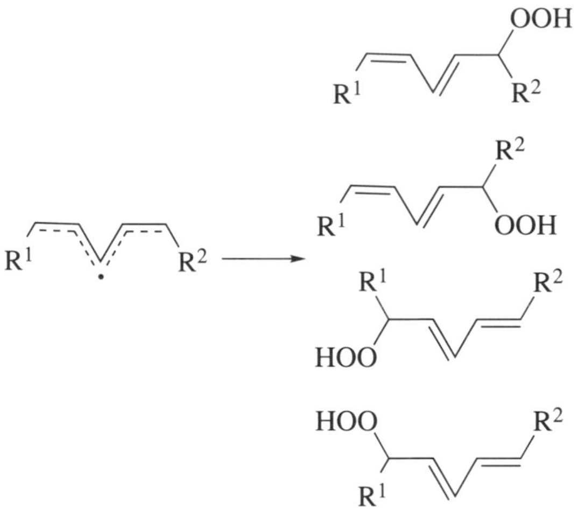

$$R^1\text{–CH=CH–}\dot{C}H\text{–CH=CH–}R^2 \longrightarrow \begin{array}{l} R^1\text{–CH=CH–CH=CH–CH(OOH)–}R^2 \\ R^1\text{–CH=CH–CH=CH–CH(OOH)–}R^2 \\ R^1\text{–CH(OOH)–CH=CH–CH=CH–}R^2 \\ R^1\text{–CH(OOH)–CH=CH–CH=CH–}R^2 \end{array} \quad (69)$$

The peroxyls produced from PUFAs undergo intramolecular addition as well as intermolecular H-atom metathesis, a reaction pathway that opens the door for lipid peroxyls to form a host of different products; the complexity of lipid peroxidation results from these multiple pathways. Arachidonic acid, for example, gives six different *trans*-, *cis*-peroxyls, four of which can undergo cyclization. Each of the cyclization pathways can lead to multiple products as is illustrated by the mixture formed in the following scheme for the peroxyl **9**:

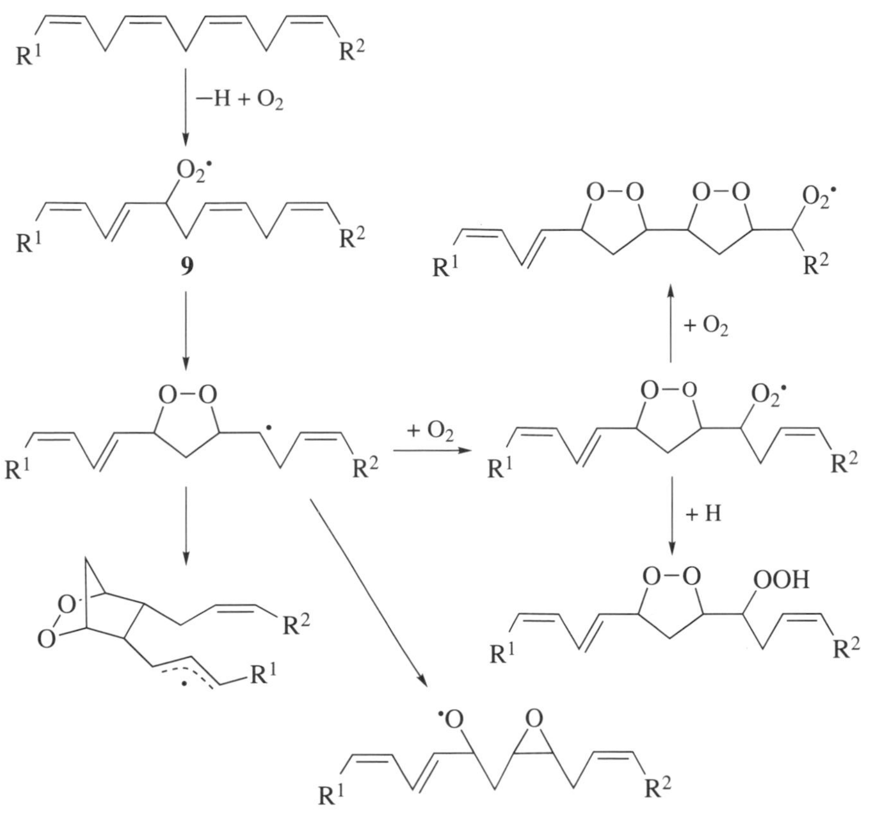

where R^1 is C_5H_{11} and R^2 is $(CH_2)_3CO_2H$.

Bowry [65] has recently proposed that autoxidation of glyceryl trilinoleate gives monohydroperoxides by intermolecular abstraction and di- and trihydroperoxides by *arm-to-arm* intramolecular hydrogen abstraction. The latter reaction increases in importance as the concentration of glyceryl trilinoleate decreases. This explains why the rate of autoxidation of glyceryl trilinoleate does not obey the rate law [5].

4 RADICAL–RADICAL REACTIONS

4.1 SELF-REACTION

4.1.1 Alkyl-, alkenyl-, aralkyl-, and aralkenylperoxyls

A few examples of the many termination rate constants ($2k_t$) and rate constants for self-reaction ($2k_t'$) that have been determined for peroxyls in solution [15] are listed in Tables 17 and 18. A perusal of these data reveals the following facts: (i) all rate constants, except $2k_t$ for $HO_2^{\cdot}$, are less that the diffusion-controlled limit of ca. 2×10^9 $M^{-1}s^{-1}$; (ii) values of $2k_t$ vary over a range of 10^5, from 1.5×10^4 $M^{-1}s^{-1}$ for isopropylbenzene to 3×10^8 $M^{-1}s^{-1}$ for 1,4-dimethylbenzene and 1.3×10^9 $M^{-1}s^{-1}$ for $HO_2^{\cdot}$ in chlorobenzene; (iii) $2k_t$ for acyclic sec-peroxyls are approximately a factor of ten smaller than $2k_t$ for primary peroxyls, while $2k_t$ for the cyclic sec-peroxyls from tetralin and cyclohexene are similar and are smaller than $2k_t$ for acyclic peroxyls; (iv) $2k_t$ for oct-1-ene, with reactive 2° hydrogens, is similar to $2k_t$ for

Table 17 Overall chain-termination rate constants for autoxidation of organic compounds in solution at 303 K

Peroxyl[a]	$10^{-6}\,2k_t$ ($M^{-1}s^{-1}$)
Hydro[b]	1260
Tetrahydrofuranyl	3
Tetrahydrothiofuranyl	3
Cyclohexenyl	5.6
p-Methylbenzyl	300
Oct-2-en-1-yl	260
2,5-Dimethylhex-3-en-2-yl	0.18
1-Phenylethyl	32
1-Tetralyl	7.6
1,1-Dimethylbenzyl	0.015
Poly(peroxystyryl)	42

[a] The parent substrate is the solvent in most cases.
[b] In chlorobenzene.

Table 18 Rate constants for self-reactions of peroxyls in solution at 303 K[a]

Peroxyl	$10^{-6}\,2k_t'$ ($M^{-1}s^{-1}$)
Methyl	370
n-Butyl	43
sec-Butyl	1.5
tert-Butyl	0.0013
2-Cyano-2-propyl	0.15
Cyclohexyl	2
1-Tetralyl	7.2
Diphenylmethyl	28
1,1-Dimethylbenzene	0.0058

[a] In hydrocarbon solvents.

1,4-dimethylbenzene because the principal chain-terminating peroxyl is the 1° oct-2-enyl-1-peroxyl; (v) $2k_t$ for poly(peroxystyryl)peroxyl is similar to $2k_t$ for 1-phenylethylperoxyl, suggesting that the size of R has little effect on $2k_t$; (vi) $2k_t$ for the tert-peroxyl from 2,5-dimethyhex-3-ene is approximately a factor of ten larger than $2k_t$ for isopropylbenzene, while $2k_t'$ for 1,1-dimethylbenzylperoxyl is approximately five times larger than $2k_t'$ for $(CH_3)_3CO_2^{\bullet}$; (vii) $2k_t'$ for 2-cyano-2-propylperoxyl is approximately two orders of magnitude larger than $2k_t'$ for tert-butylperoxyl; (viii) values of $2k_t'$ for unactivated alkylperoxyls are up to an order of magnitude smaller than $2k_t$ for peroxyls with reactive α–H atoms; (ix) replacing the α–H atom by deuterium reduces $2k_t$, e.g. $2k_t(H)/2k_t(D) = 1.36$ for diphenylmethylperoxyl and 1.37 for sec-butylperoxyl at 303 K; sec-peroxyls with activated α–C—H bonds are more reactive than those with stronger α-C—H bonds, while primary peroxyls with stronger α–C—H bonds are the most reactive.

There are few accurate determinations of the Arrhenius parameters for $2k_t$ because of the limited temperature range over which indirect studies can be made [15]. Direct studies can, however, be made over a much wider temperature range and selected Arrhenius parameters are reported in Table 19. Pre-exponential factors and activation energies are variable and it is not possible to assign differences in $2k_t$ or $2k_t'$ to differences in A_t, A_t', E_t or E_t'.

Table 19 Arrhenius parameters for self-reactions of peroxyls

Peroxyl	$\log A_t'$[a]	E_t' (kJ mol^{-1})
$(CH_3)_2CHCO_2^{\bullet}$[b]	7.12	34
$(CH_3)_2CH(CH_3)O_2^{\bullet}$	9.0	11.3
$CH_2(CH_2)_3CHO_2^{\bullet}$	10.0	13
$CH_2(CH_2CH{=}CH)CHO_2^{\bullet}$	7.8	4.2
$(CH_3)_3CO_2^{\bullet}$	9.2	35.6

[a] A_t' in units of $M^{-1}s^{-1}$.
[b] Average values from growth and decay of peroxyl in cyclopropane.

Where it has been possible to measure $2k_t$ and $2k_t'$ for the same peroxyl, e.g. 1-tetralylperoxyl, these rate constants are identical for sec-peroxyls, whereas for tert-peroxyls $2k_t$ is up to a factor of ten times larger than $2k_t'$. Thus all reactive encounters of sec-peroxyls (and presumably primary peroxyls) are terminating, while most reactive encounters of tert-peroxyls are non-terminating. This suggests that there is a reaction pathway open to primary and sec peroxyls that is not open to tert-peroxyls.

The simplest tert-alkylperoxyl, $(CH_3)_3CO_2^{\bullet}$, self-reacts above 193 K by the following terminating and non-terminating reactions:

$$2(CH_3)_3CO_2^{\bullet} \longrightarrow (CH_3)_3COOC(CH_3)_3 + O_2 \quad (70)$$

$$2(CH_3)_3CO_2^{\bullet} \longrightarrow 2(CH_3)_3CO^{\bullet} + O_2 \quad (71)$$

The ratio of the rate constants, k_{71}/k_{70}, is approximately ten and is obtained from the free-radical induced decomposition of $(CH_3)_3COOH$ [6]. The rate constant for self-reaction depends on the medium and the minimum value ($2k_t'$) is obtained in the presence of enough $(CH_3)_3COOH$ or $(CH_3)_3CH$ to convert all $(CH_3)_3CO^{\bullet}$ back to $(CH_3)_3CO_2^{\bullet}$. Under other conditions, $(CH_3)_3CO^{\bullet}$ can react with the solvent to generate solvent-derived peroxyls, e.g. $SO_2^{\bullet}$, and the $(CH_3)_3CO_2^{\bullet}$ radicals are then destroyed by reaction with these radicals. The overall rate constant, $2k_t$, therefore depends on the reaction conditions and has a maximum when the alkoxyls do not regenerate a peroxyl.

The self-reaction of other tert-alkylperoxyls is more complex because of β-scission of the alkoxyls generated in reaction (71) e.g. 1,1-dimethylbenzylperoxyl decomposes to give acetophenone and methyl, with the latter reacting with O_2 to give $CH_3O_2^{\bullet}$:

$$C_6H_5C(CH_3)_2O^{\bullet} \xrightarrow{O_2} C_6H_5C(O)CH_3 + CH_3O_2^{\bullet} \quad (72)$$

In this case, $2k_t$ is larger than $2k_t'$ because of the involvement of $CH_3O_2^{\bullet}$ in the termination.

The self-reaction of 1,1-diphenylethylperoxyl is even more complex because 1,1-diphenylethyloxyl undergoes a phenyl shift and the resulting carbon-centred radical reacts with O_2 to give 1-phenyl-1-phenoxylethylperoxyl:

$$(C_6H_5)_2C(CH_3)O^{\bullet} \xrightarrow{O_2} C_6H_5\overset{O_2^{\bullet}}{\overset{|}{C}}(CH_3)OC_6H_5 \quad (73)$$

This peroxyl self-reacts to give 1-phenyl-1-phenoxylethyloxyl, which decomposes to acetophenone and phenoxyl:

$$C_6H_5\overset{O^{\bullet}}{\overset{|}{C}}(CH_3)OC_6H_5 \longrightarrow C_6H_5C(O)CH_3 + C_6H_5O^{\bullet} \quad (74)$$

The phenoxyl radical can react with a peroxyl in the system or abstract a hydrogen atom to give phenol. In either case, $[RO_2^{\bullet}]_{ss}$ decreases and $d[O_2]/dt$ slows down. The overall termination rate constant for 1,1-diphenylethylperoxyl is therefore complex.

It is widely believed that self-reaction of primary and sec-peroxyls is more efficient than self-reaction of tert-peroxyls, from 173 K to ambient temperatures, because they have an α–H atom and can react by a non-radical, six-centre 1,5-H-atom shift (**10**) (i.e. the Russell mechanism):

$$2\,\underset{\text{H}}{\overset{|}{-\text{C}}}\text{O}_2^{\bullet} \underset{k_{-75a}}{\overset{k_{75a}}{\rightleftharpoons}} -\underset{\text{H}}{\overset{|}{\text{C}}}\text{OOOO}\underset{\text{H}}{\overset{|}{\text{C}}}- \;\underset{k_{75b}}{\overset{k_{-75b}}{\rightleftharpoons}}\; \mathbf{10} \xrightarrow{k_{75c}} -\overset{|}{\text{C}}=\text{O} + \text{O}_2 + -\overset{|}{\text{C}}\text{HOH} \qquad (75)$$

10

The main experimental evidence for this mechanism is: (i) the formation of equal yields of alcohol and ketone from self-reaction of sec-peroxyls, whereas if alkoxyls are formed then more alcohol than ketone would be expected; (ii) singlet (both $^1\Delta$ and $^1\Sigma$) oxygen and triplet ketone are generated while tert-peroxyls give ground-state oxygen; (iii) there is an isotope effect when the α–H atom is replaced by deuterium, although this evidence may be discounted since Mendenhall and Quinga [66] have found a deuterium-isotope effect of the same magnitude as that observed for peroxyls for the dismutation of caged alkoxyls.

Bennett and Summers [67] have proposed an alternative non-radical mechanism, involving the cyclic transition state **11**, for self-reaction of sec-peroxyls. Thus although equal yields of alcohol and ketone are produced at ambient temperatures, the alcohol-to-ketone ratio decreases as the temperature is reduced to a value as low as 0.06 at 173 K. Furthermore, significant yields of hydrogen peroxide are produced:

$$2\text{R}\overset{|}{\text{C}}\text{HO}_2^{\bullet} \rightleftharpoons \mathbf{11} \longrightarrow 2\text{R}\overset{|}{\text{C}}=\text{O} + \text{H}_2\text{O}_2$$

11

In 1980, Benson and coworkers [68] concluded that the thermochemistry of the self-reaction of primary and sec-peroxyls is not compatible with the cyclic transition state **10**. For sec-peroxyls the reaction to give ketone, alcohol and oxygen is exothermic enough (414 kJ mol^{-1}) to give electronically excited products, but the activation energy of < 8.4 kJ mol^{-1} for decomposition of the cyclic intermediate appears to be too low. They proposed the following chain mechanism, which involves production of a Criegee zwitterion, to account for the large values of $2k_t$ for alkylperoxyls containing α–H atoms:

Self-reaction

$$2RCH_2O_2^{\cdot} \longrightarrow RCH_2OOH + RCH{=}O^{+}{-}O^{-} \tag{76}$$

Chain reaction

$$RCHOO + RCH_2O_2^{\cdot} \longrightarrow [RCHO + RCH_2O_3^{\cdot}] \longrightarrow RCHO + RCH_2O + O_2 \tag{77}$$

Termination

$$RCH_2O^{\cdot} + RCH_2O_2^{\cdot} \longrightarrow RCH_2OH + RCHOO \tag{78}$$

However, experiments designed to trap the zwitterion have failed, e.g. the self-reaction of 1-tetralylperoxyls does not give α-hydroperoxytetralyl methyl ether in methanol, 1-chloro-1-hydroperoxytetralin in the presence of LiCl, or the ozonide in the presence of benzaldehyde. In addition, the value of $2k_t$ is not influenced by a change of solvent from the non-polar tetralin to the polar acetonitrile [69].

A milestone in our understanding of peroxyl chemistry was the discovery by Bartlett and Guaraldi [70] that $(CH_3)_3CO_2^{\cdot}$ exists in thermal equilibrium with di-tert-butyltetroxide:

$$2(CH_3)_3CO_2^{\cdot} \rightleftharpoons (CH_3)_3COOOOC(CH_3)_3 \tag{79}$$

Below 193 K, $(CH_3)_3CO_2^{\cdot}$ is so persistant that its concentration can be increased and decreased reversibly by raising and lowering the temperature with no irreversible loss of the radical. The thermodynamic parameters for this equilibrium are $\Delta S^{\circ}_{79} = -126$ to -142 J deg^{-1}mol^{-1} and $\Delta H^{\circ}_{79} = -33.5$ to -37 kJ mol^{-1}. It has subsequently been shown that other tert- and sec-peroxyls exist in reversible equilibrium with their tetroxide at temperatures below 190 K. Furthermore, within the limits of experimental error, the magnitudes of ΔS° and ΔH° are not influenced by the structure of R [15]. Irreversible radical decay before the tetroxide has completely dissociated, however, makes the accurate determination of ΔS° and ΔH° difficult. The reversible equilibrium between primary peroxyls and their corresponding tetroxides has not been demonstrated experimentally because these peroxyls decay irreversibly at much too fast a rate. Below 200 K, the rates of decay of sec-peroxyls slow down faster than would be predicted from the Arrhenius equation, because of the influence of the tetroxide on the radical concentration.

Above 158 K, tert-peroxyls decay irreversibly because the tetroxide decomposes to give two caged alkoxyls and oxygen, and a fraction of the alkoxyls combine in the solvent cage to give di-tert-alkyl peroxide:

$$\mathrm{ROOOOR} \longrightarrow [\mathrm{RO^{\bullet}\,O_2\,{}^{\bullet}OR}]_{\mathrm{cage}} \quad (80)$$

$$[\mathrm{RO^{\bullet}\,O_2\,{}^{\bullet}OR}]_{\mathrm{cage}} \longrightarrow \mathrm{ROOR} + \mathrm{O_2} \quad (81)$$

The tert-alkoxyls that escape the solvent cage:

$$[\mathrm{RO^{\bullet}\,O_2\,{}^{\bullet}OR}]_{\mathrm{cage}} \longrightarrow 2\mathrm{RO^{\bullet}} + \mathrm{O_2} \quad (82)$$

either react with the solvent to give, after reaction with O_2, solvent-derived peroxyls, or with a tert-peroxyl to give di-tert-alkyl trioxide:

$$\mathrm{RO^{\bullet}} + \mathrm{SH} \xrightarrow{\mathrm{O_2}} \mathrm{ROH} + \mathrm{SO_2^{\bullet}} \quad (83)$$

$$\mathrm{RO^{\bullet}} + \mathrm{RO_2^{\bullet}} \longrightarrow \mathrm{ROOOR} \quad (84)$$

The ratio of tert-alkoxyl radicals that escape the solvent cage to those that combine in the cage depends on the temperature and solvent viscocity. This reaction pathway plays an increasing role in the self-reaction of primary and sec-peroxyls as the temperature is increased above ambient. Thus, although sec-butylperoxyls do not give di-sec-butyl peroxide at 303 K [11], production of ethyl formate from self-reaction of 1-ethoxyethylperoxyl, benzaldehyde from 1,2-diphenylethylperoxyls, and di-1-tetralylperoxide from tetralyl-peroxyls is indicative of the intermediacy of the alkoxyls [69, 71].

4.1.2 Acyl- and aroyl peroxyls

Even though these peroxyls do not have an α–H atom, the rate constants for their self-reaction are large, e.g. $2k_t = 7.5 \times 10^6\ \mathrm{M^{-1}s^{-1}}$ for decanoylperoxyl, and $2k_t = 2.1 \times 10^8\ \mathrm{M^{-1}s^{-1}}$ for benzoylperoxyl. These radicals probably combine to give acyl and aroyl tetroxides which are believed to decompose via a six-centred cyclic transition state to give di-acyl and di-aroyl peroxides and O_2 [4]:

$$\mathrm{RC(=O)O{-}O{-}O{-}OC(=O)R} \longrightarrow \mathrm{RC(=O)OOC(=O)R} + \mathrm{O_2} \quad (85)$$

4.2 REACTIONS WITH OTHER RADICALS

4.2.1 Alkyls

The reaction of peroxyls with alkyls to give dialkyl peroxides (reaction (86)), is important in autoxidations under conditions where the concentration of O_2

in solution is too low to scavenge all alkyls:

$$RO_2^{\cdot} + R^{\cdot} \longrightarrow ROOR \quad (86)$$

The rate constants for this reaction are close to the diffusion-controlled limit, which has the effect of reducing $[RO_2^{\cdot}]_{ss}$ and $-d[O_2]/dt$. The involvement of this reaction in autoxidation depends on the resonance stabilization energy of $R^{\cdot}$, since the higher this energy then the more readily $RO_2^{\cdot}$ decomposes to $R^{\cdot}$ and O_2 (reaction (−2)). This is the reason why triphenylmethane, which loses a hydrogen atom to give the triphenylmethyl species, autoxidizes to give di-triphenylmethyl peroxide even at one atmosphere pressure of oxygen. Furthermore, triphenylmethyl hydroperoxide inhibits aralkane autoxidation because the chain-transfer reaction (14) gives triphenylmethylperoxyl which decomposes to give triphenylmethyl.

4.2.2 Alkoxyls and thiyls

We have seen above that peroxyls combine with alkoxyls to give dialkyl trioxides (reaction (84)). These trioxides are very unstable and readily decompose back to $RO_2^{\cdot}$ and $RO^{\cdot}$ (reaction (−84)). At low temperatures, this reaction can complicate KESR studies of peroxyl decay. It is also the reason why there is sometimes a momentary increase in $[RO_2^{\cdot}]_{ss}$, for photochemically generated tert-alkylperoxyls, when the light is switched off. In an interesting experiment, reaction (84) has been used to prepare $(CH_3)_3CO_2^{\cdot}$ specifically labelled with ^{17}O on the terminal oxygen. Thus if di-tert-butyl peroxide in cyclopentane (RH) saturated with ^{17}O-enriched O_2 is photolyzed at 183 K the cyclopentylperoxyls, $RO^{17}O$, $R^{17}OO$, and $R^{17}O^{17}O$, are generated. These radicals disappear completely when the light is switched off. On warming to 163 K the persistant $(CH_3)_3CO^{17}O$ is produced by the themolysis of $RO^{17}OOC(CH_3)_3$ [72].

The tert-butylperoxyl radical reacts efficiently with alkyl- and arylthiyls to give sulfanyl peroxides, compounds that are analogous to trioxides [73]:

$$(CH_3)_3CO_2^{\cdot} + RS^{\cdot} \rightleftharpoons (CH_3)_3COOSR \quad (87)$$

As with trioxides, there is a post-irradiation growth of $(CH_3)_3CO_2^{\cdot}$.

4.2.3 Phenoxyls

Alkylperoxyls add to 2,4,6-trialkylphenoxyls to give peroxycyclohexadienones, e.g. 2,6-di-tert-butyl-4-methylphenoxyl gives 2,6-di-tert-butyl-4-methyl-4-alkylperoxy-2,5-cyclohexadien-1-one, **12**, and 2,4-di-tert-butyl-6-methylphenoxyl gives the 4- and 6-adducts, **13** and **14**, respectively:

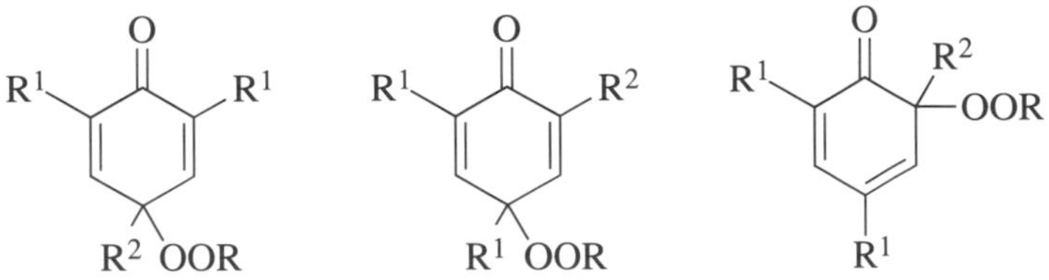

where R^1 is $C(CH_3)_3$ and R^2 is CH_3.

The reactions of $RO_2^{\bullet}$ with 2,4- and 2,6-dialkylphenoxyls gives substituted quinones because the alkylperoxy adducts are unstable, for example:

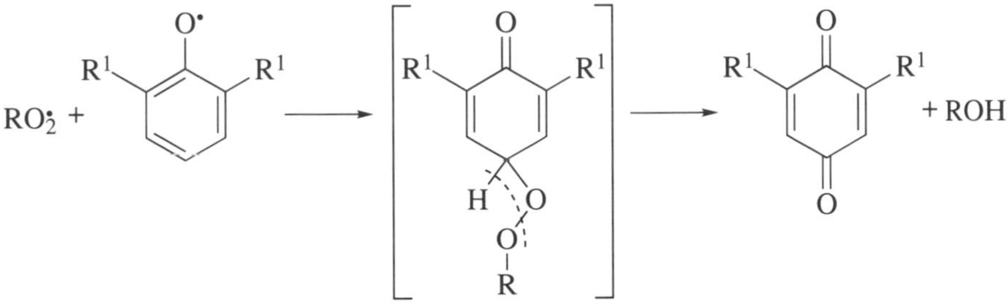

$$RO_2^{\bullet} + \text{(phenoxyl)} \longrightarrow [\text{adduct}] \longrightarrow \text{(quinone)} + ROH$$

where R^1 is $C(CH_3)_3$.

Ten products have been identified from the reaction of tert-butylperoxyls with 2,4-di-tert-butylphenoxyl. This is because the initial adducts are unstable and phenoxyl coupling occurs. Reaction of the coupled products with tert-butylperoxyls adds to the complexity of the reaction products.

1-Cyano-1,3-dimethylbutylperoxyl reacts with *d*-α-tocopherol in oxygen-saturated acetonitrile and hexane at 323 K to give 8a-(1-cyano-1,3-dimethylbutylperoxyl) α-tocopherone, **15**, 4a,5-epoxy-8a-hydroperoxy α-tocopherone, **16**, and 7,8-epoxy-8a-hydroperoxy α-tocopherone, **17** [74]. These three products account for at least 96% of the products formed. Four diastereoisomers of **15** are formed in similar amounts, thus indicating a rapid addition of $RO_2^{\bullet}$ to the tocopheroxyl:

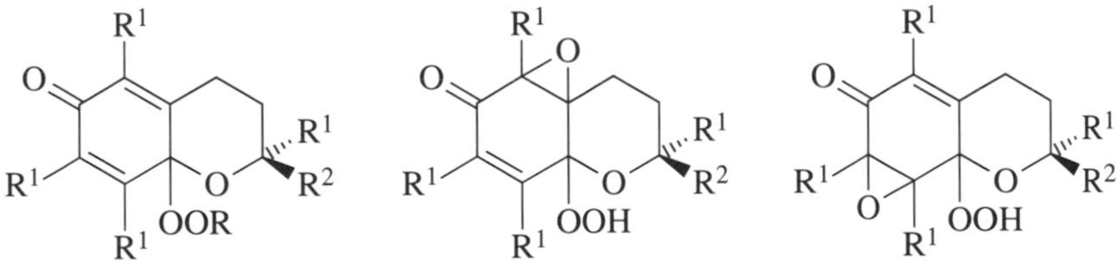

where R^1 is CH_3, R^2 is phytyl, ($C_{16}H_{33}$), and R is $C(CN)(CH_3)CH_2CH(CH_3)_2$.

There have been several determinations of the rate constants for the reactions of $RO_2^{\bullet}$ with phenoxyls [15], e.g. it has been estimated that $(CH_3)_2CHCH_2C(CH_3)(CN)O_2^{\bullet}$ adds to α-tocopheroxyl with a rate constant of 4.3×10^8 $M^{-1}s^{-1}$ in chlorobenzene at 318 K [75].

4.2.4 Aminyls

If the incipient aminyl, produced by reaction of a peroxyl with an aromatic amine (reaction (38)), is comparatively unreactive towards propagation and chain transfer it will react with a second peroxyl to give non-radical products (reaction (88)), or to give an alkoxyl and an aminoxyl by an oxygen-atom-transfer reaction (89):

$$RO_2^{\bullet} + \rangle N^{\bullet} \longrightarrow \text{non-radical products} \quad (88)$$

$$RO_2^{\bullet} + \rangle N^{\bullet} \longrightarrow RO^{\bullet} + \rangle NO^{\bullet} \quad (89)$$

with the relative importance of these two reactions depending on the structure of the peroxyl and aminyl species. Reaction (89) is, of course, a chain-transfer reaction because the alkoxyl will propagate the chain by reaction with the substrate. The occurance of reaction (89) makes inhibition of autoxidation by aromatic amines kinetically more complex than inhibition by phenols.

With regards to the influence of the structure of $RO_2^{\bullet}$ and the secondary aminyl on reaction (89), primary and sec-peroxyls are not as efficient as tert-peroxyls at converting an aminyl to its aminoxyl. This is because caged primary and sec-alkoxyls and aminoxyls can disproportionate to give a carbonyl compound and a hydroxylamine:

$$-\overset{|}{\underset{H}{C}}O_2^{\bullet} + {}^{\bullet}N(C_6H_5)_2 \rightleftharpoons -\overset{|}{\underset{H}{C}}OON(C_6H_5)_2 \quad (90)$$

$$\longrightarrow \left[-\overset{|}{\underset{H}{C}}O^{\bullet} + {}^{\bullet}ON(C_6H_5)_2\right]_{cage}$$

$$\swarrow \qquad \searrow$$

$$-\overset{|}{C}{=}O + HON(C_6H_5)_2 \qquad -\overset{|}{\underset{H}{C}}O^{\bullet} + {}^{\bullet}ON(C_6H_5)_2 \quad (91)$$

Diphenylaminyl produces large concentrations of diphenylaminoxyl because it is comparatively stable, whereas *N*-methylphenylaminyl produces low concentrations of *N*-methylphenylaminoxyl, even though the aminoxyl is produced efficiently, because it is very reactive. Only small amounts of *N*-phenylnaphthylaminyl are converted to aminoxyl because peroxyls add to the naphthyl ring rather than to the nitrogen atom. An aminoxyl is not formed from *N*, *N'*-diphenyl-*p*-phenylenediamine because the second hydrogen atom is removed in reaction (92) to give *N,N'*-diphenyl-*p*-quinonediimine as a major reaction product:

$$RO_2^{\bullet} + C_6H_5N(H)C_6H_4\dot{N}C_6H_5 \longrightarrow ROOH + C_6H_5N{=}C_6H_4{=}NC_6H_5 \quad (92)$$

Peroxyls react with primary aromatic aminyls to give mainly an alcohol and an aromatic nitroso compound:

$$RO_2^{\bullet} + C_6H_5\dot{N}H \longrightarrow \left[ROON(H)C_6H_5\right] \longrightarrow ROH + C_6H_5NO \quad (93)$$

4.2.5 Aminoxyls

As we have seen above, aminoxyls can be generated in hydrocarbon autoxidations which are inhibited by aromatic amines via reaction (89). Being free radicals, these transients might be expected to react with peroxyls and it has been shown that reaction does occur with aromatic but not with alkyl aminoxyls, e.g. 4,4'-dimethoxydiphenyl nitroxide inhibits the autoxidation of styrene, with $k_{inh} \approx 4 \times 10^3$ $M^{-1}s^{-1}$ [15]. This suggests that the peroxyl adds to the aromatic ring, for example:

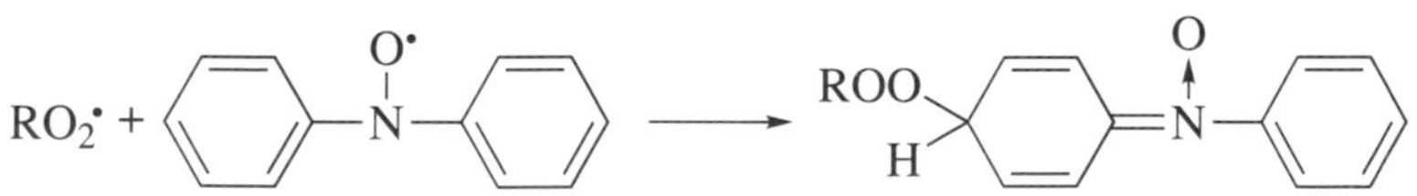

$$RO_2^{\bullet} + C_6H_5N(O^{\bullet})C_6H_5 \longrightarrow (ROO)(H)C_6H_5{=}N(\rightarrow O)C_6H_5 \quad (94)$$

5 UNIMOLECULAR DECOMPOSITION

All peroxyls decompose to give a carbon-centred radical and oxygen, i.e. the reverse of reaction (2), with the importance of this reaction depending on the strength of the bond formed between the radical $R^{\bullet}$ and oxygen and the reaction temperature:

$$RO_2^{\bullet} \longrightarrow R^{\bullet} + O_2 \quad (-2)$$

Typical bond strengths are 197 kJ mol^{-1} for H–OO, 109–121 kJ mol^{-1} for alkylperoxyls, and 54–63 kJ mol^{-1} for aralkylperoxyls and allylperoxyls [4]. The tert-alkylperoxyl $(CH_3)_3CO_2^{\cdot}$ does not decompose below 420 K, while $C_6H_5C(CH_3)_2O_2^{\cdot}$ and 1-tetralylperoxyl decompose slowly at 303 K. The rate constants for some examples of reaction (−2) [15] are listed in Table 20.

Table 20 Rate constants for β-scission of alkylperoxyls at 303 K

Peroxyl	k_{-2} (s^{-1})
1,1-Dimethylbenzyl	2
1,1-Diphenylethyl	1700
13-Peroxylinoleic acid	144
15-Peroxyaracidonic acid	152

Triphenylmethyl hydroperoxide inhibits the autoxidation of tetralin and 9,10-dihydroanthracene because the peroxyls derived from these hydrocarbons abstract the hydroperoxidic hydrogen to generate triphenylmethylperoxyl:

$$RO_2^{\cdot} + (C_6H_5)_3COOH \longrightarrow ROOH + (C_6H_5)_3CO_2^{\cdot} \quad (95)$$

Triphenylmethylperoxyl decomposes to give triphenylmethyl and oxygen:

$$(C_6H_5)_3CO_2^{\cdot} \longrightarrow (C_6H_5)_3C^{\cdot} + O_2 \quad (96)$$

and triphenylmethyl efficiently traps a chain-propagating alkylperoxyl:

$$(C_6H_5)_3C^{\cdot} + RO_2^{\cdot} \longrightarrow (C_6H_5)_3COOR \quad (97)$$

Aralkanes inhibit the high-temperature autoxidation of alkanes because the reaction of the aralkyl with O_2 is reversible and an aralkyl can terminate the chain reaction by reacting with an alkylperoxyl. Cyclohex-1,4-diene autoxidizes to give benzene and hydrogen peroxide because cyclohexadienyl reacts with O_2 to give an unstable peroxyl which decomposes to give the aromatic compound and $HO_2^{\cdot}$. The driving force of this sequence of reactions is the formation of the aromatic ring. In the case of 1,4-dihydronaphthalene the chain is propagated by 1,4-dihydronaphth-2-peroxyl, 1,4-dihydronaphth-3-peroxyl and hydroperoxyl, with the relative concentrations of each being dependent on the reaction conditions.

It has been known for some time that some unsaturated compounds autoxidize to give a complex mixture of hydroperoxides because of allylperoxyl rearrangement (reaction (20)). Furthermore, hydroperoxides formed when alkenes react with ¹Δ oxygen rearrange upon standing by a mechanism that involves the intermediacy of the allylperoxyls. For example, 5α-peroxy-3β-hydroxycholest-6-ene, which is formed by abstraction of the hydroperoxidic hydrogen from 5α-hydroperoxy-3β-hydroxycholest-6-ene, i.e. the product of reaction of cholesterol with ¹Δ oxygen, rearranges in non-polar solvents to give 7α-peroxy-3β-hydroxycholest-5-ene [76]:

The simplest mechanism that would account for allylperoxyl isomerization is unimolecular allylperoxyl decomposition, isomerization of the allyl (reaction (23)), and the readdition of O_2. Two pieces of experimental evidence do, however, argue against this mechanism: (i) retention of stereochemistry, e.g. optically pure *trans*-allylperoxyls derived from methyl oleate rearrange in a highly stereoselective process, and (ii) minimal incorporation of atmospheric oxygen, e.g. the latter is not incorporated into the hydroperoxide function when the Δ^8-10- and Δ^{10}-9-hydroperoxides produced by autoxidation of oleic acid are allowed to interconvert in the presence of $^{18}O_2$ or in the rearrangement (reaction (30)) [77]. It was suggested that allylperoxyl rearrangement proceeds through the 1,2-dioxolan-3-yl (**18**) produced by cyclization of the allylperoxyl [78]:

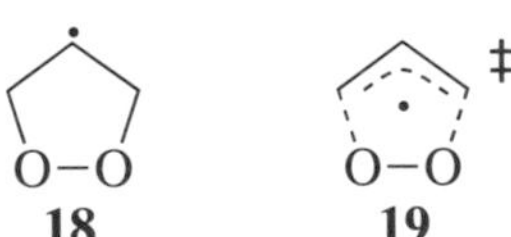

However, this cyclic carbon-centred radical has been discounted, as there is no evidence for ring-opening β-scission when the radical centre is next to a small strained ring and it is not trapped by oxygen to give peroxyl [79]. These observations prompted Porter and coworkers [80] to suggest the alternative cyclic transition state **19**.

In support of the dissociative mechanism it was also found that 7α-hydroperoxy-3β-hydroxycholest-5-ene epimerizes to give 7β-hydroperoxy-3β-hydroxycholest-5-ene via the intermediacy of the peroxyl, with atmospheric $^{18}O_2$ being incorporated into the hydroperoxy group [77]. In addition, Mills *et al.* [80] have reported that optically pure *cis*-allylperoxyls rearrange with stereoselectivity and some incorporation of labelled oxygen, and that these processes depend on the solvent viscocity. Theoretical calculations also support a dissociative mechanism over a concerted transition state.

Caldwell and Porter [81] have very recently reported the first unequivocal evidence that alkylperoxyls decompose and recombine within the solvent cage. Thus 1,1-dimethylbenzylperoxyl, specifically labelled with $^{18}O_2$ in the terminal position, undergoes internal exchange with the proximal oxygen:

$$RO^{18}O^{\bullet} \rightleftharpoons \left[R^{\bullet}\ \ O^{18}O\right] \rightleftharpoons R^{18}OO^{\bullet}$$

REFERENCES

1. F.R. Mayo, *Acc. Chem. Res.* **1**, 193 (1968).
2. K.U. Ingold, *Metal Catalalysis and Lipid Oxidation*, Discussion Papers of SIK Symposium (1968), 11; **71**.
3. L.R. Mahoney, *Angew. Chem. Int. Ed. Engl.* **8**, 547 (1969).
4. K.U. Ingold, *Acc. Chem. Res.* **2**, 1 (1969).
5. K. Adamic, D.F. Bowman and K.U. Ingold, *J. Am. Oil Chem. Soc.* **47**, 109 (1970).
6. J.A. Howard, *Adv. Free Radical Chem.* **4**, 49 (1972).
7. J. Betts, *Q. Rev. Chem. Soc.* **25**, 265 (1971).
8. J.A. Howard, in *Free Radicals*, edited by J.K. Kochi, Vol. 2, John Wiley & Sons, New York (1973) p. 16.
9. J.A. Howard, *Rubber Chem. Technol.* **47**, 976 (1974).
10. D.G. Hendry, T. Mill, J.A. Howard, L. Piszkiewicz and H.K. Eigenmann, *J. Phys. Chem. Ref. Data* **3**, 937 (1974).
11. J.A. Howard, *ACS Symp. Ser.* **69**, 413, (1978).
12. S.W. Benson and P.S. Nangia, *Acc. Chem. Res.* **12**, 223 (1979).
13. J.A. Howard, in *Frontiers of Free Radical Chemistry*, edited by W.A. Pryor, Academic Press, New York (1980), p. 237.
14. J.A. Howard, *Isr. J. Chem.* **24**, 33 (1984).
15. J.A. Howard and J.C. Scaiano, in *Radical Reaction Rates in Liquids*, edited by H. Fischer, Landolt-Börnstein, New Series, vol. 13d, Springer Verlag, Berlin (1984).
16. J.A. Howard, *Rev. Chem. Intermed.* **5**, 1 (1984).
17. N.A. Porter, *Acc. Chem. Res.* **19**, 262 (1986).
18. G.W. Burton and K.U. Ingold, *Acc. Chem. Res.* **19**, 194 (1986).
19. P. Neta, R.E. Huie and A.B. Ross, *J. Phys. Chem. Ref. Data* **19**, 413 (1990).
20. V.A. Belyakov, G. Lauterbach, W. Pritzkow and V. Voerckel, *J. Prakt. Chem.* **334**, 373 (1992).
21. G. Scott (Ed.), *Atmospheric Oxidation and Antioxidants*, Vols 1–3, 2nd Ed., Elsevier, Amsterdam (1993).
22. C.E. Boozer, G.S. Hammond, C.E. Hamilton and J.N. Sen, *J. Am. Chem. Soc.* **77**, 3238 (1955).
23. L.R. Mahoney, S. Korcek, S. Hoffman and P.A. Willermet, *Ind. Eng. Chem. Prod. Res. Dev.* **17**, 250 (1978).
24. G.W. Burton, A. Joyce and K.U. Ingold, *Arch. Biochem. Biophys.* **221**, 281 (1983).
25. G.M. Burnett and H.W. Melville, in *Techniques of Organic Chemistry,* edited by

S.L. Freiss, E.S. Lewis and A. Weissberger, Vol. 8, Part 2, John Wiley & Sons, New York (1963), Ch. 20, p. 1107.
26. C.H. Bamford and M.J.S. Dewar, *Proc. R. Soc. London A* **198**, 252 (1949).
27. J.A. Howard, in *Magnetic Properties of Free Radicals*, edited by H. Fischer and H.-K. Hellwege, Landolt-Börnstein, New Series, Vol. 9c2, Springer Verlag, Berlin (1979), p. 5' J.A. Howard in *Magnetic Properties of Free Radicals*, edited by H.Fischer and H.-K. Hellwege, Landolt-Börnstein, New Series, Vol. 17e, Springer Verlag, Berlin (1988), p.5.
28. Ya.S. Lebedev, V.F. Tsepalov and V.Ya. Shlyapintokh, *Dokl. Akad. Nauk SSSR* **139**, 1409 (1961).
29. H.W. Melville and S. Richards, *J. Chem. Soc.* 944 (1954).
30. J.E. Bennett and R. Summers, *J. Chem. Soc. Faraday Trans. 2,* **69**, 1043 (1973).
31. J.H.B. Chenier, E. Furimsky and J.A. Howard, *Can. J. Chem.* **52**, 3682 (1974).
32. J.E. Bennett, *J. Chem. Soc. Faraday Trans. 1* **83**, 1805 (1987).
33. J.R. Thomas and C.A. Tolman, *J. Am. Chem. Soc.* **84**, 4872 (1967).
34. B.S. Middleton and K.U. Ingold, *Can. J. Chem.* **45**, 191 (1967).
35. G.A. Russell, *J. Am. Chem. Soc.* **77**, 4583 (1955).
36. G.A. Russell, in *Free-Radicals*, edited by J.K. Kochi, Vol. 1, John Wiley & Sons, New York (1973), p. 275.
37. R. Harnisch, G. Lauterbach and W. Pritzkow, *J. Prakt. Chem.* **337**, 60 (1995).
38. J. Igarashi, R.K. Jensen, J. Lusztyk, S. Korcek and K.U. Ingold, *J. Am. Chem. Soc.* **114**, 7727 (1992).
39. J.A. Howard and K.U. Ingold, *Can. J. Chem.* **45**, 785 (1967).
40. D. Griller, J.A. Howard, P.R. Marriott and J.C. Scaiano, *J. Am. Chem. Soc.* **103**, 619 (1981).
41. J.R. Thomas, *J. Am. Chem. Soc.* **85**, 593 (1963).
42. J.A. Howard and T. Yamada, *J. Am. Chem. Soc.* **103**, 7102 (1981).
43. A.A. Zavitsas and C. Chatgilialoglu, *J. Am. Chem. Soc.* **117**, 10645 (1995).
44. D.D.M. Wayner, E. Lusztyk, D. Page, K.U. Ingold, P. Mulder, L.J.J. Laarhoven and H.S. Aldrich, *J. Am. Chem. Soc.* **117**, 8737(1995).
45. J.H.B. Chenier, J.-P. Tremblay and J.A. Howard, *J. Am. Chem. Soc.* **97**, 1618 (1975).
46. J.A. Howard, J.H.B. Chenier and D.A. Holden, *Can. J. Chem.* **55**, 1463 (1977).
47. R.K. Jensen, S. Korcek, M. Zinbo and J.L. Gerlock, *J. Org. Chem.* **60**, 5396 (1995).
48. J.E. Bennett, G. Bunton, A.R. Forrester and J.D. Fullerton, *J. Chem. Soc. Perkin Trans. 2* 1477 (1983).
49. R.K. Jensen, S. Korcek, L.R. Mahoney and M. Zinbo, M. *J. Am. Chem. Soc.* **101**, 7574 (1979).
50. J.A. Howard and J.H.B. Chenier, unpublished results.
51. J.L. Courtneidge, *J. Chem. Soc. Chem. Commun.* 1270 (1992).
52. A.G. Davies and B.P. Roberts, in *Free Radicals,* edited by J.K. Kochi, Vol. 1, John Wiley & Sons, New York (1973), p. 547.
53. A.G. Davies, K.U. Ingold, B.P. Roberts and R. Tudor, *J. Chem. Soc. B* 698 (1971).
54. S. Korcek, G.B. Watts and K.U. Ingold, *J. Chem. Soc. Perkin Trans. 2* 242 (1972).
55. J.A. Howard and S.B. Tong, *Can. J. Chem.* **58**, 1962 (1980).
56. A. Tkác, K. Vesely and L. Omelka, *J. Phys. Chem.* **75**, 2575 (1971); A. Tkác, K. Vesely and L. Omelka, *J. Phys. Chem.* **75**, 2580 (1971).
57. K.U. Ingold, *J. Phys. Chem.* **76**, 1385 (1972).
58. J.A. Howard, J.C. Tait, Y. Yamada and J.H.B. Chenier, *Can. J. Chem.* **59**, 2184 (1981).

59. J.A. Howard and J.C. Tait, *Can. J. Chem.* **56**, 176 (1978).
60. E.G. Jansen, P.H. Krygsman, D.A. Linday and D.L. Haire, *J. Am. Chem. Soc.* **112**, 8279 (1990).
61. E. Niki, S. Yokoi, J. Tsuchiya and Y. Kamiya, *J. Am. Chem. Soc.* **105**, 1498 (1983).
62. C. Chatgilialoglu, J.A. Howard and K.U. Ingold, *J. Org. Chem.* **47**, 4361 (1982).
63. L.R.C. Barclay, K.A. Dakin, and H.A. Zahalka, *Can. J. Chem.* **70**, 2148 (1992).
64. R.C. Mordi, J.C. Walton, G.W. Burton, L. Hughes, K.U. Ingold, D.A. Lindsay and D.J. Moffatt, *Tetrahedron* **49**, 911 (1993).
65. V.W. Bowry, *J. Org. Chem.* **69**, 2200 (1994).
66. G.D. Mendenhall and E.M.Y. Quinga, *Int. J. Chem. Kinet.* **17**, 1187, (1985).
67. J.E. Bennett and R. Summers, *Can. J. Chem.* **52**, 1377 (1974).
68. P.S. Nangia and S.W. Benson, *Int. J. Chem. Kinet.* **12**, 43 (1980).
69. A. Baignee, J.H.B. Chenier and J.A. Howard, *Can. J. Chem.* **61**, 2037 (1983).
70. P.D. Bartlett and G. Guaraldi, *J. Am. Chem. Soc.* **89**, 4799 (1967).
71. J.A. Howard and K.U. Ingold, *Can. J. Chem.* **48**, 873 (1970).
72. J.A. Howard, *Can. J. Chem.* **50**, 1981 (1973).
73. B. Mile, C.C. Rowlands, P.D. Sillman and A.J. Holmes, *J. Chem. Soc. Perkin Trans. 2* 2141 (1995).
74. D.C. Liebler, P.F. Baker and K.L. Kaysen, *J. Am. Chem. Soc.* **112**, 6995 (1990).
75. V.W. Bowry and K.U. Ingold, *J. Org. Chem.* **60**, 5456 (1995).
76. C.H. Schiesser and H. Wu, *Aust. J. Chem.* **46**, 1437 (1993).
77. N.A. Porter and J.S. Wujek, *J. Org. Chem.* **52**, 5085 (1987).
78. A.L. Beckwith, A.G. Davies, I.G.E. Davison, A. Maccoll and M. Mruzek, *J. Chem. Soc. Perkin Trans. 2* 815 (1989).
79. N.A. Porter and P. Zuraw, *J. Chem. Soc. Chem. Commun.* 1472 (1985).
80. K.A. Mills, S.E. Caldwell, G.R. Dubay and N.A. Porter, *J. Am. Chem. Soc.* **114**, 9689 (1992).
81. S.E. Caldwell and N.A. Porter, *J. Am. Chem. Soc.* **117**, 8676 (1995).

11 Electron Spin Resonance Studies of Peroxyl Radicals in Solid Matrices

CHRISTOPHER J. RHODES
John Moores University, Liverpool, UK

1 INTRODUCTION

I was pleased to be invited by Professor Alfassi to contribute a chapter to his book which deals specifically with peroxyl radicals, since it compelled me to expand my knowledge of the literature on such species; something that I had planned to do for some time. I was, of course, aware of their wide implication and importance, for example in the oxidation of hydrocarbons, in the chemistry of the atmosphere, and in the oxidative degradation of polymers and of lipids. It is, indeed, my growing interest in the involvement of free radicals in biological systems which prompted my interest in peroxyl radicals, particularly in the toxicological action of certain xenobiotics – the classic example being carbon tetrachloride, which destroys liver cells through lipid peroxidation of their membranes, initiated by $CCl_3^{\cdot}$ radicals which are rapidly converted to highly reactive $CCl_3OO^{\cdot}$ peroxyl radicals. Most of the information regarding these species has been gathered directly from electron spin resonance (ESR) spectroscopic studies, as is indeed true of all radicals, and since I am principally an ESR spectroscopist, I shall deal specifically with this aspect of the technique and what provision has been made with regard to structural and mechanistic investigations of peroxyl radicals, focussing my attention, as I was requested to, on their properties in solid matrices.

2 ELECTRON SPIN RESONANCE SPECTROSCOPY

2.1 G-VALUES

The essence of the ESR method [1] involves the irradiation of a sample containing unpaired electrons with microwave radiation, while simultaneously applying an external magnetic field (B). In the absence of the field, the two possible spin states that may be taken by unpaired electrons (denoted by the spin quantum numbers, $m_s = +1/2, -1/2$) have the same

Peroxyl Radicals. Edited by Z.B. Alfassi

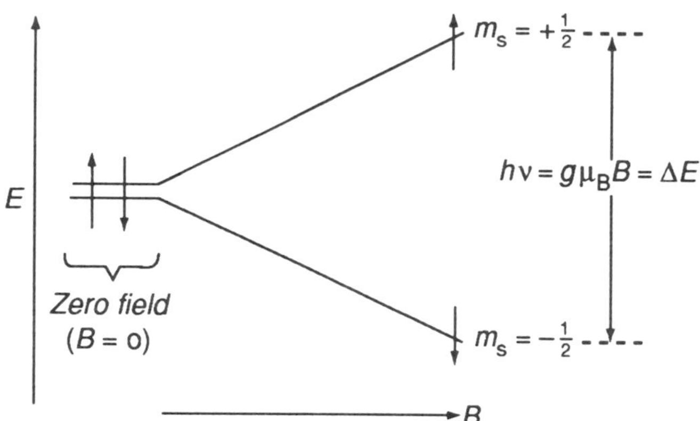

Figure 1 Schematic representation of the energy levels for an unpaired electron in an applied magnetic field *B*.

energy and are equally populated. On the application of a field *B* (Figure 1), this is no longer the case, and there is a net lowering of the energy of the state $m_s = -1/2$, and a corresponding increase in that of $m_s = +1/2$. The energy splitting (E) between the $m_s = +1/2$ and the $m_s = -1/2$ states is a function of *B* and is given by equation (1):

$$\triangle E = h\nu = g\,\mu_B B \tag{1}$$

In this equation there is a dimensionless parameter, *g*, normally called the '*g*-factor', which is structurally diagnostic, and is in fact exactly analogous to the chemical shift in nuclear magnetic resonance (NMR) spectroscopy; it increases with the unpaired electron density on 'heavy atoms', since it arises from spin–orbit coupling, where the mixing of excited states with electronic angular momentum into the ground state gives rise to a secondary field which couples with the magnetic moment of the electron as a result of its spin. For a free electron, $g = 2.0023$, but shifts, with either positive or negative values, from this can occur, depending on whether the unpaired electron orbital is coupled magnetically with a filled or with a vacant orbital, respectively. Since the term *E* in equation (1) depends on the total field experienced, i.e. the sum of the external field *B* and the secondary field (see above), *g* is seen to vary for a given measured value of *B*.

The practical consequence of the above may be illustrated by means of the radical $O_2^{-\bullet}$, in which the π^* manifold is degenerate, as shown in Figure 2. However, differential perturbation of the π^*_x and π^*_y levels by environmental effects such as hydrogen bonding can give rise to a splitting of these energy states. When the radical is oriented with its O—O *z*-axis

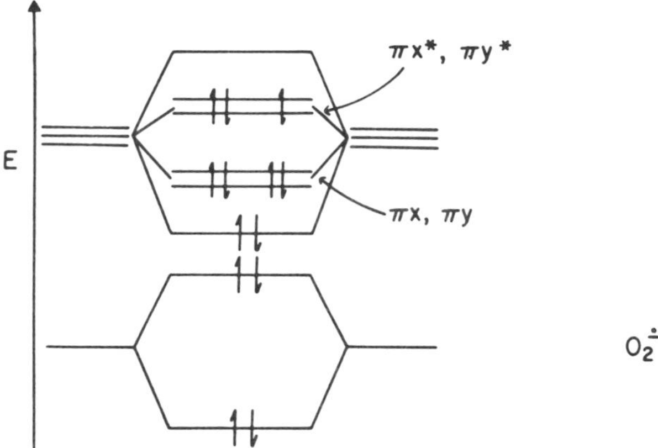

Figure 2 Schematic representation of the degenerate π^*-manifold for the $O_2^{-\cdot}$ radical anion.

(Scheme I) parallel to the applied magnetic field B, electronic angular momentum is induced about that axis, and may be considered to arise from the motion of the unpaired electron in a circular path, flowing from the π^*_x to the π^*_y level, and then back to π^*_x.

x, y, z

$R \blacktriangleright O \dot{-} O$

(I)

The ease of flow depends, therefore, on the splitting (ΔE) which provides a barrier to it. The shift (Δg) for a given orientation of the radical in the applied field B is often dominated by the contribution from a single atom, but depends on the contributions from all atoms (i) in the radical, as given by equation (2):

$$\Delta g_z = \sum_i \rho_i \lambda_i / \Delta E_i \qquad (2)$$

where E_i determines the sign of the shift (alluded to earlier) and is positive in the case of $O_2^{-\cdot}$, because the angular coupling of the singly occupied molecular orbital (SOMO) is with a filled lower-energy orbital, but is negative for coupling with a vacant higher-energy orbital as in $O_2^{+\cdot}$; λ_i and ρ_i are, respectively, the spin–orbital coupling constant and the spin density for each atom i.

The (isotropic) g-value observed from a rapidly reorienting radical, as in a liquid, differs from free-spin (2.0023) by the average of the shifts corresponding to each molecular Cartesian axis (x,y, and z) (equation (3)):

$$g(\text{isotropic}) = \frac{(\Delta g_x + \Delta g_y + \Delta g_z)}{3} + 2.0023 \quad (3)$$

In the present example, the shift associated with the O—O axis is dominant ($B(z)$), since ΔE is relatively small (it would be zero in the absence of an environmental perturbation), whereas it is larger for ($B(y)(E\pi_x\text{-}E_\sigma)$), thus leading to a weaker coupling, while for $B(x)$ there is essentially no shift since circulation about $B(x)$ does not couple π^*_x with any other orbitals.

Following the above discussion, we can now consider how a peroxyl radical, ROO$^{\cdot}$, might be formed by a *hypothetical* reaction between a proton (or other positively charged atomic centre) and the $O_2^{-\cdot}$ radical-anion unit (Figure 3). If the bonding (with the σ-2p combination) is linear, as shown first, the electrostatic effect on the x- and y-levels in both the bonding and antibonding manifolds is the same. However, an angular bonding interaction, which is shown next, will also involve the π_x levels, so that there is now a differential perturbation of the x- and y-levels in both the bonding and

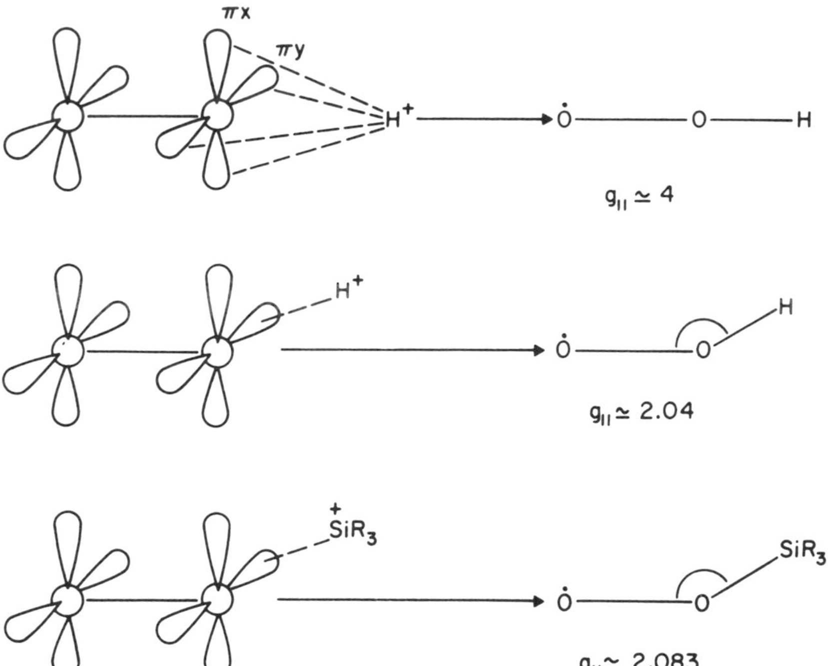

Figure 3 Schematic representation of the (hypothetical) formation of the peroxyl radical (shown as a bonding perturbation of the orbitals in the $O_s^{-\cdot}$ radical anion, in order to explain the differing $g(z)$ shifts in these species (see text).

antibonding manifolds, and the important π^* (x,y) degeneracy is lifted, thus leading to a reduced $g(z)$ shift; a typical value is ca 2.035 compared with the value (g=4) expected for complete degeneracy, while the other values are always close to $g_x = 2.003$ and $g_y = 2.008$.

2.2 HYPERFINE COUPLING

Where it can be observed, hyperfine coupling, the analogue of nuclear spin–spin coupling in NMR spectroscopy, is diagnostically the most useful feature of ESR spectroscopy, since it reveals the nature of magnetic nuclei that are present in a particular paramagnetic species (free radical). In order to appreciate this effect, the simplest starting point is the hydrogen atom H$^{\cdot}$, which consists of a unique proton interacting with a single unpaired electron. We need to consider the $m_s = -1/2$ state of the electron, since the ESR experiment measures the net absorption of energy due to the promotion of electrons from this state to the +1/2 level. Accordingly, the associated proton may take either the $m_p = +1/2$ or $-1/2$ state with respect to this, and as shown in Figure 4, the magnetism due to the intrinsic proton spin may either augment (+1/2) or detract from (−1/2) the applied magnetic field B. This means that each of the electron spin levels is split into two further levels, as shown in the figure. Free radicals are, of course, more complex than hydrogen atoms, and often contain more interacting nuclei. In the case, specifically, of peroxyl radicals, ROO$^{\cdot}$, the magnetic nuclei most usually protons which are present in the R group give rise to further splittings of the x,y, and z, g-components. As a consequence of the low natural abundance of the ^{17}O isotope, couplings from the oxygen atoms themselves are rarely detected, but this can be achieved by reacting the radical R$^{\cdot}$ with ^{17}O-enriched O_2, leading to the isotopomeric radicals

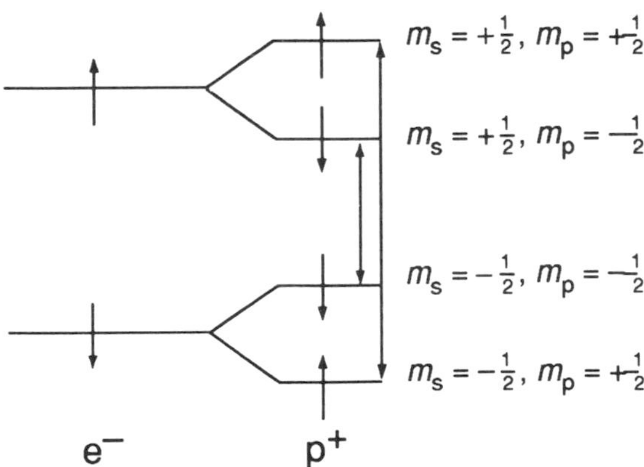

Figure 4 Magnetic energy levels for a radical containing a single proton.

R—^{17}O—O$^{\cdot}$ and R—O—^{17}O$^{\cdot}$ [2]. In this case, the two distinct ^{17}O coupling constants provide a measure of the relative spin-densities on the two oxygen atoms (*vide infra*).

3 STRUCTURAL ASPECTS

Having outlined those details relevant to the ESR spectroscopic detection of peroxyl radicals, we shall now consider the broader issue of their structure, for which a good starting point is an extensive *ab initio* investigation of the peroxyl radicals, $O_2^{-\cdot}$, $HO_2^{\cdot}$, $MeO_2^{\cdot}$, $Me_2CHO_2^{\cdot}$, and $Et_2CHO_2^{\cdot}$ [3]. Of particular interest are their equilibrium geometries, but additional parameters have been calculated, including dipole moments, harmonic vibrational frequencies and isotropic and anisotropic hyperfine couplings. For $HO_2^{\cdot}$, the bond angle, calculated by various methods, is found to be in reasonable agreement with the experimental value of 104.18 ± .1° [4–6], whereas for the hydrocarbon peroxyl radicals the C—O—O bond angles are estimated to be significantly higher, but are in accord with the value (111°) deduced for the peroxyl radical formed in irradiated isotactic polyethylene [7,8].

One further feature of the calculations is their prediction that the spin density is highest on the terminal oxygen atom, and experimental work on carbon peroxyl radicals indicates [2] that the distribution is ca. 70–61% and 30–39% on the terminal and inner oxygen atoms, respectively. A variation is observed in these numbers (Table 1) which follows the trend in electron demand of the R group; electron-withdrawing groups tend to increase the spin density on O–1, in accord with the limiting structure R—O—O$^{\cdot}$, while electron-donating groups render the distribution more symmetrical and so

Table 1 Representative ^{17}O hyperfine couplings for peroxyl radicals in glassy matrices[a]

Radical	^{17}O coupling range	
	O–1	O–2
$RCH_2OO^{\cdot}$	95.4–10.28	58.9–51.3
$R_2C(OH)OO^{\cdot}$	94.8–95.9	58.2–56.8
$R_2CHOO^{\cdot}$	94.4–94.8	59.7–58.7
$R_3COO^{\cdot}$	94.0–95.0	60.3–59.0
Halocarbon–OO$^{\cdot}$	97.5–105.0	55.5–45.2
$RSOO^{\cdot}$	81.0	64.0
$RSO_2OO^{\cdot}$	105.0–106.0	44.8–44.6
$O_2^{-\cdot}$	77.3	77.3

[a] All couplings measured in gauss $G=10^{-4}T$ [2].

favour an increased proportion of $R—O^{+\bullet}—O^-$, with the extreme case being $O_2^{-\bullet}$, which is of necessity symmetrical. The sum of the spin densities is not much changed according to the nature of R, which shows that the electronic effect is one of changing the relative spin distribution within the group, with only a minor delocalization of the spin-density on to the substituent itself. Clearly, this situation is reminiscent of the behaviour of nitroxides, in which the spin density is distributed between the O and N atoms within the N—O unit. Most studies of $RO_2^\bullet$ radicals to date have been made using glassy matrices, but we note one study [9] of a peroxyl radical produced by X-irradiation of a single crystal, i.e. that of tetralin peroxide which had been enriched in ^{17}O; the results are broadly in accord with those obtained in aqueous and alcohol media and in urea clathrates, but the ratio of the spin densities on the terminal and inner oxygen atoms is lower than that in any other carbon-atom-peroxyl radicals [2].

4 MOTIONAL BEHAVIOUR OF PEROXYL RADICALS IN SOLID MATRICES

In all chemical and biological processes, it is molecular reorientation and diffusion which permit the constituent molecules to interact with each other, and thus to undergo molecular reactions. Therefore, efforts to understand these motional phenomena are manifest and profound. The study of these aspects of reactive intermediates is generally very difficult, with the greatest success being met with free radicals, particularly by means of ESR and related kinds of spectroscopy. Indeed, stable free radicals (usually nitroxides, $R_2N–O^\bullet$) may be deliberately incorporated as functional groups into polymers, proteins (enzymes) and membranes so that ESR spectroscopy may monitor their motional behaviour and therefore that of the local molecular environment, at the behest of particular events.

Peroxyl radicals may be similarly utilized as this application is of direct relevance to those oxidation processes which they mediate. The gross changes in the ESR spectra which signify both the type and rate of molecular reorientation are shown in Figure 5, in which (a) represents the condition of very slow reorientation, in which three *g*-values are measured (*vide supra*), while (b) shows the case in which the peroxyl radical reorients rapidly, and equally, about all three axes (I); this is well illustrated in the paper by Mach *et al.* [10] which reports the detection of the tropenylperoxyl radical (II) formed

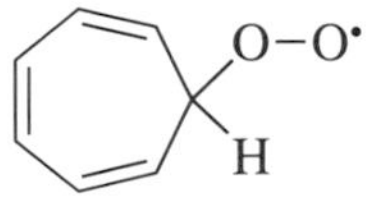

O–O•
H

(II)

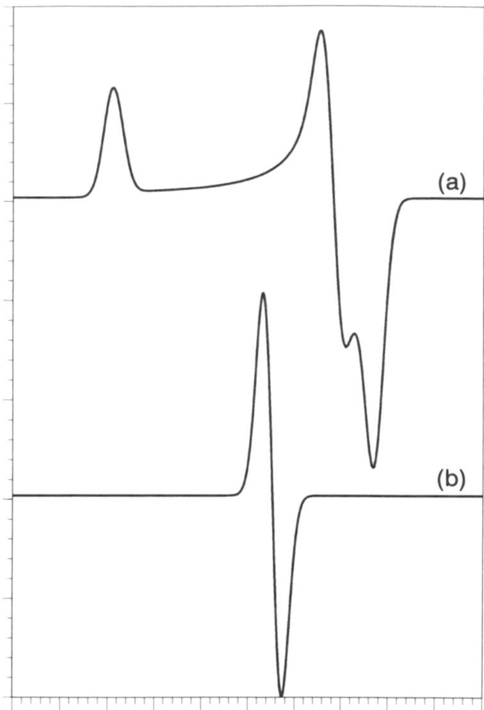

Figure 5 ESR spectra of an $RO_2^{\bullet}$ radical which is reorienting (a) very slowly and (b) very fast, with respect to the spectral time-scale.

in an adamantane matrix. As a consequence of the relatively small size of II in relation to the cavities present in solid adamantane (which is a 'rotor-solid', in which radicals have been frequently isolated to obtain sharp liquid-like ESR spectra), rapid tumbling occurs at 130 K, and the ESR spectrum is isotropic (b), while a spectrum of type (a) is recorded at 77 K, due to the fact that any reorientational motion is very slow on the ESR spectroscopic time-scale. The ESR parameters measured at 77 K ($g_1 = 2.035$, $g_2 = 2.009$, $g_3 = 2.002$; $a_1 = 8.4$, $a_2 = 9.6$, $a_3 = 5.35$ G) (1G = 10^{-4}T) are seen to average to 2.0153 and 7.78 G, which are similar to those values, i.e. 2.0157 and 7.5 G, measured at 130 K. The drop in a at the higher temperature is probably real, and reflects an increased torsional oscillation of the O—O group relative to the unique CH unit. In a less simple system, preferential reorientation about one axis may occur, so that the ESR spectrum is only partially averaged, but in a manner that reveals the preferred axis. With reference to the axis system in (I), Figure 6 indicates the effects of such motion about each of the axes shown.

When polycrystalline Ph_3CCO_2H or Ph_3CCl is irradiated with ultraviolet (UV) light at room temperature [11], the $Ph_3C^{\bullet}$ radical is formed, as signified

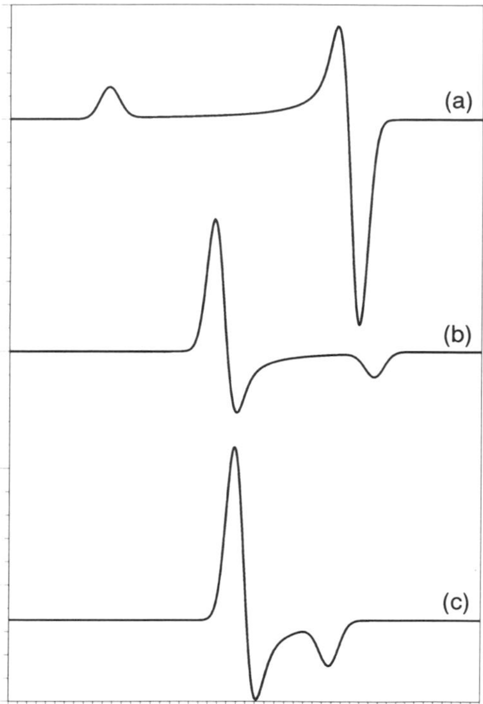

Figure 6 Partial averaging of the ESR spectra of $RO_2^{\cdot}$ radicals by preferential reorientation of the peroxyl unit about the three axes (I): (a) *x*-axis; (b) *y*-axis; (c) *z*-axis.

by a single line at $g = 2.0024$. On admission of O_2 gas to the evacuated sample, an almost symmetrical signal is produced, but at $g = 2.014$. In order to distinguish between the mechanisms in which either the whole molecule is tumbling rapidly or the peroxyl group is rotating with respect to the remainder of the molecule, the ^{17}O-enriched $Ph_3CO_2^{\cdot}$ radical was studied; this investigation showed conclusively that the latter process was dominant.

Sevilla, Champagne and Becker produced peroxyl radicals from lipids (fatty acid esters) present in their clathrates with urea [12] by exposure of the γ-irradiated material to O_2 (the peroxyl group being attached to the α-carbon atom of the fatty acid chain). Orientation and temperature dependence studies show that at room temperature the peroxyl radicals rotate or oscillate, as indeed do the carbon-centred precursor radicals, in the urea channels about their long axis. The rotational jump mechanism found for the peroxyl radicals usually follows the trigonal symmetry of the urea host channels, which are distorted at low temperatures. At room temperature, the entrapped molecules rotate freely about their long axis, and as a consequence, orientation of the samples at the 'magic-angle' (54.7°) results in isotropic couplings. ^{17}O coupling data are also reported, and are typical for hydrocarbon peroxyl radicals (Table 1). In previous related work this group also investigated [13] the motion of lipid peroxyl radicals in the neat lipids, (triarachidin and linoleic acid). The spectra were interpreted, as above, in terms of models used previously for peroxyl radicals formed in hydrocarbon polymers; for triarachidin, it was found that simulations based on the modified Bloch equations suggest a 90° jump about the chain axis with the plane of the peroxyl group perpendicular to this axis. For linoleic acid, simulations were made assuming that the principal axis system of the *g*-tensor is first rotated by 42° about the *y*-axis (I), which approximates to a rotation about the plane of the COO$^{\bullet}$ group. Good agreement with the experimental spectra was found based on a chain-axis jump of 180°, but an alternative model which involves rotation about the C—O bond by 90° jumps was also considered. The activation energy for rotation in triarachidin is four times that found for linoleic acid, i.e. ca. 3.4 kJ/mol^{-1}.

5 LIPID-PEROXYL RADICALS

The importance of this particular aspect of peroxyl radicals rests, as stated above upon their involvement in the destruction of cell membranes (and hence, of the cells themselves), by peroxidation. In relation to radiation biology and, moreover, to various disease processes, particularly heart disease, rheumatic disease, diabetes and carcinogenesis, with the latter being the final conclusion of the toxic effects of particular xenobiotic materials, the aim is to prevent, or at least to ameliorate, the related undesired physiological effects, by means of deactivating the oxygen-centred radicals which drive them. One solution is to administer 'antioxidants', with some of these being phenolic, which can readily scavenge reactive radicals, including peroxyls. Thus motivated, Sevilla and coworkers have used ESR spectroscopy in order to study the reactions of carbon-centred radicals derived from lipids with various phenolic antioxidants, such as tert-butylhydroquinone (TBHQ), n-propyl gallate (PG), butylated hydroxyanisole (BHA), butylated hydroxytoluene (BHT) and vitamin E, using the lipid as a matrix at low temperatures [14]. In all cases, these were found to react at 135 K with molecular oxygen to form peroxyl radicals, while at 170 K, the peroxyl radicals were found to react with the added antioxidants to form the corresponding phenoxyl radicals. Those radicals derived from tributyrin were found to decrease in their rate of reaction in the order, BHT > TBHQ > vitamin E > PG > BHA, while those derived from triolein and trilinolein decrease in the order, BHT > vitamin E > BHA > TBHQ. These differences are accounted for in terms of the way the antioxidants are dispersed in the different matrices. The fraction of unsaturated lipid peroxyl radicals which abstract from the antioxidant is lower than that in saturated lipids, and in the case of tributyrin the initial carbon-centred radicals can abstract directly from the antioxidant, while the more stable conjugated radicals derived from unsaturated lipids do so only to a negligible extent; the decay of all five antioxidant phenoxyl radicals in tributyrin follows second-order kinetics with an apparently common activation energy of 23 kcal mol^{-1}. The same researchers have previously determined the reactivity of radicals formed from cholesterol and its analogues by γ-irradiation [15]. In oxygen-free samples of cholesterol itself a tertiary side-chain radical (III) and an allylic radical (IV) are produced; the structure of the latter is confirmed by experiments using two cholesterol analogues, namely 7-hydroxycholesterol (V) and the selectively deuterated 7-deutero-7-hydroxycholesterol (VI), both of which produce the allylic radical after radiolysis by loss of the hydroxyl group. When oxygen is present in the samples, evidence is provided for the formation of two distinct peroxyl radicals, and these are suggested to possess differing reactivities resulting from their differing motional freedoms. It is further suggested that the products detected following radiation-induced autoxidaiton at or close to ambient temperature are consistent with the differing reactivities of these intermediary peroxyl radicals.

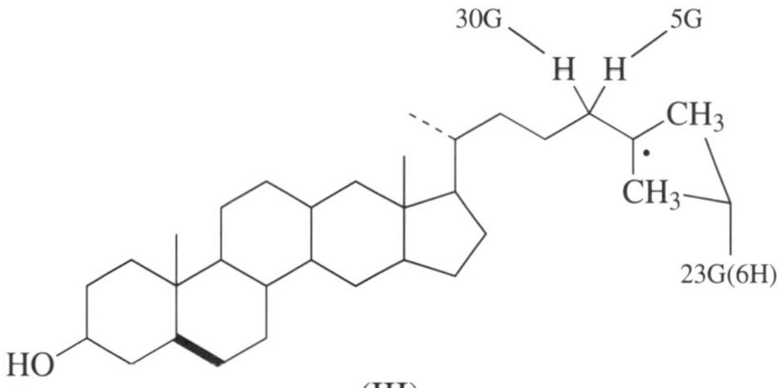

(III)

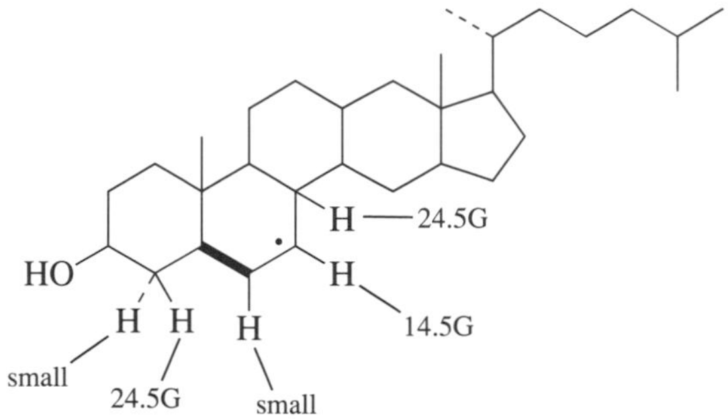

(IV)

7-OH cholesterol

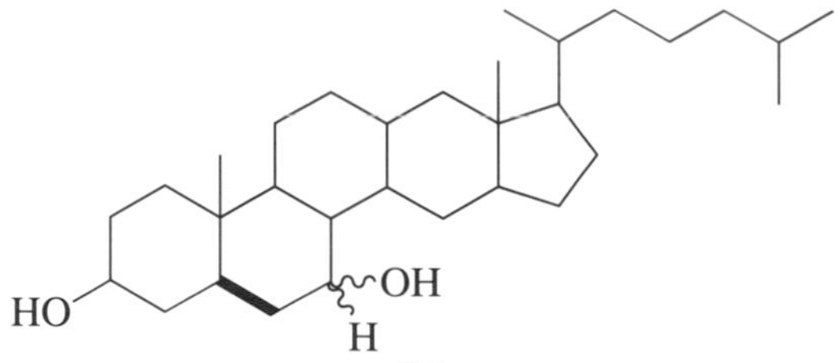

HO OH H

(V)

7-D-7-OH cholesterol

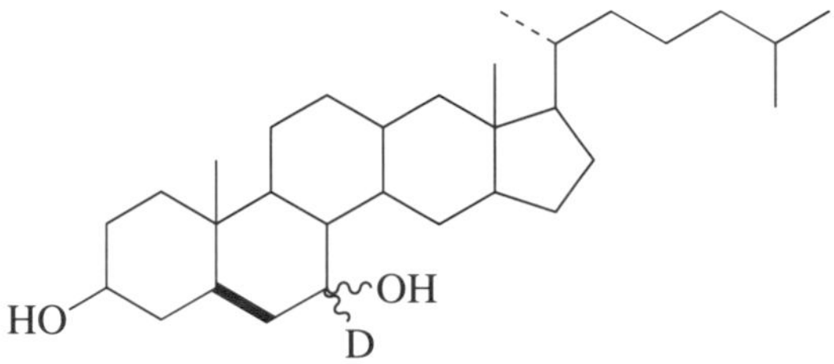

HO OH D

(VI)

The autoxidation phenomena has also been explored for the lipids tributyrin, triarachidin, triolein, triliolein, trilinolenin and linolenic acid [16]. It was found that the *unsaturated* lipids generally contained sufficient hydroperoxides such that their direct UV photolysis as neat samples at 77 K resulted in intense ESR signals; in contrast, it was necessary either to add tert-butyl hydroperoxide to the samples of *saturated* lipids prior to photolysis, or instead to use irradiation of the pure material, with the latter being found to give a more even distribution of radicals in the sample. The diffusion of molecular oxygen was found to occur in each case at a specific temperature, which was characteristic of the particular lipid being considered, when it reacts with those carbon-centred radicals produced initially and so forms the peroxyl radicals. Most interestingly, for unsaturated lipids the peroxyl radicals were found to decay on further annealing with the concomitant production of the respective allylic or pentadienyl lipid radicals, depending on the number of double bonds in the chains. We can see, then, the essential free-radical chemistry involved in the peroxidation process, i.e. a hydrogen atom is abstracted from a weak C—H bond, the resulting stabilized radical then adds molecular oxygen to form a peroxyl radical, which then abstracts a weakly bound hydrogen atom. In experiments with linolenic acid, subsequent introduction of oxygen resulted in the formation of further peroxyl radicals at low temperatures, but the signal from the pentadienyl radical was found to prevail at higher temperatures; an explanation was proposed in terms of the relative rates of oxygen migration and hydrogen-atom abstraction processes.

Later work by Sevilla and coworkers [17] concerned investigations of the kinetics of the autoxidation of triglycerides, again using ESR. It was found that following the initial production of carbon-centred radicals the following *distinct reaction stages* then occur: formation of peroxyl radicals by addition of O_2 molecules, and thus a depletion of the oxygen content of the sample; conversion of the lipid peroxyl radicals into allylic or pentadienylic radicals, by hydrogen-atom abstraction and finally, combination of these carbon-centred radicals. From the kinetics of these various stages it was concluded that the peroxidation step is controlled by O_2 diffusion, which has an apparent activation energy of 24 kJ mol^{-1} in unsaturated lipids. The subsequent H-atom-abstraction step (autoxidation cycle) depends on the lipid structure, and presumably on the relative C—H bond strengths for the atoms being abstracted, since the activation energies are 9 ± 2,34 ± 8, 88 ± 11 kJ mol^{-1}, for trilinolenin, trilinolein and triolein, yielding radicals conjugated, presumably, with three, two or one C=C double bond(s), respectively. In the final stage, these carbon-centred radicals combine at temperatures approaching the softening point of the particular lipid matrix, with a common activation energy of ca. 40 kJ mol^{-1}. For saturated lipids, the peroxyl radical signal decays following second-order kinetics, indicating a bimolecular radical combination mechanism, in contrast to the unsaturated lipid peroxyl decay, which is unimolecular; this is explained on the basis that the strength of the ROO—H bond is less than that of a C—H bond in a *saturated* molecule, thus preventing its mode of decay by H-atom abstraction.

6 REACTIVITY

As is by now clear, carbon-centred radicals react very readily with molecular dioxygen to form peroxyl radicals, $RO_2^{\cdot}$, with these lying somewhere between carbon-centred radicals and the superoxide anion in their propensity to abstract hydrogen atoms; however, until recently, very little was known about the relative reactivities of different peroxyl radicals, particularly with regard to the electronic effect of the substituent group R. Sevilla and coworkers have made considerable advances in relating the reactivity of peroxyl radicals with their electronic structures [2]. This group have found that good linear correlations are obtained between the $^{17}O_{\parallel}$ coupling from the terminal O atom in a series of peroxyls and the (-log) rate constants measured for both reduction by ascorbate and electron transfer from *N,N,N′,N′* tetramethyl-*p*-phenylenediamine; clear trends are found in both of the corresponding plots for a higher reactivity as the coupling increases. The coupling values also show good correlations with the Taft-substituent parameters, which enables predictions of the reactivities of unknown peroxyl radicals from ^{17}O couplings in more general cases. It is noteworthy that peroxyl radicals derived from halocarbons are predicted to be highly reactive; in the biological field it is well known that CCl_4 is highly toxic [18], being mediated by the formation of $CCl_3O_2^{\cdot}$ radicals, and this is in accord with the easy metabolic reduction of CCl_4, both *in vivo* and *in vitro*, and the conversion of $CCl_3^{\cdot}$ to a yet more reactive species, according to the above arguments.

7 SULFUR-PEROXYL RADICALS

It has been predicted that the hydrogen-abstracting power of oxygen-centred radicals might be related to the spin density at the terminal oxygen atom [19,20]. Observations have been made that in frozen matrices containing both $CysSO_2OO^{\cdot}$ and $CysSOO^{\cdot}$ (Cys=cysteine) the former species abstracts hydrogen atoms from the parent thiol at temperatures at which $CysOO^{\cdot}$ appears to be unreactive towards it. The correlation between the reactivity and ^{17}O hyperfine coupling [2] is in agreement with the pattern observed with ascorbate and electron transfer (see above), and the large coupling (Table 1) from the terminal O atom in $CysSO_2OO^{\cdot}$ suggests that it should be a highly reactive species (with regard to hydrogen abstraction and/or electron transfer), comparable to $CCl_3OO^{\cdot}$.

The different nature of the $RSOO^{\cdot}$ radical relative to the other peroxyl radicals being considered is of some interest; $GSOO^{\cdot}$ (G=glutathione), in pulse radiolysis studies, undergoes a relatively fast electron transfer from ascorbate (k=1.15 × 10^8 dm^3 $mol^{-1}s^{-1}$; pK_a=8.24) [21], while hyperfine coupling data predict a far lower rate [2]. It is significant that the pulse radiolysis studies show that the rate of reaction (hydrogen abstraction?) of GSH with $GSOO^{\cdot}$ is slow ($k < 10^6$ dm^3 $mol^{-1}s^{-1}$) [21], which agrees with the solid matrix investigations [2,20] that $GSOO^{\cdot}$ does not abstract from GSH in frozen samples. It may be relevant that $CysSOO^{\cdot}$ is the only peroxyl radical in which the sum of the ^{17}O couplings is less than ca. 150 G, showing that the sulfur atom delocalizes 6% of the spin-density from the O—O unit; indeed, the terminal-oxygen-atom spin density approaches that observed for the superoxide anion, which is predicted to undergo addition to ascorbate, in preference to electron transfer from ascorbate to $O_2^{-\cdot}$ [22]. In order to explain a visible absorption band at 540 nm, which is unique for peroxyl radicals, it is proposed that a significant contribution is made to the structure of $RSOO^{\cdot}$ by the charge-transfer state, RS^+—$O_2^{-\cdot}$, which would also account for the near equivalence of the ^{17}O coupling values [23]. Sevilla and coworkers [19] and Chatgilialoglu and Guerra [24] had previously resorted, independently, to *ab initio* calculations in an attempt to understand the properties of $RSSO^{\cdot}$ radicals, specifically with respect to the spin density and electronic transition energies of $MeSOO^{\cdot}$, but the results obtained did not predict a visible absorption, nor that the spin-density distribution should be other than that of a normal peroxyl radical. Indeed, Chatgilialoglu and Guerra [24] felt compelled to conclude that some species other than $RSSO^{\cdot}$, possibly $RSO^{\cdot}$, must be invoked to account for the visible absorption. In the later study [23] it was shown that the spin-density distribution in $RSOO^{\cdot}$ radicals varied significantly with the nature of the matrix in which they were isolated and with the nature of R; in a Freon matrix, for example, the coupling values for the terminal and inner oxygen atoms (^{17}O–1 and ^{17}O–2) vary for primary (79 and 62 G), secondary (84 and 57 G) and tertiary (96 and 51 G) thiyl-derived radicals, whereas in aqueous systems or in methanol, all of the thiols yield $RSOO^{\cdot}$ radicals with approximately the same couplings (80 and 62 G). In polar media, all $RSOO^{\cdot}$ species have a visible absorption at ca. 540 nm, and are found to be photosensitive, rearranging to their $RSO_2^{\cdot}$ isomers (where the latter are predicted to be more stable by 35–50 kcal mol^{-1}), and undergoing further oxygen addition to form $RSO_2OO^{\cdot}$. The degree of charge transfer ($RS^{+}OO^{-\cdot}$), and hence the variation in the ^{17}O coupling values, is proposed to be a function of the specific nucleophilic solvation of the sulfur atom by the medium, with the latter acting as an electron-pair donor; the solvent-stabilized charge-transfer state is found to be far more thermally stable than the uncomplexed radical, which is found to rearrange thermally to $RSO_2^{\cdot}$, even at 100 K. The charge-transfer state can be mimicked by *ab initio* calculations in which ions such as F^- or OH^- are allowed to associate with the sulfur atom.

Radicals of the general formula $RSO^{\cdot}$ were observed as thermal products of $RSOO^{\cdot}$ radicals after annealing to the softening point of the appropriate matrix [19]. It was proposed that these are formed by the reaction of matrix-derived $RSOO^{\cdot}$ radicals (signals from which are initially observed) and thiols, according to the following:

$$RSOO^{\cdot} + RSH \longrightarrow RSO^{\cdot} + RSOH \quad (4)$$

It is considered that matrix $ROO^{\cdot}$ (carbon-derived) peroxyl radicals are formed in this system, and initiate $RS^{\cdot}$ (reaction (5)), and hence $RSOO^{\cdot}$ formation (reaction (6)). However, such radicals might also themselves react with thiols (reaction (7)), in a step which is exoergic by ca. 58 kcal mol^{-1}, in direct competition with reaction (4).

$$ROO^{\cdot} + RSH \longrightarrow ROOH + RS^{\cdot} \quad (5)$$

$$RS^{\cdot} + O_2 \longrightarrow RSOO^{\cdot} \quad (6)$$

$$ROO^{\cdot} + RSH \longrightarrow RSO^{\cdot} + ROH \quad (7)$$

This section is concluded with reference to a review on the oxidation of simple thioethers, methionone and small methionone-containing peptides by $OH^{\cdot}$ and halogenated peroxyl radicals. The most important conclusion reached was that even for very small peptides, a pronounced dependence on the peptide structure was found in the radical yield and selectivity, in addition, of course, to the nature of the initial oxidizing radical [25].

8 SILYL-PEROXYL RADICALS

To date, there appears to be only one report of such species, namely the radical t-Bu_3Si—OO$^{\bullet}$, which was observed following radiolysis of tri-tert-butylsilane as a pure material at 77 K and after annealing to 130 K [26]. Remarkably, the initial radiolysis produced the primary t-$Bu_3SiH^{+\bullet}$ radical cation — such species normally require an inert matrix such as a Freon to stabilize them, since they react in the pure silane — along with tri-tert-butylsilyl radicals. On annealing, a 'normal' peroxyl radical signal is observed, with the typical $g(x,y,z)$ values quoted earlier, but, interestingly, an additional $g(z)$ feature at 2.083 is seen (Figure 7). The latter is most unusual for a peroxyl radical, and we propose that it is due to an increase in the Si—O—O angle from that typical of C—O—O-type radicals. As

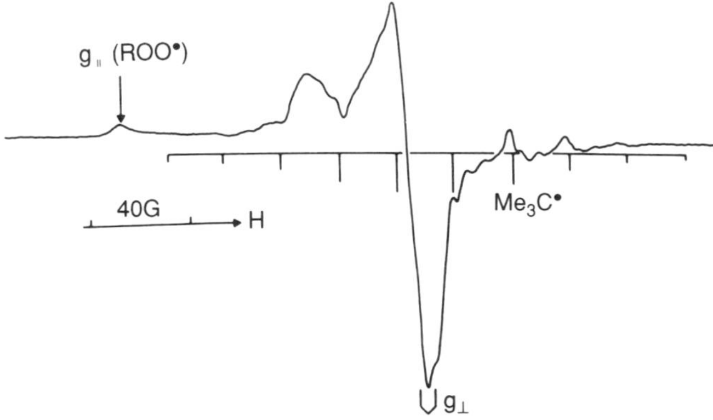

Figure 11.7 ESR spectrum showing features from the $RO_2^{\bullet}$ and t-$Bu_3SiO_2^{\bullet}$ radicals.

shown in Figure 3, in accord with our description of the relationship between the structure and the g-value for the O—O (z) direction of a peroxyl radical, an expansion of this angle will reduce the differential perturbation of the $\pi(x,y)$ manifold and so the g-shift will be greater. There is ample demonstration that the bond angles in silylamines and siloxanes are increased compared with their organic counterparts, and ESR data obtained for silylamine radical cations provide strong evidence for an appreciable π-bonding interaction between the silicon and nitrogen centres [27].

9 TRAPPING FROM THE GAS PHASE

The motivation for this kind of approach is similar to that adopted in 'spin-trapping' experiments, i.e. that the concentration of reactive radicals may be raised by isolation in an inert matrix, but with the further advantage that the radicals may be detected *directly*, rather than as spin-adducts. For example, the identity of the trapped species may not always be determined with absolute certainty, and the adducts often have a significant half-life in solution. Moreover, regarding studies of actual gas samples by ESR spectroscopy, the spectra are complicated by the strong magnetic coupling between the electron spin and the rotational angular momenta of the radicals. This has, of course, implications in atmospheric chemistry studies. Using a development of this basic approach, i.e. a numerical procedure for the deconvolution of composite spectra of matrix-isolated radicals from atmospheric samples, it has proved possible to determine the concentration of small amounts of $HO_2^{\cdot}$ and $MeC(O)O_2^{\cdot}$ in tropospheric air at concentrations as low as *40* parts per trillion [28]. Better signal-to-noise conditions are realized when a matrix of D_2O is used rather than H_2O, presumably because of the reduced dipolar broadening by the deuterons, on account of their smaller magnetic moment.

A diffusion model has been presented for $HO_2^{\cdot}$ radicals, trapped in condensates obtained by the deposition of dissociation products from microwave discharges through O_2–H_2 mixtures, on cold walls; this model is able to describe the stepwise radical recombination processes and provides the coefficient for diffusion of the radicals within the matrix which contains them [29].

In connection with the partial oxidation of hydrocarbons over bismuth oxide catalysts, Martir and Lunsford have reported that allyl radicals are formed in such reactions since they were able to matrix isolate such species from downstream of the reactor [30]; at higher oxygen flow rates, spectra which are characteristic of peroxyl radicals are produced, which might be taken to accord with speculation in the literature that allyl peroxyl radicals are surface intermediates in the conversion of propylene to acrolein [31–33]. However, it is clear that such intermediates are formed in the cooler post-catalytic volume, and it has been estimated that the dissociation equilibrium, $C_3H_5^{\cdot} + O_2 \rightleftharpoons C_3H_5O_2^{\cdot}$, lies far to the left at the temperature used under real catalytic conditions (700°C). During the partial oxidation of methane over a magnesium oxide catalyst, $CH_3O_2^{\cdot}$ radicals have been similarly detected [34]; these radicals must be intrinsically more stable, and moreover, since the reaction is carried out at lower temperatures (500°C), could be important reaction intermediates.

10 INORGANIC PEROXYL RADICALS

The superoxide ion has been observed in various inorganic media, and its ESR parameters show some variation according to the type of surface interaction, but in the main the two oxygen atoms remain equivalent, as demonstrated by ^{17}O–labelling [35]. There are a few exceptions to this, which show that there is a dominant interaction between a surface site and *one* of the oxygen atoms, which renders their ^{17}O couplings distinct, and the ESR spectrum in fact resembles that of a typical peroxyl radical. This analogy becomes closer in cases where the superoxide anion is formed at a cation with non-zero nuclear spin, because not only are normal peroxyl-type *g*-values measured, but hyperfine structure from the cation is also observed [35]. In a study of one such species, formed by the reaction of peroxides with various vanadium compounds, the ^{51}V hyperfine tensor was determined [36], the form of which shows that the spin density on the vanadium atom is negative and stems from spin polarization of the *covalent* V—O bond.

11 POLYMERS

Peroxyl radicals are key intermediates in the photo-oxidation of polymers, and so the study of these radicals in polymer matrices helps to reveal some of the mechanistic details involved, and indeed has provided part of the motivation for some of the work covered earlier, in addition, of course, to their biological significance. The reactivity of peroxyl radicals generated in polypropylene films by γ-irradiation at $-78°C$ has been measured by ESR spectroscopy for a series of samples having differing morphology and radical-decay history [37]. Spectral changes which were consistent with chain propagation being brought about by the peroxyl radicals abstracting hydrogen atoms from the polymer matrix were observed for both ^{17}O-labelled radicals in a $^{16}O_2$ atmosphere and for peroxyl radicals in oxygen-free films. Some peroxyl radicals were extremely unreactive towards propagation, and instead underwent exclusive termination reactions. These latter radicals were believed to exist in motionally restricted regions of the polymer and became the dominant component in the radical population at long decay times. The reactivity of 2-(2-hydroxy-5-vinyl)-2*H*-benzotriazole in a copolymer with methyl methacrylate has been studied by ESR spectroscopy [38], and the alkyl and peroxyl radicals monitored. The alkyl radicals were found to be generated via mechanical degradation, with the peroxyl radicals being subsequently produced by the addition of O_2.

Peroxyl radicals have also been observed in *isotactic* polypropylene, following γ-irradiation [39]; in this case, however, their decay was monitored as a function of pressure. The rate constants for their decay were measured and from these data the corresponding activation volumes were determined. The kinetic characteristics implied that the peroxyl radicals were formed preferentially in the amorphous region of the polymer. In general, much useful information on the molecular mechanism of radical reactions in the solid phase can be derived from the activation volumes. In a further study, the decay of peroxyl radicals in polypropylene also monitored by ESR spectroscopy while the associated oxidation products were simultaneously studied by means of Fourier transform infrared (FT-IR) spectroscopy [40]; both *isotactic* and *atactic* forms of the polymer were examined, and distinct differences were found. In the *atactic* polymer, $RO_2^{\cdot}$ radical self-reaction was found to predominate, and was very fast, whereas chain oxidation was prevalent in the *isotactic* polymer, together with non-terminating peroxyl self-reactions. A comparison of the peroxyl decay rates and the rates of formation of hydroperoxide groups and of β-scission products clearly indicated that $RO_2^{\cdot}$ propagation did occur, and that previous studies based solely on ESR spectroscopy were not definitive — a cautionary note on which to end this review.

12 CONCLUSIONS

I have certainly not covered every publication dealing with solid-state systems in which peroxyl radicals were found, largely because there are so many, since peroxyl radicals are often unwelcome guests in irradiated materials from which oxygen has not been excluded. However, I hope to have spanned the broad gamut of their importance, and it is clear, certainly to me, now having completed this survey, that the methods currently available for their study, particularly in relation to biology, polymer science, atmospheric reactions and catalysis, are likely to be improved upon. There are currently enormous developments in ESR methodology, particularly pulsed techniques and the use of very high frequencies, and advances should enhance this particular field, as well as many other aspects of free-radical structure and reactivity. Time will tell, as ever, what benefits these will bring!

REFERENCES

1. A Carrington and A.D. McLachlan, *Introduction to Magnetic Resonance*, Chapman and Hall, London (1979).
2. M.D. Sevilla, D. Becker and M. Yan, *J. Chem. Soc. Faraday Trans.* **86**, 3279 (1990).
3. B.H. Besler, M.D. Sevilla and P. MacNeille, *J. Phys. Chem.* **90**, 6446 (1986).
4. Y. Beers and C.J. Howard, *J. Chem. Phys.* **64**, 1541 (1976).
5. K.G. Lubic, T. Amano, H. Uehara, K. Kawaguchi and E. Hirota, *J. Chem. Phys.* **81**, 4826 (1984).
6. C.E. Barnes, J.M. Brown and H.E. Radford, *J. Mol. Spectrosc.* **84**, 179 (1980).
7. S. Shimada, A. Kotake, Y. Hori and H. Kashiwabara, *Macromolecules*, **7**, 1104 (1984).
8. S. Shimada, Y. Hori and H. Kashibara, *J. Am. Chem. Soc.* **18**, 179 (1985).
9. E. Melamud, S. Schlick and B.L. Silver, *J. Magn. Reson.* **14**, 104 (1974).
10. K. Mach, J. Novakova, V. Hanus and J.B. Raynor, *Tetrahedron* **45**, 843 (1989).

11. E. Melamud and B.L. Silver, *J. Magn. Reson.* **14**, 112 (1974).
12. M.D. Sevilla, M. Champagne and D. Becker, *J. Phys. Chem.* **93**, 2653 (1989).
13. D. Becker, J. Yanez, M.D. Sevilla, M.G. Alonso-Amigo and S. Schlick, *J. Phys. Chem.* **91**, 492 (1987).
14. J. Zhu, W.J. Johnson, C.L. Sevilla, J.W. Herrington and M.D. Sevilla, *J. Phys. Chem.* **94,** 7185 (1990).
15. C.L. Sevilla, D. Becker and M.D. Sevilla, *J. Phys. Chem.* **90**, 2963 (1986).
16. J. Yanez, C.L. Sevilla, D. Becker and M.D. Sevilla, *J. Phys. Chem.* **91**, 487 (1987).
17. J. Zhu and M.D. Sevilla, *J. Phys. Chem.* **94**, 1447 (1990).
18. T.F. Slater, K.H. Cheeseman and K.I. Ingold, *Philos. Trans. R. Soc. London B* **311**, 451 (1985).
19. S.G. Swartz, D. Becker, S. DeBolt and M.D. Sevilla, *J. Phys. Chem.* **93**, 155 (1989).
20. M.D. Sevilla, D. Becker and M. Yan, *Int. J. Radiat. Biol.* **57**, 65 (1990).
21. M. Tamba, G. Simone and M. Quintiliani, *Int. J. Radiat. Biol.* **50**, 595 (1986).
22. D.E. Cabelli and B.H. Bielski, *J. Phys. Chem.* **87**, 1809 (1983).
23. Y. Razskazovskii, A.-O. Colson and M.D. Sevilla, *J. Phys. Chem.* **99**, 7993 (1995).
24. C. Chatgilialoglu and M. Guerra, in *Sulfur-Centred Reactive Intermediates in Chemistry and Biology*, edited by C. Chatgilialoglu and K.-D. Asmus, Plenum, New York (1990), p. 31.
25. C. Schoeneich, K. Bobrowski, J. Holman and K.-D. Asmus, in *Proceedings of 5th Biennial Meeting of the International Society for Free-Radical Researc* (1991), p. 380.
26. C.J. Rhodes, *J. Organomet. Chem.* **443**, 19 (1993).
27. C.J. Rhodes, *J. Chem. Soc. Perkin Trans. 2* 235 (1992).
28. D. Mihelcic, A. Volz-Thomas, H.W. Paetz, D. Kley and M. Mihelcic, *J. Atmos. Chem.* **11**, 271 (1990).
29. O. Yu. Berezin, E.E. Antipenko and B.V. Strakhov, *Zh. Fiz. Khim.* **67**, 947 (1993).
30. W. Martir and J.H. Lunsford, *J. Am. Chem. Soc.* **103**, 3728 (1981).
31. C.R. Adams and T.J. Jennings, *J. Catal.* **2**, 63 (1963).
32. N.W. Cant and W.K. Hall, *J. Catal.* **22**, 310 (1971).
33. A.L. Dent and R.J. Kokes, *J. Am. Chem. Soc.* **92**, 6709 (1970).
34. D.J. Driscoll, W. Martir, J.-X. Wang and J.H. Lunsford, *J. Am. Chem. Soc.* **107**, 58 (1985).
35. J.H. Lunsford, *Catal. Rev.* **8**, 135 (1973).
36. M.C.R. Symons, *J. Chem. Soc. A* 1889 (1970).
37. D.J. Carlsson, C.J.B. Dobbin and D.M. Wiles, *Macromolecules* **18**, 1791 (1985).
38. E. Borsig, Z. Hlouskova, F. Szocs, L. Hrckova and O. Vogl, *Eur. Polym. J.* **27**, 841 (1991).
39. F. Szocs, *J. Appl. Polym. Sci.* **27**, 1865 (1982).
40. D.J. Carlsson, C.J.B. Dobbin and D.M. Wiles, *Macromolecules* **18**, 2092 (1985).

12 Organic Peroxy Radicals in Polymeric Systems

YASURO HORI
Nagoya Institute of Technology, Showa, Nagoya 466, Japan

1 GENERATION OF PEROXY RADICALS IN POLYMERIC SYSTEMS

1.1 OXIDATION OF POLYMERS

Peroxy radicals of polymers were first directly observed in polytetralfuoroethylene (PTFE) in 1955 by Ard, Shields and Gordy by means of the electron spin resonance (ESR) spectroscopic method [1]. These ESR spectra were observed when X-irradiated samples of PTFE were aged for several days in air or oxygen, whereas those aged in N_2 or in a vacuum were found to have the same ESR spectra as they had immediately after irradiation. Therefore these newly observed ESR spectra were assigned as being due to the peroxy radicals of PTFE.

Since this first report, the peroxy radicals of polymers and the reactions of such radicals have been the main subject in the oxidation processes of polymers. The following free-radical chain mechanism for the oxidation of a polymeric system is now generally accepted:

$$\mathrm{RH} \longrightarrow \mathrm{R}^{\cdot} \quad (1)$$

$$\mathrm{R}^{\cdot} + \mathrm{R'H} \longrightarrow \mathrm{RH} + \mathrm{R'}^{\cdot} \quad (2)$$

$$\mathrm{R}^{\cdot} + \mathrm{O_2} \longrightarrow \mathrm{ROO}^{\cdot} \quad (3)$$

$$\mathrm{ROO}^{\cdot} + \mathrm{R'H} \longrightarrow \mathrm{ROOH} + \mathrm{R'}^{\cdot} \quad (4)$$

$$\mathrm{ROO}^{\cdot} + \mathrm{X} \longrightarrow \text{non radical} \quad (5a)$$

$$\mathrm{ROO}^{\cdot} + \mathrm{R}^{\cdot}\ (\text{or } \mathrm{ROO}^{\cdot}) \longrightarrow \text{non radical} \quad (5b)$$

$$\mathrm{ROOH} \longrightarrow \mathrm{RO}^{\cdot} + {}^{\cdot}\mathrm{OH} \quad (6)$$

where $R^{\cdot}$, ROOH, $RO^{\cdot}$ and $^{\cdot}OH$ represent the alkyl radicals of polymeric systems, peroxides, alkoxy radicals and hydroxy radicals, respectively.

Peroxyl Radicals. Edited by Z.B. Alfassi

The chain reaction of steps (3) and (4) is the key reaction in the oxidation of polymers. However, the formation of the first peroxy radical in the chain reaction needs a parent radical, i.e. a free radical, $R^{\cdot}$, which is a carbon-centred radical of the polymer. This may be produced by γ-irradiation, X-irradiation, photoirradiation or a chemical reaction.

1.2 BRIEF DESCRIPTION OF POLYMER FEATURES

Many papers have been published on studies of peroxy radicals in a wide range of polymers. From these polymers, only a selection of materials will be discussed in this chapter, such as polyethylene (PE), polytetrafluoroethylene (PTFE), and polypropylene (PP). Poly(vinyl chloride) (PVC) and poly(methyl methacrylate) (PMMA) will also appear briefly.

Polymer molecules consist of linear combinations of monomer molecules, such as ethylene (CH_2CH_2) for PE, and are denoted as $(—CH_2CH_2—)_n$ or simply $—CH_2CH_2—$ for PE, $—CF_2CF_2—$ for PTFE, $—CH_2CH(CH_3)—$ for PP, $—CH_2CHCl—$ for PVC and $—CH_2C(CH_3)(COOCH_3)—$ for PMMA.

Solid polymers usually contain two regions, i.e. a crystalline region and an amorphous one. In the crystalline regions the polymer molecules are regularly folded and stacked. It is noteworthy that in the crystalline region PE molecules have a planar zigzag form and PTFE is almost planar zigzag in its structure, but PP molecules have a 1/3 helical structure. In the amorphous regions the molecules exist in random forms and usually make faster and more random motions than those in the crystalline region.

As the parent radicals for the peroxy radicals of polymers, there are two types of carbon-centred radicals of polymers, i.e. the scission-type radical and the main chain radical, which has two forms, namely alkyl and polyenyl. The scission-type radical is produced by the scission of a polymer main chain, such as $—CH_2CH_2^{\cdot}$ for PE and $—CF_2CF_2^{\cdot}$ for PTFE, and is called a chain-end radical. The alkyl radical of the main chain is also called a chain radical and as such is denoted as $—CH_2CH^{\cdot}CH_2—$ for the PE chain alkyl radical or $—CH_2C(CH_3)^{\cdot}CH_2—$ for the PP tertiary alkyl radical. The simplest radical among the polyenyl radicals is the allyl radical, such as $—CH_2CH^{\cdot}CH{=}CHCH_2—$ for PE.

The peroxy radical of a polymer is formed by the reaction of the parent radical with oxygen (reaction (3)). The formation of the peroxy radical is believed to take place very fast when an oxygen molecule reaches the parent radical. This means that the formation of the peroxy radical can be regarded as the diffusion-controlled reaction of oxygen with the parent radical. It should be noted that oxygen can diffuse easily into the amorphous regions of all polymers, and even into the crystalline regions of PTFE or PP, but it cannot diffuse into the crystalline region of PE. This indicates that peroxy radicals can be easily formed in the amorphous or crystalline regions of PTFE or PP, but not in the crystalline regions of PE.

1.3 ESR SPECTRA OF PEROXY RADICALS IN POLYMERIC SYSTEMS

The first observed ESR spectrum for the peroxy radicals of PTFE [1] is shown in Figure 1. This figure shows that the ageing in O_2 of X-irradiated PTFE changed the ESR spectrum having the hyperfine structure (bottom), due to the carbon-centred radical of PTFE, into the asymmetric spectrum having no hyperfine structure (upper), where the latter has been assigned to the presence of the peroxy radicals.

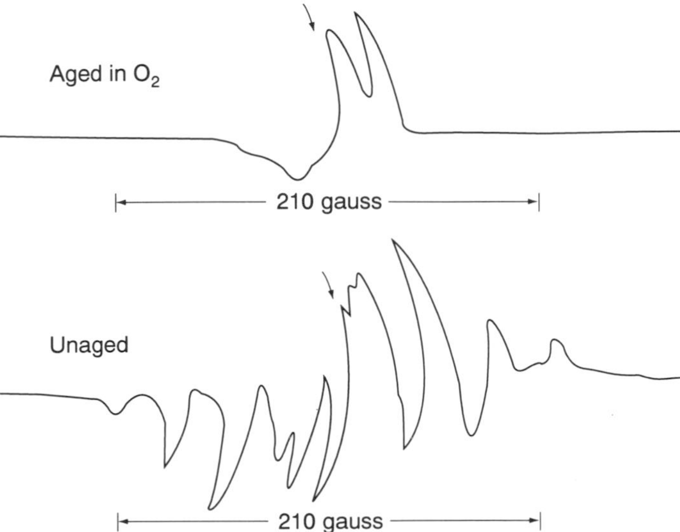

Figure 1 ESR spectra of X-irradiated PTFE before and after ageing in O_2 (temperature = 25°C, frequency = 23 GHz). The arrow indicates the position of the 2,2-di(4-t-octyl phenyl)-1-picryl hydrazyl (DPPH) resonance, $g = 2.0036$ [1].

The ESR-line shapes arising from randomly oriented particles in a fixed position were calculated by Kneubühl [2]. It has been found that it is possible to determine the three principal g-factors ($g_1 < g_2 < g_3$) with relatively high accuracy and without the need to investigate single crystals. The g-tensor of peroxy radicals in the rigid state is approximately axially symmetric, and the value of $g_{\parallel}$ is much larger than $g_{\perp}$. This indicates that $g_{\parallel} = g_3$ and $g_{\perp} \approx g_1 \approx g_2$. From theoretical considerations [3,4] the direction of $g_{\parallel}$ (= g_3) is expected to be parallel to the O—O• bond.

It should be noted here that two notation systems have been used in the literature for the principal g-factors; one is the ($g_1 < g_2 < g_3$) system while the other is the ($g_3 < g_2 < g_1$) form. In this present chapter the ($g_1 < g_2 < g_3$) system will be used, with the exception of Section 3 where the ($g_3 < g_2 < g_1$) form will be used.

There are many cases of small amounts of peroxy radicals existing in a large amount of parent radicals. In order to differentiate the spectrum of the peroxy radical from that of the parent one, the power saturation technique was developed by Ohnishi, Sugimoto and Nitta [5]. Figure 2 shows a plot of the detected signal voltage V against the square root of the microwave power P for the oxygenated radical and for the parent radical trapped in the irradiated polymer (polypropylene being used as a typical example). This demonstrates clearly the differences in saturation behaviour. If the ESR spectrometer is run at a microwave power of, say 100 mW, V for the oxygenated radicals is observed as being magnitided three times that of the parent carbon-centred radicals.

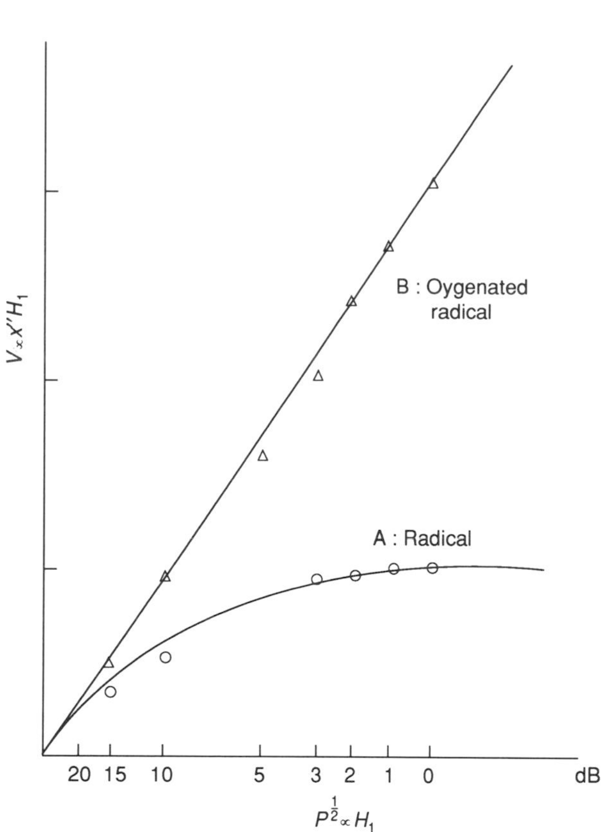

Figure 2 Saturation effect of free radicals trapped in irradiated polypropylene: (A) $-CH_2-C^{\bullet}(CH_3)-CH_2-C^{\bullet}(CH_3)_2$; (B) the oxygenated radicals produced on introduction of air to these radicals. V denotes the detected signal voltage, P is the microwave power to the cavity (0 dB corresponding to ca. 100 mW), H_1 is the rotating magnetic field, and χ'' is the magnetic susceptibility. The sample was electron-irradiated at 1 Mrad (1 rad = 10^{-2} Gy) [5].

Figure 3 shows the typical spectra of peroxy radicals which have an anisotropic *g*-tensor, where the peroxy radicals of PE in a urea–PE inclusion compound (UPEC) change their molecular motion from rigid to rotational with increasing temperature [6].

The precise principal values of the *g*-tensors of peroxy radicals can be obtained from the rigid-state ESR spectra by computer simulation. Examples

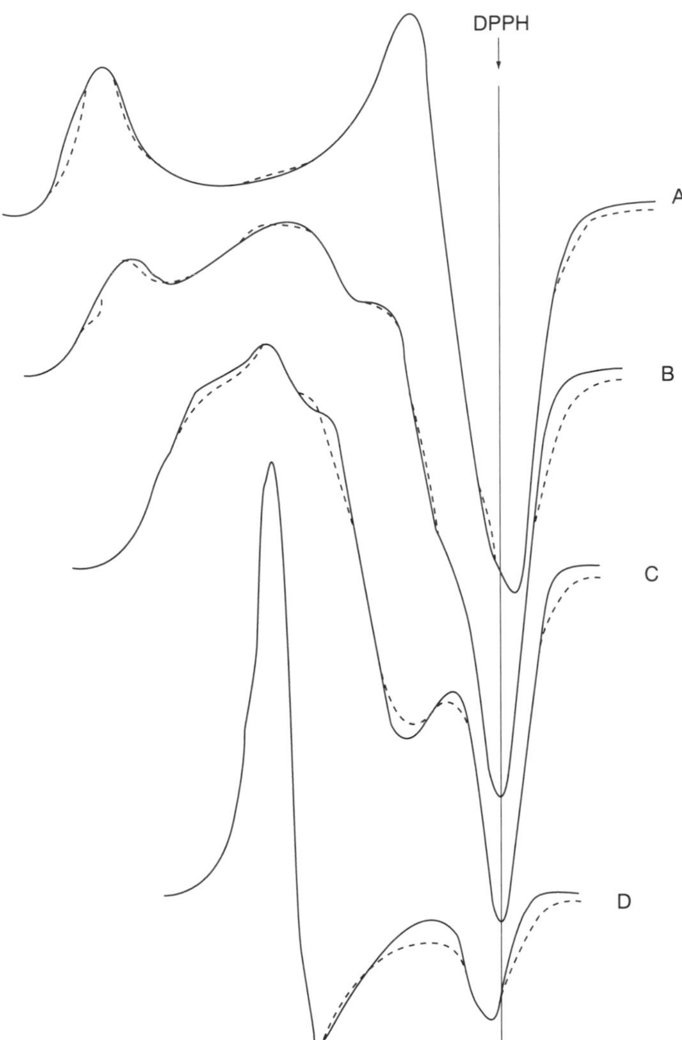

Figure 3 ESR spectra of peroxy radicals in the UPEC observed at various temperatures (- - -), plus the corresponding simulated spectra (—): (A) 113; (B) 200; (C) 249; (D) 326 K [6].

of the g-values are listed in Table 1 for some typical peroxy radicals of various polymers.

Table 1 Principal g-values for some Typical peroxy radicals

Radical	g_1	g_2	g_3	Ref.
$—CH_2CHOO^{\cdot}CH{=}CHCH_2—$	2.0024	2.0092	2.0361	7
$—CH_2CHOO^{\cdot}CH_2—$ [a]	2.0022	2.0081	2.0366	6
$—CH_2C(CH_3)^{\cdot}CH_2—$	2.0021	2.0081	2.0353	8,9
$—CF_2CFOO^{\cdot}CF_2—$	2.0023	2.0079	2.0385	10

[a] In urea–polyethylene inclusion complex.

1.4 STRUCTURES AND MOLECULAR MOTIONS OF PEROXY RADICALS IN POLYMERS

Whereas detailed discussions will be given in later sections, we shall describe briefly here the structures and motions of the peroxy radicals of some specific polymers. It has been determined by Iwasaki and Sakai [11] for PTFE and by Hori and coworkers [6,7,12] for PE that the COO group of the peroxy radical lies in a plane perpendicular to the polymer chain axis. A schematic representation of the peroxy radical in the case of the UPEC [6] is shown in Figure 4. The g_1 axis is parallel to the chain axis, while the g_3 direction is perpendicular to it. The motion of an entire COO group about the main chain is faster than the rotation of O—O about the C—O bond [6,7,11,12] in PTFE or UPEC and in the amorphous region of PE. Spectrum D in Figure 3 shows the rotation of the COO group around the chain axis. In the case of PP, similar conclusions for the structure and motion of peroxy radicals have been obtained [8], although the PP molecule has the helical structure. The COO angle at a peroxy radical site formed in irradiated isotactic PP has been determined by Shimada and coworkers [9,13] to be 111°. *Ab initio* molecular orbital calculations [14] have given the value of 111.5° for the COO angle. Moreover, a large value of the COO internal rotation barrier has been calculated. This result supports the above mentioned conclusion that rotation of the $OO^{\cdot}$ group about the C—O bond does not usually take place.

2 PEROXY RADICALS OF POLYETHYLENE (PE)

2.1 FEATURES OF PE

Ethylene monomer, $CH_2{=}CH_2$, can be polymerized into a high-density polyethylene (HDPE) by using a low-pressure technique, or into a low-

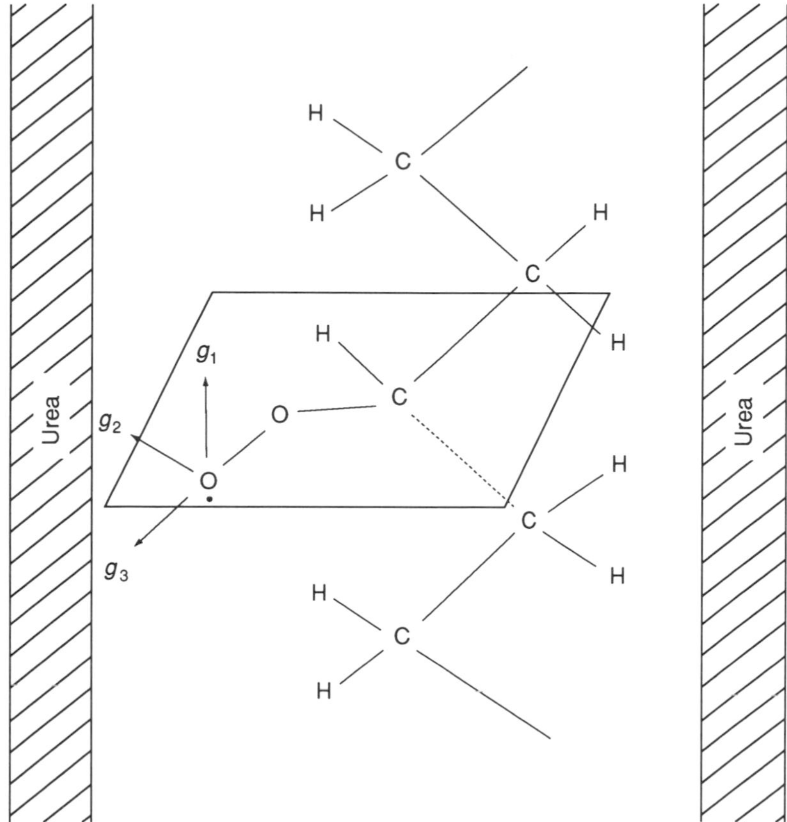

Figure 4 Schematic representation of peroxy radicals in the UPEC [6]

density polyethylene (LDPE) by using a high-pressure method; HDPE is a linear polymer. It has a density of 0.94–0.97 kg dm^{-3} and a crystallinity of 0.80–0.95. LDPE has a relatively large amount of branches and its crystallinity is 0.20–0.60. The melting point is dependent on the crystallinity and ranges from 100 (LDPE) to 135°C (HDPE).

There have been many studies on the peroxy radicals trapped in HDPE, but only a few for those in LDPE. Therefore, the word 'PE' will be used hereafter to indicate HDPE, unless mentioned otherwise.

PE has a folded crystal structure, and in the crystal, the PE molecule adopts a planar zigzag structure. It is noteworthy that it is very hard to diffuse oxygen molecules into the crystalline regions of PE on account of the fact that the molecular size of oxygen is almost equal to the size of vacancy in the PE crystallite..

2.2 CARBON-CENTRED RADICALS OF PE

The ESR spectrum of the secondary alkyl radical, —$CH_2CH^{\cdot}CH_2$— [15], was first observed and identified in γ-irradiated PE in 1958. The pattern is a six-line spectrum, arising from four equivalent β-hydrogen atoms with couplings of 3.1 mT each, and one α-hydrogen atom with the same coupling of 3.1 mT.

Another more stable radical of PE was also found trapped in irradiated PE and was assigned as an allyl radical, —$CH_2CH^{\cdot}CH{=}CH$—CH_2— [16]. This radical gives a septet spectrum, for which the line separation was estimated to be ca. 1.8 mT, with each peak of the septet split further into a doublet. This radical is very stable at room temperature.

The third radical observed in PE is a polyenyl radical, —CH_2—$CH^{\cdot}$(—$CH{=}CH$—$)_n$—CH_2—. This radical is produced in irradiated PE at large dose rates, and gives a singlet ESR spectrum [17,18].

2.3 FORMATION OF PEROXY RADICALS OF PE

The first ESR spectrum of the peroxy radical of PE was observed in 1958 by Abraham and Whiffen [19], when air was admitted to a sample of γ-irradiated PE. The same spectrum was formed directly when the sample was irradiated in air.

The behaviour of the alkyl radical when exposed to air has been studied by Loy [20]. The asymmetrical narrower line, which represents the peroxy radical, grew with time and was superimposed on the alkyl spectrum. The reaction ceased after 55–60% of the alkyl radicals had reacted.

Using the power saturation technique, clear ESR spectra of peroxy radicals in irradiated PE were observed by Ohnishi, Sugimato and Nitta [5]. The septet spectra, which are assigned to the allyl radicals, change on the introduction of air to the irradiated PE, as shown in Figure 5. The spectra observed at room temperature indicate that the peroxy singlet appeared in the region with a high *g*-factor immediately after the introduction of air (see spectrum B2), but the concentration of the peroxy radical was small, as observed in spectrum A2. From the decay curve observed at 3 dB, it is known that the peroxy radical, —$CH(OO^{\cdot})$—$CH{=}CH$—, decayed rather quickly at room temperature (half-life of ca. 20 min), although the allyl radical is very stable *in vacuo*, even at room temperature.

The change in the spectra from the allyl radicals to the peroxy radicals has been observed at various temperatures by Hori and coworkers [21], and quantitative measurements of the reaction processes of oxygen with the allyl radicals trapped in the amorphous part of PE have been made. When the peroxy radical is produced from the allyl radical in a ratio of one to one, the function $((S)(t) - S(0))$ has been found to be proportional to the concentration of the peroxy radicals, where $S(t)$ is the integral value of the power saturated ESR spectrum observed at time t.

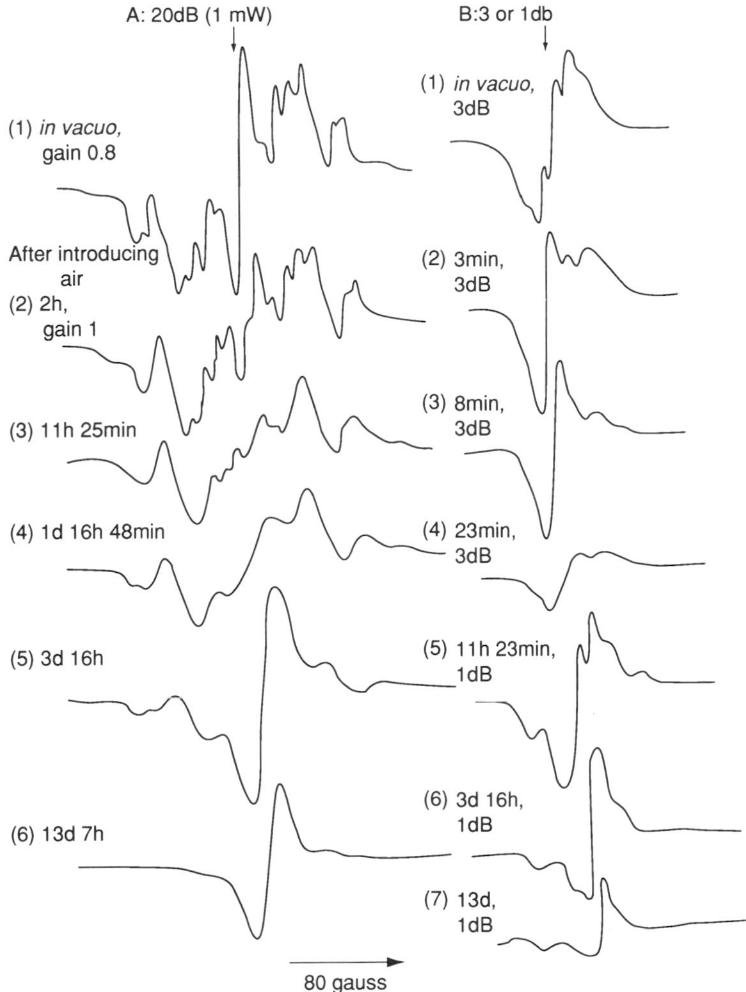

Figure 5 Changes in the ESR spectra after introduction of air to samples of PE which have been electron-irradiated at −196°C and 60 Mrad and then stored at room temprature for 2 d, showing the reaction with oxygen of the allyl radical (–CH$^{\bullet}$–CH=CH–). Series A was measured at 1mW and series B at 50 or 80 mW [5].

Examples of the results obtained at various temperatures are shown in Figure 6, where the values of $S(t)$ are plotted against the time elapsed after the introduction of the oxygen into the irradiated PE sample, i.e. the reaction time. This figure clearly shows that the peroxy radicals produced by the reaction of oxygen with allyl radicals are stable below 213 K and decay above this temperature.

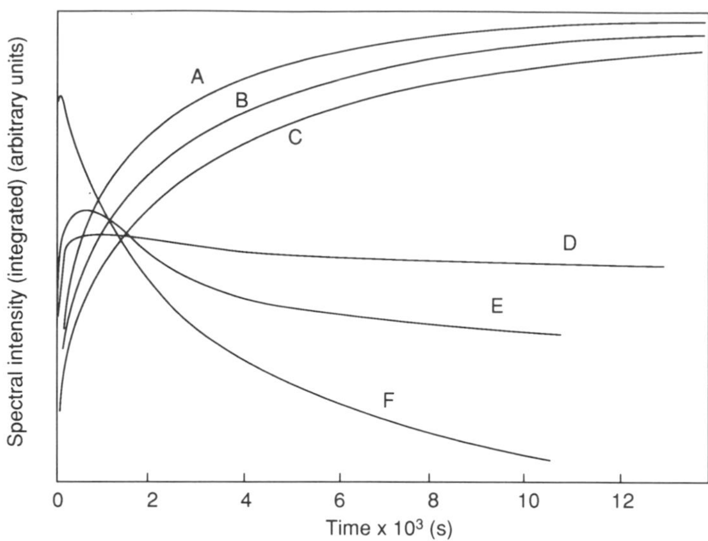

Figure 6 Changes in the spectral intensities of the ESR spectra of irradiated PE at various temperatures when 10 mW microwave power was applied: (A) 198; (B) 190; (C) 182; (D) 213; (E) 227; (F) 241 K [21].

There is an opportunity here to answer the question as to whether the formation reaction of the peroxy radical is controlled by the diffusion of oxygen. From the experimental results, the data observed at 182, 190 and 198 K, at an oxygen pressure of 41 torr, and at 201 K at 222 torr, were analyzed based on the diffusion-controlled process theory [21]. The diffusion equations were solved by the computer simulation method. The best-fitted curves to the experimental data were determined and these are illustrated in Figure 7. The excellent agreement between the simulated curves and the experimental data indicates that the reaction of oxygen with allyl radicals in the amorphous region of the linear PE at low temperatures is a diffusion-controlled process.

The diffusion coefficient (D) and the solubility constant of oxygen can be calculated from the parameters in these best-fitted simulations, and the values obtained are shown in Table 2. The comparatively low values for the diffusion coefficients at low temperatures were estimated first and are very close to the values calculated by the extrapolation of the data at higher temperatures in the non-irradiated PE [22].

By using the fact that peroxy radicals in normal PE are rather unstable and that the reaction of oxygen with allyl radical is a diffusion-controlled process, the diffusion coefficient of oxygen into the crystalline region of PE has been estimated [23]. In powdered single crystals, the allyl radicals were trapped in

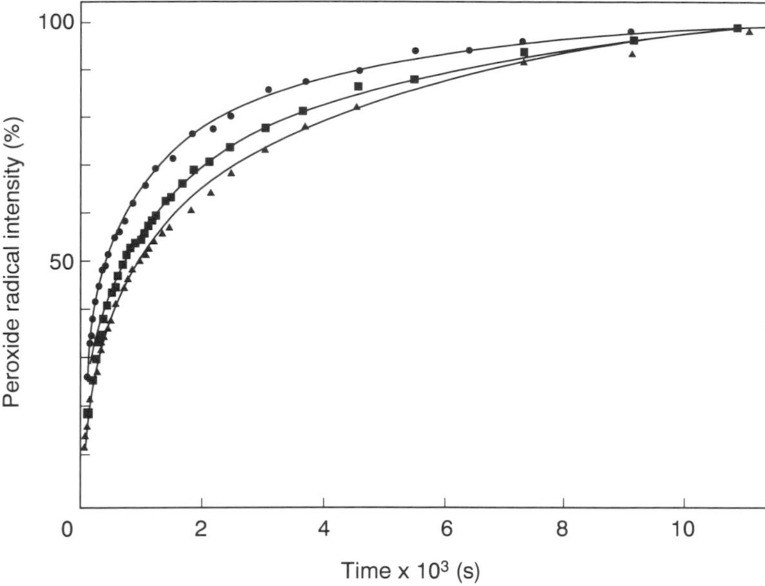

Figure 7 Concentration of peroxy radicals in irradiated PE vs. reaction time normalized at 10^4 s: (—) calculated; experimental at 198 (●), 190 (■), and 182 (▲) K [21].

Table 2 Diffusion Coefficients of oxygen into the amorphous part of PE and solubility constants estimated from the parameters obtained by computer simulation [21]

Temperature (K)	P_{O2} (torr)	N^a	$A_0/B_0{}^b$	D ($cm^2\,s^{-1}$)	Solubility constant (cm^3(STP) $cm^{-3}\,atm^{-1}$)[c]
182	41	200	0.16	2.5×10^{-11}	0.095
190	41	100	0.20	1.0×10^{-10}	0.120
198	41	55	0.25	3.3×10^{-10}	0.145
201	222	40	1.5	6.3×10^{-10}	0.165

[a] N is a parameter related to the diffusion constant.
[b] A_0/B_0 is a parameter related to the solubility constant.
[c] 1 atm = 1.013×10^5 Pa.

the pure state. After introduction of oxygen, the allyl radicals decayed with the storage time. The calculated curves were obtained by computer simulation, and the diffusion coefficients of oxygen were then determined. The order of magnitude of the diffusion coefficients in the crystalline region was found to be 10^{-16} $cm^2\,s^{-1}$ at ca. 320 K, whereas that in the amorphous region is 10^{-8} $cm^2\,s^{-1}$.

2.4 STRUCTURE AND MOLECULAR MOTION OF PEROXY RADICALS IN PE

The temperature-dependent ESR spectra of the peroxy radicals in PE have been successfully observed by Hori and coworkers [7] by using the power saturation technique and spectral subtraction. Mixtures of the peroxy and allyl radicals were prepared by reaction of the calculated amount of oxygen with the allyl radicals at 241 K for 75 min, and ESR spectra of these samples were then observed at various temperatures, using 10 mw of microwave power. The peroxy radicals disappeared after storing the samples for 2 h at

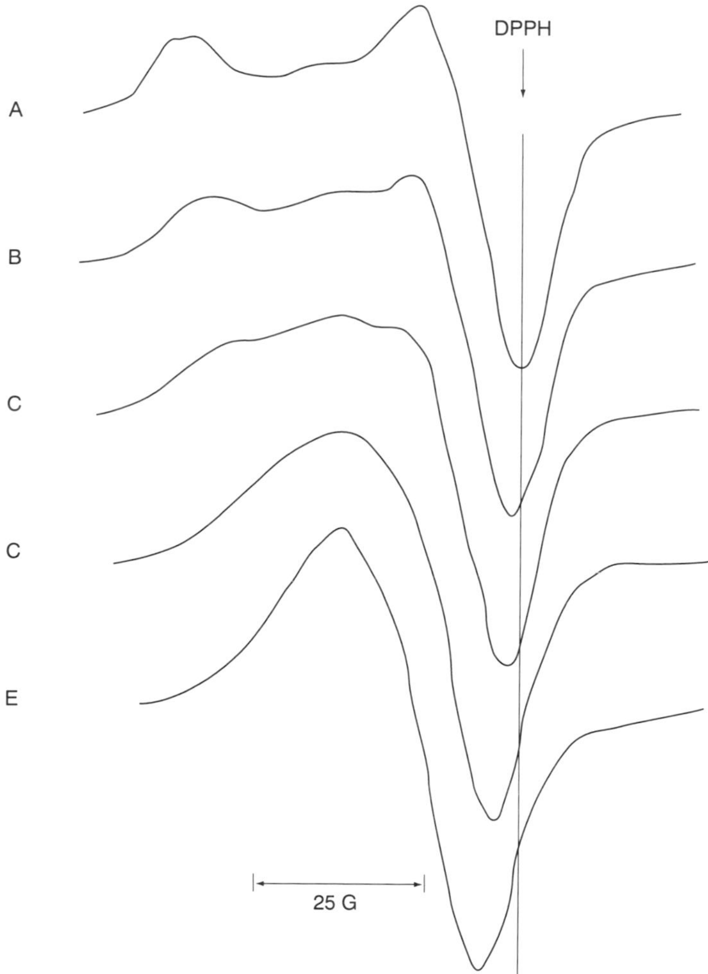

Figure 8 ESR spectra of peroxy radicals in PE obtained at various temperatures: (A) 118; (B) 168; (C) 207; (D) 257; (E) 294 K. The arrow indicates the position of the DPPH resonance, g = 2.0036 [7].

room temperature, and thus corresponding spectra from only the allyl radicals were also observed over the same temperature range.

The spectral patterns of the peroxy radicals at various temperatures were obtained by subtraction of the spectra of the allyl radicals from the superposed spectra of the mixture for the corresponding temperatures and these are illustrated in Figure 8. The spectra consist of two components, as can be seen from the Figure, in particular for spectra B and C. One component, which is due to a radical which is referred to as the A-radical, has features which were similar to the amorphous patterns observed for rigid peroxy radicals. The other component, due to the so-called B-radical, has a broad singlet-like pattern. The A-radical was readily assigned to the peroxy radical, while the B-radical was also assigned to the same radical, because of the reversibility of the spectral changes with temperature. It should be noted that both kinds of radicals exist in the same amorphous region of PE under these experimental conditions.

A computer simulation of the spectra was carried out and examples of the simulated spectra are shown in Figure 9. The method of simulation used here was as follows. It was assumed that there were two groups of peroxy radicals; one of these is rigid (the A-radical), while the other is more mobile (the B-radical). The parameters used were $g_1 < g_2 < g_3$, ΔH_1, and ΔH_2 for the A-radical, and $g_1' < g_2' < g_3'$, $\Delta H_1'$, and $\Delta H_2'$ for the B-radical, with the ratio of concentrations being [B]/[A]. A Gaussian function was used as the line-shape function.

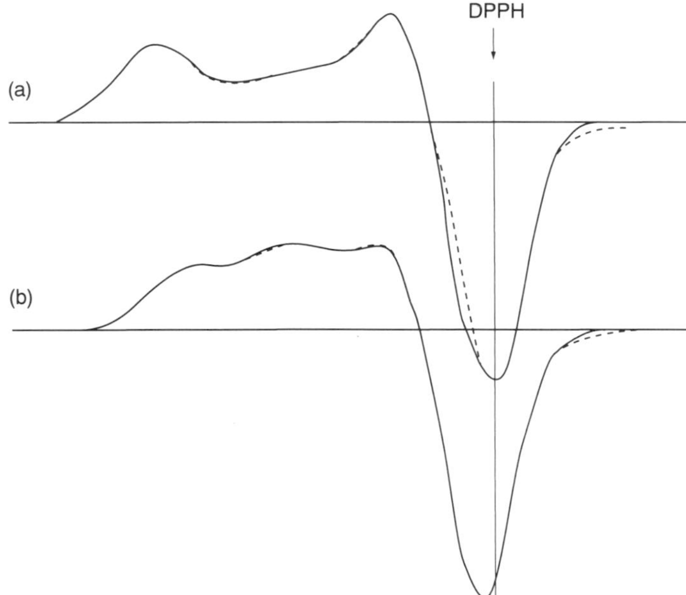

Figure 9 Comparisons of the simulated and observed spectra of peroxy radicals in PE at (a) 118 and (b) 195 K: (—) simulated patterns; (- - -) observed spectra [7].

The good agreement obtained between the simulated and experimental spectra confirmed the co-existence of two kinds of radicals, i.e. A-radicals and B-radicals. At 77 K, there was one kind of peroxy radical and the principal values of the g-tensor obtained from the simulation were g_1 = 2.0024, g_2 = 2.0092, and g_3 = 2.0361. The temperature-dependent anisotropic g-values obtained from the simulation are shown in Figure 11b for both the A- and B-radicals (see below).

Clearer temperature-dependent ESR spectra of the peroxy radicals of PE have been observed in urea-PE inclusion complexes (UPECs) by Hori and coworkers [6]. The structure of the UPEC is represented schematically in Figure 10. In the irradiated complex, the secondary alkyl radicals of PE are

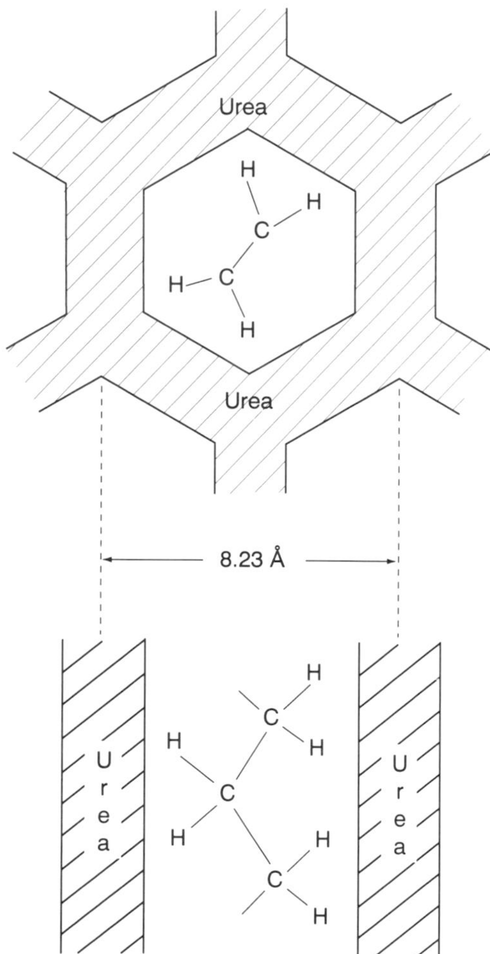

Figure 10 Schematic representation of the structure of the urea–polyethylene inclusion complex [6].

trapped in the pure state and are stable at room temperature. All of the alkyl radicals in the complex were successfully converted to peroxy radicals. The temperature dependence of the ESR spectra showed that the peroxy radicals in the UPEC are stable below 336 K, but convert into alkyl radicals above 354 K. Note that the peroxy radicals in normal PE were unstable even at 241 K.

Examples of the observed spectra of peroxy radicals in the UPEC have already been shown in Figure 3, along with the best-fitted simulated spectra. For the peroxy radicals produced from the alkyl radicals in the UPEC, the principal values of the g-tensor at 77 K were $g_1 = 2.0022$, $g_2 = 2.0081$, and $g_3 = 2.0366$.

Both of the temperature-dependent g-values for the cases of the UPEC [6] and normal PE [7] are shown in Figure 11. The common features between both cases will be discussed first.

Figure 11 clearly shows that the motional averaging of the g-values for both the A- and B-radicals becomes faster at higher temperatures. The smallest g-values for the A- and B-radicals, g_1 and g_1', are equal at all temperatures and the temperature dependence is small. However, the motional averaging between g_2' and g_3' in the B-radicals is different from that in the A-radicals at the same temperatures, and both temperature dependencies are large. In other words, the molecular motion of the B-radicals around the g_1-axis is more rapid than that of the A-radicals and there exists another (slow) motion which moves the g_1-axis. An additional result was derived for the relaxation time of the A-radicals from the Kneubühl equation [2]. The two relaxation times estimated from g_1 and g_3 of the A-radicals were clearly different; the relaxation time estimated from g_3 was shorter than that obtained from g_1. This indicates that the g_3-axis moves more rapidly than the g_1-axis.

In the polymeric chain molecule, two main motions of the peroxy radical should be considered. The first is a rotation of the OO group around the C—O bond axis, and the second is a rotation or vibration of the entire COO group around the chain axis. When rotation of the OO group around the C—O bond occurs, all of the g-values, g_1, g_2 and g_3, are averaged with the same relaxation time, because the g_3-axis is along the O—O bond direction. This conflicts with the experimental results. Therefore, the rotation of the COO group around the chain axis has been concluded to be faster and the direction of the g_1-axis should be along the chain axis. A structure for the peroxy radical as derived from the above discussion has already been shown in Figure 4 for the case of the UPEC, and the same structure is now attributed to the peroxy radical produced from the allyl radical in the amorphous region of PE.

From the above discussion, the following three important conclusions have been reached:

(1) For the structure of the peroxy radical of PE, the g_1-axis is parallel to the chain axis and the C—O—O plane is perpendicular to the chain axis.

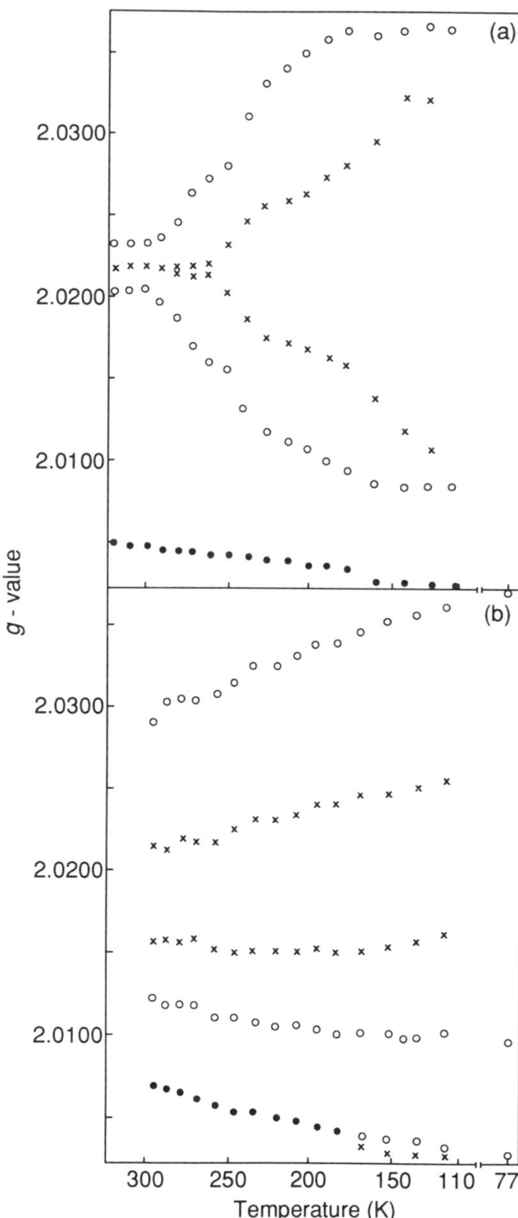

Figure 11 Changes in the anisotropic g-values of peroxy radicals with temperature for (a) the UPEC and (b) normal PE: (○) A-radical; (×) B-radical [6].

(2) The rotation or vibration of the COO group around the chain axis is faster than the rotation of the OO group around the C—O bond.
(3) There are two kinds of peroxy radicals present in PE, namely A- and B-radicals. The speeds of the molecular motions are different between these two radicals. The rotation around the chain axis for the B-radical is faster than that of the A-radical, but the speed of rotation around the C—O bond for the two radicals is similar.

Next, the differences between the peroxy radicals in the amorphous regions of normal PE and those in the UPEC will be considered.

The g_1 value for normal PE shows a much larger temperature dependence than the same quantity for the UPEC, as shown in Figure 11. This difference must be a reflection of the different situations where three-dimensional

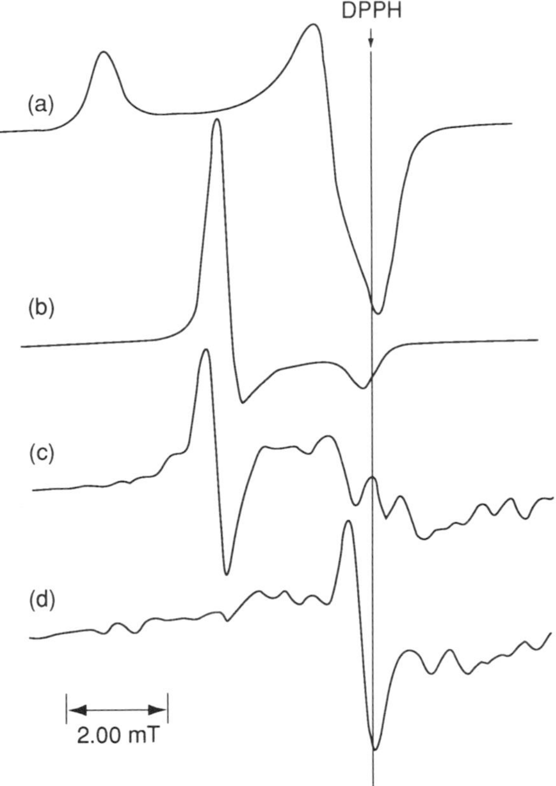

Figure 12 ESR spectra of peroxy radicals in a powdered $UP_{24}C$ sample and in a single crystal of the $UP_{24}C$: (a) powder, 77 K, 2 mW; (b) powder, room temperature, 2 mW; (c) single crystal, room temperature, 50 mW, perpendicular to, *H* *c*-axis; (d) single crystal, room temperature, 50 mW, *H* parallel to *c*-axis [12].

motion of the peroxy radicals is marginally allowed in normal PE, while in the UPEC only two-dimensional motion is possible.

Rotation around the g_1-axis occurs more rapidly in the UPEC than in normal PE at the same temperatures, as shown in Figure 11. This rotation in the complex completely averages out the g_2- and g_3-values at room temperature and gives an axially symmetric spectrum, as shown in Figure 3D, where $g_{\parallel} = 2.0048$ and $g_{\perp} = 2.00213$. This result indicates that rotation around the chain in the complex is faster than in normal PE. This may be due to the fact that the distance between the PE chain and the urea wall in the complex is larger than the distance between the nearest-neighbour PE chains in the amorphous regions of normal PE. On the other hand, the peroxy radicals in the UPEC are most stable than those in normal PE, whereas the rotation around the chain axis is faster in the complex.

Against rotation around the chain axis, it was claimed [24] that 180° rotational jumps of the OO fragment take place around the C—O bond. However, the results obtained for a single crystal of a urea-n-tetracosane complex ($UP_{24}C$) by Hori and coworkers [12] indicate conclusively that rotation around the chain axis is much more rapid than that of the rotation of the OO fragment around the C—O bond.

Angular-dependent ESR spectra of peroxy radicals trapped in single crystals of the $UP_{24}C$ have been obtained in both the *a–b* and *a–c* planes at room temperature. Figure 12 shows ESR spectra of the peroxy radical in a single crystal of the $UP_{24}C$ and in a powdered $UP_{24}C$ sample. The angular dependence of the *g*-values of the peroxy radical in the single crystal of the $UP_{24}C$ is shown in Figure 13.

These two figures show clearly that the peroxy radical of $UP_{24}C$ has an axial *g*-tensor at room temperature, and $g_{\parallel}$ is parallel to the *c*-axis, which is parallel to the chain axis in the crystal of the $UP_{24}C$. The principal values obtained for the *g*-tensor are summarized in Table 3, together with those at 77 K, which were obtained from a powdered sample. From this table it can be safely said that the g_1-direction is parallel to $g_{\parallel}$, and that g_2 and g_3 are

Table 3 Principal *g*-values for the peroxy radicals in $UP_{24}C$ [12].

Powdered $UP_{24}C^a$	$UP_{24}C$ single crystals^b
$g_1 = 2.0025$	$g_{\parallel} = 2.0044^c$
$g_2 = 2.0082$	
$g_3 = 2.0355$	
$(g_1 + g_2)/2 = 2.0218$	$g_{\parallel} = 2.0210^d$

[a] Measured at 77 K.
[b] Measured at room temperature.
[c] Parallel to *c*-axis.
[d] Perpendicular to *c*-axis.

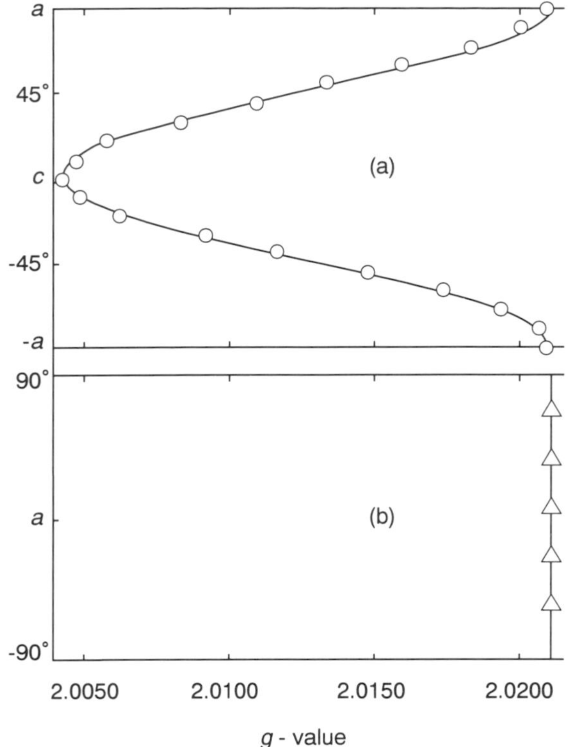

Figure 13 Angular dependence of the *g*-values of the peroxy radical in the $UP_{24}C$ at room temperature: (a) rotating in the *a–c* plane; (b) rotating in the *a–b* plane (the continuous lines represent the theoretical results [12].

averaged at room temperature to become $g_\perp$. Therefore, it is concluded that g_1 is parallel to the chain axis and that the structure of the peroxy radical in $UP_{24}C$ is truly the same as that shown in Figure 4. In order to fully conclude that there is free rotation of the peroxy radical around the chain axis, rather than 180° jumps around the bisector between the g_2- and g_3-directions, the help of other results is needed. In the urea–lauric acid ($C_{11}H_{23}$–COOH) complex [25] the acid is in the form of a dimer and is freely rotating at room temperature. By comparing the volume of the fragment of the peroxy radical with that of this dimer, it can be safely concluded that the peroxy radical is freely rotating around the chain axis at room temperature.

From the UPEC, but not the $UP_{24}C$ studies, the PE molecule was found to be freely rotating at room temperature [26]. From the similarity between the $UP_{24}C$ and UPEC, it was concluded that the peroxy radical in the UPEC is also rotating around the chain axis rather than making jumps around the C—O bond.

2.5 DECAY OF PEROXY RADICALS IN PE

Much work has been carried out on the course of the post-irradiation oxidation of PE [5,21,27–30]. The main results that have been obtained are as follows:

(1) One allyl radical can eventually produce about 12 carbonyl groups and about 5 hydroxyl groups [5]. This result indicates that the chain reaction is very important in the oxidation of PE.

(2) The hydroperoxide of PE was found to be fairly stable below a temperature of 100°C [27].

(3) The peroxy radical of PE trapped in the amorphous region has been found to decay rapidly at room temperature [21,28]; this has already been shown above in Figure 6 [21].

(4) Several methods of oxidation of PE yield isolated hydroperoxide groups, but none of these give adjacent hydroperoxide groups [29].

The conversion from peroxy radicals to alkyl radicals in the UPEC [6] has already been mentioned. There can be two possible reaction schemes for this conversion, i.e. the hydrogen-abstraction reaction (4) or the reverse of the peroxy-radical-formation reaction (3), as shown in the following:

$$\mathrm{ROO^{\cdot} + R'H \longrightarrow ROOH + R'^{\cdot}} \tag{4}$$

$$\mathrm{ROO^{\cdot} \longrightarrow R^{\cdot} + O_2} \tag{3R}$$

Reaction (3R) has been claimed to occur for the conversion of the peroxy radicals of PTFE [31]. However, for the decay mechanism of the peroxy radicals in PE, which is the true reaction, (4) or (3R)?

In order to answer this question, the decay of the peroxy radicals in the UPEC was studied by Hori, Shimada and Kashiwabara [32]. The decay was observed for two cases, namely one in which the radicals were reconverted to the alkyl radicals by annealing the samples *in vacuo*, while in the other method conversion was achieved by annealing the samples in an atmosphere of oxygen. ESR spectra showing the changes are presented in Figure 14 for samples annealed *in vacuo*. Spectrum A arises from pure peroxy radicals, while spectrum D is due to pure alkyl radicals; spectra B and C are superpositions of the peroxy and alkyl signals.

In all of the cases, after annealing the UPEC or deuterated UPEC, either *in vacuo* or in oxygen, at the various temperatures, the decay of the peroxy radicals was found to be a first-order reaction. The rate constants of these reactions were estimated and are listed in Table 4.

It has been concluded [32] that the abstraction of hydrogen atoms by the

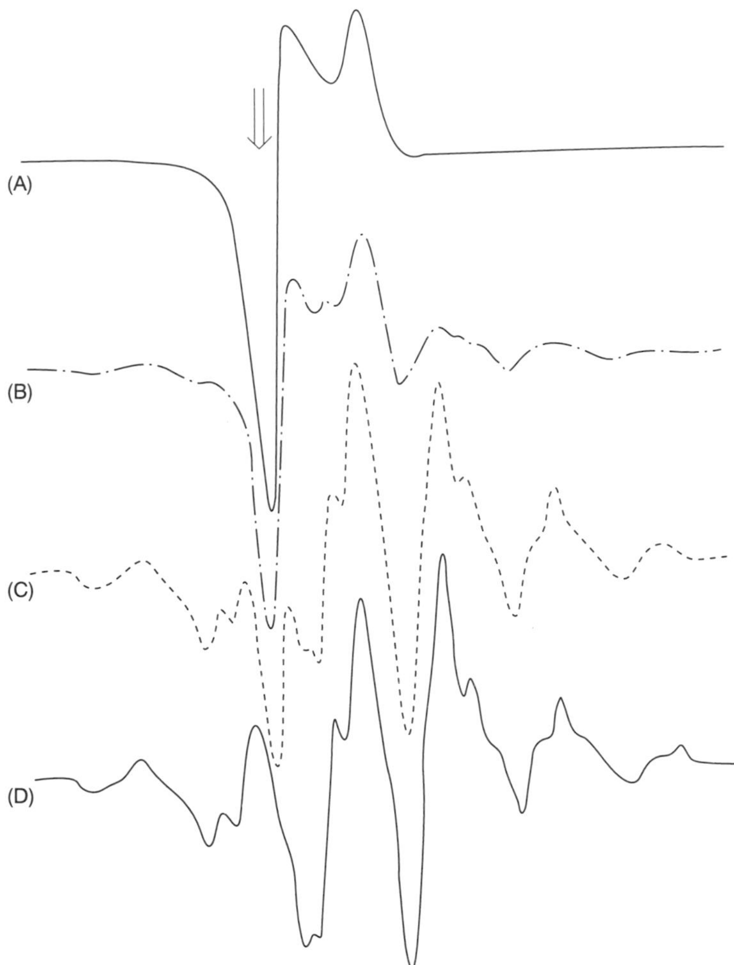

Figure 14 Changes in the ESR spectra of radicals trapped in the UPEC by annealing *in vacuo* at 361 K: (A) 0 min; (B) 21 min, × 1.25; (C) 90 min, × 2.8; (D) 210 min, × 2.8 [32].

peroxy radicals (reaction (4)) takes place in the UPEC, and that the decomposition reaction (3R) does not occur, on account of the following:

(a) As shown in Table 4, the reaction rates in O_2 are almost equal to those *in vacuo* at the same temperatures. If decomposition is a major part of the process, the reaction rate in oxygen must be different from that *in vacuo*, and perhaps even lower.

Table 4 Rate constants obtained for the decay of peroxy radicals trapped in the urea–polyethylene inclusion complex annealing at various temperatures [32].

Temperature (K)	Irradiation dose (Mrad)	Rate constant (s^{-1})	Conditions
343	11.2	5.08×10^{-5}	*In vacuo*
352	11.2	1.19×10^{-4}	*In vacuo*
361	11.2	2.80×10^{-4}	*In vacuo*
371	11.2	7.38×10^{-4}	*In vacuo*
381	11.2	1.6×10^{-3}	*In vacuo*
361	11.2	2.80×10^{-4}	*In vacuo*
361	3.0	2.59×10^{-4}	*In vacuo*
361	11.2	2.82×10^{-4}	In O_2
361	3.0	2.70×10^{-4}	In O_2
361	3.0	2.59×10^{-4}	*In vacuo*[a]

[a] Deuterated complex.

(b) Infrared (IR) spectroscopic measurements confirmed the production of oxidation products after three cycles of post-irradiation oxidation annealing at 360 K for 4 h *in vacuo*.
(c) The amount of residual gas after annealing *in vacuo* was found to be about a quarter of the amount of the peroxy radicals; such an amount indicates that decomposition is not the main part of the reaction.

It has also been established [32] that the hydrogen abstracted by the peroxy radicals comes from the hydrogen on the γ-carbon of the PE and not from the urea. The decay rate of the peroxy radicals *in vacuo* did not change when the hydrogen atoms of the urea were substituted by deuterium, as shown in Table 4. This conclusion is supported by the fact that the peroxy radicals in the complex are very stable at room temperature, even though they are rotating rapidly around the PE chain and that the decay reaction begins above the temperature at which the motion of the OO group about the C—O bond becomes rapid.

For the decay of peroxy radicals in normal PE, the intermolecular hydrogen-abstraction reaction is strongly suggested, on account of the fact that the peroxy radicals in normal PE decay at lower temperatures than those in the UPEC. At these low temperatures, the motion of the OO group about the C—O bond, which is advantageous to the intramolecular hydrogen-abstraction process, does not occur, but rotation of the COO group about the molecular chain, advantageous to intermolecular hydrogen abstraction, does take place.

3 PEROXY RADICALS OF POLYTETRAFLUOROETHYLENE (PTFE)

A reminder is given here that in this section the notation system ($g_3 < g_2 < g_1$) for the anisotropic g-factors will be used.

3.1 FEATURES OF PTFE

PTFE is produced by the polymerization of tetrafluoroethylene, $CF_2{=}CF_2$. Its density is 2.1–2.3 kg dm^{-3}, and its first-order transition temperature is 327°C, with decomposition starting at temperatures above 400°C. This polymer is very strong mechanically at low temperatures and maintains its plasticity even below 4 K. PTFE is also chemically very stable.

In its crystalline form, the PTFE molecule has a helical structure [33], consisting of a 15/7 helix above 19°C and a 13/6 helix below this transition temperature. It is noteworthy that both the 15/7 and 13/6 helix forms have almost planar zigzag structures (of 14/7 or 12/6 helices), and that oxygen molecules can diffuse into the crystalline region of PTFE.

3.2 CARBON-CENTRED RADICALS OF PTFE

Two types of fluoroalkyl radicals can be produced as PTFE radicals. One is a secondary alkyl radical (a chain-type radical), —$CF_2CF^{\cdot}CF_2$—, and the other is a primary alkyl radical (a scission-type radical), —$CF_2CF_2^{\cdot}$. Indeed, both types of radical are fairly stable at room temperature and have been observed by ESR spectroscopy.

The ESR spectrum of the secondary alkyl radical was first obtained by Rexroad and Gordy in 1959 [34]. The pattern consists of a double quintet, arising from four equivalent fluorines with couplings of 3.3 mT each and a fifth fluorine with a coupling of 9.2 mT.

The scission-type alkyl radical was clearly observed for the first time by Siegel and Hedgpeth in 1967 [35]. These propagating radicals, —$CF_2CF_2^{\cdot}$, are formed when the peroxy radicals are irradiated with ultraviolet light in vacuum, and are believed to arise from the following photoreaction:

$$—CF_2CF(OO^{\cdot})CF_2— + h\nu \longrightarrow —CF_2CF_2^{\cdot} + CF_2O + CO + CF_3CF_2— \quad (7)$$

Both CF_2O and CO were found as products of the photolysis; this reaction mechanism was also supported by the INDO calculations [36].

The alkyl radical has an ESR spectrum consisting of a triplet of triplets. In the spectrum observed at room temperature, the central triplet set is well resolved, with a splitting of 1.6 mT; the wing lines are broad, however. The g-value for this radical is 2.0038. Here it should be noted that there have been some reports in the literature [31,34] in which the central triplet was thought to be due to the hyperfine structure of the corresponding alkoxy radical.

3.3 FORMATION OF PEROXY RADICALS OF PTFE

The peroxy radical of PTFE was first identified by Ard, Shields and Gordy after the ageing in air of X-irradiated PTFE samples and were the first polymer peroxy radicals to be observed [1]. Its ESR spectrum has already been shown in Figure 1 above, and a description of the identification of this peroxy radical was also previously given.

After this first report, much work was carried out on the ESR spectra of peroxy radicals in PTFE [31,34,37–41]. The ESR spectra at 4 and 77 K were similar [39], while on the other hand, the spectrum observed at room temperature was different in shape from those observed at 4 or 77 K.

Two types of spectra were observed by Matsugashita and Shinohara in an ESR study carried out at room temperature [31,38]. One of these is asymmetric (denoted as O_I in this section), while the other showed a symmetrical singlet, situated at $g = 2.016$ (denoted as O_{II}).

The disappearance of the double quintet and the growth of O_I after admission of oxygen, and also the reverse process that occurs after re-evacuation at 150°C for 15 min, suggest that O_I is a signal for the radicals which are produced when the —$CF_2CF^{\cdot}CF_2$— radicals are oxygenated [31,38]. The likely reaction is as follows:

$$\underset{\text{(double quintet)}}{—CF_2—CF^{\cdot}—CF_2—} + O_2 \longleftrightarrow \underset{(O_I)}{—CF_2—CF(OO^{\cdot})—CF_2} \qquad (8)$$

Admission of oxygen causes this reaction to proceed towards the right, while re-evacuation at a higher temperature leads to the reverse reaction.

From the orientation dependence of the ESR spectra of O_I in stretched PTFE films, the g-values were estimated: $g_1 = 2.037$, $g_2 = g_3 = 2.0065$ [40], or $g_\perp = 2.002$ and $g_\parallel = 2.04$ [41].

On the other hand, O_{II} was assumed to represent the signal of a chain-end peroxy radical, —$CF_2OO^{\cdot}$ [31,38]; its ESR spectrum is symmetrical and has a g-value of 2.016. For the stretched PTFE film, it was also observed that the additional peroxy radical, O_{II}, gave an angle-independent ESR signal [40]. The singlet for this radical has a g-value of 2.018, which is very close to the average g-value of the above mentioned secondary chain-type peroxy radical of 2.017 (= (2.037 + 2.0065 + 2.0065)/3)). The closeness of the g-values and the features of this radical strongly suggest that this is also a peroxy radical, and in addition is moving very rapidly at room temperature.

Conclusive evidence has been presented [35] for the assignment of O_{II} to —CF_2—CF_2—$OO^{\cdot}$ radicals. The chain-end fluoroalkyl radicals were produced by the method described in Section 3.2. When oxygen was added to a PTFE system containing only the propagating primary alkyl radicals, the ESR spectrum shown in Figure 15a was observed. This spectrum is essentially a symmetric single line and is similar to O_{II}. However, if the temperature was lowered to 77 K, the spectrum shown in Figure 15b was seen. This latter spectrum was very similar to that observed for the chain-type peroxy radicals. Therefore, the spectra shown in Figure 15 can be readily assigned to the peroxy radicals of the propagating alkyl radicals, while the O_{II} spectrum is considered to arise from the propagating peroxy radicals.

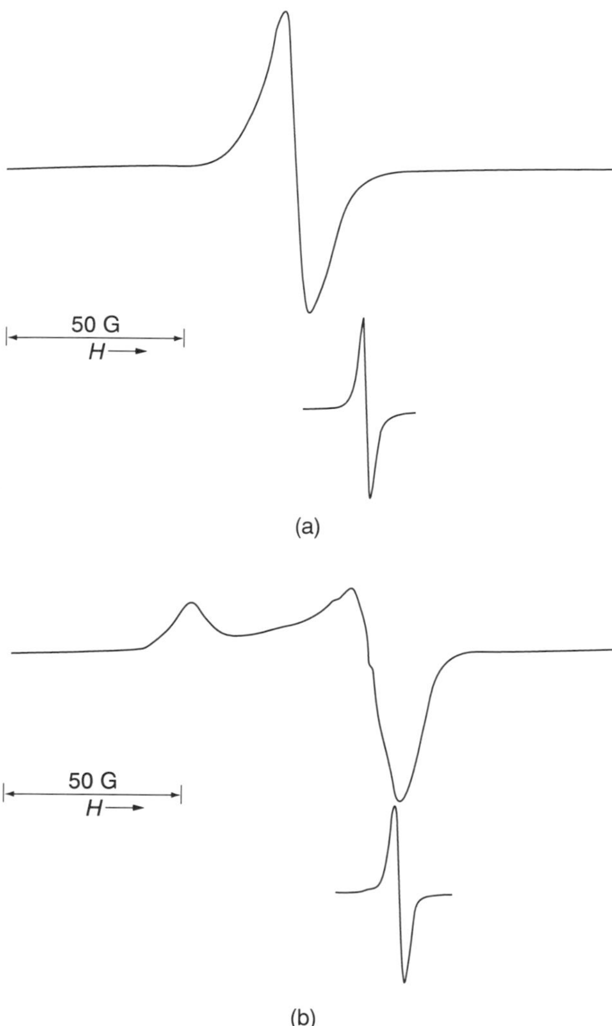

Figure 15 (a) the ESR spectrum, measured at room temperature, of PTFE which has been γ-irradiated in vacuum, oxygenated, re-evacuated, photolyzed with ultraviolet light in vacuum, and finally re-oxygenated. (b) the equivalent ESR spectrum measured at 77 K (the spectrum of DPPH, $g = 2.0036$ is shown in both cases) [35].

3.4 STRUCTURE AND MOLECULAR MOTION OF PEROXY RADICALS IN PTFE

From the ESR spectrum of O_I for stretched samples of PTFE [40,41], the OO group has been suggested to lie perpendicular to the chain axis. The ESR spectra observed at room temperature [41] also suggest rotation of the COO group about the chain axis.

Iwasaki and Sakai [11] have measured the ESR spectra of peroxy radicals in irradiated powders and oriented samples of PTFE using a K-band (24 GHz) ESR spectrometer, and determined the principal values and directions of the g-tensor, both at room temperature and at 77 K.

Figure 16 shows the variation with temperature of the ESR powder

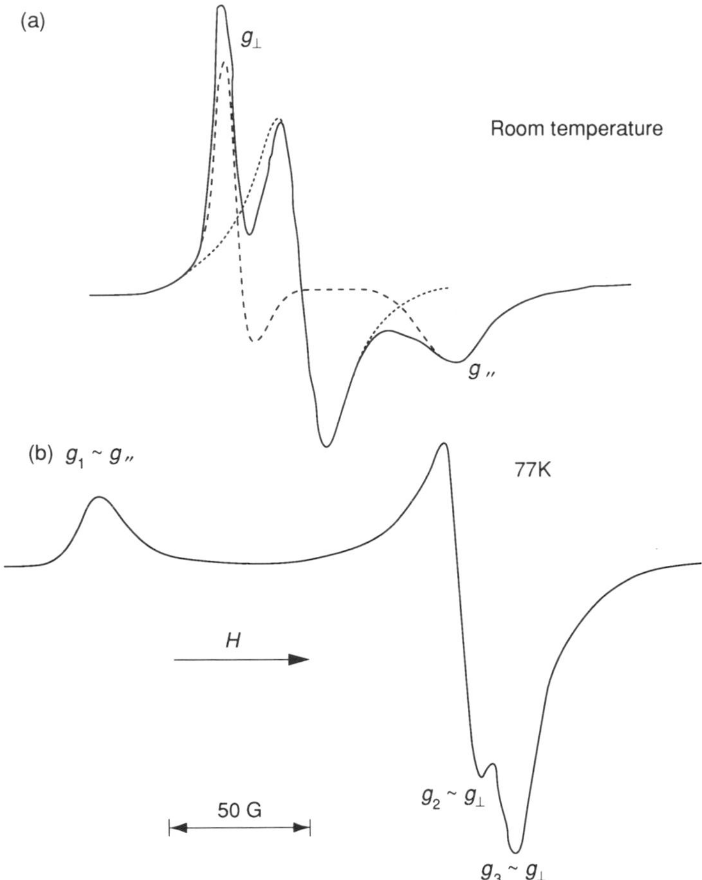

Figure 16 ESR spectra of peroxy radicals in γ-irradiated PTFE powders observed (a) at room temperature and (b) at 77 K: (...) symmetric component; (- - -) asymmetric component. Measurements were made at 24 GHz [11].

spectra, measured at 24 GHz, of peroxy radicals trapped in PTFE. The spectrum observed at 77 K consists of a single component (shown in Figure 16b). The following principal values of the g-tensor were obtained at 77 K from the line-shape of this spectrum: $g_1 = 2.038(4)$ $g_2 = 2.007(1)$, and $g_3 = 2.002(6)$; g_1 is considered to correspond approximately to $g_{\|}$, while g_2 and g_3 represent $g_{\perp}$.

On the other hand, the spectrum measured at room temperature (Figure 16a) was split into two components, as has been mentioned above [31,38,40]. The symmetric component (indicated by the dotted line) is attributed to those peroxy radicals which have enough motional freedom to average out the entire anisotropy in g. The asymmetric component (indicated by the broken line) has a characteristic line-shape arising from the axially symmetric g-tensor. From the line-shape one can obtain the principal values: $g_{\|}^{r} = 2.006(1)$ and $g_{\perp}^{r} = 2.022(1)$, where $g_{\|}^{r}$ and $g_{\perp}^{r}$ represent the apparent principal values at room temperature. The averages of the principal values of the g-tensors of the asymmetric component are 2.016(8) (room temperature) and 2.016(0) (77 K). The value of the symmetric component at room temperature is also 2.016(9). Given the following facts, (1) the temperature changes in the principal values are reversible, and (2) the average values of the g-tensor elements are nearly equal, both at 77 K and at room temperature, there is a strong indication that the apparent spectral changes at room temperature are due to complete or partial averaging of the tensor elements by molecular motion. Such motional effects should be quenched at 77 K.

The ESR spectra for stretched PTFE films were also measured at room temperature and at 77 K [11]. The room-temperature spectra obtained at various orientations of the stretch axis to the magnetic field are shown in Figure 17. In addition to a strong signal with large anisotropy, a broad weak signal can be seen at $g = 2.016$. Since this peak is nearly isotropic and its g-value is very close to the g-value of the symmetric component of the powder samples, we can again attribute this peak to the peroxy radical of PTFE which has averaged out the entire anisotropy in g.

The value of 2.005(9), which was found with the field parallel to the stretch axis, is very close to $g_{\|}^{r} = 2.006(1)$, obtained for the powder spectrum, while 2.021(3), measured for the perpendicular direction, is close to $g_{\perp}^{r} = 2.022(1)$. This means that the symmetry axis of the g-tensor is parallel to the molecular chain axis at room temperature.

The angular dependence of the spectra was also measured at 77 K [11]. Typical spectra obtained for both the parallel and perpendicular directions are shown in Figure 18. In order to eliminate the contribution from randomly oriented radicals, the powder spectrum obtained at 77 K was subtracted from the spectra of the oriented films. The residual spectra are indicated by the dotted lines in Figure 18. The parallel spectrum of —$CF_2CF(OO^{\bullet})CF_2$— thus obtained, has a narrow symmetric peak at $g = 2.003(1)$, while the perpendicular spectrum still extends from $g = 2.005(1)$ to 2.038(3).

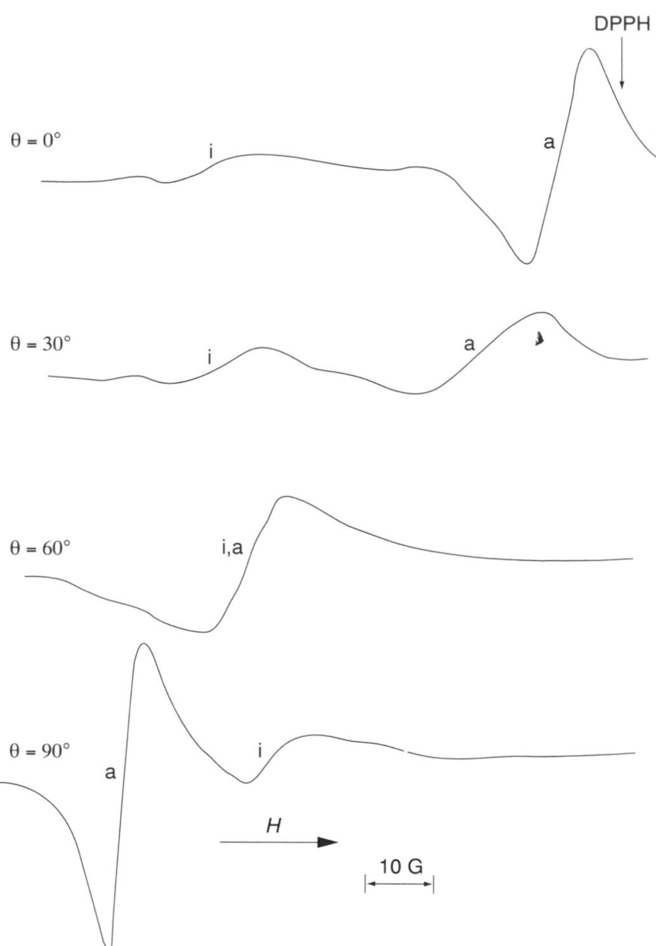

Figure 17 Angular dependence of the room-temperature ESR spectra of peroxy radicals in γ-irradiated oriented films of PTFE, where the angles indicated are the field directions measured from the stretch axis; i and a represent the isotropic and anisotropic components, respectively. Measurements were made at 24 GHz [11].

Therefore, the direction of the maximum principal value is perpendicular to the stretch axis. This means that the direction of the O—O bond is perpendicular to the molecular chain axis. It was concluded that the COO group lies in the plane perpendicular to the molecular chain axis, as shown in Figure 19 [11]. Iwasaki and Sakai could not conclude definitely that the g_3-direction is parallel to the chain direction (as shown in Figure 19) and considered that alternative directions for g_2 and g_3 might be possible.

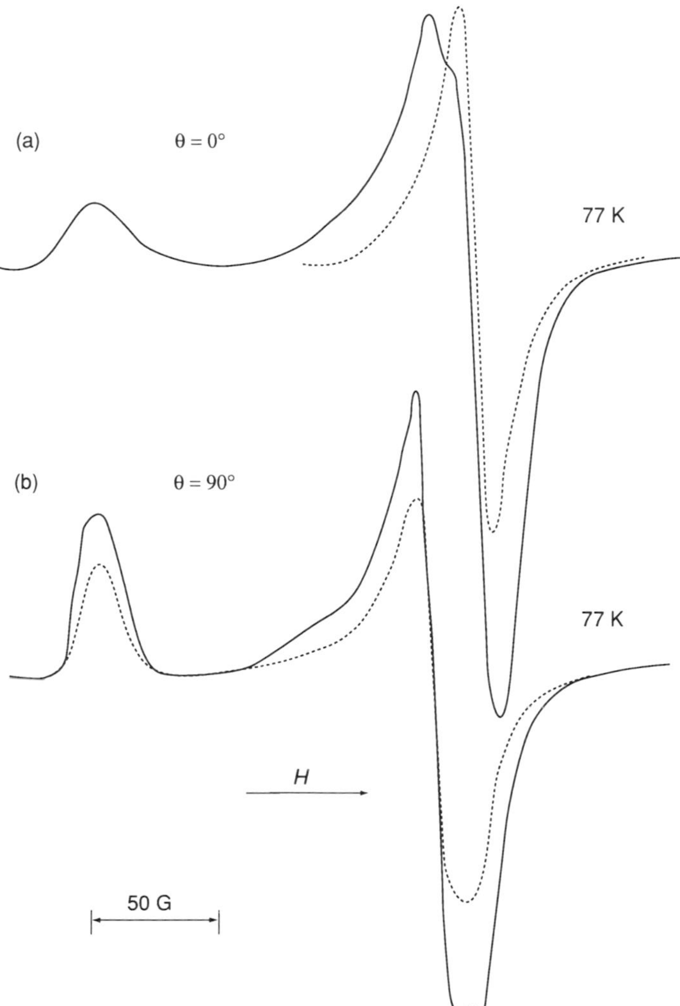

Figure 18 ESR spectra at 77 K of peroxy radicals in γ-irradiated oriented films of PTFE observed with the field (a) parallel and (b) perpendicular to the stretch axis; (...) curves obtained by subtracting from the continuous curves the contributions to the powder spectra due to randomly oriented radicals. Measurements were made at 24 GHz [11].

Let us suppose that very rapid motion around the C—O bond occurs at room temperature; the *g*-tensor should then be axially symmetric about this bond, and consequently the direction of $g_{\perp}{}^{r}$ becomes perpendicular to the chain axis. This result contradicts the experimental findings. On the other hand, if rapid motion around the chain axis takes place, the *g*-tensor should

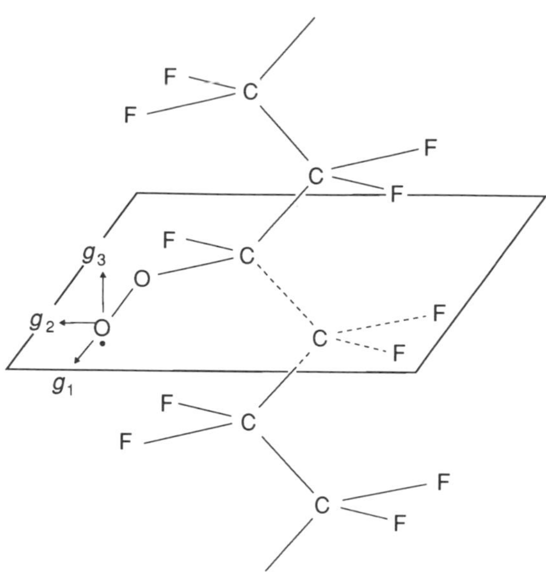

Figure 19 Structure of peroxy radicals in γ-irradiated PTFE showing the principal directions of the g-tensor at 77 K; note that alternative directions for g_2 and g_3 may be possible [11].

be axially symmetric about the molecular chain axis, which is in agreement with experiment. A rotation around the chain axis, rather than rotation around the O—O bond, has therefore been concluded.

Temperature variations in the line-shapes of the ESR spectra from the two types of peroxy radicals, namely chain and end radicals, of PTFE have been investigated by Moriuchi *et al.* [10]. The peroxy chain radicals were produced by the usual procedure, while the peroxy end radicals were produced by the method of Siegel and Hedgpeth [35]. The temperature changes in the ESR spectra of the peroxy chain radicals are shown in Figure 20.

From this figure the conclusion has been reached that the peroxy chain radical is freely rotating about the chain axis at room temperature. Moreover, it has been concluded that the g_3-direction is parallel to the chain axis, because its position remains almost unchanged during the temperature changes.

On the other hand, for the peroxy end radicals the position of the main peak, corresponding to g_3 at low temperatures, did not stay unchanged, but shifted gradually to the lower-field side with increasing temperature, with the line-shape appearing finally as a narrow singlet above 240 K. This temperature variation indicates that the molecular motion of the peroxy end radical is a three-dimensional random motion which averages out the three principal values of the g-tensor.

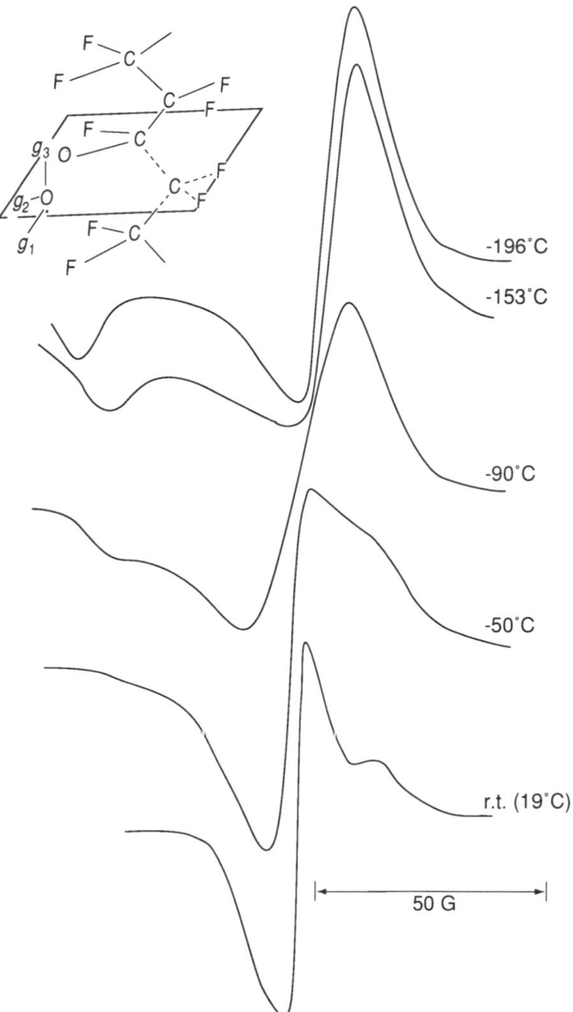

Figure 20 Temperature changes in the ESR spectra observed for the peroxy chain radicals of PTFE; the structure of the species shows the directions of the principal axes of the *g*-tensor [10].

More precise studies [42] of temperature-dependent ESR spectra have shown the following features. The peroxy propagating radical undergoes rotation about the polymer chain axis above −120°C, with additional rotation about the C—O bond beginning to take place above −40°C. However, the peroxy chain radical has the same rotation about the chain axis above −20°C, and any additional rotation about the C—O bond is not observed below the decomposition temperature.

The existence of the peroxy radical of PTFE has been definitively proved by Che and Tench [43,44], based on observations of ESR lines from two oxygen nuclei both labelled with ^{17}O. The peroxy chain radical, —$CF_2CF(O^1O^{2\cdot})CF_2$—, has $a_{\parallel}(1) = 4.0$ and $a_{\parallel}(2) = 8.9$ mT at 300 K, while the peroxy propagating radical, —$CF_2CF_2(O^1O^{2\cdot})$, has $a_{iso}(1) = 1.3$ and $a_{iso}(2) = 2.65$ mT at the same temperature. At 77 K, both types of peroxy radicals have the same parameters, i.e. $a_1(1) = 4.6$ and $a_1(2) = 10.7$ mT. These results indicated that the p orbital is parallel to the chain direction and that 70% of the spin is on the O^2 atom with 30% on the O^1 atom. Temperature changes of the hyperfine parameters of the ^{17}O support the proposals that the peroxy chain radical undergoes rotation around the polymer chain axis and the peroxy propagating radical has three-dimensional random motion at room temperature.

As an application of using the peroxy radicals of PTFE, the degree of orientation in elongated PTFE films for two directions have been estimated by Shimada, Hori and Kashiwabara [45]. The rigid peroxy radicals, —$CF_2CF(OO^\cdot)CF_2$—, were cleanly trapped in the crystalline region of the elongated PTFE film at a stretch ratio of 3.6.

Angular-dependent spectra were obtained, with examples being shown in Figure 21. The remarkable angular dependence which is observed indicates that orientation of the polymer chains occurred along the stretching direction. Moreover, further features of the angular dependent of the observed ESR spectra suggest that there were biaxial orientations. From comparisons of the experimental spectra with the simulated spectra, the degree of orientation was estimated as follows: $f_\alpha = 0.45 \pm 0.02$ and $f_\alpha' = 0.10 \pm 0.02$, where f_α and f_α' are the degrees of orientation of the polymer chains in the stretching direction and the film plane, respectively. In addition, it has been observed that even in a copolymer of tetrafluoroethylene–hexafluoropropylene a chain-axis rotation of the peroxy chain radical, –$CF_2CF(OO^\cdot)CF_2$–, takes place at room temperature [46].

3.5 DECAY OF PEROXY RADICALS IN PTFE

The effects of various gases on the oxygenated radicals of PTFE have been reported by Rexroad and Gordy [34]. When an atmosphere of H_2 gas was admitted to the samples, the signal from the oxygenated radical disappeared within minutes. A possible mechanism for this effect is that H_2 reacts with the radical to form a hydroperoxide and atomic hydrogen, and that the atomic hydrogen then proceeds to escape through the lattice. When ionized air or nitric oxide (NO) were admitted to the samples, the signal for the oxygenated radical disappeared. These effects are not surprising, since radicals present in ionized air or the NO radical can destroy an oxygenated PTFE radical by direct addition.

After re-evacuation of oxygenated samples at 150°C for 15 min, the signals

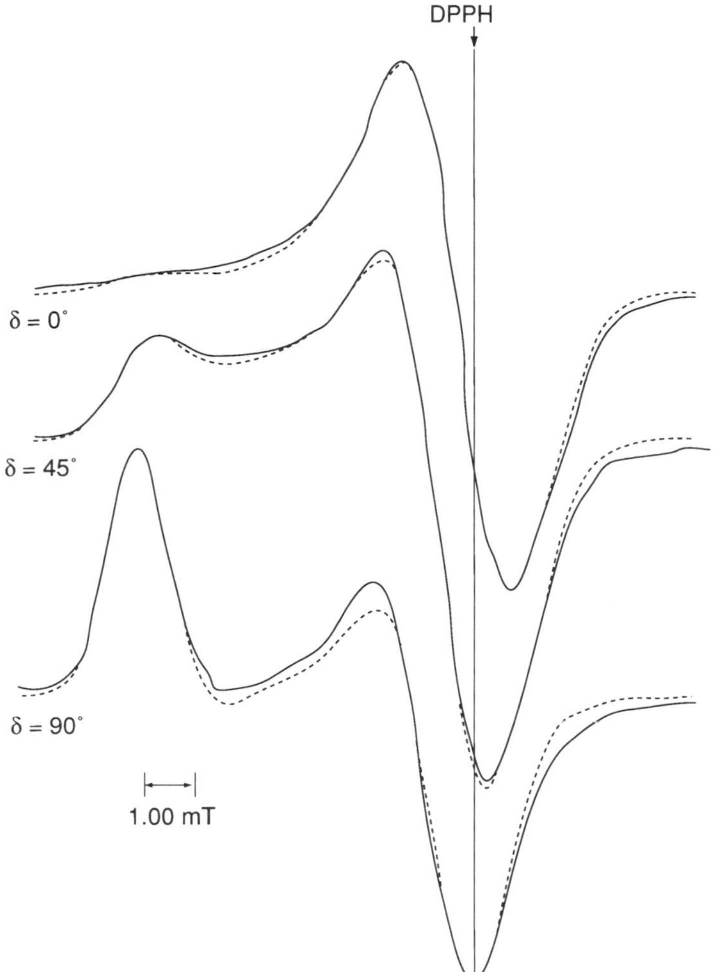

Figure 21 Comparison of experimental ESR spectra of the peroxy radicals of PTFE with simulated spectra for different rotation angles (δ) recorded at 77 K: (—) experimental data; (- - -) simulated data, for $f_\alpha = 0.44$ and $f_\alpha{}' = 0.10$ [45].

of the oxygenated radicals partially disappeared and the signals of the double quintets from the secondary fluorocarbon radicals were recovered [31,38]. The signal lost in this change will be referred to as O_1. Further evacuation of the same sample at 250°C for 15 min led to the signal of the oxygenated radicals disappearing completely. The signal lost during this change was a symmetrical singlet, situated at $g = 2.016$ (denoted at O_{II}).

When the oxygenated sample was exposed to N_2 or CO_2, O_I disappeared and the double quintets were recovered at temperatures of 50–150°C; O_{II}

disappeared at 250°C in the same manner as when the sample was evacuated. In H_2, at 50–150°C, O_I also disappeared, but the double quintets were not recovered.

For the reactions between the peroxy radical of PTFE and the various gases, the following reactions can be proposed:

$$\underset{(O_I)}{—CF_2CF(OO^{\cdot})CF_2—} \longrightarrow \underset{\text{(double quintet)}}{—CF_2CF^{\cdot}CF_2—} + O_2 \quad (9)$$

$$—CF_2CF(OO^{\cdot})CF_2— + H_2 \longrightarrow —CF_2CF(OOH)CF_2— + H^{\cdot} \quad (10)$$

This last reaction was thought to occur in H_2 at room temperature, but thus has not been confirmed. If the reverse reaction (9) occurs and then the secondary fluorocarbon radical decays in H_2 [34], the PTFE peroxy radical might then decay in H_2. If this is the case there will be no hydroperoxides produced.

When the peroxy radicals decay in H_2 at room temperature hydroperoxides should be present, because of the following facts: 89% of the oxygenated radicals were still remaining after 15 min, with 75% remaining after 24 h after re-evacuation at room temperature [38], but within minutes after the admission of H_2 the signal from the oxygenated radicals disappeared [34]. These two results taken together indicated that ca. 90% of the oxygenated radicals converted to hydroperoxides at room temperature when H_2 was admitted.

We will consider next whether $—CF_2CF(OOF)CF_2—$ actually exists or not. In other words, does the following reaction take place?

$$—CF_2CF(OO^{\cdot})CF_2— + -CF_2CF_2CF_2— \longrightarrow —CF_2CF(OOF)CF_2— + —CF_2CF^{\cdot}CF_2— \quad (11)$$

Such a reaction could also explain the fact of the disappearance of O_I and the growth of the double quintets by re-evacuation at 150°C, in the same way that reaction (9) can do. In the case of hydrocarbon polymers other than PTFE, reactions similar to (11) should take place, rather than the corresponding reverse actions (c.p. reaction (9)). However, in PTFE reaction (11) may not take place. If reaction (11) occurs in oxygen, a radical site would migrate and thus cause decay of the oxygenated radicals. However, such a decay has not been observed. Thus it has been concluded that reaction (11) does not take place; this might also suggest instability of $—CF_2CF(OOF)CF_2—$.

The effect of CO, C_2H_4, SO_2, and $^{17}O_2$ on the ESR spectra of the two types of peroxy radicals of PTFE has been investigated by Olivier, Marachi and Che [47]. The peroxy chain radicals appear chemically stable up to 100°C, while under the same experimental conditions the peroxy propagating radicals do react. From changes in the ESR spectra, combined with mass spectrometric analysis, the following reactions have been shown to take place:

$$R{-}CF_2{-}OO^{\bullet} + 2\,CO \longrightarrow R{-}CF_2^{\bullet} + 2\,CO_2 \qquad (12)$$

$$R{-}CF_2{-}OO^{\bullet} + SO_2 \longrightarrow R{-}CF_2{-}SO_2^{\bullet} + O_2 \qquad (13)$$

$$(R{-}CF_2{-}OO^{\bullet}) + (C_2H_4) \longrightarrow \text{diamagnetic compound} \qquad (14)$$

By using ^{17}O labelled peroxy radicals, it has also been shown that no alkoxy radicals are produced in the above reaction schemes.

4 PEROXY RADICALS OF POLYPROPYLENE (PP)

4.1 FEATURES OF PP

This polymer is produced by the polymerization of propylene, $CH_2{=}CH(CH_3)$. There are three types of PP which have three different configurations, namely isotactic PP (i-PP), syndiotactic PP (s-PP), and atactic PP (a-PP), with the latter being completely amorphous. Nearly 100% of commercially produced PP is in the isotactic form, and so the use of 'PP' in the following discussion will refer to this form, unless mentioned otherwise.

The moleculars in crystalline PP have a helical structure (3/1helix) [33]; this crystalline form has a melting temperature of 180°C.

Oxygen molecules are able to diffuse easily into both the crystalline and amorphous regions of PP. This can be adequately explained by the fact that the molecular size of oxygen is smaller than the size of vacancy in the PP crystallites.

4.2 CARBON-CENTRED RADICALS OF PP

Three types of alkyl radicals can be trapped in PP by irradiation at low temperatures; namely the primary alkyl radical, $—CH_2CH(CH_2^{\bullet})CH_2—$, the secondary alkyl radical, $—CH^{\bullet}CH(CH_3)—$, and the tertiary alkyl radical, $—CH_2C(CH_3)^{\bullet}CH_2—$. The ESR spectrum observed at room temperature consists of one component with 17 well resolved lines and has been assigned as being due to the tertiary alkyl radical, $—CH_2C(CH_3)^{\bullet}CH_2—$ [48]. These radicals are very stable at room temperature. In this radical, the four β-protons are equivalent pairs, and are denoted as follows: $—CH^1H^2—C(CH^0_3)^{\bullet}—CH^1H^2—$.The spectrum of this radical is very nearly isotropic at room temperature, with isotropic hyperfine coupling parameters of $a_0 = 2.1$, $a_1 = 0.9$, and $a_2 = 4.3$ mT.

4.3 FORMATION OF PEROXY RADICALS OF PP

ESR spectra of the peroxy radicals of PP were first reported by Ohnishi, Sugimoto and Nitta [49]. On the introduction of air to irradiated PP, the ESR spectrum changed immediately to the typical asymmetric singlet of the peroxy radical (see Figure 22). This result suggests that almost all of the carbon radicals of PP were oxidized to form peroxy radicals immediately after the introductin of air. This behaviour is very different to that observed for PE. In PP, the oxygen molecules can easily diffuse into the crystalline regions of the polymer, and therefore all of the carbon radicals can react with oxygen at a very fast reaction rate.

The spectrum of the peroxy radicals obtained immediately after the introduction of air has been found to consist of two components [50]. The typical amorphous pattern resulting from the anisotropic *g*-tensor was

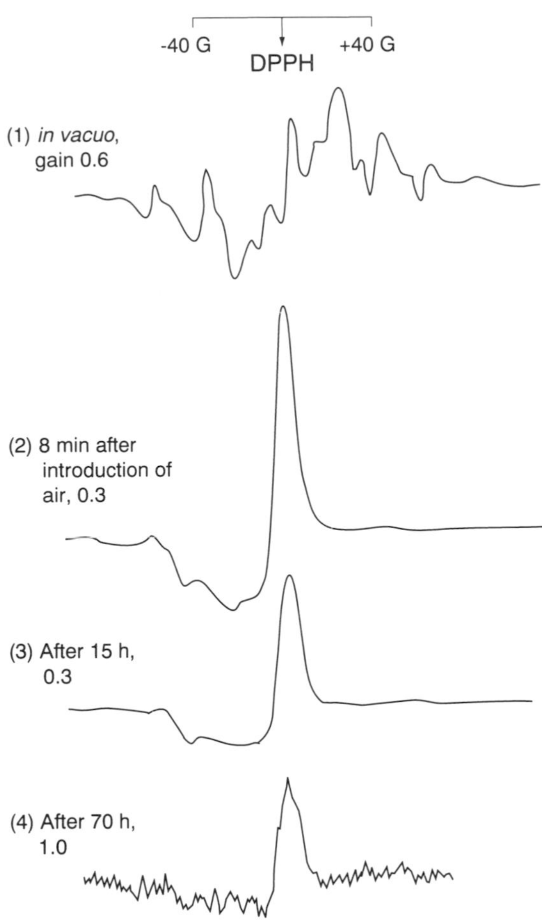

Figure 22 Changes in the ESR spectra after introduction of air to PP electron-irradiated at −78°C and 1 Mrad, showing the reaction of the PP carbon-centred radical with oxygen [49].

observed after annealing in air at 70°C; the components of the axial symmetric g-tensor, i.e. $g_{\parallel} = 2.034$ and $g_{\perp} = 2.004$, were obtained from the amorphous pattern [50].

4.4 STRUCTURE AND MOLECULAR MOTION OF PEROXY RADICALS IN PP

Early work strongly suggested that the O—O bond of the peroxy group in stretched PP is perpendicular to the draw axis [50,51]. Furthermore, it was pointed out [52] that there are two types of peroxy radicals present, with one having more motional freedom than the other.

In 1984, the temperature-dependent ESR spectra of the peroxy radicals resulting from the tertiary alkyl radicals trapped in i-PP and a-PP were obtained, and the molecular motions, along with the structure of the peroxy radicals, were extensively discussed by Hori and coworkers [8].

The ESR spectra of the peroxy radicals in i-PP observed at 4 and 77 K were identical. This result indicates that peroxy radicals at 77 K do not have any motion. The spectra observed at higher temperatures consist of two components, arising from the rigid peroxy radicals and the mobile peroxy radicals. All of the observed spectra were reconstructed by means of computer simulation and the changes in the anisotropic g-values with temperatures were obtained. The simulation method used here was essentially the same as that described in Section 2.4. Some typical temperature-dependent ESR spectra of the peroxy radicals trapped in i-PP are shown in Figure 23, along with the corresponding simulated spectra. The agreement between the observed spectra and the simulated ones is excellent.

The temperature dependence of the g-values obtained from the simulation are plotted in Figure 24 for the peroxy radicals in i-PP. For the peroxy radicals in a-PP, the ESR spectra and temperature dependence of the g-values showed similar features to those in Figures 23 and 24, respectively. However, for a-PP the line-shapes were broader and the molecular motion occurs at lower temperatures when compared with the radicals in i-PP.

The values of g_1 and g_1' for PP are different at higher temperatures , as seen in Figure 24, whereas g_1 and g_1' for PE have the same values at all temperatures [6,7,32]. This difference between g_1 and g_1' was confirmed by the direct observation of two peaks in the spectra by means of a K-band ESR spectrometer [8].

As can be seen for the A-radical in Figure 24, g_3 (maximum g-value) decreased and g_1 (minimum g-value) increased slightly with increasing temperature. This result can be explained by assuming that only a small vibration of the O—O group around the C—O bond occurs in the A-radical. The motion of the B-radical evidently becomes greater or faster with increasing temperature, as shown in Figure 24.

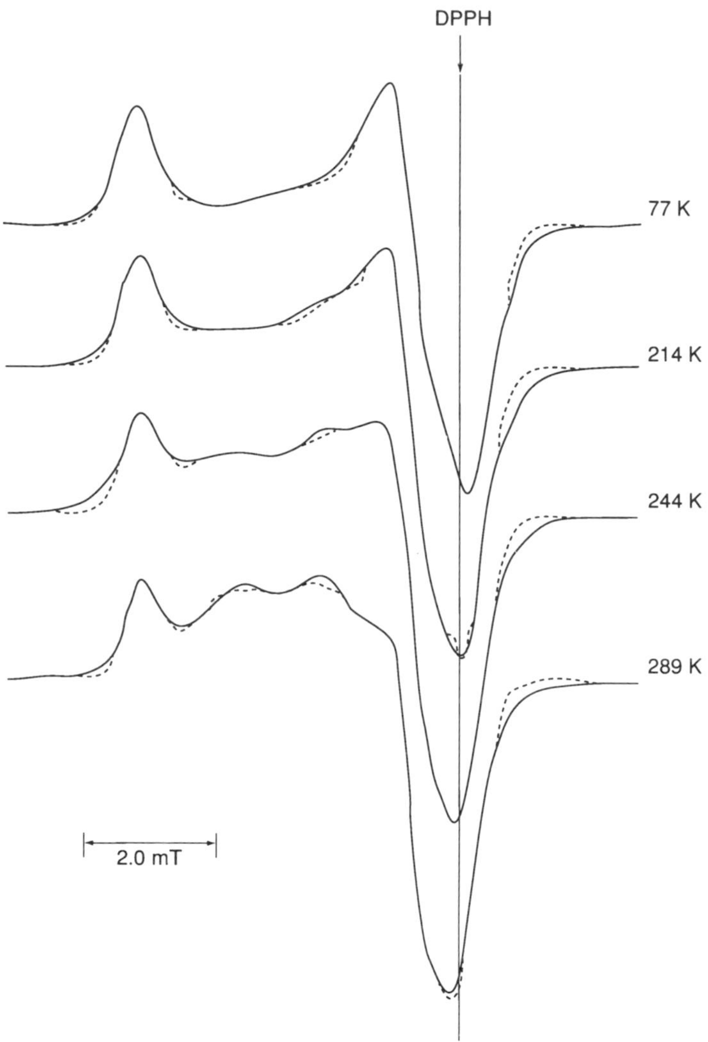

Figure 23 Some examples of the temperature-dependent ESR spectra of the peroxy radicals trapped in i-PP, plus the corresponding simulated spectra: (—) experimental data; (- - -) simulated data [8].

The structure of the peroxy radical of PP may be analogous to that of PE, particularly from the viewpoint of its micro-conformation, and thus the C—O—O plane can be thought of as being perpendicular to the C^1–C^3 direction, when the radical is described as ($\sim C^1H_2$–C^2(Me)($OO^{\cdot}$)–$C^3H_2\sim$). The angle between g_1 and the chain axis is then estimated to be ca. 35° from the helical structure of PP. The C—O—O rotation around the chain axis,

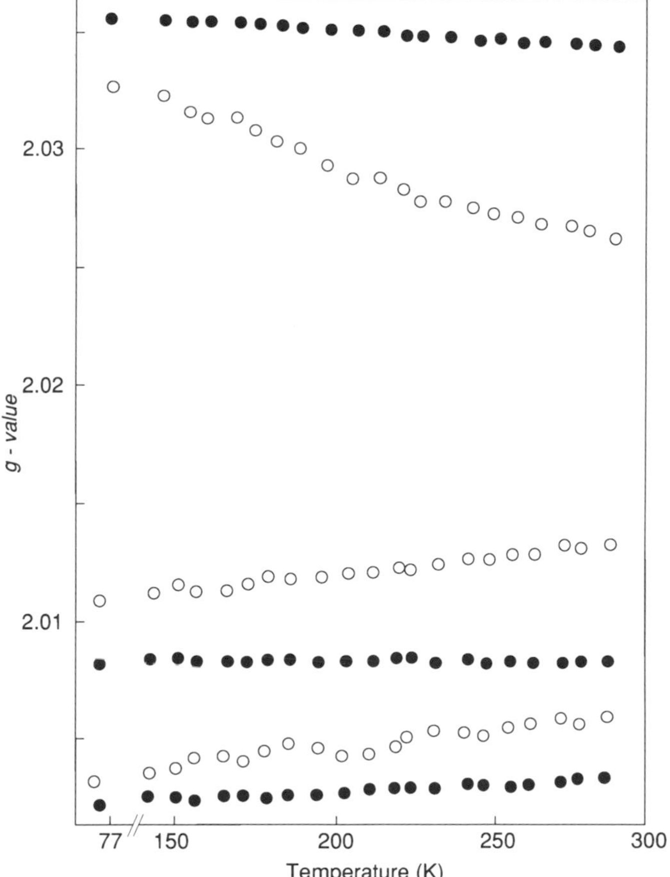

Figure 24 Changes in the g-values of the peroxy radicals trapped in i-PP with temperature: (●) A-radical; (○) B-radical [8].

combined with the inclination of the g_1-direction to the chain axis, could explain the temperature dependencies of g_1', g_2', and g_3', and also the difference between g_1 and g_1' mentioned above.

It has therefore been concluded that the C—O—O plane is perpendicular to the C^1—C^3 direction, and the mobile peroxy radicals are rotating or rotationally vibrating around the chain axis.

The similarity between the temperature dependencies of the g-values of the peroxy radicals in i-PP and a-PP indicates that the peroxy radicals in the atactic form have the same structure to those in the isotactic form and are also rotating about the chain axis.

In order to determine a more precise structure for the peroxy radicals of

PP, angular-dependent ESR spectra of peroxy radicals trapped in an elongated film of i-PP were investigated by Shimada and coworkers [9,53]. The rigid peroxy radicals (A-radicals) from the tertiary alkyl radicals were cleanly trapped in the crystalline region of elongated PP by using the method described in the next section [8,54]. A very marked angular dependence was observed for the spectrum measured at 77 K (shown in Figure 25). A computer simulation of the angular-dependent ESR spectra was also carried out, and this is also presented in the same figure. The profiles of the experimental spectra show good agreement with those of the calculated spectra, as shown in Figure 25 for the orientation function, $f = 0.968$, with $\lambda_1 = 39°$ and $\lambda_3 = 68°$.

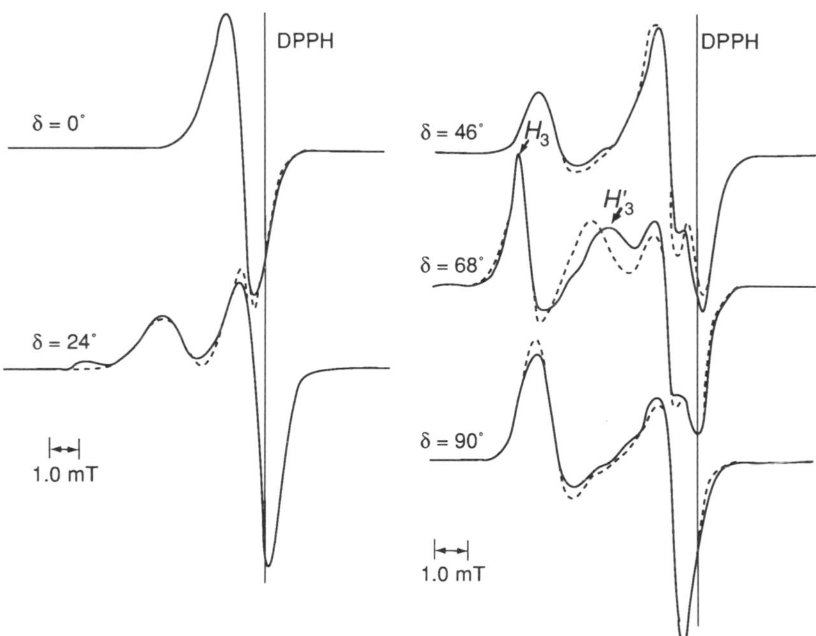

Figure 25 Angular dependencies of (—) observed and (- - -) calculated ESR spectra at 77 K of the peroxy radicals trapped in i-PP upon the rotation angle (δ) between the direction of elongation and that of the magnetic field [9].

The values of λ_1 and λ_3 have established two important angles in the structure of the peroxy radical, which is schematically illustrated in Figure 26. The (C—O—O) bond angle has also been obtained, with its value being supported by an *ab initio* calculation [14], with the latter giving a value of 111.5°.

Next, the dihedral angle, ϕ (Figure 26), was determined to be 55.5°; this value is close to 60°, corresponding to the *gauche*-conformation. It should be

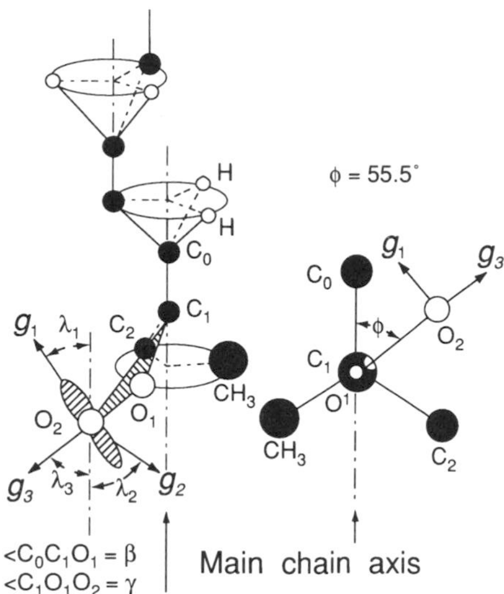

Figure 26 Schematic representation of the peroxy radical of i-PP, where λ_1, λ_2, and λ_3 are the angles between the principal directions of the *g*-tensor and the polymer chain axis, g_1 is perpendicular to the C—O—O plane, and g_3 is the direction of the O—O bond [9].

noted that in the *gauche*-conformation the C—O—O plane is perpendicular to the C_0—C_2 direction (described as the C^1—C^3 direction in the above), with the O—O and C_1—Me directions being in the *trans*-form. This perpendicularity is similar for both PE and PTFE, with the *trans*-form being reasonable from the view point of conformational energy. The deviation of 4.5° from the *gauche*-conformation may be caused by the 3_1 helical structure of PP.

This work has been extended to determine the structure of the mobile peroxy radical (the B-radical) in i-PP [13]. Angular-dependent ESR spectra were observed at 77 K for all of the radicals immediately after the production of the peroxy radicals, and for the rigid radicals after annealing at 313 K for 40 h. The two series of angular-dependent spectra produced were found to be different to each other.

The angular-dependent ESR spectra of the mobile peroxy radicals at 77 K were obtained by subtracting the ESR spectra measured after annealing from those obtained before annealing; values of $f = 0.536$, $\lambda = 24°$, and $\lambda_3 = 73°$ were obtained. These values of λ_1 and λ_3 correspond to $\phi = -103°$ and again a bond angle C—O—O of 111°; the value of ϕ is fairly close to $-120°$, corresponding to a skew conformation, which indicates the *cis*-form. The small value of $f(0.536)$ suggests that the mobile radicals reside in molecularly disordered sites, such as defects in the crystals.

Angular-dependent ESR spectra of the peroxy radicals in the non-crystalline regions of elongated PP have also been successfully measured [55]. The shape of the spectra change with the annealing temperature, and from the spectral simulation, the degree of orientation was calculated for the various annealing temperatures. A skew conformation has been identified for all of the peroxy radicals in the non-crystalline regions. For the most mobile radicals, which decay at the lowest temperatures, the smallest value of f was found to be 0.31. The orientation function, f, of the mobile radicals has been found to increase with the increase in annealing temperature, or with the decay of the peroxy radicals. These results clearly show that the less stable radicals have the lower degree of orientation. In another words, the more mobile peroxy radicals exist in the more disordered structures.

4.5 DECAY OF PEROXY RADICALS IN PP

As before, the chain-reaction scheme can be represented as follows:

$$R^{\cdot} + R'H \longrightarrow RH + R'^{\cdot} \quad (2)$$

$$R^{\cdot} + O_2 \longrightarrow ROO^{\cdot} \quad (3)$$

$$ROO^{\cdot} + R'H \longrightarrow ROOH + R'^{\cdot} \quad (4)$$

In the post-irradiation oxidation of PP, the chain length was postulated to be 3.4 [49], because 3.4 ketone groups per carbon radical were observed by infrared spectroscopy.

The reversibility [50] of the formation of the peroxy radicals suggests that the above chain reaction does take place. Such a reaction process was also supported by the oxygen consumption of samples containing peroxy radicals.

The isothermal decay of the peroxy radicals in a vacuum, measured by Eda and coworkers [52], showed a rapid decay within the first hour, followed by two further periods of relatively slow decay. It has been suggested that the radicals exhibiting rapid decay undergo different molecular motions to the other peroxy radicals.

Conclusive results for the decay of peroxy radicals trapped in irradiated i-PP were obtained by Hori and coworkers [54]. The decay of the tertiary peroxy radical, $—CH_2—CH(CH_3)(OO^{\cdot})—CH_2—$, trapped in the crystalline region was studied in air at various temperatures from 284 to 309 K by ESR spectroscopy. A typical example of the dependence of the ESR spectra on reaction time, measured at 301 K, is shown in Figure 27. Spectrum (d), obtained after annealing at 363 K, represents a typical amorphous pattern for the rigid peroxy radicals. The spectra observed at all stages of the reaction consist of two components, which arise from the mobile and the immobile fractions. It was previously known that the peak found at the midfield of the spectra arise from the mobile peroxy radicals. Since these peaks decreased

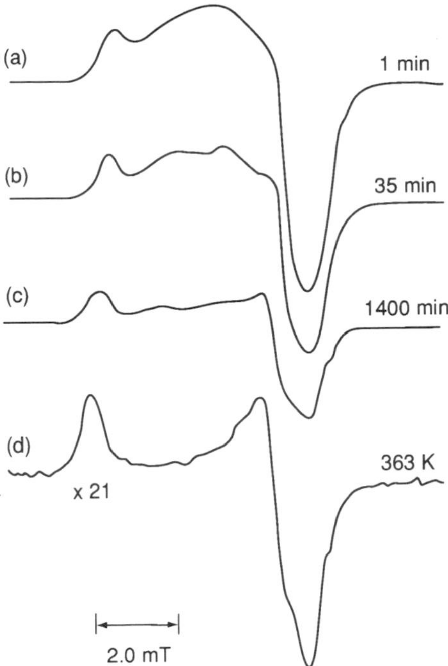

Figure 27 Changes in the ESR spectra of peroxy radicals in i-PP for the indicated oxidation times at 301 K (a–c). Spectrum (d) was obtained at 301 K after annealing at 363 K for 15 min [54].

with reaction time (Figure 27), it was proposed that the decay of the peroxy radicals is mainly due to the mobile fraction of these radicals.

In order to examine this hypothesis, all of the spectra observed during the course of the decay at 301 K were computer simulated by assuming two components, i.e. mobile and immobile. The relative concentrations of the mobile and immobile radicals, and the total concentrations obtained from the simulation, are plotted against the reaction time in Figure 28. One characteristic feature of this figure is that the concentration of immobile radicals is almost constant throughout the decay process. These radicals are thought not to participate in the decay reaction. Therefore, it was again concluded that the decay of the peroxy radicals is due mainly to the decay of the mobile radicals, with the immobile radicals playing no role in the decay process.

We next need to consider which reaction mechanism can best explain the peroxy radical decay. The non-participation of the immobile radicals in the decay process has been taken into account when using the following equation to analyze the data obtained [54]:

$$C(t)/C_0 = f(t) + X_0 \qquad (15)$$

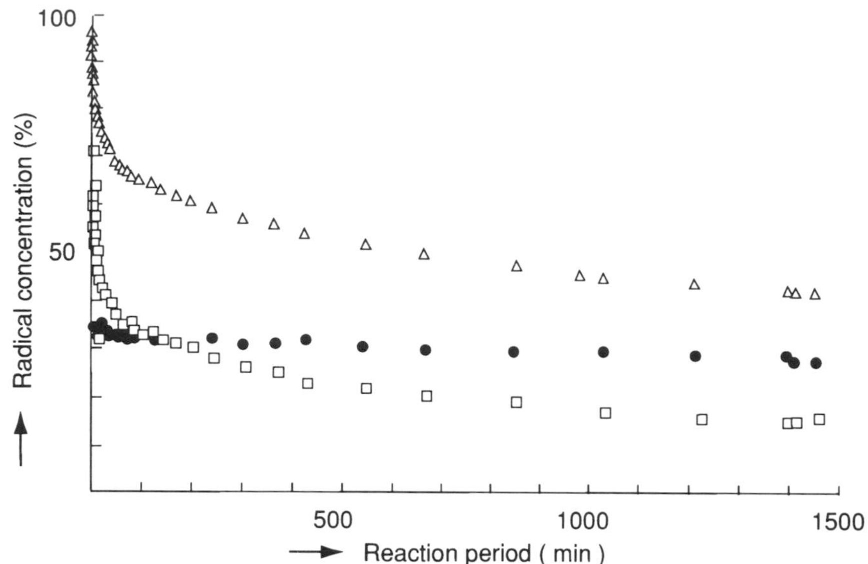

Figure 28 Changes in the peroxy radical concentration in i-PP with decay time at 301 K, where concentrations of the mobile and immobile radicals were calculated from the ratios obtained by simulation: (△) total concentration; (□) mobile radical; (●) immobile radical [54].

where X_0 is a parameter representing the fraction of immobile peroxy radicals and $f(t)$ is the decay curve of the mobile radicals. In order to determine $f(t)$, five prospective reaction mechanisms were examined: two first-order mechanisms, with one composed of two reactions, i.e. a fast reaction and a slow reaction, and the other single reaction, two second-order mechanisms of the same types as for the firsr-order case, and a diffusion-controlled mechanism. The diffusion-controlled reaction is described by the equation:

$$f(t) = (1 - X_0)/(1 + At^{1/2} + Bt) \tag{16}$$

These five cases were examined by least-squares fitting procedures. The diffusion-controlled mechanism, which has three parameters, gave the best fit to the data, although four adjustable parameters are available in the calculations for the cases of the two rate constants.

Therefore, it has been concluded that the diffusion-controlled mechanism can best explain the peroxy radical decay. Curves obtained by assuming such a mechanism are shown in Figure 29 for the decay at various temperatures.

The parameters obtained from this calculation are listed in Table 5, where it can be seen that the value of X_0 decreases with increasing temperature. This agrees with the results obtained from the temperature-dependent ESR

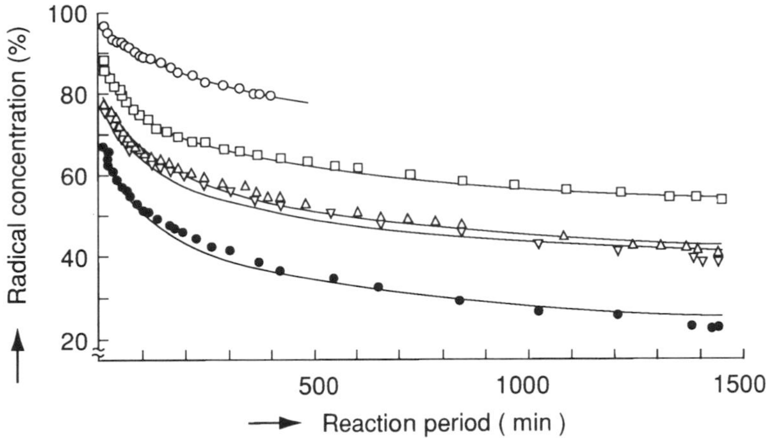

Figure 29 Change of peroxy radical concentration in i-PP with decay time at various temperatures: (○) 284; (□) 294; (△, ▽) 301; (●) 309 K. The solid lines are theoretical curves obtained by assuming a diffusion-controlled mechanism; parameters used are given in Table 5 [54].

spectra [8] where the component due to the immobile fraction also decreases with increasing temperature. The correlation between X_0 and the immobile radical fraction further supports the above mentioned conclusion that the immobile peroxy radicals play no role in the decay reaction.

Both immobile and mobile peroxy radicals should exist in the crystalline region of PP, on account of the fact that the parent tertiary alkyl radicals of PP reside in the crystalline region. Therefore the locality of the peroxy radical decay should be solely within the crystalline region.

The value of A in equation (16) increases with temperature, with B having a very low value (Table 5). These parameters can be used to obtain the diffusion coefficient, D, from equation, $\pi D/2r^2_0 = (B/A)^2$ [56]. A value

Table 5 Least-squares fitting parameters for the equation, $C(t)/C_0 = (1 + At^{1/2} + Bt) + X_0$, used to describe the diffusion-controlled mechanism for peroxy radical decay[a]

Parameter	Temperature (K)				
	284	294	301a	301b	309
A ($\times 10^{-1}$)	0.11	0.73	0.86	0.93	1.32
B ($\times 10^{-6}$)	1×10^2	<1	<1	<1	<1
X_0	0.0	0.39	0.25	0.25	0.12

[a] Time in min.

of $< 1 \times 10^{-12} s^{-1}$ which corresponds to $D < 1.5 \times 10^{-24}$ $cm^2 s^{-1}$ if r_0 is assumed to be 5 Å was estimated at 309 K,. indicating that the peroxy radicals diffuse very slowly in crystalline PP. The temperature dependence of A should thus show the temperature dependence of the diffusion coefficient.

In the decay reaction of the peroxy radicals of PP, intermolecular hydrogen abstraction has been suggested by considering that the O—O group of the B-radical rotates around the chain axis [8] at the same temperature at which the B-radicals decay [54]. This interchain hydrogen abstraction process has been supported by the model proposed for the peroxy radical [13]. Surprisingly, the distances between the half-charge of the radical spin and the hydrogen atoms of the interchain are of the same order of magnitude when compared with those of the hydrogen atoms of the intrachain methyl group. These interchain distances become shorter when rotation of the C—O—O group occurs about the chain axis. Therefore, it is concluded that the peroxy radicals should abstract the intermolecular hydrogens.

In the stretched PP, the concentration of rigid peroxy radicals has also been found to remain unchanged throughout the decay process [55]. This result indicates that even in the elongated PP film the rigid peroxy radicals play no role in the decay process of the peroxy radicals [54].

In the decay of ^{17}O-labelled peroxy radicals of PP [57,58], very little $PP^{17}O_2^{\cdot}$ to $PP^{16}O_2^{\cdot}$ conversion was observed when the $PP^{17}O_2^{\cdot}$ were stored at 23°C under $^{16}O_2$. This result also supports the conclusion [54] that the rigid peroxy radicals play no role in the decay reaction of peroxy radicals.

Further information was also obtained from this labelling study [58]. After a brief period in normal oxygen at 23°C, all reactions were stopped by cooling to 77 K, and an ESR spectrum was recorded.

The ^{17}O hyperfine structure was extensively reduced and the new line-shape resembled that of the $PP^{16}O_2^{\cdot}$ line-shape at 77 K. This result indicates that the following reaction sequence takes place:

$$ROO^{\cdot} + R'H \longrightarrow ROOH + R'^{\cdot} \tag{4}$$

$$R'^{\cdot} + RH \longrightarrow R'H + R^{\cdot} \tag{2}$$

$$R^{\cdot} + O_2 \longrightarrow ROO^{\cdot} \tag{3}$$

The decay behaviour of peroxy radicals in two PP films, unoriented and bioriented [59], has been found to be similar to that observed for the powder [54], i.e. a rapid decrease of the mobile peroxy radicals, and a slow decay of the peroxy radicals which are trapped in the immobile phase. The decay kinetics of these PP films were better described by a bimolecular reaction, controlled by a diffusion mechanism.

4.6 OXIDATION OF PP

The autoxidation of PP has been extensively studied by Chien and coworkers [60–62]. Oxidized PP (PPH) was prepared from a chlorobenzene-soluble fraction of commercial PP. More than 90% of the hydroperoxides contained in the PPH had been determined as being intramolecularly hydrogen bonded, with all of the hydroperoxide groups present in sequences with lengths of two or greater. The intramolecular-hydrogen abstraction during oxidation could account for the formation of these neighbouring hydroperoxides, although in post-irradiation oxidation interchain abstraction should occur in crystalline PP.

The autoxidation of PP inhibited by 2,6-di-tert-butyl-*p*-cresol (AH) was examined [60] by using PPH samples as the starting materials. Using various concentrations of AH, the oxygen consumption and the radical concentration were monitored continuously, and the hydroperoxide and stabilizers were analyzed after the autoxidation was terminated by quenching. There was a well defined induction period during which [ROOH] decreased rapidly, along with [A$^\bullet$] and [AH]. At the end of the induction period, [ROOH], [ROO$^\bullet$], and $-d[O_2]/dt$ increased rapidly until steady-state values were obtained for all of these. The steady-state [ROO$^\bullet$] is approximately equal to that found in experiments carried out with no stabilizers being present.

From these results, it is suggested that the following reactions occur in the autoxidation of PP, in addition to the previously mentioned chain reactions: (1–5) (see Section 1.1):

$$\mathrm{ROOH} \longrightarrow \mathrm{RO}^\bullet + \mathrm{OH}^\bullet \tag{16}$$

$$\mathrm{RO}^\bullet(\mathrm{OH}^\bullet) + \mathrm{RH} \longrightarrow \mathrm{ROH}(\mathrm{H_2O}) + \mathrm{R}^\bullet \tag{17}$$

$$\mathrm{ROO}^\bullet + \mathrm{AH} \longrightarrow \mathrm{ROOH} + \mathrm{A}^\bullet \tag{18}$$

$$\mathrm{ROO}^\bullet + \mathrm{A}^\bullet \longrightarrow \text{products} \tag{19}$$

The existence of an induction period indicates that reactions (18 and 19) are faster than reaction (4). The steady-state concentration of ROOH clearly shows that the hydroperoxides decompose into the oxidation products, and that this decomposition occurs above 85°C [62].

For the mechanism of the decomposition of ROOH, a unimolecular homolytic cleavage reaction (16) [62] has been suggested, since the decomposition of PPH yields a maximum of ca. 1.8 radicals. A different mechanism, i.e. disproportionation (reaction (20)), has been suggested by Zolotova and Denisov [29]:

$$2\mathrm{ROOH} \longrightarrow \mathrm{RO}^\bullet + \mathrm{ROO}^\bullet + \mathrm{H_2O} \tag{20}$$

Two forms of PPH, i.e. PPH_a with many adjacent hydroperoxide groups and PPH_i with isolated hydroperoxide groups, were prepared. The PPH_i decay to free radicals is slower than that of PPH_a, i.e. the adjacent hydroperoxide groups decompose more readily than the isolated ones. Thus the decay of the adjacent hydroperoxide groups (reaction (20)) is the main reaction of degenerate chain branching in PP oxidation. PPH_i have been found to decay via reaction (16). It is interesting to note that the reaction rates of PPH_i in both the solid phase and in solution are almost equal [29]; consequently, the unimolecular decomposition rate of the PP hydroperoxides does not depend on the phase of the polymer.

In the photo-oxidation of solid PP [63,64] only a small proportion (4.5%) of the radicals will escape secondary cage recombination, but these will carry the bulk of the oxidation [64]; an average kinetic chain length (KCL) of 10^4 propagation steps was estimated. This large KCL may make radical scavenging effective in stabilization. The average KCL for all radicals which escape the primary cage was estimated to be 800 without any scavengers, but only 2.3 with scavengers, which adds only a probability of 0.025 for termination to each propagation.

From the pressure dependence of the decay of the peroxy radicals [65] it was confirmed that decay occurs in the amorphous region. From studies of various types of PP [66–68], values of $D(O_2) = 8 \times 10^{-8}\ cm^2\,s^{-1}$ and $D(PPO_2^{\cdot}) < 1 \times 10^{-17}\ cm^2\,s^{-1}$ were obtained, compared with $D(PPO_2) < 1.5 \times 10^{-24}\ cm^2\,s^{-1}$ obtained previously [54]. In solution, the decay of peroxy radicals of a-PP has been concluded to follow second-order kinetics.

In order to study the long-term oxidation of PP, oxygen absorption measurements, in conjunction with hydroperoxide analysis and ESR spectroscopy, were carried out over periods greater than a year by Geuskens and Nedelkos [69]. Just after irradiation, the migration of mobile peroxy radicals is responsible for the oxidation, as already stated [54]. After about 200 h, this mechanism cannot operate any longer because all of the peroxy radicals are now immobile. However, the rates of oxygen absorption and of hydroperoxide production remained constant for more than a year. This second stage of post-irradiation oxidation of PP is initiated by the slow decomposition of clustered hydroperoxides, which proceeds even in the dark at room temperature. It was suggested that 5% of peroxy radicals escape the cage, with a kinetic chain length (KCL) of 20 steps being estimated.

In various ethylene–propylene copolymers [70], the ethylene unit has been found to be more easily photo-oxidized than the propylene unit. Fourier-transform infrared (FT-IR) spectra indicated, that linear low-density polyethylene (LLDPE) generates isolated hydroperoxides, whereas i-PP contains amounts of associated hydroperoxides. The corresponding features for other copolymers will depend upon the ratio of ethylene to propylene.

5 PEROXY RADICALS OF OTHER POLYMERS

5.1 PEROXY RADICALS OF POLY(VINYL CHLORIDE) (PVC)

This polymer is produced by the polymerization of vinyl chloride, $CH_2{=}CHCl$. The crystallinity of PVC is very low and thus it can be considered to be an amorphous polymer. The structure of PVC contains syndiotactic, isotactic and atactic configurations.

The ESR pattern of PVC irradiated in vacuum has a spread of 12.0 mT but does not show any hyperfine structure [49,71–73], which could be considered to be due to the presence of polyenyl radicals with short conjugation lengths [49]. A septet signal was also found [73] resulting from the allyl radicals, and this decayed rapidly at room temperature.

Formation of the peroxy radicals of PVC has been observed [71,72]. After exposure of irradiated samples to air at −78°C, the peroxy signal reached a maximum amplitude after ca. 6000 s; this signal was relatively stable. The formation of the peroxy radicals was found to be (oxygen)-diffusion controlled, which was confirmed by the fact that the ratio of surface areas (ca. 4.9) was close to the ratio of the decay reaction rates at −78°C (4.4) in experiments carried out for two samples showing different surface areas.

At temperatures above 25°C, the formation of peroxy radicals was virtually complete after 24 s of exposure to air. Subsequent spectra showed a continuing decrease of the peroxy radical signal with no observable change in the line-shape. Following evacuation, after 210 s of exposure to air, a series of spectra showed the reverse reaction. These results are consistent with an autoxidative scheme in which the rate controlling step of the decay is the abstraction of hydrogen from the polymer substrate to leave an alkyl radical.

A comparison of the rates of peroxy radical decay in air and *in vacuo* gave an average chain length of 1.5 per radical, although simultaneous gas and free-radical measurements indicated that three molecules of oxygen are consumed per peroxy radical. The increase of the carbonyl group of the ketone and the decay of the polyenyl radicals were observed simultaneously [49]. Comparison of the carbonyl concentration with that of the polyenyl radicals suggested that ca. 1.6 ketone groups were eventually produced per carbon-centred radical. On the other hand, when irradiation was carried in the presence of air, the oxidation of PVC produced approximately ten times as many ketone groups as that found for the post-irradiation oxidation process.

5.2 PEROXY RADICALS OF POLY(METHYL METHACRYLATE) (PMMA)

This polymer is produced from methyl methacrylate, $CH_2{=}C(CH_3)$-$(COOCH_3)$. Almost all of the commercially produced PMMA is in the atactic form, and thus in the following discussion 'PMMA' will refer to this form; A-PMMA is completely amorphous.

There are three types of carbon-centred radicals of PMMA, namely the primary radical I, —CH_2—$C(CH_3)(COOCH_3)^{\cdot}$, which gives the well known quintet–quartet ESR spectrum, a second primary radical II, —CH_2—$C(CH_3)(COOCH_3)$—$CH_2^{\cdot}$, which gives a triplet, and a secondary radical III, —$C(CH_3)(COOCH_3)$—$CH^{\cdot}C(CH_3)(COOCH_3)$—, which gives a doublet in its ESR spectrum. The scission-type radical I is stable and is usually only observed in irradiated PMMA.

When air was introduced to γ-irradiated PMMA powder which contained the scission-type radical I, the spectrum suddenly changed to a superposition of signals from free radicals and oxygenated radicals, probably $ROO^{\cdot}$, which appears as an asymmetric singlet [74]. The decay of the free radicals in the presence of air was faster than that in vacuum, which indicates that the peroxy radical is less stable than the scission-type radical I.

Temperature-dependent ESR spectra of peroxy radicals in irradiated PMMA have been observed [75]. Although the observed spectra were not of high quality, because of the instability of the peroxy radicals, the observed *g*-shift suggests that the peroxy radicals in PMMA move faster at higher temperatures.

The PMMA radicals prepared mechanically by the grinding PMMA for short periods of time *in vacuo* at 77 K reacted with oxygen at 113 K [76]. The formation of diamagnetic dimers of the peroxy radicals, i.e. ROO—OOR, was suggested. With increasing temperature, these dimers were found to decompose back into the single peroxy radicals.

REFERENCES

1. W.B. Ard, H. Shields and W. Gordy, *J. Chem. Phys.* **23**, 1727 (1955).
2. F.K. Kneubühl, *J. Chem. Phys.* **33**, 1074 (1960).
3. D.W. Ovenall, *J. Phys. Chem. Solids* **21**, 309 (1961).
4. T. Ichikawa, M. Iwasaki and K. Kuwata, *J. Chem. Phys.* **44**, 2979 (1966).
5. S. Ohnishi, S. Sugimoto and I. Nitta, *J. Polym. Sci. (A)* **1**, 605 (1963).
6. Y. Hori, S. Shimada and H. Kashiwabara, *Polymer* **18**, 1143 (1977).
7. Y. Hori, S. Shimada and H. Kashiwabara, **18**, 567 (1977).
8. Y. Hori, Y. Makino and H. Kashiwabara, *Polymer* **25**, 1436 (1984).
9. S. Shimada, A. Kotake, Y. Hori and H. Kashiwabara, *Macromolecules* **17**, 1104 (1984).
10. S. Moriuchi, M. Nakamura, S. Shimada, H. Kashiwabara and J. Sohma, *Polymer* **11**, 630 (1970).
11. M. Iwasaki and Y. Sakai, *J. Polym. Sci. (A–2)* **6**, 265 (1968).
12. Y. Hori, S. Aoyama and H. Kashiwabara, *J. Chem. Phys.* **75**, 1582 (1981).
13. S. Shimada, Y. Hori and H. Kashiwabara, *Macromolecules* **18**, 170 (1985).
14. B.H. Besler, M.D. Sevilla and P. MacNeille, *J. Phys. Chem.* **90**, 6446 (1986).
15. B. Smaller and M.S. Matheson, *J. Chem. Phys.* **28**, 1169 (1958).
16. S. Ohnishi, Y. Ikeda, M. Kashiwagi and I. Nitta, *Polymer* **2**, 119 (1961).
17. S. Ohnishi, Y. Ikeda, S. Sugimoto and I. Nitta, *J. Polym. Sci.* **47**, 503 (1960).
18. E.J. Lawton, J.S. Balwit and R.S. Powell, *J. Chem. Phys.* **33**, 405 (1960).

19. R.J. Abraham and D.H. Whiffen, *Trans. Faraday Soc.* **54**, 1291 (1958).
20. B.R. Loy, *J. Polym. Sci.* **44**, 341 (1960).
21. Y. Hori, S. Shimada and H. Kashiwabara, *Polymer* **18**, 151 (1977).
22. A.S. Michaels, H.J. Bixler, *J. Polym. Sci.* **50**, 413 (1961).
23. Y. Hori, Z. Fukunaga, S. Shimada and H. Kashiwabara, *Polymer* **20**, 181 (1979).
24. S. Schlick and L. Kevan, *J. Am. Chem. Soc.* **102**, 4622 (1980).
25. Y. Chatani, H. Anraku and Y. Taki, *Mol. Cryst. Liq. Cryst.* **48** 219 (1978).
26. Y. Hori, T. Tanigawa, S. Shimada and H. Kashiwabara, *Polym. J.* **13**, 293 (1981).
27. J.C.W. Chien, *J. Polym. Sci. (A–1)* **6**, 375 (1968).
28. T. Seguchi and N. Tamura, *J. Phys. Chem.* **77**, 40 (1973).
29. N.V. Zolotova and E.T. Denisov, *J. Polym. Sci. (A–1)* **9**, 3311 (1971).
30. L.A. Davis, C.A. Pampillo and T.C. Chiang, *J. Polym. Sci. Polym. Phys. Edn* **11**, 841 (1973).
31. T. Matsugashita and K. Shinohara, *J. Chem. Phys.* **35**, 1652 (1961).
32. Y. Hori, S. Shimada and H. Kashiwabara, *Polymer* **20**, 406 (1979).
33. H. Tadokoro, *Structures of Polymers* Kagaku Dojin, Kyoto (1976) (in Japanese).
34. H.N. Rexroad and W. Gordy, *J. Chem. Phys.* **30**, 399 (1959).
35. S. Siegel and H. Hedgpeth, *J. Chem. Phys.* **46**, 3904 (1967).
36. A. Noguchi, S. Kondo and M. Kuzuya, *Chem. Pharm. Bull.* **42**, 1 (1994).
37. S. Ohnishi, M. Kashiwagi, Y. Ikeda and I. Nitta, *Isotop. Radiat. Jpn* **1**, 210 (1958).
38. T. Matsugashita and K. Shinohara, *J. Chem. Phys.* **32**, 954 (1960).
39. G.A. Almanov, L. Ya. Dzhavakhishvili and G.D. Ketiladze, *Opt. Spectrosc.* **12**, 446 (1962).
40. H. Tanaka, A. Matsumoto and N. Goto, *Bull. Chem. Soc. Jpn* **37**, 1128 (1964).
41. D.W. Ovenall, *J. Phys. Chem. Solids* **26**, 81 (1965).
42. D. Olivier, C. Marachi and M. Che, *J. Chem. Phys.* **71**, 4688 (1979).
43. M. Che and A.J. Tench, *J. Polym. Sci. Polym. Lett. Edn* **13**, 345 (1975).
44. M. Che and A.J. Tench, *J. Chem. Phys.* **64**, 2370 (1976).
45. S. Shimada, Y. Hori and H. Kashiwabara, *Polym. J.* **16**, 539 (1984).
46. S. Schlick, W. Chamulitrat and L. Kevan, *J. Phys. Chem.* **89**, 4278 (1985).
47. D. Olivier, C. Marachi and M. Che, *J. Chem. Phys.* **72**, 3348 (1980).
48. P.B. Ayscough and S. Munari, *J. Polym. Sci. (B)* **4**, 503 (1966).
49. S. Ohnishi, S. Sugimoto and I. Nitta, *J. Polym. Sci. (A)* **1**, 625 (1963).
50. H. Fischer, K.-H. Hellwege and P. Neudörfl, *J. Polym. Sci. (A)* **1**, 2109 (1963).
51. R.P. Gupta, *J. Phys. Chem.* **68**, 1229 (1964).
52. B. Eda, K. Nunome and M. Iwasaki, *J. Polym. Sci. Polym. Lett. Edn* **7**, 91 (1969).
53. H. Kashiwabara, S. Shimada and Y. Hori, *Radiat. Phys. Chem.* **37**, 511 (1991).
54. Y. Hori, S. Shimada and H. Kashiwabara, *J. Polym. Sci. Polym. Phys. Edn* **22**, 1407 (1984).
55. S. Shimada, Y. Hori and H. Kashiwabara, *Macromolecules* **21**, 979 (1988).
56. S. Shimada, Y. Hori and H. Kashiwabara, *Polymer* **22**, 1377 (1981).
57. J. Reuben and B.H. Mahlman, *J. Phys. Chem.* **88**, 4904 (1984).
58. D.J. Carlsson, C.J.B. Dobbin and D.M. Wiles, *Macromolecules* **18**, 1793 (1985).
59. I.L.J. Dogué, N. Mermilliod and F. Genoud, *J. Polym. Sci. Polym. Chem. Edn.* **32**, 2193 (1994).
60. J.C.W. Chien and C.R. Boss, *J. Polym. Sci. (A–1)* **5**, 1683 (1967).
61. J.C.W. Chien, E.J. Vandenberg and H. Jabloner, *J. Polym. Sci. (A–1)* **6**, 381 (1968).
62. J.C.W. Chien and H. Jabloner, *J. Polym. Sci. (A–1)* **6**, 393 (1968).
63. A. Garton, D.J. Carlsson and D.M. Wiles, *Macromolecules* **12**, 1071 (1979).
64. A. Garton, D.J. Carlsson and D.M. Wiles, *Makromol. Chem.* **181**, 1841 (1980).
65. F. Szöcs, *J. Appl. Polym. Sci.* **27**, 1865 (1982).

66. A. Faucitano, A. Buttafava, F. Martinotti, V. Comincioli and F. Gratani, *Polym. Photochem.* **7**, 483 (1986).
67. A. Faucitano, A. Buttafava, F. Martinotti, V. Comincioli and P. Bortolus, *Polym. Photochem.* **7**, 491 (1986).
68. A. Faucitano, A. Buttafava, F. Martinotti, P. Bortolus and V. Comincioli, *J. Polym. Sci. Polym. Chem. Edn.* **25**, 1517 (1987).
69. G. Geuskens and G. Nedelkos, *Makromol. Chem.* **194**, 3349 (1993).
70. R.P. Singh, R. Mani, S. Sivaram, J. Lacoste and J. Lemaire, *Polymer* **35**, 1382 (1994).
71. Z. Kuri, H. Ueda and S. Shida, *J. Chem. Phys.* **32**, 371 (1960).
72. B.R. Loy, *J. Phys. Chem.* **65**, 58 (1961).
73. R.E. Michel, *J. Polym. Sci. (A–2)* **10**, 1841 (1972).
74. S. Ohnishi and I. Nitta, *J. Polym. Sci.* **38**, 451 (1959).
75. D. Suryanarayana and L. Kevan, *J. Phys. Chem.* **86**, 2042 (1982).
76. J. Pilar and K. Ulbert, *J. Polym. Sci. Polym. Phys. Edn* **16**, 1973 (1978).

13 The Reactions of HO_2/O_2^- Radicals in Aqueous Solution

DIANE E. CABELLI
Brookhaven National Laboratory, New York, USA

1 INTRODUCTION

This chapter contains a description of the physical properties and reactivity of the superoxide radical, O_2^-, and its conjugate acid, the perhydroxyl radical, HO_2, in aqueous solution. The physical properties of superoxide/perhydroxyl radicals are addressed first, with a description of the ultraviolet (UV) spectra, thermodynamics and electron spin resonance (ESR) parameters. The rate of spontaneous dismutation and the acid–base equilibrium of these radicals are also described. The remainder of the chapter concerns the reactivity of HO_2/O_2^- radicals with a variety of substrates. The discussion begins with the reactions of some well studied organic molecules with HO_2/O_2^- radicals i.e., Nitro Blue Tetrazolium, ascorbic acid/ascorbate and nicotinamide–adenine dinucleotide (NADH) and lactate dehydrogenase (LDH)-bound NADH, prefaced by a brief description of spin-traps and the reactivity of the nitrone spin-trap, 5,5-dimethyl-pyrroline-*N*-oxide (DMPO). Then, the reactivities of metal-centered porphyrins, metal ions and complexes, specifically manganese, iron and copper complexes, and superoxide dismutases (SODs) with HO_2/O_2^- are addressed. The chapter concludes with a description of very recent studies of the reactions of nitric oxide (NO) with HO_2/O_2^- radicals, a topic of great interest with the recent identification of NO as a neurotransmitter *in vivo*.

For a very comprehensive and detailed description of the chemistry of HO_2/O_2^- in both aqueous and non-aqueous solutions, the reader is referred to a recently published two-volume series by Afanas'ev [1], and the references given therein. In addition, selected subjects of topical interest in aqueous [2] and none aqueous [3] HO_2/O_2^- chemistry have been recently reviewed.

2 PROPERTIES OF HO_2/O_2^- RADICALS

2.1 SPECTRAL AND KINETIC PROPERTIES

The acid-base equilibrium between HO_2 and O_2^- is well established in

Peroxyl Radicals. Edited by Z.B. Alfassi

aqueous solution, equilibrium (1,-1) [4,5]:

$$HO_2 \rightleftharpoons O_2^- + H^+ \ (K_1 = 1.6 \times 10^{-5}\ \text{mol dm}^{-3}) \qquad (1,-1)$$

Both of these radicals absorb in the UV region with absorption maxima at 245 nm ($\epsilon_{245\ nm}$ = 2350 dm^3 mol^{-1} cm^{-1}) and 225 nm ($\epsilon_{225\ nm}$ = 1400 dm^3 mol^{-1} cm^{-1}) for O_2^{-1} and HO_2, respectively; the UV spectra of O_2^{-1} and HO_2 are given in Figure 1. It is important to note that in the pH range 3–7, the observed absorbance is a composite of that due to both HO_2 and O_2^-. As the extinction coefficients of these radicals are significantly different, the acid/base equilibrium constant of 1.6×10^{-5} mol dm^{-3} (pK_a = 4.8) must be taken into account when calculating the respective concentrations of HO_2 and O_2^- radicals in this pH range.

HO_2/O_2^- radicals undergo disproportionation to H_2O_2 and O_2 via a pH-dependent mechanism that involves equilibrium (1) and reactions (2) and (3):

$$HO_2 + HO_2 \longrightarrow H_2O_2 + O_2 \qquad (2)$$

$$HO_2 + O_2^- \xrightarrow{H^+} H_2O_2 + O_2 \qquad (3)$$

$$O_2^- + O_2^- \longrightarrow \text{no reaction} \qquad (4)$$

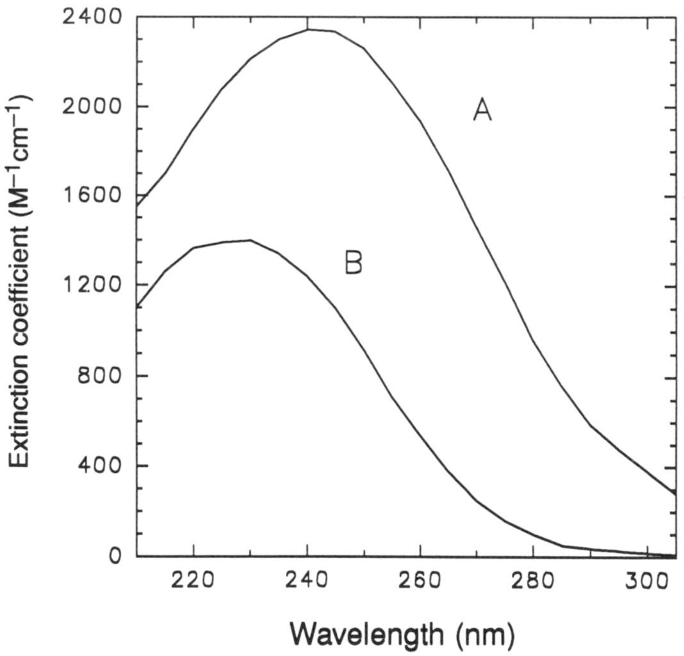

Figure 1 UV/Vis spectra of (A) O_2^- (pH > 7) and (B) HO_2 (pH < 2.5) radicals in aqueous solution

A kinetic equation can be derived where $[HO_2/O_2^-]_{tot} = [HO_2] + [O_2^-]$, and $-d[HO_2/O_2^-]_{tot}/dt = 2k_2[HO_2]^2 + 2k_3[HO_2][O_2^-]$:

$$\frac{-d[HO_2/O_2^-]tot}{[HO_2/O_2^-]_{tot}^2} = k_{obs}, \text{dm}^3\,\text{mol}^{-1} = \frac{2(k_2 + k_3K_1/[H^+])}{(1 + K_1/[H^+])^2} \qquad (5)$$

Using the acid–base equilibrium constant for these radicals, values of $k_2 = 8.3 \times 10^5\,\text{dm}^3\,\text{mol}^{-1}\,\text{s}^{-1}$ and $k_3 = 9.7 \times 10^7\,\text{dm}^3\,\text{mol}^{-1}\,\text{s}^{-1}$ can be calculated by fitting the experimental measurement of the observed rates of disproportion at varying pH to this equation. As seen in Figure 2, the slope of k_{obs} versus pH is -1 above pH 6, and the lack of deviation from this line indicates that, even at pH 13, any contribution from reaction (4) is negligible.

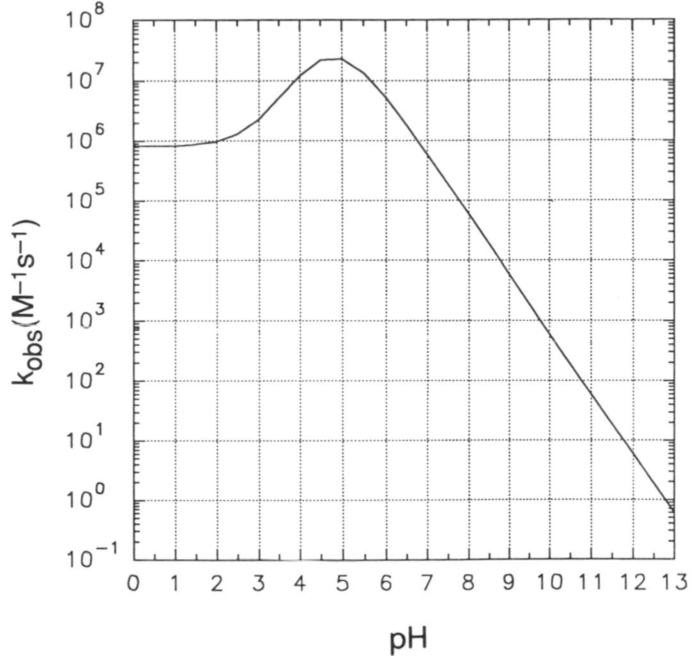

Figure 2 The calculated rate of decay of HO_2/O_2^- radicals as a function of pH using equation (5) and the rate and equilibrium constants reported in Ref. 5 for reactions (2) and (3) and equilibrium (1,-1).

2.2 THERMODYNAMIC PARAMETERS

Thermodynamic parameters have been measured for HO_2 and O_2^- under a variety of conditions and compilations of these can be found elsewhere [6,7]. The reduction potentials of HO_2 and O_2^- radicals in aqueous solution under 1.0 atmosphere pressure of O_2 (versus Normal Hydrogen Electrode

(NHE)) are E° (O_2/O_2^-) = 0.33 V and E° (H^+, O_2/HO_2) = −0.05 V. The oxidation potentials for these radicals are E° ($O_2^-,H^+/HO_2^-$) = 1.03 V and E° ($HO_2,H^+/H_2O_2$) = 1.45 V.

2.3 ESR SPECTRA

The ESR spectrum of HO_2 has been observed [8] in a rapid mix–flow system (room temperature, 0.8 M H_2SO_4) where Ce^{IV} sulfate and H_2O_2 react to yield HO_2 and Ce^{III}. The ESR spectrum of HO_2 has a line-width of 27 G and a g-value of 2.016. A large number of adducts of metal complexes with HO_2 have also been observed in ESR experiments [9]. The ESR spectrum of O_2^-, however, has only been observed in irradiated ice [10], as line-broadening due to rapid spin relaxation in aqueous solution occurs here.

3 GENERATION OF HO_2/O_2^- RADICALS

3.1 PULSE AND STEADY-STATE RADIOLYSIS

Perhaps the most efficient method for the generation and study of O_2^-/HO_2 radicals is through the use of radiolysis. Upon radiolysis of water, either by a electron beam from an accelerator or by $^{60}Co\gamma$-radiolysis, the following species are generated [11]:

$$H_2O \rightsquigarrow OH\ (2.75),\ e_{aq}^-\ (2.65),\ H\ (0.65),\ H_2\ (0.7),\ H_2O_2\ (0.45) \quad (6)$$

where the values in brackets are *G*-values, i.e. the number of species generated per 100 eV of energy dissipated in water. In the presence of O_2 and formate, the primary radicals OH, H and e_{aq}^- react via reactions (7–10) and equilibrium (1,-1) to yield HO_2/O_2^- radicals with rate constants [12] of $k_7 = 1.9 \times 10^{10}$, $k_8 = 1.2 \times 10^{10}$, $k_9 = 3.2 \times 10^9$ and $k_{10} = 3 \times 10^9$ dm^3 mol^{-1} s^{-1}:

$$^{e}aq^- + O_2 \longrightarrow O_2^- \quad (7)$$

$$H + O_2 \longrightarrow HO_2 \quad (8)$$

$$OH + HCO_2^- \longrightarrow CO_2^- + H_2O \quad (9)$$

$$CO_2^- + O_2 \longrightarrow O_2^- + CO_2 \quad (10)$$

The magnitudes of these rate constants are such that in the radiolysis of an oxygen-saturated (1.2 mM O_2) aqueous solution containing 1 mM formate, the HO_2/O_2^- radicals are virtually completely formed by the end of the first microsecond. Upon radiolysis with either electron beams or γ-rays, H_2O_2 is formed with a yield that is ca. 10% that of HO_2/O_2^- (see equation 6). This is particularly important to note in studies involving metal complexes, as reactions between H_2O_2 and metals can complicate the mechanisms under study.

Formate and oxygen can be used to convert the primary radicals to HO_2/O_2^- radicals over the entire pH range. In the alkaline pH region, however, it is equally satisfactory to use alcohols [13] as scavengers for the OH radical:

$$OH + CH_3CH_2OH \longrightarrow CH_3{}^{\cdot}CHOH + H_2O \quad (11)$$

$$CH_3{}^{\cdot}CHOH + O_2 \longrightarrow CH_3CH(O_2^{\cdot})OH \quad (12)$$

$$CH_3CH(O_2^{\cdot})OH + OH^- \longrightarrow CH_3CHO + O_2^- + H_2O \quad (13)$$

In general, it is advisable to use a sufficient concentration of the radical scavengers O_2 and formate/ethanol such that reactions (7–13) will have gone virtually to completion within the first microsecond after the pulse in pulse radiolysis experiments. As discussed earlier, this occurs with concentrations of > 1 mM formate. As the rate constants [12] for reactions (11–13) are also quite high, this is not a limitation of the use of alcohols instead of formate.

The most common method of detection of transients generated by pulse radiolysis is UV/Vis spectroscopy. Examples of absorbance versus time traces measured at 260 nm by pulse radiolysis of an oxygen-saturated aqueous solution of formate at pH 5.4 and 8.25 are shown in Figure 3 (curves B and D).

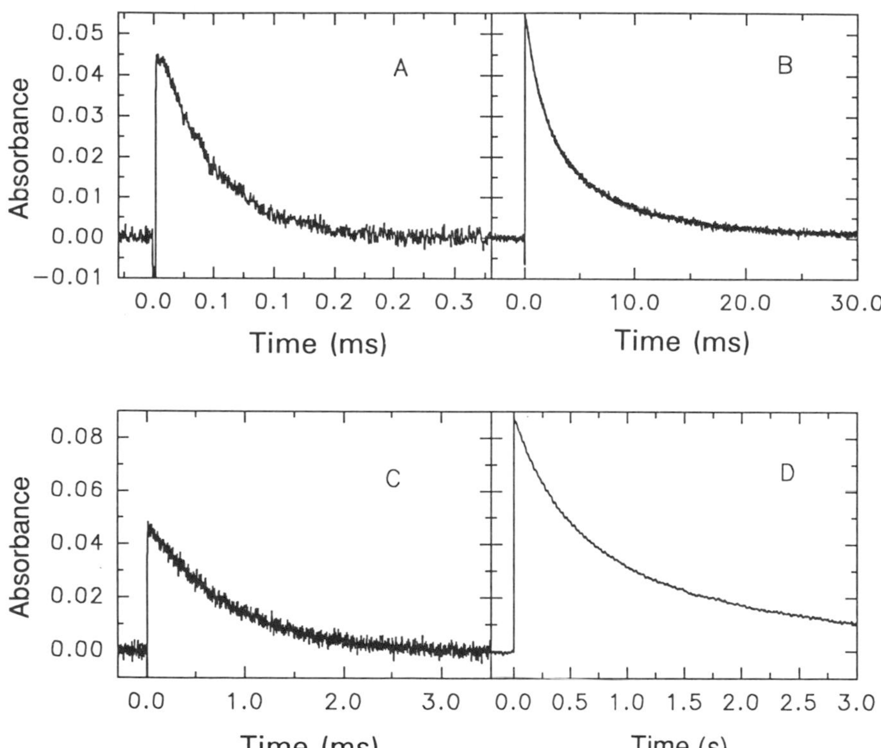

Figure 3 Disappearance of O_2^-(260 nm) measured by pulse radiolysis of an O_2-saturated aqueous solution in the presence of 10 mM formate and 50 mM phosphate: (A) 3.57 μM $CuSO_4$, pH 5.57, 13 μM O_2^-; (B) no $CuSO_4$, pH 5.4, 16 μM O_2^-; (C) 3.57 μM $CuSO_4$, pH 8.25, 11 μM O_2^-; (D) no $CuSO_4$, pH 8.25, 23 μM O_2^-.

These are contrasted with curves A and C in the same figure, which are absorbance versus time traces of the same solutions, but with the addition of 3.57 μM Cu^{2+} ion, which is known to catalytically dismutate O_2^- radicals very efficiently (see Section 4.2.3). Direct measurement of the rate of disappearance of the absorbance of HO_2/O_2^- radicals in the presence of a substrate has led to the measurement of hundreds of rate constants for the reactions of HO_2/O_2^- radicals with a wide variety of substrates. These rates constants have been compiled into a database that can be obtained through the National Institute of Science and Technology (NIST), Gaithersburg, USA [14].

3.2 PHOTOCHEMICAL GENERATION OF HO_2/O_2^- RADICALS

It is also possible to generate O_2^-/HO_2 radicals by photolysis of an aqueous solution. One method is via the vacuum-UV photolysis of water at alkaline pH [15–18]. Here, the energy of the photons from photolysis converts the water molecules into H and OH radicals:

$$H_2O \xrightarrow{h\nu} H + OH \tag{14}$$

Vacuum-UV photolysis in the presence of O_2 and formate/alcohols will yield HO_2/O_2^- radicals in a similar fashion to that described earlier for the radiolytic generation of HO_2/O_2^- radicals, where the subsequent reactions of the H and OH radicals are given by reactions (8–10).

Another method for the photolytic generation of O_2^-/HO_2 radicals was described by McDowell and coworkers [19]. Here, ketones, such as benzophenone, acetone, etc., are photoexcited to the n-π^* state. This photoexcited state is then quenched by reaction with primary and secondary alcohols. The photoexcitation of a ketone had the advantage that it is carried out at much higher wavelengths than the vacuum-UV photolysis that is necessary to split H_2O in to $^{\bullet}OH$ and $H^{\bullet}$, thus allowing the use of mercury lamps or even sunlight:

$$(CH_3)_2CO \xrightarrow{h\nu} {}^3(CH_3)_2CO \tag{15}$$

$${}^3(CH_3)_2CO + CH_3CH_2OH \longrightarrow (CH_3)_2{}^{\bullet}COH + CH_3{}^{\bullet}CHOH \tag{16}$$

$$O_2 + (CH_3)_2{}^{\bullet}COH \longrightarrow (CH_3)_2C(O_2^{\bullet})OH \tag{17a}$$

$$O_2 + CH_3{}^{\bullet}CHOH \longrightarrow CH_3C(O_2^{\bullet})HOH \tag{17b}$$

$$OH^- + (CH_3)_2C(O_2^{\bullet})OH \longrightarrow O_2^- + (CH_3)_2\,CO + H_2O \tag{18a}$$

$$OH^- + CH_3C(O_2^{\bullet})HOH \longrightarrow O_2^- + CH_3CHO + H_2O \tag{18b}$$

Finally, the photolysis of H_2O_2 also yields OH radicals [21] (reaction 19). Here, the photolysis wavelength is much more accessible as H_2O_2 has a very broad absorption band that extends as far into the red as ca. 360 nm, with mercury lamps irradiating at 253.7 nm:

$$H_2O_2 \xrightarrow{h\nu} 2OH \tag{19}$$

The OH radicals generated here can then undergo the reactions described earlier (9 and 10). This method [22,23] was used very successfully to measure the reaction of O_2^- with ascorbate.

Another physical method by which O_2^-/HO_2 radicals have been generated is the sonolysis of oxygen-saturated aqueous solution containing formate [24,25]. Sonolysis, like vacuum-UV photolysis, produces only $H^{\bullet}$ and $^{\bullet}OH$ radicals. The generation of HO_2/O_2^- radicals from the $H^{\bullet}$ and $^{\bullet}OH$ radicals then occurs by reactions (8–10).

3.3 CHEMICAL GENERATION OF HO_2/O_2^- RADICALS

3.3.1 KO_2

Studies using KO_2 as a method for generating O_2^- radicals are generally carried out in aprotic solution [26], due to the problem of rapid O_2^- disproportionation in aqueous solution. In aprotic solvents such as dimethyl sulfoxide (DMSO), however, KO_2 and NaO_2 are only sparingly soluble. The use of crown ethers to solubilize KO_2 or NaO_2 allows the preparation of ca. 0.1 M O_2^- solutions.

A very innovative method was developed [27] involving the use of a triple-mixer stopped-flow spectrophotometer in which millimolar concentrations of O_2^- can be reacted in a predominantly aqueous environment with a substrate and the kinetics and spectral characteristics monitored on a stopped-flow time-scale. Here, KO_2 is dissolved in DMSO, where the O_2^- is quite stable, and then rapidly mixed with 10 parts of water. The resultant solution is again rapidly mixed with 10 parts of water containing a substrate. The stopped-flow time-scale is fast enough that the O_2^- is still stable, even at only a moderately alkaline pH, and is amenable to kinetic studies.

It is also possible to use solutions of KO_2 dissolved directly in aqueous solution [28,29] at a pH above 9. These studies are most effectively carried out at temperatures near freezing and in the presence of metal chelators such as ethylenediameinetetraacetate (EDTA) and diethylenetriaminetetraace (DTPA) to further stabilize the superoxide radical.

3.3.2 Enzymatic Generation by Xanthine/Xanthine Oxidase

By far the most common method of generation of HO_2/O_2^- radicals is by using the enzymatic generating system, xanthine/xanthine oxidase. The widespread use of this method is driven by the great interest in the effect of

O_2 in biological systems and the convenience of this method of generating O_2^-. However, the disadvantages are that the radical is generated in nanomolar concentrations per minute and the kinetics are generally determined by competition methods and not direct observations.

The overall mechanism here involves the oxidation of xanthine by xanthine oxidase [30–34]. The reduced xanthine oxidase then reacts with oxygen to produce O_2^- via a stepwise mechanism. Ultimately, two O_2^- radicals are generated for each molecule of xanthine oxidase that is reduced.

4 REACTIVITY OF HO_2/O_2^- RADICALS

The reactivity of HO_2/O_2^- radicals in aqueous solution is discussed here initially with regards to organic molecules. Then, inner-sphere and outer-sphere electron-transfer reactions of these radicals with metal complexes are explored. The various mechanisms by which superoxide dismutases, naturally occurring enzymes that function to remove superoxide in aerobic living systems, react with O_2^- are presented. Finally, some very current research involving measuring the rate constants and defining the mechanisms for the reaction of HO_2/O_2^- with NO, a molecule that has become pre-eminent in discussions of neurotransmission, as well as having relevance in air pollution schemes, is discussed.

4.1 ORGANIC COMPOUNDS

In aqueous solution, HO_2/O_2^- radicals are not very reactive towards many organic compounds; an exception to this is quinones [1], which react very rapidly with O_2^-. One class of compounds that have been studied extensively is amino acids, as the reactivity of HO_2/O_2^- radicals with the building blocks of proteins is essential to any discussion of superoxide toxicity in living systems. Measurements of the rate constants [35] were made for the reactions of amino acids with HO_2/O_2^- radicals at pH extremes; no detectable reactions were observed. There has been one report of a relatively low rate constant [36] for the reaction of O_2^- with cysteine, as measured in ^{60}Co γ-ray studies at pH 3.1–7. At the pH extremes of 1 and 10, the amino acids are fully protonated or fully deprotonated, respectively (see Scheme I), and the reactivities of the zwitterionic forms of the amino acids with O_2^- radicals were not measured. However, deprotonation of the amine moiety occurs at pH 9–10, while protonation of the carboxylate group takes place at pH 2–4, so even if the zwitterionic form of the amino acids did react, the rate

$$^{+}H_3N{-}CH(R){-}CO_2R \xrightleftharpoons{H^+} {}^{+}H_3N{-}\underset{\text{zwitterion}}{CH(R){-}CO_2^-} \xrightleftharpoons{H^+} H_2N{-}CH(R){-}CO_2^-$$

Scheme I

constants for the reactions of HO_2/O_2^- with the zwitterions could not be greater than 10–10^2 $dm^3\ mol^{-1}$.

The reactivity of HO_2/O_2^- with spin-traps is a very important topic as these compounds are used extensively as assays for the presence of HO_2/O_2^- radicals. This subject has been reviewed elsewhere [37,38]. However, it is worth describing here the interactions of 5,5-dimethyl-pyrroline-*N*-oxide (DMPO), the most common spin-trap for O_2^-, with HO_2/O_2^-, in order to illustrate the advantages and drawbacks of the assay:

$$O_2^- + \text{DMPO} \xrightarrow{H^+} \text{DMPO—OOH} \tag{20}$$

$$HO_2 + \text{DMPO} \longrightarrow \text{DMPO—OOH} \tag{21}$$

The reaction was studied [39] between pH 5–9, and the data yielded rate constants of $k_{20} = 10\ dm^3\ mol^{-1}\ s^{-1}$ and $k_{21} = 6.6 \times 10^3\ dm^3\ mol^{-1}\ s^{-1}$. It is interesting to note that at high pH, where the reaction is between O_2^- and DMPO, the product is the same protonated radical-adduct. The DMPO—OOH is identified by the characteristic ESR signature [38] of $a^N = 14.2$, $a\beta^H = 11.3$ and $a\gamma^H = 1.25$ G. A drawback of using this as an assay for O_2^- is that the rate constants for reactions (20) and (21) are quite low, so very high concentrations of DMPO must be used. As a result of these low rate constants (k_{20} and k_{21}), the spontaneous dismutation of O_2^- is likely to be a major complication of any quantitative study, except at a very high pH. In addition, the lifetime of the DMPO—OOH adduct is relatively short at ambient temperature (the half-life was observed to be ca. 60 seconds at pH 7) [40]. In spite of these caveats, spin-trapping has been used very successfully in studies involving O_2^-, particularly for *in vivo* and biological systems.

This section will describe the reactivity of HO_2/O_2^- radicals with three specific organic compounds in which the mechanistic details are well defined or that illustrate particularly interesting kinetic features. The systems described here are Nitro Blue Tetrazolium (NBT^{2+}), ascorbic acid/ascorbate and NADH in contrast with LDH-bound NADH.

4.1.1 Nitro Blue Tetrazolium

Nitro Blue Tetrazolium (NBT^{2+}), 2,2′-di-*p*-nitrophenyl-5,5′-diphenyl-3,3′-(3,3′-dimethoxy-4,4′-diphenylene)ditetrazolium dichloride, is a widely used indicator for the presence of O_2^-. The spectral properties in both non-aqueous and aqueous solution and the mechanism by which the tetrazolium is reduced are well established [41–44]. The parent tetrazolium has an absorption maximum in the UV region (ϵ_{257} nm $= 6.1 \times 10^4\ dm^3\ mol^{-1}\ cm^{-1}$) while the two-electron reduced monoformazan (MF^+), absorbs strongly in the visible region ($\epsilon_{530\ nm} = 2.34 \times 10^3\ dm^3\ mol^{-1}\ cm^{-1}$). The rate constant for the reduction of NBT^{2+} by O_2^- was measured [45] by pulse radiolysis, and gave $k_{22} = 3 \times 10^4$ dm^3 mol^{-1} s^{-1}:

$$O_2^- + NBT^{2+} \rightleftharpoons O_2 + NBT^{+\cdot} \quad (22,\text{-}22)$$

$$NBT^{+\cdot} + NBT^{+\cdot} \rightleftharpoons NBT^{2+} + MF \quad (23,\text{-}23)$$

$$MF + H_2O \longrightarrow MF^+ + OH^- \quad (24)$$

The complete mechanism for the formation of MF^+ involves the disproportionation of the one-electron reduced radical of NBT^{2+}, $NBT^{+\cdot}$, followed by the hydrolysis of MF to MF^+. Equilibrium (23) has been studied and the forward rate constant was measured in aqueous solution [45]; $k_{23} = 3.1 \times 10^9$ dm^3 mol^{-1} s^{-1}. In aqueous solution, this is observed as a undirectional process because of the rapid rate of hydrolysis. It is written as an equilibrium based on the analogy to studies carried out in non-aqueous systems, where the forward and reverse reactions are observed.

The importance of the reverse of equilibrium (22) was suggested by Auclair, Torres and Hakim [46]. The rate constant of this reaction was measured by pulse radiolysis at high O_2 concentrations, and gave $k_{-22} = 1.1 \times 10^5$ dm^3 mol^{-1} (unpublished results). The use of NBT^{2+} in assays for O_2^- activity has inherent problems as, at the steady-state concentration of O_2^- produced enzymatically (usually nanomolar) relative to the concentration of O_2 (0.25–1.2 nM), the concentration of the radical intermediate $NBT^{+\cdot}$ may not be sufficiently large as to allow reaction (23) to compete effectively with reaction (-22).

While NBT^{2+} is somewhat soluble in water, MF^+ is only sparingly soluble (ca. 5 μM). In addition, NBT^{2+} is a ditetrazolium and, therefore, the entire mechanistic cycle described above will occur at the other tetrazolium centre:

$$MF^+ + O_2^+ \rightleftharpoons MF^\cdot + O_2 \quad (25,\text{-}25)$$

$$MF^\cdot + MF^\cdot \rightleftharpoons MF^+ + DF^- \quad (26,\text{-}26)$$

$$DF^- + H_2O \longrightarrow DF + OH^-$$

As seen here, the ultimate species formed in the mechanism is diformazan (DF), which is insoluble in water.

4.1.2 Ascorbic acid/ascorbate

Ascorbic Acid is a biological reductant which is known to behave synergistically with vitamin E to prevent oxidative damage in biological systems [47,48]. The rate of reaction of ascorbic acid (AH_2) and its anion, ascorbate (AH^-) (Scheme II) with HO_2 and its anion, O_2^-, was observed to vary widely with pH [23,49,50].

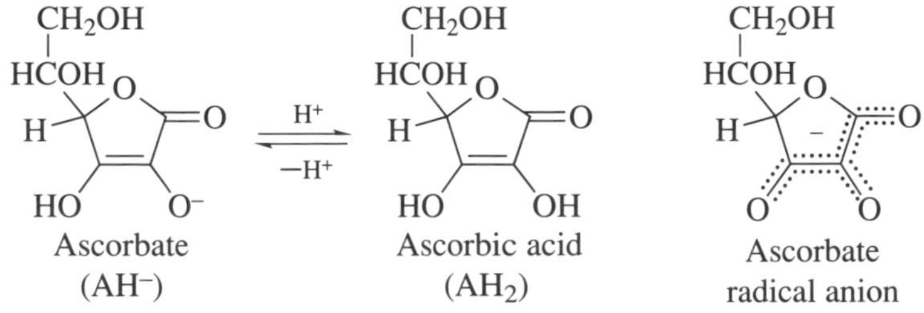

Ascorbate (AH^-) $\underset{-H^+}{\overset{H^+}{\rightleftharpoons}}$ Ascorbic acid (AH_2) — Ascorbate radical anion

Scheme II

This system was studied in detail between pH 0.5–11, using both the pulse radiolysis and stopped-flow techniques [44] in order to cover this large pH range. The observed second-order rates as a function of pH, where the k_{obs} $dm^3\ mol^{-1}\ s^{-1}$ was calculated by dividing the observed pseudo-first-order rate by the total ascorbic acid/ascorbate concentration, can be fitted to an equation involving the four possible reactions of HO_2/O_2^- with AH_2/AH^- and the acid–base equilibria of these species, i.e. reactions (28–31) and the equilibria (1,-1) and (32,-32) (see Figure 4):

$$AH_2 + HO_2 \longrightarrow A^{-\bullet} + H_2O_2 + H^+ \tag{28}$$

$$AH_2 + O_2^- \longrightarrow A^{-\bullet} + H_2O_2 \tag{29}$$

$$AH^- + HO_2 \longrightarrow A^{-\bullet} + H_2O_2 \tag{30}$$

$$AH^- + O_2^- \longrightarrow \text{product(s)} \tag{31}$$

$$AH_2 \rightleftharpoons AH^- + H^+\ (pK_a = 4.3) \tag{32,-32}$$

$$k_{obs}\,(dm^3\,mol^{-1}\,s^{-1}) = \frac{2k_{28} + 2k_{29}K_{32}/[H^+] + 2k_{30}K_1/[H^+] + 2k_{31}K_{32}K_1/[H^+]^2}{(1 + K_{32}/H^+])(1 + K_1/H^+])} \tag{33}$$

The factor of 2 is introduced into this equation as the reactions of the ascorbate free radical ($A^{-\bullet}$) with both HO_2 and O_2^- are very rapid; $k(HO_2 + A^{-\bullet}) = 5 \times 10^9$ $dm^3\ mol^{-1}\ s^{-1}$ and $k(O_2^- + A^{-\bullet}) = 2.6 \times 10^8$ $dm^3\ mol^{-1}\ s^{-1}$. Hence, the stoichiometry is that two radicals are consumed for every ascorbate oxidized. The rate constants for reactions (28–31) were calculated as $k_{28} = 1.6 \times 10^4$ $dm^3\ mol^{-1}\ s^{-1}$, $k_{31} = 5 \times 10^4$ $dm^3\ mol^{-1}\ s^{-1}$ and $(k_{30} + 0.36k_{29}) = 1.2 \times 10^7$ $dm^3\ mol^{-1}\ s^{-1}$. It is kinetically impossible to separate the two rate constants k_{29} and k_{30}; a composite rate representing a fixed ratio of the two rate constants is the only quantitative measure of these reactions. An interesting feature here is that although the ascorbate free radical can be generated by the oxidation of AH_2/AH^- by radicals such as Br_2^- and is stable over a very wide pH range, upon oxidation of ascorbate by O_2^- at pH > 8, the $A^{-\bullet}$ species was not observed.

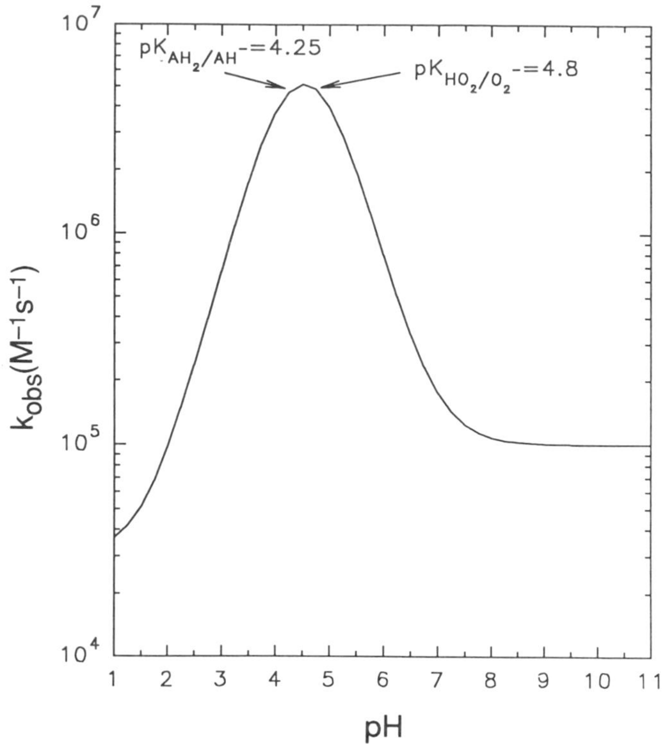

Figure 4 The calculated bimolecular rate of reaction of ascorbic acid/ascorbate with HO_2/O_2^- using equation (33) and the rate constants obtained by fitting the observed rates to the mechanism described by reactions (28–31) and equilibria (1,-1) and (32,32) (data taken from Ref. 50).

4.1.3 NADH and LDH-bound NADH

Nicotinamide-adenine dinucleotide (NADH) in its reduced form is a hydrogen-carrying coenzyme. The rate constant for the reaction of NADH with O_2^- is vanishingly small, with an upper limit of $k_{34} < 27$ dm^3 mol^{-1} s^{-1} [22,51]. In contrast, the rate constant for the reaction of HO_2 with NADH is significantly larger, i.e. $k_{35} = 1.8 \times 10^5$ dm^3 mol^{-1} s^{-1} [22,51]. However, the kinetic behaviour in this system changes markedly when NADH is bound to the enzyme, lactate dehydrogenase (LDH). A chain mechanism was proposed [52,53] that involves the initial formation of a bound NADH and the reaction of the bound LDH–NADH with O_2^- with a rate constant of $k_{37} = 1 \times 10^5$ dm^3 mol^{-1} s^{-1}, a rate constant that is significantly faster than that of the unbound NADH with O_2^-. The chain carrier is the reaction of the bound NAD$^\bullet$ with O_2:

$$O_2^- + \text{NADH} \longrightarrow HO_2^- + \text{NAD}^\bullet \tag{34}$$

$$HO_2 + NADH \longrightarrow H_2O_2 + NAD^{\cdot} \quad (35)$$

$$NADH + LDH \rightleftharpoons LDH\text{-}NADH \quad (36,-36)$$

$$O_2^- + LDH\text{-}NADH \longrightarrow HO_2^- + LDH\text{-}NAD^{\cdot} \quad (37)$$

$$LDH\text{-}NAD^{\cdot} + O_2 \longrightarrow LDH\text{-}NAD^+ + O_2^- \quad (38)$$

$$LDH\text{-}NAD^+ \rightleftharpoons LDH + NAD^+ \quad (39,-39)$$

This reaction is very efficient, i.e. the rate constant [54] for the reaction of free $NAD^{\cdot}$ with O_2 is 2×10^9 dm^3 mol^{-1} s^{-1}. A significant difference in stability constants between the LDH complexes of NADH and NAD^+ was observed, with the LDH–NADH complex favoured by a factor of 100/1 over the complex between LDH and NAD^+.

4.2 METAL IONS AND COMPLEXES

The reactivities of metal ions and metal complexes with HO_2/O_2^- radicals have been studied extensively. This was fueled in part by the discovery of superoxide dismutases, which are enzymes containing metals at the active sites that function to dismutate superoxide at near-diffusion-controlled rates *in vivo*. With the exception of a few studies of the reactivity of metal complexes such as chromium and ruthenium [5] with HO_2/O_2^- radicals, the majority of research in this area has focused on iron, copper manganese complexes, not coincidentally these being the metals at the active sites of superoxide dismutases. Complexes containing these three metals exhibit complex and varied chemistry with HO_2/O_2^- radicals, a chemistry that is, for the most part, dependent upon the nature of the ligand. These three metals are discussed separately below, followed by a discussion of superoxide dismutases and preceded by a discussion at the reactions of porphyrins with HO_2/O_2^- radicals.

4.2.1 Porphyrins

Research in the effectiveness of porphyrins for the catalytic dismutation of O_2^- in aqueous solution has focused mainly on Fe(II)/Fe(II)- and Mn(II)/Mn(III)-containing porphyrins [55–60]. One Co(III)-containing porphyrin, $Co^{III}TMPyP^{5+}$, (where TMPyP = tetrakis (4-*N*-methyl-pyridyl)porphine) was reported [56,59] to have a substantial rate constant for the reduction of O_2^-, i.e. $k(Co^{III}TMPyP^{5+} + O_2^-) = 1.4 \times 10^7$ and 1×10^5 dm^3 mol^{-1} s^{-1} at pH 5.6 and 10.1, respectively, while porphyrins that contain Ni(II), Zn(II) and Cu(II) at the active site [56] have been shown to be unreactive to O_2^-.

As seen in Table 1, the Mn(II)- and Mn(III)-containing porphyrins react with O_2^- by an outer-sphere mechanism. Although the oxidation of Mn(II)-

Table 1 Rate constants[a] for the reactions of O_2^- with various manganese- and iron-containing porphyrins[b]

M^{II} Porphyrin	k_{calc} ($dm^3\ mol^{-1}\ s^{-1}$)	M^{III} porphyrin	k_{calc} ($dm^3\ mol^{-1}\ s^{-1}$)	Ref.
$Mn^{II}TMPyP^{4+}$	$(2.5–3.3) \times 10^9(6.7–9.3)^c$	$Mn^{III}TMPyP^{5+}$	$(3.5–5.1) \times 10^7(5.6–9.3)^{c,d}$	55,56
$Mn^{II}TAPP^{4+}$	$(5.6–20) \times 10^8(6.7–9.3)^c$	$Mn^{III}TAPP^{5+}$	$(1.5–13) \times 10^6(5.6–9.3)^{c,d}$	55,56
$Mn^{II}PFP^{4+}$	$(5–9) \times 10^9(6.7–9.3)$	$Mn^{III}PFP^{5+}$	$1.7 \times 10^7(6.7)$	55
			$4.0 \times 10^5(9.3)$	55
$Fe^{II}PFP^{4+}$	$3.7 \times 10^8(5.8)$	$Fe^{III}PFP^{5+}$	$7.5 \times 10^8(5.8)$	58
			$7.6 \times 10^7(7.9)$	58
$Fe^{II}TMPyP(His)_2^{4+}$	$3.1 \times 10^6(8.0)$	$Fe^{III}TMPyP(His)_2^{5+}$	$1.2 \times 10^6(8.0)$	57
$Fe^{II}TMPyP(Im)_2^{4+}$	$3.8 \times 10^6(8.0)$	$Fe^{III}TMPyP(Im)_2^{5+}$	$1.0 \times 10^6(8.0)$	57
$Fe^{II}TMPyP(CN)_2^{4+}$	$3.1 \times 10_6(10.2)$	$Fe^{III}TMPyP(CN)_2^{5+}$	$2 \times 10^6(8.0)$	57
		$Fe^{III}TMPyP^{5+}$	$2 \times 10^9(5.6–8.0)$	57
		$Fe^{III}TMPyP(O_2^-)^{4+}$	$2.3 \times 10^9(8.1)$	57
		$Fe^{II}TPPS^{3-}$	$(1.2–8) \times 10^6(5.6–8.0)$	56
		$Fe^{III}TPPS(Im)_2^{3-}$	$3 \times 10^5(9.7)$	59
		$Fe^{III}TPPS(BSA)^{3-}$	$(5–6) \times 10^5(9.7–10.1)$	59

[a] Figures in parentheses (after *K*) represent pH (range) of measurements.
[b] TMPyP, tetrakis (4-*N*-methylpyridyl)porphine; TAPP, tetrakis-4-*N,N,N*-trimethylamion)phenylporphine; TPPS, tetrakis(*p*-sulfonatophenyl)porphine; PFP, α,α,α,β-tetrakis(*N*-methylisonicotinomidophenyl)porphine; lm, imidazole; His, histidine; CN, cyano; BSA, bovine serum albumin.
[c] Ref. 55.
[d] Ref. 56.

porphyrins by O_2^- is generally quite rapid ((10^8–10^{10}) $dm^3\ mol^{-1}\ s^{-1}$) the reduction of Mn(III)-porphyrins is slower by two to three orders of magnitude. It is the slower rate constant that governs the catalytic efficiency of any species for the dismutation of O_2^- so these compounds are clearly not very efficient as superoxide dismutases.

The reactions of O_2^- with both Fe(II)- and Fe(III)-containing porphyrins occur by somewhat more complex mechanisms. Iron-containing porphyrins have been shown to react either via outer-sphere electron transfer with O_2^- or to form an intermediate adduct with the radical. The determining factor in which behaviour is observed is the structure of the particular porphyrin and the redox potential of the metal when bound in this way. Specifically, studies of the reactivity of substituted Fe(II,III)TMPyP species with O_2^- suggest that the ligands bound in the axial position on the porphyrin seem to govern the mechanism of the reaction. If these ligands are H_2O, the mechanism is inner-sphere, while when the water is substituted with imidazole, histidine or CN^-, the mechanism changes to an outer-sphere pathway (see Table 1). In contrast to the manganese-containing porphyrins discussed above, the rate constants observed for the reduction of Fe(III)-containing porphyrins by O_2^- are very similar to those for the corresponding reoxidation reactions.

4.2.2 Iron

The reactivity of iron complexes with O_2^-/HO_2 has probably received the most attention of all metal complexes, due, in large part, to the implication of O_2^- in observed deleterious processes and aging in aerobic biological systems. Given the relatively unreactive behaviour of O_2^- towards many biologically important substrates, oxygen-initiated damage in living systems was initially attributed erroneously to the Haber–Weiss reaction [62]. This latter reaction is the purported production of OH radicals, which are very powerful and relatively non-specific oxidizing species, from the reaction between peroxide and superoxide (reaction (40)) which was originally postulated by Haber and Weiss [63].

$$H_2O_2 + O_2^- \longrightarrow OH + OH^- + O_2 \tag{40}$$

Subsequent studies, most conclusively by the radiolytic generation of the superoxide radical in the presence of high concentrations of peroxide in very clean systems [64], have put an upper limit on the rate constant for this reaction of $k_{40} < 0.03\ dm^3\ mol^{-1}\ s^{-1}$.

Given the lack of any direct reaction of H_2O_2 with O_2^- in aqueous solutions, the current theory for the source of oxygen-derived damage in biological systems is that it is metal mediated [65]. The damage is thought to occur via a Fenton-type reaction [66], resulting in the formation of OH radicals:

$$M^{n+} + H_2O_2 \longrightarrow M^{(n+1)+} + OH + OH^- \quad (41)$$

$$M^{(n+1)+} + O_2^- \longrightarrow M^{n+} + O_2 \quad (42)$$

In this mechanism, O_2^- serves as a biological reductant. A problem with the mechanism as written above is that the OH radical is so reactive that it is hard to imagine it diffusing very far from the site where it is produced, particularly in a cellular environment, without first reacting with a nearby substrate. However, the suggestion that the metal is complexed at a specific site [67], such as DNA, where OH-radical reactions in the vicinity of the metal complex lead to significant damage, is quite powerful. The site-specific production of OH radicals may be particularly important in DNA cleavage, as will be discussed later.

The rate constants for the reactions of Fe(II) and Fe(III) with HO_2/O_2^- radicals have been measured for the various aquo and hydroxy complexes formed in aqueous solution [68–70]:

$$Fe(II) + HO_2 \xrightarrow{H^+} Fe(III) + H_2O_2 \ (k_{43} = 1.2 \times 10^6 \ dm^3 \ mol^{-1} \ s^{-1}) \quad (43)$$

$$Fe(II) + O_2^- \xrightarrow{2H^+} Fe(III) + H_2O_2 \ (k_{44} = 1 \times 10^7 \ dm^3 \ mol^{-1} \ s^{-1}) \quad (44)$$

$$Fe(III) + HO_2 \longrightarrow Fe(II) + O_2 + H^+ \ (k_{45} < 10^3 \ dm^3 \ mol^{-1} \ s^{-1}) \quad (45)$$

$$Fe(III) + O_2^- \longrightarrow Fe(II) + O_2 \ k_{46} = 1.5 \times 10^8 \ dm^3 \ mol^{-1} \ s^{-1} \quad (46)$$

The addition of formate and sulfate does not significantly affect these values. At low pH, a somewhat more complex mechanism for the HO_2/Fe(II) system was proposed [71–73] that involves both mononuclear and binuclear iron complexes with the ultimate formation of Fe(III):

$$Fe(II) + HO_2 \longrightarrow [Fe(III)(HO_2^-)] \quad (47)$$

$$[Fe(III)(HO_2^-)] + Fe(II) \rightleftharpoons [Fe(III)(HO_2^-)Fe(II)] \quad (48,-48)$$

$$[Fe(III)(HO_2^-)] \longrightarrow Fe(III) + HO_2^- \quad (49)$$

$$[Fe(III)(HO_2^-)Fe(II)] \longrightarrow Fe(III) + Fe(II) + HO_2^- \quad (50)$$

The high-oxidation-state species, ferrate, $K_2Fe(VI)O_4$, can be made synthetically and is marginally stable in very alkaline solution. The pK_as for protonation of $Fe(VI)O_4^{2-}$ have been measured and it has been found that only the protonated species, $HFe(VI)O_4^-$, and $H_2Fe(VI)O_4$ react with O_2^-.

$$HFe(VI)O_4^- + O_2^- \longrightarrow Fe(V)O_4^{3-} + O_2 \quad (51)$$

$$Fe(V)O_4^{3-} + O_2^- \longrightarrow Fe(IV)O_4^{4-} + O_2 \quad (52)$$

The rate constants for the reduction of the high-valent iron species, $HFe(VI)O_4^-$, by O_2^-, as well as the reduction of $Fe(V)O_4^{3-}$ by O_2^-, have been measured [74], giving values of $k_{51} = 1.7 \times 10^7$ dm^3 mol^{-1} s^{-1} and $k_{52} = 1 \times 10^7$ dm^3 mol^{-1} s^{-1}, respectively.

The reactions of polyaminocarboxylate (PAC) complexes of Fe(II) and Fe(III) with HO_2/O_2^- have been studied extensively and in detailed studies [75] of the Fe(II)–EDTA complex, a modified mechanism to that described above was observed. Here, the O_2^- radical reduces the Fe(III)–EDTA complex but reoxidation occurs by a transient O_2^- adduct of Fe(II) EDTA (reaction 54). Conversion of this complex of Fe(III) EDTA occurs via equilibrium (55,-55).

$$\text{Fe(III)–EDTA} + O_2^- \longrightarrow \text{Fe(II)–EDTA} + O_2 \qquad (53)$$

$$\text{Fe(II)–EDTA} + O_2^- \longrightarrow [\text{Fe(III)–EDTA}\ (O_2^{2-})] \qquad (54)$$

$$\text{Fe(III)–EDTA} + H_2O_2 \overset{-2H^+}{\rightleftharpoons} [\text{Fe(III)–EDTA}\ (O_2^{2-})] \qquad (55,\text{-}55)$$

The reaction between O_2^-/HO_2 and cytochrome c (cyt c) a haem-containing protein, has received much attention, in part due to the widespread use of cytochrome c as an assay for superoxide radical reactivity. An advantage in the use of cytochrome c as such an assay is that, at 550 nm, the difference in extinction coefficient between the reduced and the oxidized cytochrome c [76,77] is 2.1×10^4 dm^3 mol^{-1} cm^{-1}. The magnitude of this molar extinction coefficient allows accurate quantitative measurements at a very low $[O_2^-]$.

The rate constant for reaction (56) was measured [78] in the presence of a metal chelator in order to eliminate any contribution to the observed rate by trace metal (usually copper) contamination, giving $k_{56} = 2.6 \times 10^5$ dm^3 mol^{-1} s^{-1} at pH 7.8. The rate constant for the reoxidation of Fe(II)–cyt c by O_2^- was shown to be negligible [79]. Studies carried out under a wide variety of experimental conditions have confirmed a pK_a for the protein at pH $\approx$ 9.2, which is likely to be the result of a conformational change, with a decrease in the observed rate at higher pH, being consistent with a very low rate constant of the high-pH form with O_2^-:

$$O_2^- + \text{Fe(III)–cyt c} \longrightarrow O_2 + \text{Fe(II)–cyt c} \qquad (56)$$

$$O_2^- + \text{Fe(II)–cyt c} \xrightarrow{2H^+} H_2O_2 + \text{Fe(III)–cyt c} \qquad (57)$$

The rate constant for the reduction of cytochrome c by O_2^- is slower than would be expected for an outer-sphere electron-transfer process as, by comparison, the reduction of cytochrome c by CO_2^- occurs with a rate constant of 1×10^9 dm^3 mol^{-1} s^{-1}. It is thought that the most likely mechanism [80] involves the transfer of an electron to an acceptor site on the protein and then internal electron transfer through the protein to the Fe(III)-containing haem.

4.2.3 Copper

The rate constants [13] for the reduction of copper ions/copper hydroxide complexes by HO_2/O_2^- are very high; $k_{58} = 1 \times 10^9$ dm^3 mol^{-1} s^{-1} and $k_{59} = 1 \times 10^{10}$ dm^3 mol^{-1} s^{-1}.

The reoxidation steps [13] are somewhat slower, i.e. $k_{60} = 1 \times 10^6$ dm^3 mol^{-1} s^{-1} and $k^{61} = (5\text{–}8) \times 10^9$ s^{-1}:

$$HO_2 + Cu^{2+} \longrightarrow Cu^+ + O_2 + H^+ \tag{58}$$

$$O_2^- + Cu^{2+} \longrightarrow Cu^+ + O_2 \tag{59}$$

$$HO_2 + Cu^+ \xrightarrow{H^+} Cu^{2+} + H_2O_2 \tag{60}$$

$$O_2^- + Cu^+ \xrightarrow{2H^+} Cu^{2+} + H_2O_2 \tag{61}$$

The magnitude of these rate constants are such that copper contamination of either water or reagents is a major cause of incorrect kinetic data in the study of O_2^- reactivity with substrates, or in the measurement of the O_2^- lifetime in aqueous solution. Once the copper is complexed, however, both the rates constant and mechanisms change markedly. In general, PAC metal chelators (EDTA, DTPA, etc.) are used to complex out adventitious copper in water when carrying out studies of the reactivity of O_2^-. The advantage of the use of PACs as chelators for copper in these systems is that the rate constants of the Cu^{2+}–PAC complexes with O_2^- are generally negligible, while the rate constants for the reaction of the Cu^+–PAC complexes with O_2 to regenerate the corresponding complexes Cu^{2+} are quite fast.

The initial studies [81–83] of the reactivity of copper(II) amino acid complexes with O_2^- radicals demonstrated that the rate constants in these systems could be quite substantial, but the kinetic parameters were measured only over limited pH ranges. Two very detailed studies of the reactivity of Cu(II)–histidine complexes [84] and Cu(II)–arginine (Arg) complexes [85] with HO_2/O_2^- radicals as a function of pH demonstrated that his high activity was only over a very narrow pH range and that only complex(es) that had an open site available for O_2^- addition, or an exchangeable water, could react at any substantial rate. This is seen in Figure 5, where the observed rate of disappearance was measured as a function of pH in the presence of 100 μM Cu(II)-complex solutions. When the observed rates are corrected for the known distribution of the Cu^{2+} complexes in solution at each pH, it is easily demonstrated that for both systems, only one of the specific complexes in solution is reacting at a substantial rate with O_2^-. In addition, the latter study involving the reactivities of Cu(II)–arginine complexes with O_2^- showed the

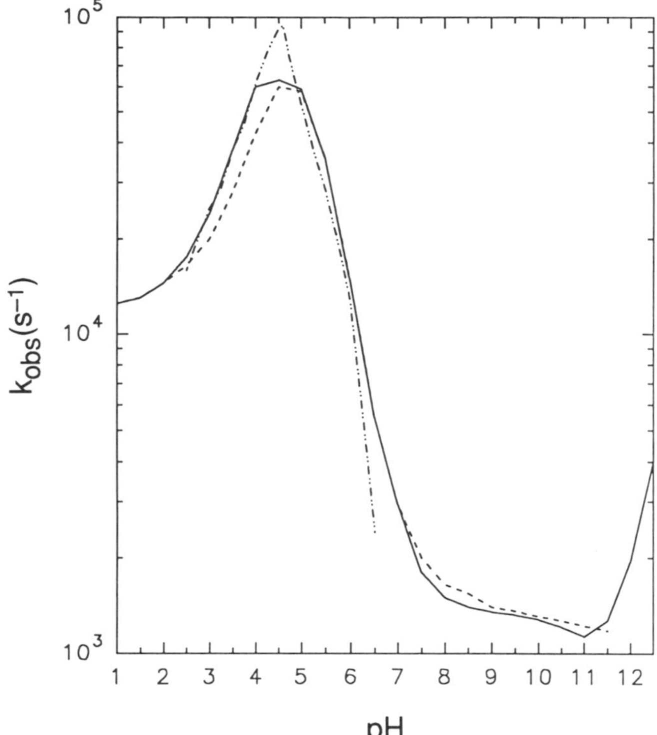

Figure 5 Pseudo-first-order reaction rates for the disappearance of O_2^-/HO_2 Radicals in the presence of copper complexes as a function of pH: (—-) observed rate in the presence of 100 μM Cu^{2+}/10 mM Arginine, 0.1 M formate, O_2 saturated solution; (–) calculated rate, fitting observed data to the known stability constants of the various Cu^{2+}-arginine complexes in aqueous solution (data taken from Ref. 85); (–..–..–) calculated rate in the presence of 20.8 μM Cu^{2+}/0.1 mM histidine scaled to 100 μM Cu^{2+}, assuming the observed rate is proportional to $[Cu^{2+}]$ (data taken from Ref. 85).

existence of a transient $Cu(O_2)^+$–arginine complex by pulse radiolysis studies carried out in 1.2–140 mM O_2:

$$Cu(II)ArgH^{2+} + O_2^- \rightleftharpoons [Cu(II)ArgH^{2+}(O_2^-)] \rightleftharpoons Cu(I)\text{–}Arginine + O_2 \quad (62)$$

DNA cleavage was observed in the presence of the Cu(II)–o-phenanthroline complex and peroxide [86–88]. This provoked a great deal of research into the reactions between O_2^- and Cu(II)–phenanthroline complexes. The chemistry [89] that occurs in aqueous solutions of these complexes in the presence of O_2^- involves the reduction and reoxidation of Cu(II)–phenanthroline, as well as the formation of a transient complex,

Cu(II)(O_2^-)–phenanthroline (reactions 63–66), where the phenanthroline ligand is written here as L:

$$\mathrm{Cu(II)L} + O_2^- \rightleftharpoons \mathrm{Cu(I)L} + O_2 \quad (63,\text{-}63)$$

$$\mathrm{Cu(I)L} + O_2^- \xrightarrow{2H^+} \mathrm{Cu(II)L} + H_2O_2 \quad (64)$$

$$\mathrm{Cu(I)L} + O_2 \rightleftharpoons \mathrm{Cu(I)}(O_2)/\mathrm{Cu(II)}(O_2^-)\mathrm{L} \quad (65,\text{-}65)$$

$$\mathrm{Cu(I)}(O_2)/\mathrm{Cu(II)}(O_2^-)\mathrm{L} + \mathrm{Cu(I)L} \xrightarrow{2H^+} 2\mathrm{Cu(II)L} + H_2O_2 \quad (66)$$

The rate constants [90] for reduction of a variety of substituted Cu(II)–phenanthroline complexes were all shown to be very rapid (ca. 1×10^8 dm^3 mol^{-1} s^{-1}). DNA cleavage in this system [91] seems likely to occur by a Fenton-type reaction between Cu(I)–phenanthroline and H_2O_2:

$$\mathrm{Cu(I)L} + H_2O_2 \longrightarrow \mathrm{Cu(II)L} + \mathrm{OH}/\mathrm{Cu(II)}{\cdot\cdot}(\mathrm{OH})\mathrm{L}/\mathrm{Cu(III)L} + \mathrm{OH}^- \quad (67)$$

Cu(II)–phenanthroline has been shown to bind to DNA efficiently and cleavage occurs by either a free or bound OH-radical- (or possibly a Cu(III)-complex-) produced site specifically. The question of bound versus free OH radicals or even the formation of a Cu(III) transient in systems such as these is the subject of much debate and is beyond the scope of this present chapter.

4.2.4 Manganese

The manganese ion in aqueous solution, Mn_{aq}^{2+} or the various hydroxylated forms, is only soluble and stable at low to neutral pH values. In contrast, upon complexation, manganese(II) can be studied over a broad pH range. The cumulative data [92,99] from studies of the reactions of a number of Mn(II) complexes with HO_2/O_2^- radicals has led to a general mechanism for the reactivity of Mn(II) complexes with O_2^-:

$$\mathrm{Mn(II)L} + O_2^- \rightleftharpoons \mathrm{Mn(II)}(O_2^-)\mathrm{L} \quad (68,\text{-}68)$$

$$\mathrm{Mn(II)}(O_2^-)\mathrm{L} \xrightarrow{H^+/H_2O} \mathrm{Mn(III)L} + H_2O_2 \quad (69)$$

$$\mathrm{Mn(II)}(O_2^-)\mathrm{L} + O_2^- \xrightarrow{2H^+} \mathrm{Mn(II)L} + O_2 + H_2O_2 \quad (70)$$

$$\mathrm{Mn(II)}(O_2^-)\mathrm{L} + H^+ \rightleftharpoons \mathrm{Mn(II)}(HO_2)\mathrm{L} \quad (71,\text{-}71)$$

At low pH, where the Mn(II)(HO_2) complex is formed, the mechanism is similar to that described earlier for the Fe(II)L + HO_2 system. A binuclear HO_2-bridged complex is observed with the ultimate formation of Mn(III) complexes:

$$Mn(II)L + HO_2 \longrightarrow Mn(II)(HO_2)L \quad (72)$$

$$Mn(II)(HO_2)L + Mn(II)L \rightleftharpoons [LMn(II)(HO_2)Mn(II)L] \quad (73,-73)$$

$$[LMn(II)(HO_2)Mn(II)L] \longrightarrow Mn(III)L + HO_2^- + Mn(II) \quad (74)$$

$$Mn(II)(HO_2)L \longrightarrow Mn(III)L + HO_2^- \quad (75)$$

As can be seen in the mechanism described by the above reactions and equilibria (68–75), the superoxide radical forms an adduct with many Mn^{II} complexes, with either subsequent oxidation of the transient adduct to Mn^{III} complexes and peroxide, or with a reaction of the adduct to yield an oxidized substrate and manganese (II). It should be noted that although the metal–HO_2/O_2^- adducts are written with the metal and radicals retaining their original oxidation states, the actual oxidation states are unknown. The formation of an oxidized metal–peroxo adduct such as $Mn(III)(HO_2^-)$ is as equally likely as a metal–perhydroxyl adduct ($Mn(II)(HO_2)$) in these systems. The rate constants for the reactions of Mn(II) complexes with HO_2/O_2^- are very ligand dependent (see Table 2). However, the mechanism and rate constant for the reaction of the transient intermediate, $Mn(II)(O_2^-)L$, seems to be the most sensitive to the nature of the ligand. This transient adduct is not observed when the ligand (L) is pyrophosphate, $P_2O_7^{4-}$, but instead the formation of Mn(III)–pyrophosphate is observed with the same rate constant as the disappearance of O_2^-. A likely mechanism is that the loss of peroxide is so fast in this system that the transient O_2^- adduct is never observed. In contrast, when the ligand is triethylenetetraaminehexaacetate (TTHA), protonation of the $Mn(II)(O_2^-)$TTHA complex is so slow that at alkaline pH, the transient is observed to disappear via a second-order process, which is attributable to the spontaneous dismutation of O_2^- (reaction (3)) via equilibrium (68,-68).

In order to turn this mechanism into a catalytic cycle, the reaction of O_2^- with Mn(III)L must be included:

$$Mn(III)L + O_2^- \longrightarrow Mn(II)L + O_2 \quad (76)$$

The reactions between Mn^{III} complexes and O_2^-/HO_2 radicals occur with rate constants that also vary with the ligand. The behaviour here is in great contrast with that found in manganese superoxide dismutase (MnSOD) (see Section 4.3). Reduction of manganese(III) in the enzyme is very facile and occurs without any significant transient formation. In contrast, the dismutation of O_2^- by various complexes of Mn(II) occurs with relatively low catalytic rate constants (ca. 1×10^6 dm^3 mol^{-1} s^{-1}) and as the oxidation of Mn(II) is facile, the catalysis must be limited by the rate constant for reduction of Mn(III) by O_2^-. This same feature was discussed in section 4.2.1 with regards to the rate constants for reduction of Mn(III)–porphyrins by O_2^-.

Table 2 Rate constants for the reactions of Mn(II)/Mn(III) complexes with HO_2/O_2^- radicals[a]

Ligand[b]	k_{68} ($dm^3\ mol^{-1}\ s^{-1}$)	k_{69} (s^{-1})	k_{76} ($dm^3\ mol^{-1}\ s^{-1}$)	Ref.
NTA	4×10^8(4.5)[c]	1.8×10^3(4.5)[c]	1.2 μ 10^7(6.0)[d]	95,97
	1.2×10^8(5.5)	9×10^1(5.5)		97
EDTA	3×10^7(4.5)[c]	3×10^3(4.5)[c]	5×10^4(10.0)[e]	97,98
	7.5×10^6(5.5)	3×10^3(5.5)		97
CyDTA			7.2×10^5(9.2)	98
TTHA	1.75×10^5(6–8.3)	9×10^4[f]		
$P_2O_7^{4-}$	2.6×10^7(6.5)[g]	$>2 \times 10^4$[h]		93
	1.3×10^7(7.3)[g]			100
Formate	4.6×10^7(4–7)	4(4–7)		93
Sulfate	5.4×10^7(2.8–5.6)	<1.4[i]		93
Phosphate	5×10^7(2–6)	2×10^3[j]		94
EDDA	3×10^7(7.1)	50(7.1)	$3 \times 10^4\ dm^3\ mol^{-1}$[k]	99

[a] Figures in parenthesis represent pH (range) of measurements.
[b] NTA, nitriloacetate; EDTA, ethylenediaminetetraacetate; CyDTA, diaminecyclohexane-*N,N,N′,N′*-tetraacetate; TTHA, triethylenetetraaminehexaacetate; EDDA, ethylenediaminediacetate.
[c] Ref. 97.
[d] Ref. 95.
[e] Ref. 98.
[f] This is the protonation reaction and the rate constant given is second-order and proportional to the H^+ concentration of the solution.
[g] No $Mn(II)(O_2^-)$ complex is observed in this system.
[h] This is a lower limit based on the concentration of Mn(II)–pyrophosphate used, k_{68}, with no transient being observed.
[i] Here, the reaction is ($Mn(II)(O_2^-)$ sulfate + O_2^- $\xrightarrow{H^+}$ Mn(II)sulfate + O_2 + H_2O_2).
[j] This is an observed first-order rate; the mechanism probably involves H_2O because of the pH independence of this rate.
[k] This mechanism involves the formation of a transient adduct between Mn(III)EDDA and O_2^- with the equilibrium constant given, followed by a first-order rate of 55 s^{-1}.

The reaction of Mn(II)EDDA (EDDA = ethylenediaminediacetate) with O_2^- was studied in detail by pulse radiolysis [99]. The initial reaction between O_2^- and Mn(II)EDDA resulted in a transient adduct that resembled those described earlier (see Table 2):

$$\mathrm{Mn(II)EDDA} + O_2^- \rightleftharpoons \mathrm{Mn(III)\ EDDA}(O_2^{2-}) \quad (77)$$

$$\mathrm{Mn(III)EDDA}(O_2^{2-}) + H_2O \xrightarrow{H^+} \mathrm{Mn(III)EDDA} + H_2O_2 + OH^- \quad (78)$$

$$\mathrm{Mn(III)EDDA} + O_2^- \rightleftharpoons \mathrm{Mn(III)}(O_2^-)\mathrm{EDDA} \quad (79,-79)$$

$$\mathrm{Mn(III)}(O_2^-)\mathrm{EDDA} \longrightarrow \mathrm{Mn(II)EDDA} + O_2 \quad (80)$$

The reaction of Mn(III)EDDA with O_2^-, however, was shown to exhibit saturation kinetics, indicating that the mechanism also involved formation of a transient adduct between these two species.

4.3 SUPEROXIDE DISMUTASES (SODS)

Superoxide dismutases are enzymes that catalytically dismutate superoxide into hydrogen peroxide and oxygen [101]. They are relatively small proteins whose active sites contain the redox-active metals, Cu, Mn and Fe. MnSOD and FeSOD are structurally distinct from the CuZnSOD complexes while still having many structural features in common [102].

The mechanism by which CuZnSOD dismutases superoxide radicals involves the alternate reduction and reoxidation of the Cu^{II} (resting state) centre with rate constants that are virtually diffusion controlled and identical for both reactions [103–105], $k_{81} = k_{82} = 2 \times 10^9$ dm^3 mol^{-1} s^{-1}:

$$Cu^{2+}ZnSOD + O_2^- \longrightarrow Cu^{+}ZnSOD + O_2 \quad (81)$$

$$Cu^{+}ZnSOD + O_2^- \underset{}{\overset{2H^+}{\rightleftharpoons}} Cu^{2+}ZnSOD + H_2O_2 \quad (82,-82)$$

Two novel features exhibited by this enzyme relative to many water-soluble copper complexes are (1) the rate constants for reactions (81) and (82) are relatively pH independent over a broad pH range (pH 5–10) and (2) the magnitude of the rate constants are electrostatically enhanced [106–112] because of the charges around the active site channel. It has been possible to both calculate [113–115] and construct [107–112] enzymes in which these rates constants have been modified as a result of changing the electrostatic field (by changing specific amino acid residues) at or near the active site. An example of this is the dramatic change in rate constants [111,116] observed upon modification of the arginine residue found in human CuZnSOD at position 143. This amino acid is in the active site channel and has been implicated in theoretical studies in the stabilization of a O_2^-–Cu^{2+} transient [117]. Chemical modification [111] of this arginine in bovine CuZnSOD (Arg 141 here) indicated an electrostatic control of the catalytic rate constant. In subsequent pulse radiolysis studies [116] of human CuZnSOD, Arg 143, having a net positive charge, was changed via site-directed mutagenesis to a histidine (+ charge), an isoleucine (neutral), and glutamic (− charge) and aspartic (− charge) acids. As can be seen in Figure 6, the enzyme loses approximately an order of magnitude in the rate constant for O_2^- dismutation as the charge at this site is reduced from positive to neutral, and another order of magnitude as it is reduced further to a negative charge. Of interest is the pH effect on the modification of the glutamic and aspartic acid, where the pK_a of the acid clearly visible. At low pH, where the acid is protonated and the net charge on the residue is now neutral, the rate

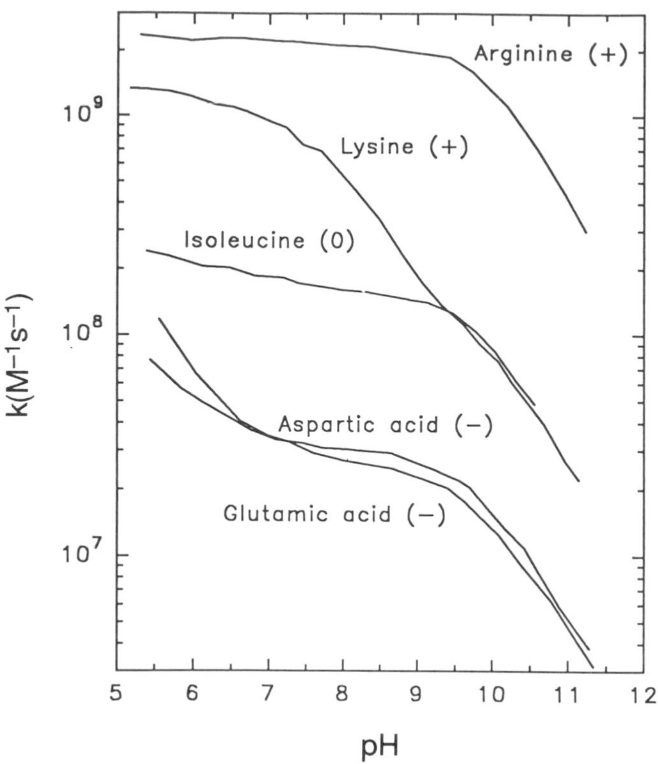

Figure 6 The observed catalytic rate constants as a function of pH for the disappearance of O_2^- radicals in the presence of human Cu,ZnSOD modified at position 143 (data taken from Ref. 116).

constant increases to become commensurate to that of the isoleucine-modified enzyme.

It has also been possible to modify the human CuZnSOD at amino acids adjacent to the active site such that the dismutation rate was accelerated significantly over that of the native enzyme [118]. Interestingly, this increase in activity comes with a price, namely a loss of stability of the enzyme to metal retention at higher pH. These studies have shown how genetic techniques which have been developed only recently, when used together with theoretical and physical methods, can help to determine the structure/function relationship in these enzymes.

Both MnSOD and FeSOD react with O_2^- by more complicated reaction mechanisms [119–123] tha CuZnSOD. In both of these systems, transient complexes between the enzyme and superoxide can be observed. The overall catalytic mechanism again involves the alternate reduction and reoxidation of the metal centre but, with both MnSOD and FeSOD, the oxidation states of the metals are +3/+2, with the resting state of the system being the +3 oxidation state:

$$M^{III}SOD + O_2^- \rightleftharpoons M^{III}(O_2^-)SOD \longrightarrow M^{II}SOD + O_2 \quad (83)$$

$$M^{II} + O_2^- \rightleftharpoons M^{II}(O_2^-)SOD \xrightarrow{H^+} M^{III}SOD + H_2O_2 \quad (84)$$

The very generalized mechanism as described above will accommodate the experimental data accumulated in studies of FeSOD in the presence of O_2^-. However, the observed kinetic data for MnSOD suggests a more complicated mechanism, involving either two parallel pathways for the reaction of O_2^- with the reduced enzyme, as described by McAdam *et al.* [121], or an isomerization of the bound O_2^- as described by Bull *et al.* [120] (see Schemes III and IV).

$$Mn^{3+}SOD + O_2^- \longrightarrow Mn^{2+}SOD + O_2$$
$$Mn^{2+}SOD + O_2^- \xrightarrow{2H^+} Mn^{3+}SOD + H_2O_2$$
$$Mn^{2+}SOD + O_2^- \longrightarrow [Mn^{2+}SOD(O_2^-)]$$
$$[Mn^{2+}SOD(O_2^-)] \xrightarrow{2H^+} Mn^{3+}SOD + H_2O_2$$

Scheme III

$$Mn^{3+}SOD + O_2^- \rightleftharpoons [Mn^{3+}SOD(O_2^-)] \longrightarrow Mn^{2+}SOD + O_2$$
$$Mn^{2+}SOD + O_2^- \rightleftharpoons [Mn^{2+}SOD(O_2^-)] \xrightarrow{2H^+} Mn^{3+}SOD + H_2O_2$$
$$[Mn^{2+}SOD(O_2^-)] \rightleftharpoons [Mn^{2+}SOD(O\text{–}O^-)]$$

Scheme IV

The isomerization, as shown in Scheme IV, is likely to be an isomerization of the bound dioxygen radical to a more stable conformation which then must reisomerize to the initial conformation to allow protonation and release of the bound peroxide. There is precedence for the isomerization of dioxygen from side-on to end-on in dioxygen-metal complexes [124] in the literature.

The mechanistic details described above for these three enzymatic systems are strikingly similar to those described earlier (see Sections 4.2.2–4.2.4) for complexes of these same three metal ions in aqueous solution. One difference between CuZnSOD and most copper complexes is that in the former system, the rate constant for reoxidation of the Cu^+ZnSOD by O_2 is very slow [125], while in the copper complexes this is generally a very facile reaction. This feature enhances the catalytic destruction of O_2^- in the enzymatic system. It is fascinating that enzymes have evolved in aerobic systems that allow the high rate constants for the reactivity of the metal complexes with O_2^- to be preserved (and even enhanced), while adding a large degree of flexibility to the conditions under which the enzymes operate with regards to the pH independence of the rate constants and the very strong binding constants of the metal ions to the enzymatic chelating system.

4.4 NITRIC OXIDE

With the recent identification [126] of NO as the endothelium-derived relaxing factor (EDRF), and the discovery that it functions as a neurotransmitter, as well as being involved in numerous other processes *in vivo*, the challenge of measuring the rate constants for the reactions between NO and HO_2/O_2^- has received much attention [127–130]. These rate constants were determined by using both pulse radiolysis [127,128,130] and flash photolysis [129] techniques. The pulse radiolysis studies [127,130] involved irradiation of solutions containing NO_2^- and formate in the presence of O_2, where reactions (85–88) are all very rapid:

$$NO_2^- + e_{aq}^- \longrightarrow NO_2^{-} \tag{85}$$

$$NO_2^- + H_2O \longrightarrow NO + 2OH^- \tag{86}$$

$$OH + HCO_2^- \longrightarrow CO_2^- + H_2O \tag{87}$$

$$CO_2^- + O_2 \longrightarrow O_2^- + CO_2 \tag{88}$$

$$O_2^- + NO \longrightarrow ONO_2^- \tag{89}$$

$$HO_2 + NO \longrightarrow ONO_2H \tag{90}$$

$$ONO_2^- + H^+ \rightleftharpoons ONO_2H \tag{91,-91}$$

By varying the quantity of nitrite and/or O_2, the ratio of O_2^-/NO can be adjusted and the reaction can then be measured under pseudo-first-order conditions. Both of the rate constants reported [127,130] for reaction (83) are essentially identical, i.e. $k_{89} = 4.0 \times 10^9$ dm^3 mol^{-1} s^{-1}.

The rate constant reported for the reaction between HO_2 and NO was also very similar [130] to k_{89}, i.e. $k_{90} = 2.1 \times 10^9$ dm^3 mol^{-1} s^{-1}. The rate constant as measured by flash photolysis [129] is in the same range, namely $k_{89} = 6.7 \times 10^9$ dm^3 mol^{-1} s^{-1}. The reaction between NO and O_2 is reasonably rapid and the mechanism is complex, thus precluding accurate determinations of k_{89} and k_{90} in air/O_2-saturated solutions containing NO. The pK_a of the resultant peroxynitrite, equilibrium (91,-91) has also been determined [131,132], having a value of 6.8.

The ultimate fate of the $ONOO^-$ species *in vivo*, namely whether the $ONOO^-$ is dismutated by superoxide dismutases, behaves as a peroxide and is involved in Fenton-like chemistry, or is itself reactive enough to account for any observed toxic effects via nitration reactions, is not yet well established. This subject has been extensively reviewed elsewhere [133] and is outside the scope of this present chapter.

ACKNOWLEDGEMENTS

This work was performed at Brookhaven National Laboratory, under contract DE-ACO2-76CHOOOO16 with the US Department of Energy, and supported by its Division of Chemical Sciences, Office of Basic Energy Sciences.

REFERENCES

1. I.B. Afanas'ev, *Superoxide Ion: Chemistry and Biological Implications*, CRC Press, Boca Raton (1991).
2. B.H.J. Bielski and D.E. Cabelli, *Int. J. Radiat. Biol.* **59**, 291.
3. D.T. Sawyer, *Oxygen Chemistry*, Oxford University Press, New York (1991).
4. B.H.J. Bielski, *Photochem. Photobiol.* **28**, 645 (1978).
5. B.H.J. Bielski, D.E. Cabelli, R.L. Arudi and A.B. Ross, *J. Phys. Chem. Ref. Data* **14**, 1041 (1985).
6. D.M. Stanbury, *Adv. Inorg. Chem.* **33**, 69 (1989).
7. P. Wardman, *J. Phys. Chem. Ref. Data* **18**, 1367 (1989).
8. E. Saito and B.H.J. Bielski, *J. Am. Chem. Soc.* **83**, 4467 (1961).
9. G. Czapski, *Annu. Rev. Phys. Chem.* **22**, 171 (1971).
10. B.H.J. Bielski and J.M. Gebicki, *Atlas of Electron Spin Resonance Spectra* Academic Press, New York (1967), pp. 292, 419 and 643.
11. H.A. Schwarz, *J. Chem. Ed.* **58**, 101 (1981).
12. G.V. Buxton, C.L. Greenstock, W.P. Helman and A.B. Ross, *J. Phys. Chem. Ref. Data* **17**, 513 (1988).
13. C. von Sonntag, *The Chemical Basis of Radiation Biology*, Taylor and Francis, London (1987) p. 31–93.
14. A.B. Ross, W.G. Mallard, W.P. Helman, G.V. Buxton, R.E. Huie and P. Neta, *NDRL-NIST Solution Kinetics Database*, Version 2 NIST Standard Reference Data, NIST, Gaithersburg (1994).
15. J.H. Baxendale, *Radiat. Res.* **17**, 312 (1962).
16. B.H.J. Bielski and J.M. Gebicki, *J. Am. Chem. Soc.* **104**, 796 (1982).
17. R.A. Holroyd and B.H.J. Bielski, *J. Am. Chem. Soc.* **100**, 5796 (1978).
18. J.A. Ghormley and C.J. Hochanadal, J.F. Riley and J.W. Boyle, *Oak Ridge National Laboratory Report ORNL-3994*, (1966), p.40.
19. M.S. McDowell, A. Bakac and J.H. Espenson, *Inorg. Chem.* **22**, 847 (1983).
20. E. Lee-Ruff, *Chem. Soc. Rev.* **6**, 195 (1977).
21. H.A. Schwarz, *J. Phys. Chem.* **66**, 255 (1962).
22. A. Nadezhdin and H.B. Dunford, *J. Phys. Chem.* **83**, 1957 (1979).
23. A. Nadezhdin and H.B. Dunford, *Can. J. Chem.* **57**, 3017 (1979).

24. B. Lippitt, J.M. McCord and I. Fridovich, *J. Biol. Chem.* **247**, 4688 (1982).
25. A. Henglein and C. Kormann, *Int. J. Radiat. Biol.* **48**, 251 (1985).
26. J.S. Valentine, A.R. Miksztal and D.T. Sawyer in *Methods of Enzymology*, edited by L. Packer and A.N. Glazer, Vol.186, Academic Press, New York (1984) p.171.
27. C. Bull, G.J. McClune and J.A. Fee, *J. Am. Chem. Soc.* **105**, 5290 (1983).
28. S.L. Marklund, *J. Biol. Chem.* **251**, 7504 (1976).
29. S.L. Marklund in *CRC Handbook of Methods for Oxygen Radical Research*, edited by R.A. Greenwald, CRC Press, Boca Raton (1985) p. 249.
30. I. Fridovich, *J. Biol. Chem.* **245**, 4053 (1970).
31. R.F. Anderson, R. Hille and V. Massey, *J. Biol. Chem.* **261**, 15870 (1986).
32. R. Hille and V. Massey, *J. Biol. Chem.* **256**, 9090 (1981).
33. A.G. Porras, J.S. Olson and G. Palmer, *J. Biol. Chem.* **256**, 9096 (1981).
34. J.S. Olson, D.P. Ballou, G. Palmer and V. Massey, *J. Biol. Chem.* **249**, 4350 (1974).
35. B.H.J. Bielski and G.G. Shiue in *Oxygen Free Radicals and Tissue Damage*, Ciba Foundation Symposium 65, Excerpta Medica (1979) p. 43.
36. J.P. Barton and J.E. Packer, *Int. J. Radiat. Phys. Chem.* **2**, 159 (1970).
37. E. Finkelstein, G.M. Rosen and E.J. Rauckman, *Arch. Biochem. Biophys.* **200**, 1 (1980).
38. G.R. Buettner and R.P. Mason, in *Oxygen Radicals in Biological Systems, Part B. Oxygen Radicals and Antioxidants*, edited by L. Packer and A.N. Glazer, Methods in Enzymology, Vol. 186, Academic Press, New York (1990) p.127.
39. E. Finkelstein, G.M. Rosen and E.J. Rauckman, *J. Am. Chem. Soc.* **102**, 4994 (1980).
40. G.R. Buettner and L.W. Oberley, *Biochem. Biophys. Res. Commun.* **83**, 69 (1978).
41. F.P. Altman, *Histochemistry*, **38**, 155 (1974).
42. M.J. Eadie, J.H. Tyrer, J.R. Kukums and W.D. Hooper, *Histochemistry*, **21**, 170 (1970).
43. C. Auclair and E. Voisin, in *CRC Handbook of Methods for Oxygen Radical Research*, edited by R.A. Greenwald, CRC Press, Boca Raton (1985) p.123.
44. A.W. Nineham, *Chem. Rev.* **55**, 355 (1955).
45. B.H.J. Bielski, G.G. Shiue and S.J. Bajuk, *J. Phys. Chem.* **84**, 830 (1980).
46. C. Auclair, M. Torres and J. Hakim, *FEBS Lett.* **89**, 26 (1978).
47. J.E. Packer, T.E. Slater and R.L. Willson, *Nature* **278**, 737 (1979).
48. A.L. Tappel, *Geriatrics* **23**, 97 (1968).
49. M. Nishikimi, *Biochem. Biophys. Res. Commun.* **63**, 463 (1975).
50. D.E. Cabelli and B.H.J. Bielski, *J. Phys. Chem.* **87**, 1809 (1983).
51. E.J. Land and A.J. Swallow, *Biochem. Biophys. Acta* **234**, 34 (1971).
52. B.H.J. Bielski and P.C. Chan, *J. Biol. Chem.* **251**, 3841 (1976).
53. B.H.J. Bielski and P.C. Chan, *J. Am. Chem. Soc.* **102**, 1713 (1980).
54. J.A. Ransohoff, *Radiat. Phys. Chem.* **18**, 239 (1981).
55. D. Weinraub, P. Levy and M. Faraggi, *In. J. Radiat. Biol.* **50**, 649 (1986).
56. P. Peretz, D. Solomon, D. Weinraub and M. Faraggi, *In. J. Radiat. Biol.* **42**, 449 (1983).
57. D. Solomon, P. Peretz and M. Faraggi, *J. Phys. Chem.* **86**, 1842 (1982).
58. M. Faraggi, D. Weinraub and P. Peretz, *Int. J. Radiat. Biol.* **49**, 951 (1986).
59. R.F. Pasternack and W.R. Skowronek, *J. Inorg. Biochem.* **11**, 261 (1979).
60. R.W. Pasternack and B. Halliwell, *J. Am. Chem. Soc.* **101**, 1026 (1979).
61. Y. Ilan, J. Rabani, I. Fridovich and R.W. Pasternack, *Inorg. Nucl. Chem. Lett.* **17**, 93 (1981).

62. I, Fridovich, *Annu. Rev. Biochem.* **44**, 147 (1975), and reference therein.
63. F. Haber and J. Weiss, *Naturwissen Schaften* **20**, 948 (1932).
64. J. Weinstein and B.H.J. Bielski, *J. Am. Chem. Soc.* **101**, 58 (1979).
65. S.D. Aust, L.A. Morehouse and C.E. Thomas, *J. Free Radical Biol. Med.* **1**, 3 (1985).
66. H.J.H. Fenton and H. Jackson, *J. Chem. Soc.* **75**, 1 (1899).
67. G. Czapski, J. Aronovich, A. Samuni and M. Chevion, in *Oxy-Radicals and their Scavenging Systems*, edited by G. Cohen and R.A. Greenwald, Vol.1, Elsevier, New York (1983), p.111.
68. W.G. Barb, J.H. Baxendale, P. George and K.R. Hargrave, *Trans. Faraday Soc.* **47**, 462 (1951).
69. W.G. Barb, J.H. Baxendale, P. George and K.R. Hargrave, *Trans. Faraday Soc.* **47**, 591 (1951).
70. J.D. Rush and B.H.J. Bielski, *J. Phys. Chem.* **89**, 5062 (1985).
71. G.G. Jayson, B.J. Parsons and A.J. Swallow, *J. Chem. Soc. Faraday Trans. 1* **68**, 2053 (1972).
72. G.G. Jayson, B.J. Parsons and A.J. Swallow, *J. Chem. Soc. Faraday Trans. 1* **69** 236 (1973).
73. G.G. Jayson, B.J. Parsons and A.J. Swallow, *J. Chem. Soc. Faraday Trans. 1* **69** 1079 (1972).
74. J.D. Rush, Z. Zhao and B.H.J. Bielski, *Free Radical Res. Commun.* (1996), in press.
75. C. Bull, G.J. McClune and J.A. Fee, *J. Am. Chem. Soc.* **105**, 5290 (1983).
76. V. Massey, *Biochim. Biophys. Acta* **34**, 255 (1959).
77. B.F. van Gelder and E.C. Slater, *Biochim. Biophys. Acta*, **58**, 593 (1962).
78. E.J. Land and A.J. Swallow, *Arch. Biochem. Biophys.* **145**, 365 (1971).
79. H. Seki, Y.A. Ilan, Y. Ilan and G. Stein, *Biochem. Biophys. Acta* **440**, 573 (1976).
80. J. Butler, W.H. Koppenol and E. Margoliash, *J. Biol. Chem.* **B257** 10789 (1982).
81. R. Brigelius, R. Spottl, W. Bors, E. Lengfelder, M. Saran and U. Weser, *FEBS Lett.* **47**, 72 (1974).
82. R. Brigelius, H.-J. Hartmann, W. Bors, M. Saran, E. Lengerfelder and U. Weser, *Hoppe-Seyler's Z. Physiol. Chem.* **356**, 739 (1975).
83. D. Klug-Roth and J. Rabani, *J. Phys. Chem.* **80**, 588 (1976).
84. J. Weinstein and B. Bielski, *J. Am. Chem. Soc.* **102**, 4916 (1980).
85. D.E. Cabelli, J. Holcman and B.H.J. Bielski, *J. Am. Chem. Soc.* **109**, 3665 (1987).
86. D.R. Graham, L.E. Marshall, K.A. Reich and D.A. Sigman, *J. Am. Chem. Soc.* **102**, 5419 (1980).
87. B.G. Que, K.M. Downey and A.G. So, *Biochemistry*, **19**, 5987 (1980).
88. L.E. Marshall, D.R. Graham, K.A. Reich and D.A. Sigman, *Biochemistry* **20**, 244 (1981).
89. S. Goldstein, G. Czapski, *J. Amer. Chem. Soc.* **105**, 7276 (1983).
90. S. Goldstein and G. Czapski in *Superoxide and Superoxide Dismutase in Chemistry, Biology and Medicine*, edited by G. Rotilio, Elsevier, New York (1985), p.64.
91. S. Goldstein and G. Czapski, *J. Am. Chem. Soc.* **108**, 2244 (1986).
92. S. Baral, C. Lume-Pereira, E. Janata and A. Henglein, *J. Phys. Chem.* **90**, 6025 (1986).
93. D.E. Cabelli and B.H.J. Bielski, *J. Phys. Chem.* **88**, 3111 (1984).
94. D.E. Cabelli and B.H.J. Bielski, *J. Phys. Chem.* **88**, 6291 (1984).
95. W.H. Koppenol, F. Levine, T.L. Hatmaker, J. Epp and J.D. Rush, *Arch. Biochem. Biophys* **251**, 594 (1986).

96. M. Pick-Kaplan and J. Rabani, *J. Phys. Chem.* **80**, 1840 (1976).
97. J. Lati and D. Meyerstein, *J. Chem. Soc. Dalton Trans.* 1105 (1978).
98. J. Stein, J.P. Fackler Jr, G.J. McClune, J.A. Fee and L.T. Chan, *Inorg. Chem.* **18**, 3511 (1979).
99. J.D. Rush and Z. Maskos, *Inorg. Chem.* **29**, 897 (1990).
100. F. Goetz and E. Lengfelder in *Oxy-Radicals and their Scavenging Systems*, edited by G. Cohen and R.A. Greenwald, Vol.1, Elsevier, New York (1983) p.228.
101. J.M. McCord and I. Fridovich, *J. Biol. Chem.* **244**, 6049 (1969).
102. W. Beyer, J. Imlay and I. Fridovich, *Prog. Nucleic Acid Res. Mol. Biol.* **40**, 221 (1991).
103. E.M. Fielden, P.B. Roberts, R.C. Bray, D.J. Lowe, G.N. Mautner, G. Rotilio and L. Calabrese, *Biochem. J.* **129**, 49 (1974).
104. D. Klug, J. Rabani and I. Fridovich, *J. Biol. Chem.* **247** 4839 (1972).
105. C. Bull and J.A. Fee, *J. Am. Chem. Soc.* **107**, 3295 (1985).
106. W.H. Koppenol in *Oxygen and Oxy-Radicals in Chemistry and Biology*, edited by M.A. Rodgers and E.L. Powers, Academic Press, New York (1981) p.671.
107. J.A. Tainer and E.L. Getzoff, J.S. Richardson and D.C. Richardson, *Nature* (London) **306**, 284 (1983).
108. E.D. Getzoff, J.A. Tainer, P.K. Weiner, P.A. Kollman, J.S. Richardson and D.C. Richardson, *Nature* (London) **306**, 287 (1983).
109. D.P. Malinowski and I. Fridovich, *Biochemistry* **18**, 5909 (1979).
110. L. Banci, I. Bertini, D.E. Cabelli, R.A. Hallewell, C. Luchinat and M.S. Viezzoli, *Free Radical Res. Commun.* **12–13,** 239 (1991).
111. A. Cudd and I. Fridovich, *J. Biol. Chem.* **257**, 11443 (1982).
112. E. Argese, P. Viglino, G. Rotilio, M. Scarpa and A. Rigo, *Biochemistry* **26**, 3224 (1987).
113. J. Sines, S. Allison, a. Wierzbicki and J.A. McCammon, *J. Phys. Chem.* **94**, 959 (1990).
114. J. Sines, S. Allison and J.A. McCammon, *Biochemistry*, **29**, 9403 (1990).
115. I. Klapper, R. Hagstrom, R. Fine, K. Sharp and B. Honig, *Proteins* **1**, 47 (1986).
116. C.L. Fisher, D.E. Cabelli, J.A. Tainer, R.A. Hallewell and E.D. Getzoff, *Proteins* **19,** 24 (1994).
117. R. Osman and H. Basch, *J. Am. Chem. Soc.* **106**, 5710 (1984).
118. E.D. Getzoff, D.E. Cabelli, C.L. Fisher, H.E. Parge, M.S. Viezzoli, L. Banci and R.A. Hallewell, *Nature* (London) **358**, 347 (1992).
119. J.A. Fee and C. Bull, *J. Biol. Chem.* **261**, 1300 (1986).
120. C. Bull, E.C. Niederhoffer, T. Yoshida and J.A. Fee, *J. Am. Chem. Soc.* **113**, 4069 (1991).
121. M.E. McAdam, R.A. Fox, F. Lavelle and E.M. Fielden, *Biochem. J.* **165**, 71 (1977); M.E. McAdam, R.A. Fox, F. Lavelle and E.M. Fielden, *Biochem. J.* **165**, 81 (1977).
122. M. Pick, J. Rabani, F. Yost and I. Fridovich, *J. Am. Chem. Soc.* **96**, 7329 (1974).
123. F. Fe Lavelle, M.E. McAdam, E.M. Fielden, P.B. Roberts, K. Puget and A.M. Michelson, *Biochem. J.* **161**, 3 (1977).
124. J.S. Valentine, *Chem. Rev.* **73**, 235 (1973).
125. P. Viglino, M. Scarpa, F. Coin, G. Rotilio and A. Rigo, *Biochem. J.* **237**, 305 (1986).
126. R.M. Palmer, A.G. Ferrige and S. Moncada, *Nature* (London) **327**, 524 (1987).
127. K. Kobayashi, M. Miki and S. Tagawa, *J. Chem. Soc. Dalton Trans.* 2885 (1995).
128. M. Saran, C. Michel and W. Bors, *Free Radical Res. Commun.* **10**, 221 (1990).
129. R.E. Huie and S. Padmaja, *Free Radical Res. Commun.* **18**, 195 (1993).

130. S. Goldstein and G. Czapski, *Free Radical Biol. Med.* **19**, 505 (1995).
131. W.H. Koppenol, J.J. Moreno, W.A. Pryor, H. Ischiropoulos and J.S. Beckman, *Chem. Res. Toxicol.* **5**, 834 (1992).
132. T. Logager and K. Sehested, *J. Phys. Chem.* **97**, 6664 (1993).
133. J.S. Beckman, *J. Dev. Physiol.* **15**, 53 (1991).

14 Heteroatom Peroxyl Radicals

HEINZ-PETER SCHUCHMANN
Max-Plank-Institut für Strahlenchemie, Mülheim an der Ruhr, Germany

and

CLEMENS VON SONNTAG
Max-Plank-Institut für Strahlenchemie, Mülheim an der Ruhr, Germany

1 INTRODUCTION

In the course of the oxidation of carbon compounds in the presence of dioxygen, peroxyl radicals are found to play a vital role. The preponderance of carbon atoms in organic compounds has tended to make carbon-centred peroxyl radicals rank foremost, and therefore heteroatom peroxy radicals can only be said to occupy a minority position among this type of radical. Moreover, as a class they are set apart from the carbon-centred ones by the fact that many elements, in contrast to carbon, can allow a rearrangement at the peroxyl-carrying atom with loss of the peroxyl function, accompanied by a change of the coordination number and valence state of the heteroatom. In these cases, this will tend to make heteroatom peroxyl radicals more unstable than their carbon-centred analogues. The basis of heteroatom-centred radicals with different possible valence states of the heteroatom implies the existence of different classes of peroxyl radicals based on the same element. Such complexities notwithstanding, many of the reaction of heteroatom peroxyl radicals are nevertheless reminiscent of those of their carbon-centred prototypes.

Peroxyl-radical structures do not only exist for non-metallic elements, but may also be formulated in the case of transition metals (superoxo complexes, cf. Refs. 1 and 2). However, owing to the multiplicity of valence states in these elements and the ease with which they can alter, the peroxyl radical structure describes only incompletely the properties of such metal–oxygen complexes. The interplay of the various oxidation states of these metal–oxygen complexes is of the widest significance in biochemistry and catalysis, and these superoxo complexes justifiably fall outside the scope of this present survey.

This short review deals with non-transition-metal heteroatom peroxyl radicals of organic compounds. Such peroxyl radicals have to date only been discovered for a rather limited series of elements, i.e. of the main groups VI, V, and IV. Nevertheless, it is a fair bet that most, if not all, heteroatom-centred

Peroxyl Radicals. Edited by Z.B. Alfassi

radicals will form peroxyl radicals on contact with dioxygen even though these species may in many cases be very unstable, thus precluding their identification. With regard to synthetic procedures, the chemical use of the heteroatom peroxyl function is not yet widespread, although the field has the potential to become another bridge between organic and inorganic chemistry.

2 MAIN GROUP VI ELEMENTS

2.1 OXYGEN PEROXYLS

Trioxyl radicals, $RO_3^{\cdot}$ arise through H-atom abstraction from the organohydrotrioxide function [3–5]. Trioxyl is unstable with respect to its decomposition into alkoxyl and O_2 (reaction (1)), even at low temperatures, since the breaking of this bond is exothermic by several tens of kJ mol^{-1} [6–8]. In this context, it is interesting to consider the prototype $HOOO^{\cdot}$ radical, which in aqueous solution is in equilibrium with the ozonide radical anion $O_3^{-\cdot}$ (reactions (2,-2); pK_a = 6.15: [9]).

$$ROOO^{\cdot} \longrightarrow RO^{\cdot} + O_2 \qquad (1)$$

$$HOOO^{\cdot} \rightleftharpoons OOO^{-\cdot} + H^+ \qquad (2,-2)$$

The ozonide radical anion can be generated in alkaline solution by the addition of the strongly nucleophilic radical, $O^{-\cdot}$, to dioxygen [10] (reaction (3)); evidently, in $^-O—OO^{\cdot}$, the bond strength is positive. This situation reverses upon protonation, when the strength of this bond becomes negative and cleavage into $^{\cdot}OH$ and O_2, analogues to reaction (1), takes place [9]:

$$O^{-\cdot} + O_2 \longrightarrow O_3^{-\cdot} \qquad (3)$$

2.2 SULFUR PEROXYLS

The free-radical chemistry of organic sulfur compounds, thiols in particular, has attracted considerable interest, particularly among radiation chemists and radiation biologists (cf. Refs 11 and 12, and references therein), on account of their ability to reduce reactive free radicals, such as the OH radical, but also many other less reactive species including carbon-centred radicals. In this way, thiol compounds can confer protection against radiative damage. Chemists researching the atmosphere study the complex web of photodegradation and oxidation reaction pathways undergone by air-borne organic sulfur compounds, particularly dimethyl sulfide which is a ubiquitous trace constituent of the atmosphere. These processes, the knowledge of whose elementary steps is important for mechanistic modelling, also involve the participation of sulfur peroxyl radicals (see Refs 13 and 14), and references therein).

While in the absence of dioxygen, thiyl radicals, $RS^{\bullet}$, which result from the free-radical-induced oxidation of the thiol, or the photocleavage of a sulfide or disulfide, mainly undergo dimerization to the disulfide, they are oxidized to the thiylperoxyl radicals, $RSOO^{\bullet}$ in the presence of O_2 (reaction (4)) [15–26]. From about a decade ago, the question has increasingly been posed as to the chemical behaviour and fate of $RSOO^{\bullet}$. It has become apparent that as a consequence of the transformation of thiylperoxyl, other sulfur-function-based radicals are also formed, among these sulfonylperoxyl, $RS(O_2)OO^{\bullet}$ [19,25], which is a good H-atom abstractor [19,27–29], and may contribute to the oxygen-enhancement effect observed in the damage to DNA by ionizing radiation [27].

The kinetics of the formation and transformation reactions of $RSOO^{\bullet}$ derived from mercaptoethanol have been studied recently [25,26] in aqueous solution. At room temperature, $k_4(= 2.2 \times 10^9\ dm^3\ mol^{-1}\ s^{-1})$ is diffusion controlled, but the reverse reaction is also fast ($k_{-4} = 6.2 \times 10^5\ s^{-1}$), which implies the existence of an equilibrium situation; the stability constant, k_4/k_{-4} is $3.5 \times 10^3\ dm^3\ mol^{-1}$ at room temperature. These values are very similar to those obtained in the case of glutathione [22]. A thermal rearrangement to the thermodynamically more stable [19,21,30] sulfonyl radical (reaction(5)) takes place with a rate constant of ca. $2 \times 10^3\ s^{-1}$ [25] ($1.4 \times 10^3\ s^{-1}$ [26]). It was found in frozen aqueous solutions that this rearrangement can also be induced by ultraviolet (UV) radiation [19]. The sulfonyl radical can be further oxidized (reactions (6)) to sulfonylperoxyl $RS(O_2)OO^{\bullet}$ [19,25]. Electron spin resonance (ESR) parameters of some thiylperoxyl radicals are collected together in Ref. 31. Only about half of the spin density is at the terminal O-atom, while the inner O-atom carries ca. 40%, and the S-atom 6 to 8%. Thiylperoxyl is remarkable for its optical absorption in the visible region, occurring near 550 nm [15,24]. This enables it to be monitored easily, despite its low absorption coefficient (ϵ is of the order of $(2–4) \times 10^2\ dm^3\ mol^{-1}\ cm^{-1}$), since other transients in these systems show absorptions only at shorter wavelengths. Thiylperoxyl may undergo oxygen-atom transfer (reaction (7)) to reducing compounds such as the parent thiol [19,24], perhaps via an adduct, $RSOO\dot{S}(H)R$ [32]; k_7 is of the order of $2 \times 10^6\ dm^3\ mol^{-1}\ s^{-1}$ [25] ($1.4 \times 10^6\ dm^3\ mol^{-1}\ s^{-1}$ [26]).

$$R{-}S^{\bullet} \underset{(4)/(-4)}{\overset{O_2}{\rightleftharpoons}} R{-}S{-}O{-}O^{\bullet} \xrightarrow{(5)} R{-}\overset{O}{\overset{\|}{\underset{\underset{O}{\|}}{S^{\bullet}}}}$$

$$R{-}S{-}O{-}O^{\bullet} \xrightarrow[(7)]{RSH} R{-}\overset{O}{\overset{/\!/}{S^{\bullet}}} + R{-}S{-}OH$$

$$R{-}SO_2^{\bullet} \xrightarrow[(6)]{O_2} R{-}\overset{O}{\overset{\|}{\underset{\underset{O}{\|}}{S}}}{-}O{-}O^{\bullet}$$

In contrast to the thiyl radicals, thiyl radical cations such as those obtained through oxidation of thioureas (reaction (8)) are unreactive toward O_2 on the pulse radiolysis time-scale [33].

$$(R_2N)_2C{=}S + {}^{\bullet}OH \xrightarrow{(8)} {}^{\oplus}R_2N{=}C(NR_2){-}S^{\bullet} + OH^{\ominus}$$

The low stability of the thiylperoxyl radical with respect to the reverse of its formation is already apparent from the relatively high value of the rate constant k_{-4}. In more quantitative terms, experimental values for the S—OO$^{\bullet}$ bond strength of 11 kcal mol^{-1} [23], or ca. 7 kcal mol^{-1} [25] have been reported. Quantum chemical calculations are in support of such low values [21,30,34].

The thiylperoxyl radical from dithiothreitol (DTT) shows a further way of expelling the peroxyl function [35]. This consists of the elimination of superoxide from the deprotonated form of the thiylperoxyl radical, upon closure of the disulfide bond (reaction (9)).

$$\text{DTT}(S{-}O{-}O^{\bullet})(S^{\ominus}) \xrightarrow{(9)} \text{ox-DTT}(S{-}S) + O_2^{\bullet\ominus}$$

The oxygenated DTT system exemplifies some of the complexities of sulfur peroxyl free-radical chemistry. Under alkaline conditions, in addition to the conventional thiylperoxyl-type species, RSOO$^{\bullet}$, a peroxo–thiolate radical dianion is apparently produced through the addition of superoxide to the thiolate (reaction (10)) and may be an intermediate in an $O_2^{-\bullet}$-catalyzed oxidation of DTT under these conditions. Contrary to what one might expect, this process does not give rise to H_2O_2, but the oxidation product has the stoichiometry shown in reaction (11) [35].

$$\text{DTT}(S^{\ominus})(SH) + O_2^{\bullet\ominus} \xrightarrow{(10)} \text{DTT}(\overset{\ominus}{S}{\dot{-}}O{-}O^{\ominus})(SH)$$

$$2\ \underset{\textbf{DTT}}{\text{DTT}(SH)(SH)} + O_2 \xrightarrow{(11)} \underset{\textbf{ox-DTT}}{\text{ox-DTT}(S{-}S)} + 2\,H_2O$$

Under acidic conditions, the oxidation proceeds via a different chain mechanism (reactions (12–14)). Here, disulfide formation is accompanied by the production of an equivalent amount of H_2O_2 (reaction (15)) [36].

$$\text{DTT-S-O-O}^{\bullet} \xrightarrow[(12)]{} \text{ox-DTT (cyclic S–S)} + HO_2^{\bullet}$$

$$HO_2^{\bullet} + \text{DTT (SH, SH)} \xrightarrow[(13)]{} \text{DTT (S}^{\bullet}\text{, SH)} + H_2O_2$$

$$\text{DTT (S}^{\bullet}\text{, SH)} + O_2 \xrightarrow[(14)]{} \text{DTT (S-O-O}^{\bullet}\text{, SH)}$$

$$\textbf{DTT} + O_2 \xrightarrow[(15)]{} \textbf{ox-DTT} + H_2O_2$$

Under near-neutral conditions, the oxidation is essentially non-chain in nature. This is because the prerequisite for the existence of a chain process, i.e. the occurrence side-by-side of both $O_2^{-\bullet}$ and thiolate, is no longer fulfilled in this pH range where the thiolate form does no longer exist, while the $HO_2^{\bullet}$, a moderately potent hydrogen abstractor in contrast to $O_2^{-\bullet}$, does not yet exist in sufficient concentration.

After $RSOO^{\bullet}$, the next-higher-oxygenated peroxyl homologue is sulfinylperoxyl, $RS(O)OO^{\bullet}$. Firm evidence for the emergence (reaction (16)) of this kind of species in the course of the radiolysis-induced oxidation of thiol systems in lacking, although some evidence has been found that sulfinic acid is formed under certain conditions in low yield. This may be due to a low reactivity of the sulfinyl radical with O_2. In fact, it has been claimed that the structurally similar alkoxysulfinyl radical, $RO\dot{S}O$, is essentially unreactive toward O_2 [37]. Nevertheless, there are indications that methylsulfinylperoxyl is one of the sulfur peroxyl radical species that play a role in the gas-phase oxidation of thiyl radicals [13,14].

$$R{-}\dot{S}({=}O) + O_2 \xrightarrow[(16)]{} R{-}S({=}O){-}O{-}O^{\bullet}$$

Apart from the oxidation of thiyl there exist other pathways to the higher-valent sulfur peroxyl radicals. 2-Nitrobenzenesulfinylperoxyl has been generated from the parent sulfinyl chloride by reacting the latter with superoxide (cf. reaction (17)) in acetonitrile. This sulfinylperoxyl radical is able to oxidize sulfides to the corresponding sulfoxides in excellent yields [29,38]. The reaction of superoxide with the acid chloride is an example of a wider class of nucleophilic substitution reactions by superoxide in aprotic solvents [39].

$$\text{R}-\text{S}(=\text{O})\text{Cl} + \text{O}_2^{\bullet\ominus} \xrightarrow{(17)} \text{R}-\text{S}(=\text{O})-\text{O}-\text{O}^{\bullet} + \text{Cl}^{\ominus}$$

The highest-valent member of the oxo–sulfur peroxyl radical homologues is sulfonylperoxyl, $RS(O_2)OO^{\bullet}$. Its presence in radiolytic systems has been verified by ESR spectroscopy [19]. Under the conditions of a relatively high $[O_2]/[RSH]$ ratio, this radical predominates. It is expected to be the precursor of the corresponding sulfonic acid which is observed as the major radiolysis product under these conditions [25]. There is evidence that $RS(O_2)OO^{\bullet}$ is able to undergo H-atom abstraction reactions [28,29,40], similar to the structurally related peroxysulfate radical, $SO_5^{-\bullet}$ [41]. This would lead to peroxysulfonic acid. However, in the mercaptoethanol system [25], acid formation appears to be prompt, while the decomposition of the peroxysulfonic acid into the sulfonic acid is expected to be relatively slow under these conditions if the behaviour of the structurally related compound, Caro's acid, is any guide [42,43]. At present, the mechanism of the sulfonic-acid formation is not clear.

A representative of these species, the 2-nitrophenylsulfonyl peroxyl radical has been generated from the parent 2-nitrobenzenesulfonic acid chloride through the displacement of chloride ion by superoxide in acetonitrile (reaction (18), cf. Refs. 29 and 44), a reaction which can be used for the epoxidation of olefins and arenes [45,46], to which it transfers an oxygen atom (reaction (19)). The 2-nitrobenzenesulfonate radical, $RS(O_2)O^{\bullet}$, which is left behind, is reduced to the sulfonate, e.g. by $O_2^{-\bullet}$ (reaction (20)) or H-atom abstraction. The corresponding arylsulfinyl, $RS(O)OO^{\bullet}$, reacts in a like manner [44]. The fact that $RS(O_2)OO^{\bullet}$ is more reactive than thiylperoxyl with respect to H-atom abstraction (thiylperoxyl prefers to react with thiol by O-atom transfer, reaction (7), rather than by H-abstraction, which would impart a pronouned chain character to the mercaptoethanol/O_2 free-radical chemistry, although, however, this is not the case [47] is perhaps a reflection of the finding that 70% of the spin density resides on the terminal O-atom, similar to the case of typical carbon peroxyl radicals, but in contrast to the ca. 50% in the case of $RSOO^{\bullet}$ [20,31].

The foregoing sulfur peroxyl radicals have in common the fact that they are physically or formally derived ultimately from thiols or thiyl radicals, as they contain only one organic function, i.e. the structural element R—S(O_X). Distinct from these are the peroxyl radicals derived from sulfuranyl radicals, $R_3S^{\bullet}$. At present, information on this class of sulfur peroxyl radicals seems restricted to the dialkylhydroxysulfuranylperoxyl radicals, $R_2(HO)SOO^{\bullet}$ which

$$R{-}S(=O)_2{-}Cl + O_2^{\bullet -} \xrightarrow{(18)} R{-}S(=O)_2{-}O{-}O^{\bullet} + Cl^{-}$$

$$NO_2{-}C_6H_4{-}S(=O)_2{-}O{-}O^{\bullet} + \rangle C{=}C\langle \xrightarrow{(19)} NO_2{-}C_6H_4{-}S(=O)_2{-}O^{\bullet} + \rangle C(O)C\langle$$

$$NO_2{-}C_6H_4{-}S(=O)_2{-}O^{\bullet} + O_2^{\bullet -} \xrightarrow{(20)} NO_2{-}C_6H_4{-}S(=O)_2{-}O^{-} + O_2$$

are produced when sulfides R_2S react with the OH radical in the presence of O_2 (reactions (21,22)) [13,14,48,49]. These peroxyl radicals are short-lived and are transformed into the sulfoxides by superoxide elimination (reaction (23)) [49]. Dimethylhydroxysulfuranylperoxyl has been suspected to be a precursor in the gas-phase formation of dimethysulfone (reaction (24)) [50]. This

$$R_2S + {}^{\bullet}OH \xrightarrow{(21)} HO{-}S^{\bullet}R_2$$

$$HO{-}S^{\bullet}R_2 + O_2 \xrightarrow{(22)} HO{-}SR_2{-}O{-}O^{\bullet} \quad \left(HO{-}S^{\bullet}R_2\langle{}^{O}_{O}|\right)$$

$$HO{-}SR_2{-}O{-}O^{\bullet} \xrightarrow{(23)} R_2S{=}O + O_2^{\bullet -} + H^{+}$$

$$HO{-}S^{\bullet}R_2\langle{}^{O}_{O}| \xrightarrow{(24)} O{=}SR_2{=}O + {}^{\bullet}OH$$

reaction could perhaps proceed via a 'side-on' peroxo–complex sulfur radical that might coexist with the sulfuranylperoxyl form [49].

Trialkylsulfuranyl peroxyl radicals, $R_3SOO^{\bullet}$, have to date apparently not been studied, even though the approach to generate them from the reaction of superoxide with trialkylsulfonium might be a promising route.

2.3 SELENIUM AND TELLURIUM PEROXYLS

There is circumstantial evidence that the equilibrium (25) lies strongly to the left [51]. The reversible formation of selenylperoxyl, $RSeOO^{\bullet}$, has been hypothesized to constitute a side reaction in the diselenide-catalyzed oxidation of glutathione [52]. Otherwise, the literature appears empty on this and other selenoperoxyl species.

$$R{-}Se^{\bullet} + O_2 \underset{(-25)}{\overset{(25)}{\rightleftharpoons}} R{-}Se{-}O{-}O^{\bullet}$$

The situation is similar in the case of tellurium. It seems that the affinity of $RTe^{\bullet}$ and $RTeTe^{\bullet}$ to oxygen must be very low, since the photolysis of dibenzyl ditelluride in the presence of O_2 produces elemental Te (cf. Ref. 53).

3 MAIN GROUP V ELEMENTS

3.1 NITROGEN PEROXYLS

While OH-radical-induced formation of thiyl radicals from thiols in the presence of O_2 provides an easy, essentially universal gateway to thiylperoxyl radicals, because of the preferential abstraction of the thiol H-atom, this approach is mostly impractical in the case of primary and secondary amines (except for ammonia), since amino hydrogen is usually more firmly bound and radical attack would predominantly materialize at the organic moiety of the amine [54]. This is in contrast to the situation regarding the UV photolysis if such amines where cleavage of the N—H bond if often the preferred primary process [55,56].

Generated by OH-radical attack, the prototype aminylperoxyl, $NH_2OO^{\bullet}$, has been observed upon radiolysis of ammonia in oxygenated aqueous solutions (reactions (26)) [57,58] where, in contrast to the conclusions reached in gas-phase [59,60] and theoretical [61] studies, the rate constant of its formation from $^{\bullet}NH_2$ and O_2 is moderately fast (reaction (27)), having a value of $3 \times 10^7\ dm^{-3}\ mol^{-1}\ s^{-1}$ [57]. Interestingly, it is eventually transformed into peroxynitrite $ONOO^-$, apparently by a reaction with superoxide [58]:

$$NH_3 + {}^{\bullet}OH \longrightarrow {}^{\bullet}NH_2 + H_2O \tag{26}$$

$$^{\bullet}NH_2 + O_2 \longrightarrow H_2NOO^{\bullet} \tag{27}$$

Alkyl-substituted aminyl radicals may be generated by photolysis of the corresponding tetrazenes (reaction (28)) [62], or in the case of amines with a weakened N—H bond such as 2,2,6,6-tetramethylpiperidines, by H-atom abstraction at the nitrogen [63]. Apparently the reaction of the corresponding aminyl radical with O_2 (reaction (29)) is fast on the ESR time-scale, even at cryogenic temperatures [62]. This is not in contradiction with data from a laser flash photolysis study where the formation of $Ph_2NOO^{\bullet}$ from $Ph_2N^{\bullet}$ could not be observed, implying a rate constant below 1×10^7 dm^3 mol^{-1} s^{-1} [64]. The ensuing aminylperoxyl radical is converted into the corresponding nitroxide [62], perhaps through reactions (30) or (31) [63]. The ESR data, as well as quantum chemical computation, indicate that the structure of the aminylperoxyl function resembles that of carbon-based peroxyl radicals [63]:

$$R_2N{-}N{=}N{-}NR_2 \xrightarrow{h\nu} 2\,R_2N^{\bullet} + N_2 \tag{28}$$

$$R_2N^{\bullet} + O_2 \longrightarrow R_2NOO^{\bullet} \tag{29}$$

$$R_2NOO^{\bullet} + R_2N^{\bullet} \longrightarrow 2\,R_2NO^{\bullet} \tag{30}$$

$$2\,R_2NOO^{\bullet} \longrightarrow R_2NO^{\bullet} + O_2 \tag{31}$$

3.2 PHOSPHORUS PEROXYL

The first radicals of the phosphoranylperoxyl, $X_4POO^{\bullet}$, type to be identified, and kinetically studied by ESR spectroscopy in a systematic fashion, were produced by the addition of alkoxyl radicals to alkyl phosphites (X = OR) in the presence of O_2 (reactions (32 and 33)); the alkoxyl radicals were generated by photolysis of di-t-butylperoxide in alkane solutions at low temperatures [65–67]. These peroxyl radicals oxidize the trialkyl phosphite (reaction (34)), giving rise to trialkyl phosphate and phosphoranyl oxyl which undergoes scission (reaction (35)) into trialkyl phosphate and alkoxyl, thus completing the reaction chain [65].

$$(RO)_3P + {}^{\bullet}OR \xrightarrow{(32)} (RO)_4P^{\bullet} \xrightarrow[(33)]{O_2} (RO)_4P{-}O{-}O^{\bullet}$$

$$(RO)_4P{-}O{-}O^{\bullet} \xrightarrow[(34)]{P(OR)_3} (RO)_3P{=}O + (RO)_4P{-}O^{\bullet}$$

$$(RO)_4P{-}O^{\bullet} \xrightarrow{(35)} {}^{\bullet}OR + (RO)_3P{=}O$$

Phosphoranylperoxyl radicals of a mixed substitution derived from the addition of t-butoxyl [67] or t-butylperoxyl [68] to triphenylphosphine have also been observed.

Another pathway to these peroxyl radicals is through the oxidation of phosphine or phosphite by an arene diazonium salt to give the radical cation (reaction (36)). This reacts with alcohol (reaction (37)), giving rise to the alkoxyl radical adduct, which then adds O_2 (reaction (38)). Reactions (39)–(41) then complete the chain which, similar to that represented by reactions (32)–(35), leads to phosphine oxide or alkyl phosphate as the final oxidation product [69]. A partitioning of the unpaired-spin density, with 60% on the terminal oxygen, has been inferred for these peroxyl radicals [67].

$$R_3P + ArN_2^{\oplus} \xrightarrow{(36)} R_3P^{\bullet\oplus} + Ar^{\bullet} + N_2$$

$$R_3P^{\bullet\oplus} \xrightarrow[(37)]{R'OH} R_3P^{\bullet}(OR') + H^{\oplus}$$

$$R_3P^{\bullet}(OR') \xrightarrow[(38)]{O_2} R_3P(OR')\text{–O–O}^{\bullet}$$

$$R_3P(OR')\text{–O–O}^{\bullet} \xrightarrow[(39)]{R_3P} R_3P{=}O + R_3P(OR')\text{–O}^{\bullet}$$

$$R_3P{=}O + R_3P(OR')\text{–O}^{\bullet} \xrightarrow{(40)} R_3P{=}O + R'O^{\bullet}$$

$$R'O^{\bullet} \xrightarrow[(41)]{R_3P} R_3P^{\bullet}(OR')$$

Similar to the production of phenylsulfonylperoxyl-type radicals (see above), the diphenylphosphinic peroxyl, $R_2P(O)OO^{\bullet}$, is accessible through chloride displacement from diphenylphosphinic acid chloride by superoxide in solvents such as acetonitrile (reaction (42)), and like its sulfonyl peroxyl homologues acts as a strong oxidant [44,70]. The dialkoxy-substituted radical, $(RO)_2P(O)OO^{\bullet}$, has been produced from diethyl chlorophosphate in a like manner and shows similar reactivity [71]. The presence of phosphate peroxyl radicals, generated in dimethylsulfoxide from nucleotides or organic phosphates by the action of superoxide through the same type of ester cleavage [39], has been suspected to be the reason for the enhancement of base release from nucleotides as compared to nucleosides, or from nucleosides in non-aqueous systems, in the presence of organic phosphates as compared to their absence [72].

$$R_2P(=O)Cl + O_2^{\bullet\ominus} \xrightarrow{(42)} R_2P(=O)\text{–O–O}^{\bullet} + Cl^{\ominus}$$

There are apparently no reports as yet on tetraalkyl (or aryl) phosphoranyl peroxyl radicals, $R_4POO^{\cdot}$, in the literature. The reaction of superoxide with the corresponding phosphonium compound might prove to be an interesting one to study in this respect.

3.3 ARSENIC PEROXYLS

Evidence has been presented for the formation of the arsoranyl peroxyl species, $R_3(RO)AsOO^{\cdot}$, from the addition of t-butoxyl to triphenylarsine or trimethylarsine in the presence of O_2; however, kinetic and product information for the subsequent reactions of these peroxyl radicals is not yet available [67,73].

The dimethyl-substituted arsinyl peroxyl, $R_2AsOO^{\cdot}$, was detected by ESR spectroscopy as a reaction product of dimethylarsine and oxygen, and was assumed to play a major role in arsenic-mediated DNA damage [74]. It has been observed that arsenic-induced DNA damage is strongly enchanced in the presence of superoxide [75]. While no mechanism has been proposed, one may suspect that, similar to what has been found regarding the reactivity of superoxide versus certain sulfonic and phosphonic acid derivaties (see above), a similar displacement reaction, perhaps of an ester function from metabolically esterfied dimethylarsinic acid, could take place under formation of the dimethyl arsinyl peroxyl (reaction (43)), which could be a hydrogen-atom abstractor (reaction (44)) of an activity similar to the organosulfonyl peroxyl or diorganophosphinyl peroxyl.

$$R_2As(=O)X + O_2^{\cdot\ominus} \xrightarrow{(43)} R_2As(=O)O{-}O^{\cdot} + X^{\ominus}$$

$$R_2As(=O)O{-}O^{\cdot} + RH \xrightarrow{(44)} R_2As(=O)O{-}OH + R^{\cdot}$$

4 MAIN GROUP IV ELEMENTS

4.1 SILICON PEROXYL

Triorganosilylperoxyl radicals have been prepared by the photolysis of the hydride in oxygen-saturated organic solvents and studied by ESR spectroscopy [76,77]. At low temperatures they exist in equilibrium with the tetroxide, but above 233 K are unstable and decay by first- or lower-order kinetics. The decay products have not been identified. The decomposition of triorganosilyl hydrotrioxides, produced by the ozonation of the corresponding silane at low temperatures, may also give rise to triorganosilylperoxyl radicals, presumably via H-atom abstraction from the silane by various radicals in this system (reactions (45 and 46)). The silyltrioxyl radical, $R_3SiOOO^•$, may also be present in this system and could be formed through H-atom abstraction from the hydrotrioxide [78].

$$R_3Si{-}H + X^• \xrightarrow{(45)} R_3Si^• + XH$$

$$R_3Si^• + O_2 \xrightarrow{(46)} R_3Si{-}O{-}O^•$$

The autoxidation of tris(trimethylsilyl)silane occurs by way of the tris(trimethylsilyl)silyl-peroxyl radical, $[(CH_3)_3Si]_3SiOO^•$ (strictly speaking, this is not an organo-heteroatom-peroxyl radical in the foregoing sense), which rearranges via the 1,3-shift of a trimethylsilyl group to the terminal peroxyl oxygen (reaction (47)), the driving force being the high affinity of silicon toward oxygen [79]. This kind of shift seems much less likely in the case of triorganosilylperoxyl radicals, as the driving force would be smaller. The peroxyl oxygen atoms in silylperoxyl radicals are magnetically inequivalent, but with a higher spin density at the terminal oxygen-atom than that shown by carbon-centred peroxyl [77].

$$(Me_3Si)_3Si{-}O{-}O^• \xrightarrow{(47)} (Me_3Si)_2\dot{S}i{-}O{-}O{-}SiMe_3$$

4.2 GERMANIUM PEROXYL

Triorganogermylperoxyl radicals have been produced (reaction (48)) in oxygen-saturated hydrocarbon solution via the reaction of t-BuO$^•$ with the corresponding hydride and studied by ESR spectroscopy [77,80]. The rate constant for the addition of O_2 to $(n\text{-}Bu)_3Ge^•$ (in benzene) is smaller than that of its homologue, $(n\text{-}Bu)_3Sn^•$ (2.5 vs. 7.5×10^9 dm^3 mol^{-1} s^{-1}) [64]. Triorganogermylperoxyl exists in equilibrium with the corresponding tetroxide and at temperatures below—60°C is persistent. Under these conditions the equilibrium favours the dimer. A small activation energy is suggested for the dimerization reaction. At higher temperatures these peroxyl radicals decay in a unimolecular process where the products are not known. The thermolysis (near 500 K) of trimethylgermane in the gas phase in the presence of O_2 has been studied recently [81]. This is a chain reaction whose propagating steps are reactions (48) to (52).

$$(CH_3)_3Ge^{\bullet} + O_2 \xrightarrow{(48)} (CH_3)_3Ge-O-O^{\bullet}$$

$$(CH_3)_3Ge-O-O^{\bullet} \xrightarrow[(49)]{Me_3GeH} (CH_3)_3Ge-O-OH$$

$$(CH_3)_3Ge-O-OH \xrightarrow{(50)} (CH_3)_3Ge-O^{\bullet} + {}^{\bullet}OH$$

$$(CH_3)_3Ge-O^{\bullet} \xrightarrow[(51)]{Me_3GeH} (CH_3)_3Ge-OH + (CH_3)_3Ge^{\bullet}$$

$${}^{\bullet}OH \xrightarrow[(52)]{Me_3GeH} (CH_3)_3Ge^{\bullet} + H_2O$$

$$(CH_3)_3Ge-OH \xrightarrow[(53)]{2\,x} (CH_3)_3Ge-O-Ge(CH_3)_3 + H_2O$$

The final product, hexamethylgermoxane, is formed by the condensation of two molecules of trimethylgermanol (reaction (53)). Organogermylperoxyl radicals show an unpaired-spin distribution similar to that observed in the silylperoxyl radicals [77].

4.3 TIN PEROXYLS

It has been shown by ESR spectroscopy that triorganostannylperoxyl radicals do not exist in a reversible equilibrium with the tetraoxide, in contrast to most other peroxyl radicals. In addition the two oxygen atoms are magnetically equivalent and the peroxy group becomes a bidentate ligand (reaction (54)) [76,77,82]. The autoxidation of triorganotin hydride to the hydroxide probably involves the triorganostannylperoxyl radical as an intermediate [83]. It has been reported that organostannylperoxyl can displace trialkylstannyl radicals from hexaalkylditin with the formation of hexaalkylstannyl peroxide (reaction (55)) [84].

$$R_3Sn^{\bullet} + O_2 \xrightarrow{(54)} R_3Sn^{\bullet}\langle O_2 \rangle \quad (54)$$

$$R_3Sn^{\bullet}\langle O_2 \rangle + R_3Sn{-}SnR_3 \xrightarrow{(55)} R_3Sn{-}O{-}O{-}SnR_3 + R_3Sn^{\bullet} \quad (55)$$

The biocidal properties of triorganostannyl halides are thought to be mediated by the stannylperoxyl radical [85,86]. These compounds are very lipophilic and are proposed to be potent initiators of lipid peroxidation. It seems that they are activated by the superoxide anion $O_2^{-\bullet}$, perhaps by a halide displacement (reaction (56)) similar to the displacement reactions discussed above in connection with organo-oxosulfur and organo-oxophosphorus halides, whereby the stannylperoxyl radicals are formed. *In-vitro* ESR studies of such systems in the presence of excess stannyl halide have shown the formation of a ditin peroxo radical cation (reaction (57)) [86].

$$R_3Sn{-}X + O_2^{\bullet\ominus} \xrightarrow{(56)} R_3Sn^{\bullet}\langle O_2 \rangle + X^{\ominus} \quad (56)$$

$$R_3Sn^{\bullet}\langle O_2 \rangle + R_3Sn{-}X \xrightarrow{(57)} \left[R_3Sn{-}O{-}O{-}SnR_3\right]^{\bullet\oplus} + X^{\ominus} \quad (57)$$

4.4 LEAD PEROXYLS

Triorganoplumbylperoxyl radicals, generated from triorganolead hydride through H-abstraction by t-BuO$^{\bullet}$ in oxygen-saturated cyclopentane, were shown by ESR spectroscopy to be stable up to the boiling point of the solvent cyclopropane, and even at low temperatures there was no evidence for a peroxyl radical–tetraoxide equilibrium [76], and one might suspect the existence of a bidentate structure similar to the tin case. Triorganoplumbyl peroxyl radicals have also been obtained by the photolysis of cyclopentadienyltriorganolead in the presence of O_2 [87].

REFERENCES

1. R.A. Sheldon, in *The Activation of Dioxygen and Homogeneous Catalytic Oxidation*, edited by D.H.R. Barton, A.E. Martell and D.T. Sawyer, Plenum Press, New York (1993), p.9.

2. A. Bakac, S.L. Scott, J.H. Espenson and K.R. Rodgers, *J. Am. Chem. Soc.* **117**, 6483 (1995).
3. F. Kovac and B. Plesnicar, *J. Am. Chem. Soc. 101*, 2677 (1979).
4. L.G. Kulak, Ye. M. Kuramshin, S.S. Zlotskii and D.L. Rakhmankulov, *Izv. Vyssh. Ucheb. Zaved., Khim. Khim. Tekhnol.* **31**, 3 (1988).
5. J. Koller, M. Hodoscek and B. Plesnicar, *J. Am. Chem. Soc.* **112**, 2124 (1990).
6. S.W. Benson and R. Shaw in *Organic Peroxides*, edited by D. Swern, Vol.1, John Wiley & Sons, New York (1970), p.105.
7. A.C. Baldwin, in *The Chemistry of Functional Groups: Peroxides*, edited by S. Patai, John Wiley & Sons, Chichester (1983) p.97.
8. J.S. Francisco and I.H. Williams, *Int. J. Chem. Kinet.* **20**, 455 (1988).
9. R.E. Bühler, J. Staehelin and J. Hoigné, *J. Phys. Chem.* **88**, 2560 (1984).
10. G.V. Buxton, *Trans. Faraday Soc.* **65**, 2150 (1969).
11. C. von Sonntag, *The Chemical Basis of Radiation Biology*, Taylor and Francis, London (1987).
12. P. Wardman and C. von Sonntag, in *Biothiols. A: Monothiols and Dithiols, Protein Thiols, and Thiyl Radicals*, edited by L. Packer. Methods in Enzymology, Vol. 251, Academic Press, San Diego (1995), p.31.
13. F. Yin, D. Grosjean and J.H. Seinfeld, *J. Atmos. Chem.* **11**, 309 (1990).
14. F. Yin, D. Grosjean, R.C. Flagan and J.H. Seinfeld, *J. Atmos. Chem.* **11**, 365 (1990).
15. G.G. Jayson, D.A. Stirling and A.J. Swallow, *Int. J. Radiat. Biol.* **19**, 143 (1971).
16. M. Tamba, G. Simone and M. Quintiliani, *Int. J. Radiat. Biol.* **50**, 595 (1986).
17. J. Mönig, K.-D. Asmus, L.G. Forni and R.L. Willson, *Int. J. Radiat. Biol.* **52**, 589 (1987).
18. M. Tamba, *Z. Naturforsch.* C **44**, 857 (1989).
19. M.D. Sevilla, D. Becker and M. Yan, *Int. J. Radiat. Biol.* **57**, 65 (1990).
20. M.D. Sevilla, D. Becker and M. Yan, *J. Chem. Soc. Faraday Trans.* **86**, 3279 (1990).
21. Ch. Chatgilialoglu and M. Guerra, in *Sulfur-Centered Reactive Intermediates in Chemistry and Biology*, edited by Ch. Chatgilialoglu and K.-D. Asmus, Plenum Press, New York (1990), p.31.
22. M. Tamba and P. O'Neill, *J. Chem. Soc. Perkin Trans. 2* 1681 (1991).
23. A.A. Turnipseed, S.B. Barone and A.R. Ravishankara, *J. Phys Chem.* **96**, 7502 (1992).
24. Y.V. Razskazovskii and M.Y. Mel'nikov, *Sov. J. Chem. Phys (Engl. Transl.)* **10**, 148 (1992).
25. X. Zhang, N. Zhang, H.-P. Schuchmann and C. von Sonntag, *J. Phys. Chem.* **98**, 6541 (1994).
26. M. Tamba, A. Torregiani and O. Tubertini, *Radiat. Phys. Chem.* **46**, 596 (1995).
27. D. Becker, S. Summerfield, S. Gillich and M.D. Sevilla, *Int. J. Radiat. Biol.* **65**, 537 (1994).
28. Y. Razskazovskii and M.D. Sevilla, *Int. J. Radiat. Biol.* **69**, 75 (1996).
29. Y.H. Kim, S.C. Lim, and K.S. Kim, *Pure Appl. Chem.* **65**, 661 (1993).
30. D. Laasko, C.E. Smith, A. Goumri, J.-D.R. Rocha and P. Marshall, *Chem. Phys. Lett.* **227**, 377 (1994).
31. A.R. Bassindale and J.N. Iley, in *Supplement S: The Chemistry of Sulphur-containing Functional Groups*, edited by S. Patai and Z. Rappoport, John Wiley & Sons, Chichester (1993), p.245.
32. K.-D. Asmus and C. Schöneich, in *Oxidative Damage and Repair*, edited by K.E.J. Davies, Pergamon Press, New York (1991), p.226.
33. W.-F. Wang, M.N. Schuchmann, H.-P. Schuchmann and C. von Sonatag, unpublished results.

34. M.L. McKee, *Chem. Phys. Lett.* **211**, 643 (1993).
35. N. Zhang, H.-P. Schuchmann and C. von Sonntag, *J. Phys. Chem.* **95**, 4718 (1991).
36. M. Lal, R. Rao, X.-W. Fang, H.-P. Schuchmann and C. von Sonntag, unpublished results.
37. Y.V. Razskazovskii, M.V. Roginskaya and M.Y. Mel'nikov, *J. Phys. Chem.* **98**, 12003 (1994).
38. Y.H. Kim and D.C. Yoon, *Tetrahedron Lett.*, **29,** 6453 (1988).
39. D.T. Sawyer and J.S. Valentine, *Acc. Chem. Res.* **14**, 393 (1981).
40. Y.H. Kim, K.S. Kim and H.K. Lee, *Tetrahedron Lett.* **30**, 6357 (1989).
41. P. Neta, R.E. Huie and A.B. Ross, *J. Phys. Chem. Ref. Data* **17**, 1027 (1988).
42. D.L. Ball and J.O. Edwards, *J. Am. Chem. Soc.* **78**, 1125 (1956).
43. E.A. Betterton and M.R. Hoffmann, *J. Phys. Chem.* **92**, 5962 (1988).
44. Y.H. Kim, S.C. Lin, M. Hoshino, Y. Otsuka and T. Ohishi, *Chem. Lett.* 167 (1989).
45. Y.H. Kim and B.C. Chung, *J. Org. Chem.* **48**, 1562 (1983).
46. H.K. Lee, K.S. Kim, J.C. Kim and Y.H. Kim, *Chem. Lett.* 561 (1988).
47. N. Zhang, H.-P. Schuchmann and C. von Sonntag, unpublished results.
48. C. Schöneich, A. Aced and K.-D. Asmus, *J. Am. Chem. Soc.* **115**, 11376 (1993).
49. C. Schöneich and K. Bobrowski, *J. Phys. Chem.* **98**, 12613 (1994).
50. I. Barnes, V. Bastian and K.H. Becker, *Int. J. Chem. Kinet.* **20**, 415 (1988).
51. O. Ito, *J. Am. Chem. Soc.* **105**, 850 (1983).
52. J. Chaudière, O. Courtin and J. Leclaire, *Arch. Biochem. Biophys.* **296**, 328 (1992).
53. H.K. Spencer and M.P. Cava, *J. Org. Chem.* **17**, 2937 (1977).
54. A.S. Nazran and D. Griller, *J. Am. Chem. Soc.* **105**, 1970 (1983).
55. J.G. Calvert and J.N. Pitts, Jr., *Photochemistry*, John Wiley & Sons, New York (1966).
56. C. von Sonntag and H.-P. Schuchmann, *Ad. Photochem.* **10**, 59 (1977).
57. B.G. Ershov, T.L. Mikhailova, A.V. Gordeev and V.I. Spitsyn, *Dok. Phys. Chem.* **300**, 506 (1988).
58. T.L. Mikhailova and B.G. Ershov, *Russ. Chem. Bull.* **42**, 235 (1993).
59. C.R.C. Lindley, J.G. Calvert and J.H. Shaw, *Chem. Phys. Lett.* **67**, 57 (1979).
60. R. Lesclaux, *Rev. Chem. Intermed.* **5**, 347 (1984).
61. C.F. Melius and J.S. Binkley, in *The Chemistry of Combustion Processes*, edited by T.M. Sloane, ACS Symposium Series, Vol. 249, American Chemical Society, Washington, DC (1984), p.103.
62. J.R. Roberts and K.U. Ingold, *J. Am. Chem. Soc.* **95**, 3228 (1973).
63. A. Faucitano, A. Buttafava, F. Martinotti and P. Bortolus, *J. Phys. Chem.* **88**, 1187 (1984).
64. B. Maillard, K.U. Ingold and J.C. Scaiano, *J. Am. Chem. Soc.* **105**, 5095 (1983).
65. A.G. Davies, D. Griller and B.P. Roberts, *J. Chem. Soc. Perkin Trans. 2* 993 (1972).
66. G.B. Watts and K.U. Ingold, *J. Am. Chem. Soc.* **94**, 2528 (1972).
67. J.A. Howard and J.C. Tait, *Can. J. Chem.* **56**, 2163 (1978).
68. E. Furimsky and J.A. Howard, *J. Am. Chem. Soc.* **95**, 369 (1973).
69. S. Yasui, M. Fujii, C. Kawano, Y. Nishimura, K. Shioji and A. Ohno, *J. Chem. Soc. Perkin Trans. 2* 177 (1994).
70. S.C. Lim and Y.H. Kim, *Heteroatom Chem.* 261 (1990).
71. M. Miura, M. Nojima and S. Kusabayashi, *J. Chem. Soc. Chem. Commun.* 1352 (1982).
72. H. Yamane, N. Yada, E. Katori, T. Mashino, T. Nagano and M. Hirobe, *Biochem. Biophys. Res. Commun.* **142**, 1104 (1987).

73. E. Furimsky, J.A. Howard and J.R. Morton, *J. Am. Chem. Soc.* **94**, 5932 (1972).
74. K. Yamanaka, M. Hoshino, M. Okamoto, R. Sawamura, A. Hasegawa and S. Okada, *Biochem. Biophys. Res. Commun.* **168**, 58 (1990).
75. K. Rin, K. Kawaguchi, K. Yamanaka, M. Tezuka, N. Oku and S. Okada, *Biol. Pharm. Bull.* **18**, 45 (1995).
76. J.E. Bennett and J.A. Howard, *J. Am. Chem. Soc.* **94**, 8244 (1972).
77. J.A. Howard, J.C. Tait and S.B. Tong, *Can. J. Chem.* **57**, 2761 (1979).
78. B. Plesnicar, J. Cerkovnik, J. Koller and F. Kovac, *J. Am. Chem. Soc.* **113**, 4946 (1991).
79. Ch. Chatgilialoglu, A. Guarini and G. Seconi, *J. Org. Chem.* **57**, 2207 (1992).
80. J.A. Howard and J.C. Tait, *Can. J. Chem.* **54**, 2669 (1976).
81. P.G. Harrison and D.M. Podesta, *Organometallics* **13**, 1569 (1994).
82. J.A. Howard and J.C. Tait, *J. Am. Chem. Soc.* **99**, 8349 (1977).
83. R.A. Jackson, in *Advances in Free-Radical Chemistry*, edited by G.H. Williams, Vol. 3, Logos Press, New York (1974), p. 231.
84. Y.A. Aleksandrov and B.A. Radbil, *J. Gen. Chem. USSR* **36** 562 (1966).
85. B.H. Gray, M. Porvaznik, C. Flemming and L.-F. Lee, *Toxicology* **47**, 35 (1987).
86. J.A. Rivera, S.C. Cummings and D.A. Macys, *Chem. Res. Toxicol.* **5**, 698 (1992).
87. A.G. Davies, J.A.-A. Hawari, C. Gaffney and P.G. Harrison, *J. Chem. Soc. Perkin Trans.* **2**, 631 (1982).

NOTE ADDED IN PROOF

Dialkoxyl-substituted sulfuranylperoxyl radicals can be produced in a low-temperature aqueous LiCl glass by one-electron reduction of dialkylsulfites through radiolysis and subsequent oxygenation with O_2 upon annealing [88]. A model of the atmospheric oxidation of dimethyl sulfide involving various sulfur peroxyl radicals has been reported [89]. It seems that the precursor dimethylhydroxysulfuranyl peroxyl radical, if it is formed at all, must be very unstable with respect to the decay into dimethylhydroxysulfuranyl and oxygen [90,91] The enthalpies of formation of trioxyl radicals $ROOO^•$ have been computed *ab initio*. For R = alkyl, the $RO{-}OO^•$ bond dissociation energy is reported to be slightly negative, near -1 kJ mol^{-1}, whereas for R = H it is reported to be slightly positive, near $+1$ kJ mol^{-1} [92].

REFERENCES

88. Y. Razskazovskii and M.D. Sevilla, *J. Phys. Chem.* **100**, 4090 (1996).
89. S.B. Barone, A.A. Turnipseed and A.R. Ravishankara, *Faraday Discuss.* **100,** 39 (1996).
90. R.B. Barone, A.A. Turnipseed and A.R. Ravishankara, *J. Phys. Chem.* **100,** 14694 (1996).
91. A.A. Turnipseed, S.B. Barone and A.R. Ravishankara, *J. Phys. Chem.* **100,** 14703 (1996).
92. P.T.W. Jungkamp and J.H. Seinfeld, *Chem. Phys. Lett.* **257,** 15 (1996).

15 Peroxy Radicals and the Atmosphere

T. J. WALLINGTON
Research Laboratory, Ford Motor Company, Dearborn, USA

and

O. J. NIELSEN
Research Center, Ford Motor Company, Aachen, Germany

1 INTRODUCTION

Peroxy radicals play a key role in atmospheric chemistry. They are intimately involved in the formation and destruction of ozone and in the photo-oxidation of all organic compounds in the atmosphere. HO_2 is the dominant peroxy radical in the atmosphere; its altitude profile, calculated by Brasseur and Solomon using a one-dimensional radiative convective photochemical model [1], is shown in Figure 1. HO_2 radicals participate in reactions which remove ozone and provide a natural regulation of stratospheric ozone levels.

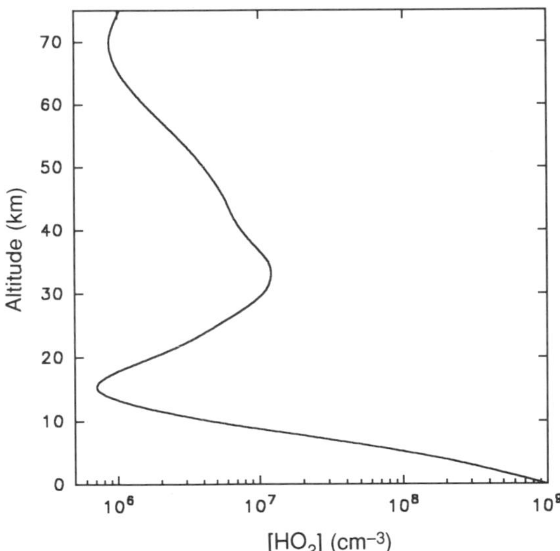

Figure 1 Altitude profile for HO_2 calculated by using a one-dimensional radiative convective photochemical model for mid-latitudes at equinox [1]

Peroxyl Radicals. Edited by Z.B. Alfassi

In the troposphere, organic peroxy radicals and HO_2 radicals undergo reactions which lead to ozone formation. In urban areas, peroxy radicals are part of the chemical cycles that produce unwanted photochemical air pollution, i.e. 'smog'. By virtue of the formation of longer-lived compounds such as peroxyacetyl nitrate, ($CH_3C(O)O_2NO_2$), peroxy radicals are involved in the long-range transport of nitrogen oxides in the atmosphere. Self-reaction of HO_2 radicals forms hydrogen peroxide, which is incorporated into cloud water and provides a major oxidant supply for aqueous-phase atmospheric chemistry. In this chapter we shall discuss the formation and measurement of peroxy radicals in the atmosphere and their role in gas-phase atmospheric chemistry.

2 SOURCES OF PEROXY RADICALS

The driving force for most of the chemistry that occurs in the atmosphere is the formation of hydroxyl (OH) radicals via the photolysis of ozone to form $O(^1D)$ atoms which react with water vapour:

$$O_3 + h\nu\ (\lambda < 320\,\text{nm}) \longrightarrow O(^1D) + O_2(^1\Delta_g) \tag{1}$$

$$O(^1D) + H_2O \longrightarrow 2\,OH \tag{2}$$

The flux of ultraviolet (UV) light, O_3, and H_2O vapour combine to give a potent source of OH radicals. The dominant loss of hydroxyl radicals is the reaction with CO and organic compounds such as CH_4, with both reactions producing peroxy radicals. Reaction with CO converts OH into HO_2 radicals, while reaction with CH_4 leads to the formation of methylperoxy radicals. In thc following reactions, 'M' denotes an inert third body which carries away the excess energy liberated when the chemical bond is formed:

$$OH + CO \longrightarrow H + CO_2 \tag{3}$$

$$H + O_2 + M \longrightarrow HO_2 + M \tag{4}$$

$$OH + CH_4 \longrightarrow H_2O + CH_3 \tag{5}$$

$$CH_3 + O_2 + M \longrightarrow CH_3O_2 + M \tag{6}$$

In 760 torr of air at 296 K the effective bimolecular rate constants for reactions (4) and (6) are both $1.2 \times 10^{-12}\,\text{cm}^3\,\text{molecule}^{-1}\,\text{s}^{-1}$ [2]. At sea level, the atmospheric O_2 concentration is $5.2 \times 10^{18}\,\text{mol cm}^{-3}$ and the lifetime of H atoms and CH_3 radicals with respect to reaction with O_2 is $0.16\,\mu\text{s}$. There are no other losses of H atoms or CH_3 radicals that compete with reactions (4) and (6) in the troposphere and any reaction which gives H atoms or alkyl radicals is a source of peroxy radicals. Atmospheric sources of H atoms and alkyl radicals include the photolysis of HCHO and other aldehydes, and ketones, e.g.

$$HCHO + h\nu \longrightarrow H + HCO \tag{7}$$

$$HCO + O_2 \longrightarrow HO_2 + CO \tag{8}$$

the reaction of OH, NO_3, and Cl radicals with organic compounds, e.g.

$$OH + CH_2{=}CH_2 + M \longrightarrow HOCH_2CH_2 + M \tag{9}$$

$$HOCH_2CH_2 + O_2 + M \longrightarrow HOCH_2CH_2O_2 + M \tag{10}$$

and the reaction of ozone with alkenes. Peroxynitric acid and peroxynitrates (e.g. peroxy acetyl nitrate, PAN) are reservoirs of peroxy radicals and their thermal decomposition gives peroxy radicals:

$$HO_2NO_2 + M \longrightarrow HO_2 + NO_2 \tag{11}$$

$$CH_3C(O)O_2NO_2\ (PAN) + M \longrightarrow CH_3C(O)O_2 + NO_2 \tag{12}$$

The R—O_2 bond strength is sufficiently strong (typically $\approx 30\,kcal\,mol^{-1}$) that thermal decomposition is of no importance. Once the R—O_2 bond is formed in the peroxy radical it is not broken in the sequence of reactions that convert hydrocarbons into CO_2 and H_2O. The concentration of peroxy radicals in ambient air at ground level is typically of the order of 5–50 pptv (10^8–$10^9\,cm^{-3}$), of which the bulk is HO_2 and CH_3O_2. HO_2 radicals are most abundant and comprise approximately 20–70% of the peroxy radicals.

3 MEASUREMENT OF PEROXY RADICALS IN THE ATMOSPHERE

The methods available to measure peroxy radicals in the atmosphere have been reviewed by Cantrell *et al.* [3]. The accurate measurement of trace levels of reactive radical species under ambient conditions is a difficult task. Reliable methods for the ambient monitoring of peroxy radicals are now emerging, although there are often significant differences (factors of 2) between the results from various techniques when intercomparisons are performed.

3.1 TROPOSPHERIC MEASUREMENTS

Several methods have been used to successfully measure peroxy radical levels in the troposphere. The direct technique developed by Mihelcic and coworkers [4–7] relies on the technique of electron spin resonance (ESR) spectroscopy combined with matrix isolation. In the ESR technique, a strong magnetic field is used to cause a splitting of the energy levels associated with the spin of an unpaired electron. Microwave radiation can then be used to induce a transition between these levels. The absorption of the microwave radiation is recorded as a function of the magnetic field strength. The electron spin interacts with neighbouring nuclear spins and thus hyperfine splitting appears in the spectra. Calibrated reference spectra of known radical species are compared to those obtained by sampling ambient air to provide identification and quantification. Numerical computer-fitting procedures enable the 'composite radical' ESR spectrum obtained by sampling ambient air to be deconvoluted to reveal the contribution of individual peroxy radicals. Figure 2 shows an example of such a process.

The drawback in using the ESR technique for monitoring peroxy radicals in ambient air is that it is not practical to maintain the large precise magnetic field needed for ESR measurements in typical remote field locations. This problem is circumvented by cryogenic condensation of 10–20 l samples of air in a D_2O matrix on a cold finger; the sampling time is ca 30 min. Such samples can be kept for weeks at a temperature of 77 K while they are transported back to the laboratory for analysis. Cryogenic condensation is also necessary to increase the radical concentration in the samples and improve the signal-to-noise ratio of the measurements. The advantage of the ESR–Matrix method is that it is an absolute technique.

The majority of data concerning the concentration of peroxy radicals in the atmosphere comes from the 'chemical amplification' technique first proposed by Cantrell and Stedman [8] and subsequently developed by Cantrell and co-workers [9–12] and Hastie *et al.* [13]. In this method, HO_2 radicals are monitored indirectly by monitoring NO_2 formation via reaction (13) when ambient air is added to a gas flow containing NO. Amplification of the NO_2 formation is achieved by adding CO to the gas flow such that the following chain reaction is set up:

$$HO_2 + NO \longrightarrow OH + NO_2 \qquad (13)$$

$$OH + CO \longrightarrow H + CO_2 \qquad (3)$$

$$H + O_2 + M \longrightarrow HO_2 + M \qquad (4)$$

A similar chain reaction can also be achieved if reactive organic compounds such as ethene or propene are used in place of CO [14]. The chain reaction converts HO_2 and OH radicals into NO_2 which can then be measured by using a luminol chemiluminescence instrument [15,16]. In the troposphere, the HO_2 radical concentration exceeds that of OH by a factor of 100–1000 [17], so the extra signal produced by the OH radicals does not interfere with the measurement of HO_2. In addition to HO_2, the chemical amplifier technique also detects alkyl peroxy radicals such as CH_3O_2 and $CH_3C(O)O_2$,

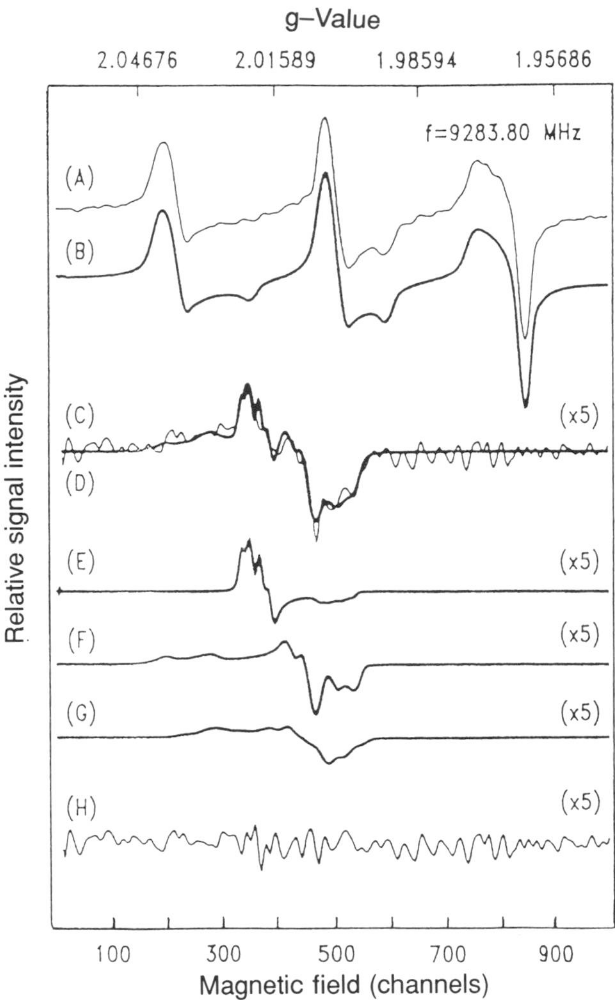

Figure 2 ESR spectra reported by Mihelcic *et al.* [7], illustrating the detection of peroxy radicals in ambient air: (A) sample of ambient air; (B) NO_2 reference spectrum; (C) result of stripping NO_2 features from A ($\times 5$); (D) Composite spectrum of NO_3, HO_2, and ΣRO_2 radicals evaluated by a computer-fitting routine to give the best fit to C; (E,F,G) reference spectra of NO_3, HO_2, and ΣRO_2 radicals, respectively; (H) C-D (reproduced with permission from Ref. 7, copyright Kluwer Academic Publishers (1993)).

because these react with NO to give NO_2 and, more importantly, are converted into HO_2 radicals:

$$CH_3O_2 + NO \longrightarrow CH_3O + NO_2 \tag{14}$$

$$CH_3O + O_2 \longrightarrow HCHO + HO_2 \tag{15}$$

Thus the chemical amplifier technique actually measures the sum of HO_2 and RO_2. The amplification factor, or number of NO_2 molecuels formed per peroxy radical sampled, is typically 50–200 and is limited by termination reactions which remove radicals from the system, such as:

$$OH + NO + M \longrightarrow HONO + M \tag{16}$$

$$HO_2 + \text{wall} \longrightarrow \text{radical loss} \tag{17}$$

$$HO_2 + NO_2 + M \longrightarrow HO_2NO_2 + M \tag{18}$$

To discriminate against compounds like O_3 that are present in the atmosphere and can oxidize NO to NO_2, the chemical amplifier is operated in a mode where the CO flow is periodically replaced with a flow of N_2. When the CO flow is switched off, the amplification ceases and the detected NO_2 falls to a background level. The NO_2 concentration recorded by the instrument is then modulated as the CO flow is switched on and off; typical data are shown in Figure 3. The peroxy radical concentration is calculated by dividing the amplitude of the NO_2 signal modulation by the ‘chain length’ or ‘amplification factor’, where the latter is defined as average number of NO_2 molecules formed by a peroxy radical.

The main difficulty associated with the chemical amplification technique is the determination of the ‘amplification factor’, which provides the absolute calibration of the instrument. In principal, if the rates and mechanisms of all of the chemical reactions occurring in the system are known, then this factor can be calculated analytically or via a chemical modelling approach. This was the method used by Cantrell and Stedman [8] when they first proposed the use of the chemical amplifier technique. However, this modelling approach assumes a complete understanding of the chemistry in the system, in particular the importance of chain termination reactions such as (16–18) which, in practice, may be rather difficult to measure. A more satisfactory approach is to use a well characterized source of peroxy radicals to calibrate the instrument. The design and implementation of such sources is not straightforward, but it now appears that two sources are becoming reliable. The first technique involves passing a H_2O_2/air flow over a heated wire. Pyrolysis of H_2O_2 then produced HO_2 radicals. The HO_2 concentration in the flow is determined by mixing with a gas stream of NO and observing the conversion of NO into NO_2 [9]. The second technique involves passing calibrated concentrations of peroxy acetyl nitrate (PAN) in air diluent

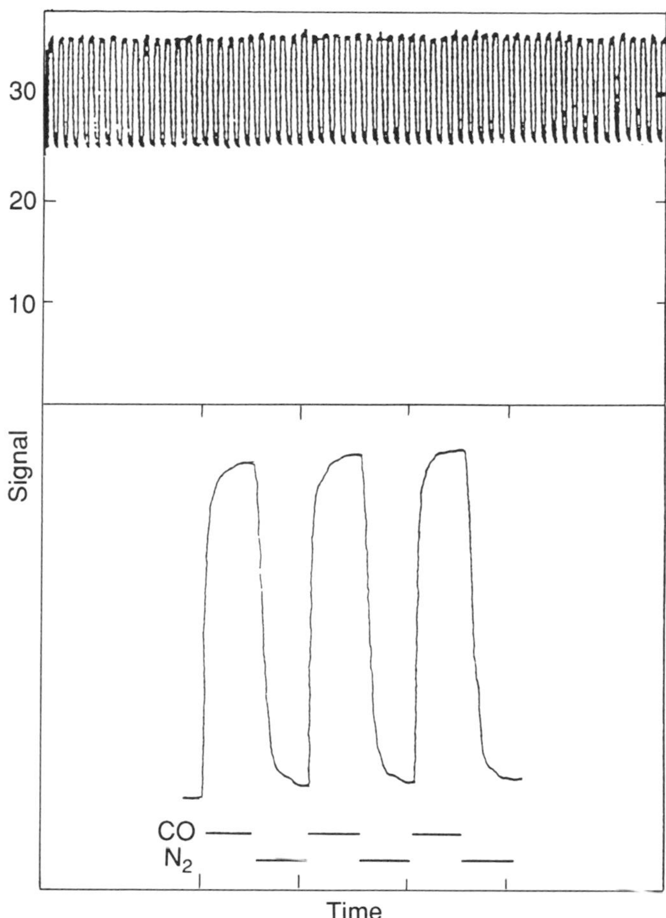

Figure 3 Typical signals obtained by the chemical amplifier technique, recorded over a total time period of 1 h (top) and 3 min (bottom) [18]. The switching between flows of CO (amplification on) and N_2 (amplification off) is indicated in the bottom figure (reprinted with permission from Ref. 3; copyright American Chemical Society (1993)).

through a heated tube. [13]. PAN decomposes rapidly when heated to give acetylperoxy radicals which are then converted into HO_2 radicals. In the most recent use of the chemical amplifer technique, both H_2O_2 and PAN calibration sources were used and shown to give indistinguishable results [19].

One final complication of the chemical amplifer technique must be noted. As discussed above, this technique measures the sum of the HO_2 and RO_2 radical concentrations. The instrument is sensitive to RO_2 radicals because they are converted into HO_2 radicals via reaction such as (14) and (15).

However, conversion of RO_2 into HO_2 is not 100% efficient because some RO_2 radicals are lost via channel (19b) to give alkyl nitrates;

$$RO_2 + NO \longrightarrow RO + NO_2 \tag{19a}$$

$$RO_2 + NO + M \longrightarrow RONO_2 + M \tag{19b}$$

and, more importantly, some alkoxy radicals, RO, are lost via reaction with NO to give alkyl nitrites (RONO). The overall efficiency with which alkylperoxy radicals, RO_2, are converted into HO_2 varies for different RO_2 radicals and is typically in the range 85–93% [11]. To place the response of chemical amplifer systems on an absolute basis, the mix of HO_2 and RO_2 radicals in the ambient air which is sampled must be evaluated by using a modelling approach. If this is not carried out, the total peroxy concentration obtained by dividing the amplitude of the NO_2 modulation by the amplification factor may be approximately 10% too low.

Finally, the 'missing oxidant' method can be used to provide information on ambient peroxy radical concentrations. In the absence of peroxy radical chemistry the following reactions dominate the formation and loss of odd oxygen (O and O_3) in the troposphere:

$$NO_2 + h\nu\ (\lambda < 420\,\text{nm}) \longrightarrow NO + O \tag{20}$$

$$O + O_2 + M \longrightarrow O_3 + M \tag{21}$$

$$NO + O_3 \longrightarrow NO_2 + O_2 \tag{22}$$

The NO_2 photolysis rate, J_{NO_2}, in the lower atmosphere (troposphere) depends on the cloud cover but is typically in the range $(0.3–1) \times 10^{-2}\,s^{-1}$, giving a lifetime of NO_2 with respect to photolysis of 2–6 min [17]. In one atmosphere pressure of air, reaction (21) has an effective bimolecular rate constant of $k_{21} = 1.5 \times 10^{-14}\,cm^3\ molecule^{-1}\,s^{-1}$ [2], $[O_2] = 5.2 \times 10^{18}\,cm^{-3}$, and O atoms have a lifetime of 13 μs with respect to reaction (21). Finally, $k_{22} = 1.8 \times 10^{-14}\,cm^3\ molecule^{-1}\,s^{-1}$, and assuming a constant O_3 concentration of 30 ppbv ($7 \times 10^{11}\,cm^{-3}$), NO then has a lifetime of 1.3 min with respect to reaction (22). As a result of the rapidity of reactions (20–22) a photostationary steady-state condition is set up where NO and O atoms are created and destroyed continually but maintain a steady concentration, i.e. $d[O]/dt = d[NO]/dt = 0$. By using this steady-state assumption one can derive the following relationship which can be used to predict the ozone concentration from the values of NO and NO_2 [20]:

$$[O_3] = \frac{J_{NO_2}[NO_2]}{k_{22}[NO]} \tag{23}$$

When measurements of J_{NO_2}, $[NO_2]$, [NO], and $[O_3]$ are performed in the field, the observed ozone level is often less than that calculated from the right-hand side of the above equation [10,21]. In other words, to maintain the observed $[NO_2]/[NO]$ mixing ratio there must be some species, other than O_3, that is capable of oxidizing NO to NO_2 which is not accounted for (i.e. missing) in the simple mechanism of reactions (20–22). This 'missing oxidant' is peroxy radicals. In the presence of peroxy radicals, additional conversion of NO to NO_2 occurs via the reaction of peroxy radicals with NO:

$$RO_2\ (\text{or } HO_2) + NO \longrightarrow RO\ (\text{or } OH) + NO_2 \tag{24}$$

When reaction (24) is included in the steady-state analysis, the predicted ozone concentration becomes:

$$[O_3] = \frac{J_{NO_2}[NO_2]}{k_{22}[NO]} - \frac{k_{24}[RO_2]}{k_{22}} \tag{25}$$

Rearrangement of this equation gives:

$$[RO_2] = \frac{k_{22}}{k_{24}}\left(\frac{J_{NO_2}[NO_2]}{k_{22}[NO]} - [O_3]\right) \tag{26}$$

The term in the brackets is known as the 'missing oxidant'. For HO_2 and CH_3O_2, under atmospheric conditions of temperature and pressure, the rate constant ratio $k_{22}/k_{24} = 0.002$, and multiplication of the 'missing oxidant' by this factor gives $[RO_2]$. The missing-oxidant method has one major limitation; many different parameters need to be measured simultaneously and accurately, and the final result is very sensitive to small uncertainties in these parameters. As a result, the missing oxidant method provides a measure of $[RO_2]$ which is uncertain by typically a factor of two. Despite this limitation, the missing oxidant method provides a useful comparison with the results from the other methods. As discussed by Cantrell *et al.* [10,18], data obtained by using this method are generally consistent with results from the chemical amplifier technique. Such agreement suggests that the general factors which control atmospheric peroxy radical concentrations are reasonably well understood.

3.2 STRATOSPHERIC MEASUREMENTS

There are relatively few measurements of stratospheric peroxy radical conentrations and no diurnal profiles. Mihelcic *et al.* [22] and Helten *et al.* [23] have used the ESR matrix isolation technique to measure HO_2 radical concentrations at an altitude of 18–34 km. Anderson *et al.* [24] and Stimpfle *et al.* [25] have used *in situ* chemical conversion of HO_2 to OH radicals via reaction with NO, with the resulting OH radicals being detected by resonance fluorescence or laser induced fluorescence at 309 nm. Emission in the far infrared region at 142–147 cm^{-1} from thermally populated rotational energy levels of HO_2 radicals has been used by Traub, Johnson and Chance [26] to measure HO_2 concentrations in the stratosphere using a balloon platform. Comparison of the observed emission spectrum with laboratory calibrated spectra enables the derivation of HO_2 radical concentrations in the stratosphere. Finally, ground-based mm-wave spectroscopy has been employed by de Zafra *et al.* [27] to measure HO_2 levels at 35–70 km. The results obtained by using far infrared emission and mm-wave spectroscopy are consistent with predictions from current chemical models, such as that given by Brasseur and Solomon [1] (see Figure 1). The results obtained from the older studies by using matrix isolation and the *in situ* chemical conversion techniques are somewhat greater.

4 TYPICAL CONCENTRATIONS OF PEROXY RADICALS IN THE ATMOSPHERE

Figure 4 shows the peroxy radical (HO_2 + RO_2) concentrations measured over a 19 d period in rural Alabama by Cantrell *et al.* [11]. The peroxy radical concentration profiles show pronounced diurnal behaviour as expected for a reactive trace radical species whose formation is driven by photochemistry. Peak peroxy radical concentrations occur around noon and are of the order of 150 pptv (3.7×10^9 cm^{-3}). During the night the peroxy radical concentrations fall to about 10 pptv (2.5×10^8 cm^{-3}); night-time peroxy radical chemistry is discussed in Section 7 below. Similar diurnal profiles but with lower maxima of 30–40 pptv have been observed in monitoring experiments conducted at Idaho Hill, Colorado [28] and at the Mauna Loa Observatory, Hawaii [29].

Weissenmayer, Burrows and Schupp [30] have used the chemical amplifier technique to measure peroxy radical concentrations in the marine boundary layer during a cruise in the north and south Atlantic oceans. The maximum local noontime peroxy radicals concentrations were generally 10–15 pptv in the north Atlantic and 5–12 pptv in the south Atlantic. Behmann, Weissenmayer and Burrows have measured daytime peroxy radical concentrations of 3–35 pptv at a coastal site in Brittany, France [31].

Madronich and Calvert [32] have performed a detailed modelling study of the peroxy radical chemistry in low NO_x environments such as those that occur in the planetary boundary layer in remote marine locations and over the Amazon basin. Figures 5 and 6 show the results of their computer simulations. In marine environments where emissions of hydrocarbons and other organic compounds is low, the total peroxy radical concentration varies between noontime values of 25–30 pptv and night-time values of 5 pptv, with HO_2 contributing 50% of the daytime radicals. In the Amazon basin scenario, the organic peroxy radical concentrations are much larger, with maxima of the order of 150 pptv and minima of 10–50 pptv. The concentration of HO_2 radicals in the Amazon simulation was approximately 20% of that of the organic peroxy radicals.

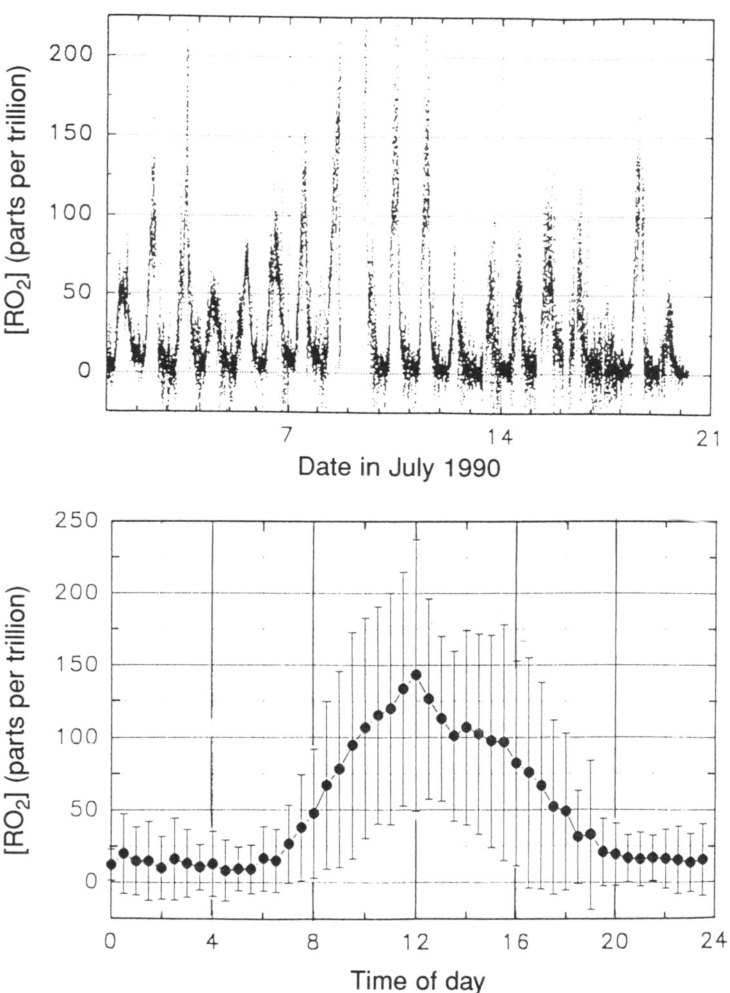

Figure 4 (Top) peroxy radical concentrations (HO_2 + RO_2) measured over a 19 d period in rural Alabama in July 1990 by Cantrell *et al.* [11] using the chemical amplifier technique. (Bottom) 30 min average radical concentrations derived from data in top figure; error bars indicate variability of the data for each time period (reproduced with permission from Ref. 11; copyright American Geophysical Union (1993)).

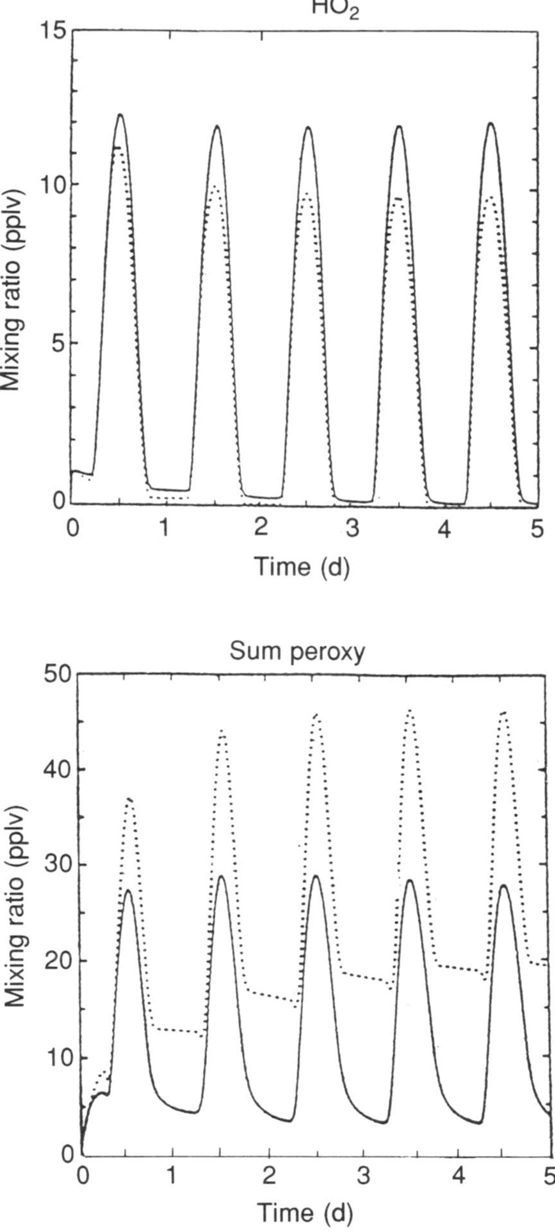

Figure 5 HO_2 (top) and total RO_2 (bottom) concentrations obtained in a simulation of a remote marine planetary boundary layer environment, with (continuous lines) and without (dotted lines) peroxy radical permutation reactions (reproduced with permission from Ref. 32; copyright American Geophysical Union (1990)).

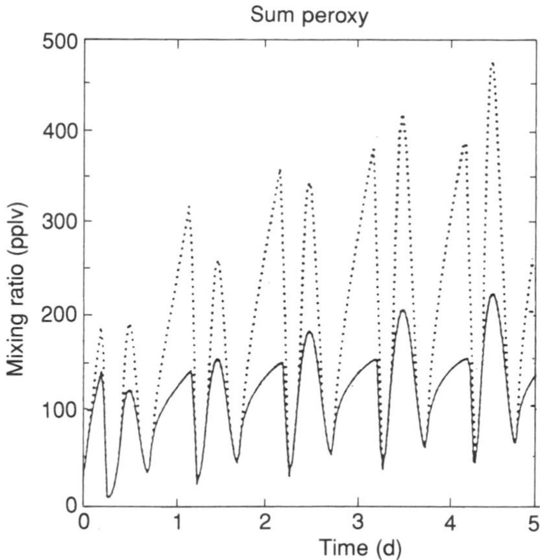

Figure 6 Total organic peroxy radical concentrations obtained in a simulation of the Amazon planetary boundary layer environment, with (continuous lines) and without (dotted lines) peroxy radical permutation reactions (reproduced with permission from Ref. 32; copyright American Geophysical Union (1990)). Permutation reactions are discussed in Section 9.

5 HO_2 RADICALS AND STRATOSPHERIC OZONE

Ozone absorbs strongly in the wavelength region from 240–290 nm and its presence in the stratosphere protects the biosphere from damaging UV solar radiation. Absorption of UV radiation by ozone heats the stratosphere and is in large measure responsible for establishing the temperature structure of the atmosphere [33]. In 1930, Chapman [34] provided the first rational explanation for the formation of the layer of ozone-rich air in the stratosphere at 15–50 km. The reactions in the Chapman scheme include the following:

$$O_2 + h\nu\ (\lambda < 242\,\text{nm}) \longrightarrow + O \quad (27)$$

$$O + O_2 + M \longrightarrow O_3 + M \quad (21)$$

$$O_3 + h\nu \longrightarrow O + O_2 \quad (1)$$

$$O + O_3 \longrightarrow 2O_2 \quad (28)$$

With increasing altitude, the O_2 concentration decreases, while the UV flux increases. Hence, with increasing altitude the rate of O-atom (and thereby

O_3) production increases to a maximum, where the product of [O_2] and the photon flux is maximized, and then decreases at higher altitudes. For several decades, the Chapman mechanism was considered to provide a good account of the formation and fate of stratospheric ozone. However, when accurate rate data for reaction (28) became available it was realized that the Chapman mechanism substantially overestimates the stratospheric ozone levels. Additional loss mechanisms for ozone were sought and found in the form of catalytic ozone destruction cycles. One of the first cycles to be identified involved OH and HO_2 radicals [35]:

$$OH + O_3 \longrightarrow HO_2 + O_2 \quad (29)$$

$$HO_2 + O \longrightarrow OH + O_2 \quad (30)$$

giving the net reaction:

$$O + O_3 \longrightarrow 2O_2 \quad (31)$$

The net effect (31) of reactions (29) and (30) is the same as reaction (28), thus augmenting the ozone loss. It is now well established that there are many similar catalytic cycles:

$$X + O_3 \longrightarrow XO + O_2 \quad (32)$$

$$XO + O \longrightarrow X + O_2 \quad (33)$$

giving the net reaction:

$$O + O_3 \longrightarrow 2O_2 \quad (34)$$

where X is H, OH, NO, Cl, and Br [33]. Figure 7 shows the relative importance of the various XO_x, cycles as a function of altitude given by Wayne [33], and illustrates the importance of HO_2 radicals in stratospheric ozone chemistry.

In addition to their direct importance in controlling the stratospheric ozone levels as discussed above, HO_2 radical chemistry provides links between the catalytic cycles. Reactions (13) and (18) link the HO_x and NO_x cycles:

$$HO_2 + NO \longrightarrow OH + NO_2 \quad (13)$$

$$HO_2 + NO_2 + M \longrightarrow HO_2NO_2 + M \quad (18)$$

Reaction (13) influences the balance between OH and HO_2, and reaction (18) ties up both HO_2 and NO_2 and so hinders the efficiency of both the HO_x and NO_x cycles. The reaction of HO_2 with ClO radicals to give HOCl and O_2 provides a link between the HO_x and ClO_x cycles.

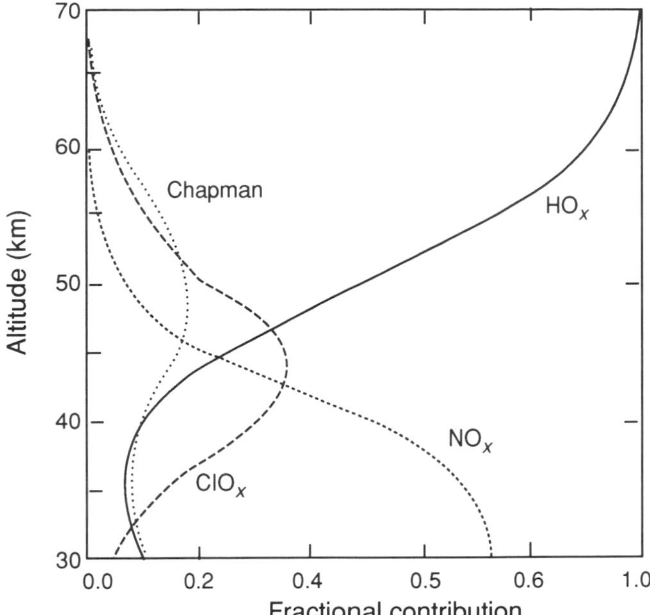

Figure 7 Fraction of (odd) oxygen loss as a function of altitude attributable to the Chapman mechanism and HO_x, ClO_x, and NO_x catalytic cycles (adapted from Wayne [33]).

6 PEROXY RADICALS AND TROPOSPHERIC OZONE

Although only 10% of atmospheric ozone resides in the troposphere (0–10 km altitude) it has a profound impact on the tropospheric chemistry. Photolysis of tropospheric ozone followed by the reaction of $O(^1D)$ atoms with H_2O vapour (reactions (1) and (2)) provides a source of OH radicals. These radicals react with almost everything emitted into the atmosphere. The generation of OH radicals is the mechanism by which the atmosphere cleanses itself. Only compounds such as chlorofluorocarbons (CFCs), Halons, and N_2O, which are inert towards OH radical attack, survive transport through the troposphere into the stratosphere where they can damage the ozone layer. An understanding of the processes which control tropospheric ozone levels is essential in any assessment of the global oxidizing capacity of the atmosphere. On regional and local scales, the photochemical production of ozone can be a serious and costly problem. Understanding the mechanisms responsible for the formation of tropospheric ozone is clearly a necessary first step in the design and implementation of measures to reduce its impact.

There are two sources of tropospheric ozone. The first of these is transport from the stratosphere in meteorological events known as 'tropospheric folding', in which a layer of stratospheric air is entrained in the tropospheric air flow and mixed into the troposphere. The second is the set of peroxy radical reactions which oxidize NO to NO_2, e.g. in the oxidation of CO:

$$OH + CO \longrightarrow H + CO_2 \qquad (3)$$

$$H + O_2 + M \longrightarrow HO_2 + M \qquad (4)$$

$$HO_2 + NO \longrightarrow OH + NO_2 \qquad (13)$$

$$NO_2 + h\nu\ (\lambda < 420\,\text{nm}) \longrightarrow NO + O \qquad (13)$$

$$O + O_2 + M \longrightarrow O_3 + M \qquad (21)$$

giving the net reaction:

$$CO + 2O_2 \longrightarrow CO_2 + O_3 \qquad (35)$$

and in the oxidation of CH_4:

$$OH + CH_4 \longrightarrow H_2O + CH_3 \qquad (5)$$

$$CH_3 + O_2 + M \longrightarrow CH_3O_2 + M \qquad (6)$$

$$CH_3O_2 + NO \longrightarrow CH_3O + NO_2 \qquad (14)$$

$$CH_3O + O_2 \longrightarrow HCHO + HO_2 \qquad (15)$$

$$HO_2 + NO \longrightarrow OH + NO_2 \qquad (13)$$

$$NO_2 + h\nu\ (\lambda < 420\,\text{nm}) \longrightarrow NO +)\ \text{(twice)} \qquad (20)$$

$$O + O_2 + M \longrightarrow O_3 + M\ \text{(twice)} \qquad (21)$$

giving the net reaction:

$$CH_4 + 4O_2 \longrightarrow H_2O + 2O_3 + HCHO \qquad (36)$$

The formaldehyde formed in the oxidation of CH_4 can either react with OH or photolyze, leading to further NO_2 formation and ozone production:

$$HCHO + OH \longrightarrow H_2O + HCO \qquad (37)$$

$$HCHO + h\nu \longrightarrow H + HCO \qquad (7)$$

$$HCO + O_2 \longrightarrow HO_2 + CO \qquad (8)$$

$$H + O_2 + M \longrightarrow HO_2 + M \qquad (4)$$

$$HO_2 + NO \longrightarrow OH + NO_2\ \text{(twice)} \qquad (13)$$

Under favourable conditions, the atmospheric oxidation of one CH_4 molecule can lead to the formation of four molecules of ozone. Recent modelling calculations suggest, that on a global scale, transport from the stratosphere accounts for ca. 10% of tropospheric ozone, while peroxy radical chemistry is the source of the remaining 90% [36]. Clearly, peroxy radicals play an important role in tropospheric ozone chemistry.

In addition to the above reactions in which peroxy radicals form ozone, they are also involved in reactions which compete with ozone formation. While HO_2 radicals react with NO to give NO_2 exclusively, RO_2 radicals react with NO via two pathways:

$$RO_2 + NO \longrightarrow RO + NO_2 \qquad (19a)$$

$$RO_2 + NO + M \longrightarrow RONO_2 + M \qquad (19b)$$

Pathway (19b) gives alkyl nitrates which are much less reactive than NO or NO_2, and so sequester NO_x (NO and NO_2). Reaction channel (19b) represents a loss of both radicals and NO_x from the system and hence slows down the photochemical chain reactions that form ozone. In general, the relative importance of the nitrate-producing channel increases with the size of 'R' [37] e.g. with R = CH_3, $k_{19b}/(k_{19a} + k_{19b}) < 0.005$ [38], while for R = t-butyl, $k_{19b}/k_{19a} + k_{19b}) = 0.18$ [39]). For C_1–C_3 hydrocarbons the formation of nitrates is of minor importance, while for C_4 and above, nitrate formation is significant.

Peroxy radicals also react with NO_2 to give peroxy nitrates with rate constants (which depend on the nature of 'R') of $k_{38} = (1–10) \times 10^{-12}\,cm^3\,molecule^{-1}\,s^{-1}$ at room temperature and atmospheric pressure (see Chapter 7 for further details):

$$RO_2 + NO_2 + M \longrightarrow RO_2NO_2 + M \qquad (38)$$

For HO_2 and alkylperoxy radicals the RO_2—NO_2 bond is relatively weak (22–25 kcal mol^{-1}) and the peroxy nitrate decomposes rapidly to reform the reactants [40]. For R = H, CH_3, and C_2H_5, the lifetime of RO_2NO_2 at room temperature and $-10°C$ lie in the ranges 0.1–10 and 10–1000 s, respectively.

$$RO_2NO_2 + M \longrightarrow RO_2 + NO_2 + M \qquad (38)$$

However, peroxy acyl nitrates ($RC(O)O_2NO_2$), such as $CH_3C(O)O_2NO_2$, $C_2H_5C(O)O_2NO_2$, and $C_6H_5C(O)O_2NO_2$, have RO_2—NO_2 bonds which are somewhat stronger than their alkyl counterparts (28 kcal mol^{-1} for $CH_3C(O)O_2NO_2$). This slight increase in bond strength has a dramatic effect on their atmospheric chemistry. The lifetime of $CH_3C(O)O_2NO_2$ at room temperature is 0.6 h, while at $-10°C$ it is 13 d. The lifetimes of peroxy acyl nitrates with respect to thermal decomposition are sufficiently long at the low temperatures characteristic of the upper troposphere to allow long-range transport of NO_x, and yet short enough at the warmer temperatures of the lower troposphere to release NO_2, which can then participate in photochemical reactions [41–43].

Peroxy acetyl nitrate (PAN) is the most abundant peroxy acyl nitrate and is an important component of photochemical smog in urban areas. PAN is largely responsible for the eye irritation experienced in smog episodes, and because of its phytotoxic nature can damage agricultural crops. Other members of the peroxyl acyl nitrate family include peroxy propyl nitrate, $C_2H_5C(O)O_2NO_2$, and peroxy benzoyl nitrate, $C_6H_5C(O)O_2NO_2$, which are found at levels ranging from a few percent to 30% of that of PAN [43]. The levels of PAN found in ambient air vary widely; the highest levels are found in polluted urban air in the afternoon and can range up to 30 ppbv, while levels as low as a few pptv are observed in cleaner air masses [44].

In remote locations, where NO concentrations are low, reactions (39) and (40) become important. These reactions compete with NO for the available peroxy radicals:

$$RO_2 + HO_2 \longrightarrow ROOH + O_2 \tag{39a}$$

$$RO_2 + HO_2 \longrightarrow R'CHO + H_2O + O_2 \tag{39b}$$

$$HO_2 + O_3 \longrightarrow OH + 2O_2 \tag{40}$$

The reaction of unsubstituted alkylperoxy radicals like CH_3O_2 and $C_2H_5O_2$ with HO_2 leads to the formation of hydroperoxides in essentially 100% yields [45,46]. In contrast, substituted peroxy radicals (e.g. CH_2FO_2 and $CH_3OCH_2O_2$) appear to react with HO_2 via two pathways to give hydroperoxide and carbonyl products [47,48]. The net effect of reaction (39) is to convert two reactive species into a relatively unreactive hydroperoxide or carbonyl compound. Reaction (39) slows down the free-radical-driven photochemical oxidation reactions and reduces the formation of ozone. At ambient temperature, HO_2 radicals react 4300 times more rapidly with NO than with ozone [2]. Tropospheric ozone levels are typically 20–40 ppbv in remote areas and 80–200 ppbv in polluted environments, while NO levels range from 1–10 pptv in remote areas to 10–200 ppbv in polluted environments. Reaction (40) is of no importance in and near polluted environments, but is important in remote areas.

The role of peroxy radicals in tropospheric ozone chemistry in remote areas depends on the relative NO and O_3 concentrations. Under conditions of low NO_x, reactions (13) and (40) compete for HO_2 radicals:

$$HO_2 + NO \longrightarrow OH + NO_2 \tag{13}$$

$$HO_2 + O_3 \longrightarrow OH + 2O_2 \tag{40}$$

At ambient temperature, the rate constant ratio $k_{13}/k_{40} = 4300$ [2]. Hence, the rates of reactions (13) and (40) become equal when $[O_3]/[NO] = 4300$. If we assume $[O_3] = 30$ ppbv (typical for a remote location), then for $[NO] < 7$ pptv, HO_2 radical chemistry leads to ozone loss, while for $[NO] > 7$ pptv, HO_2 radical chemistry leads to ozone formation. The calculation above provides a simple illustration of the effect of HO_2 radicals on tropospheric ozone chemistry. In discussions of the net effect of atmospheric photochemistry on tropospheric ozone and its dependence on NO_x ($NO + NO_2$) levels, other issues need consideration. For example, reaction (40) only accounts for ca. 30% of the ozone loss, with the rest being mainly attributable to photolysis, followed by reaction (2) [49]. In addition, the chemistry of other peroxy radicals (i.e. CH_3O_2) needs consideration (CH_3O_2 radicals react with NO but do not react appreciably with O_3). The balance between photochemical production and loss of ozone has been investigated by Liu *et al.* [49] and Jenkin [50] as a function of $[NO_x]$ by using box-model calculations. In the simulations, the NO concentration was 25–40% of the NO_x. It was found that the balance point occurs at $[NO_x] = 40–60$ pptv ($[NO] = 10–25$ pptv). With more NO_x, the net effect of atmospheric photochemistry is ozone formation, while for lower $[NO]_x$ the net effect is ozone loss.

In polluted air masses typical of continental and urban areas, the NO levels are high enough to dominate the loss of peroxy radicals. Volatile organic compounds (VOCs) are present at high concentrations in polluted air and RO_2 radical chemistry then assumes great importance. Ozone levels in such environments are controlled by a number of factors, including the rate at which peroxy radicals are formed and the extent to which they form alkyl nitrates and peroxynitrates which sequester NO_x. The latter two factors are determined by the quantity and identity of the organic compounds which are present. There is great practical interest in establishing the relationship between emissions of VOCs and NO_x into urban air masses and the formation of secondary air pollutants such as ozone. Over the past three decades, great strides have been made in understanding the fundamental chemistry which is in operation in the formation of oxidants in urban air. As described by Jeffries [51], it is now recognized that complex non-linear feedback processes relate VOCs and NO_x emissions to ozone levels. While the complexity of the situation defies easy solution, recognition of the need to reduce VOCs and/or NO_x emissions has led to new technology and new approaches which reduce the emission of these compounds. The automobile industry provides a good example of the progress made on this front. In the 1960s (prior to control), the tailpipe emissions of a typical automobile in California were 8.8 g of hydrocarbons and 3.6 g of NO_x per mile. With the help of catalyst technology, tailpipe emissions from new vehicles in California in 1994 have been reduced to 0.25 g of hydrocarbons and 0.4 g of NO_x per mile [52].

Regulations in the USA and Europe pertaining to the emission of reactive organic compounds into the atmosphere are moving away from standards in which only the total organic mass is considered, towards standards which take into account the different reactivity of different compounds. This trend in regulations has generated a considerable interest in the relative ranking of organic compounds in terms of their ability to contribute to ozone formation in urban areas. Calculation of scales of 'reactivity factors' [53] or 'photochemical ozone creation potentials' (POCPs) [54–56] for organic compounds have received considerable prominence. These scales reflect the rate at which the organic compound gives peroxy radicals, together with the efficiency with which these peroxy radicals contribute to, or hinder, ozone formation.

7 NIGHT-TIME PEROXY RADICAL CHEMISTRY

Sunset marks the end of photolytically driven reactions in the atmosphere, but it does not necessarily mark the end of peroxy radical chemistry. As the sunlight disappears, the $NO—NO_2—O_3$ photostationary steady state relaxes and the concentration of NO collapses and approaches zero within a few minutes. In the absence of NO, the nitrate radical, NO_3, becomes important; these radicals are formed by the reaction of NO_2 with O_3. NO_3 radicals react with a variety of organic species via H-atom abstraction and/or addition mechanisms to generate peroxy radicals [57–61]:

$$NO_2 + O_3 \longrightarrow NO_3 + O_2 \tag{41}$$

$$NO_3 + HCHO \longrightarrow HNO_3 + HCO \tag{42}$$

$$HCO + O_2 \longrightarrow HO_2 + CO \tag{8}$$

$$NO_3 + CH_2{=}CH_2 + M \longrightarrow CH_2CH_2NO_3 + M \tag{43}$$

$$CH_2CH_2NO_3 + O_2 + M \longrightarrow OOCH_2CH_2NO_3 + M \tag{44}$$

Reaction of O_3 with alkenes provides important night-time source of peroxy radicals [62]. As discussed by Platt *et al.* [58], in air masses that contain a particular mix of reactive organics, oxides of nitrogen, and ozone, the NO_3 radical chemistry can sustain chain reactions involving peroxy radicals. The key reaction in this system is that of HO_2 with NO_3, which is relatively rapid ($k_{45} = 3 \times 10^{-11}\,cm^3\,molecule^{-1}\,s^{-1}$ at 298 K) and produces OH radicals in a substantial yield [63–65]:

$$HO_2 + NO_3 \longrightarrow OH + NO_2 + O_2 \tag{45}$$

OH radicals then react with organic compounds leading to regeneration of peroxy radicals. Other peroxy radicals such as CH_3O_2 can also react with NO_3

via a similar mechanism [66–69]:

$$CH_3O_2 + NO_3 \longrightarrow CH_3O + NO_2 + O_2 \quad (46)$$

The CH_3O radicals produced in this reaction then react with O_2 to produce HO_2 radicals which, in turn, can react with NO_3 to give OH. Reactions (45) and (46) are the night-time equivalent of reactions (13) and (14). Mihelcic *et al.* [7] have reported an anti-correlation between the measured RO_2 and NO_3 concentrations in ambient air at Schauinsland, Germany, which is consistent with reactions (45) and (46) being important. Using modelling calculations, Platt *et al.* [58] noted the possibility that in some locations the night-time peroxy radical concentrations may actually exceed those encountered in the day. However, all measurements to date show diurnal peroxy radical profiles that are more in keeping with those shown in Figures 4 and 5, with night-time levels well below those in the daytime. Slow radical–radical reactions control the night-time peroxy radical concentrations. Hence, even if high peroxy radical concentrations were observed, this would not necessarily indicate that substantial degradation of organic compounds was occurring. As noted by Lightfoot *et al.* [40], at the present time there is no observational data which shows that night-time chemistry makes a major contribution to the atmospheric oxidation of organic compounds. The importance of peroxy radical chemistry in the night-time atmosphere remains unclear.

8 PEROXY RADICALS AND HYDROPEROXIDE FORMATION

In remote areas with low NO levels, the self-reaction of HO_2 radicals is an important loss of HO_2. This self-reaction gives hydrogen peroxide and is the only known gas-phase source of this important oxidant.

$$HO_2 + HO_2 \longrightarrow H_2O_2 + O_2 \quad (47)$$

H_2O_2 is highly soluble and is rapidly incorporated into cloud water. The importance of aqueous H_2O_2 in oxidizing SO_2 into H_2SO_4 is well established [70–72]. The relevant reactions are:

$$SO_2 + H_2O \rightleftharpoons H^+ + HSO_3^- \quad (48)$$

$$HSO_3^- + H_2O_2 \rightleftharpoons HSO_4^- + H_2O \quad (49)$$

$$HSO_4^- + H^+ \longrightarrow H_2SO_4 \quad (50)$$

Sulfuric acid, H_2SO_4, is an important component of acidic precipitation. The majority of the H_2SO_4 that is found in rain water is produced by the oxidation of SO_2 by H_2O_2 [72]. The evaporation of cloud droplets which contain sulfuric acid gives a sulfate aerosol; this aerosol is effective in scattering light and plays an important role in the earth's radiation balance and is important in discussions of potential global climate change [73]. Approximately 70% of the sulfate aerosol in the earth's atmosphere originates from the oxidation of SO_2 by H_2O_2 [72].

9 PERMUTATION REACTIONS OF ORGANIC PEROXY RADICALS IN THE TROPOSPHERE

Madronich and Calvert [32] have shown that permutation reactions (self- and cross-reactions) of organic peroxy radicals play an important role in tropospheric chemistry under low NO_x conditions. The inclusion of such reactions in models of planetary boundary layer chemistry for marine and Amazon basic conditions has four effects. First, the predicted organic peroxy radical concentrations are reduced, thus leading to a suppression of the total (organic + HO_2) peroxy radical levels (see Figures 5 and 6). Secondly, because the reaction of organic peroxy radicals with HO_2 is an important loss of HO_2 radicals, the inclusion of permutation reactions somewhat *increases* the predicted HO_2 concentrations (see Figure 5). Thirdly, permutation reactions increase the loss rate of $CH_3C(O)O_2$ radicals in the model and hence decrease the lifetime of PAN (resulting in faster release of NO_x). Fourthly, the permutation reactions provide a substantial gas-phase source of alcohols and organic acids. Inclusion of permutation reactions in current models of atmospheric chemistry is hampered by a general lack of kinetic and mechanistic data for such reactions, and research is currently underway to provide this data.

10 CONCLUSIONS

Peroxy radicals are important intermediates in the atmospheric degradation of all organic compounds. In the troposphere, peroxy radicals are involved in chemistry that leads to ozone formation, while in the stratosphere, reactions involving HO_2 lead to ozone loss. It is becoming increasingly apparent that emissions from human activities are perturbing the global atmospheric environment. To fully understand the environmental impact of such emissions it is necessary to understand the relevant gas-phase peroxy radical chemistry.

ACKNOWLEDGEMENTS

We thank Chris Cantrell and Geoff Tyndall (National Center for Atmospheric Research) for helpful discussions and preprints of some of their recent work.

REFERENCES

1. G. Brasseur and S. Solomon, *Aeronomy of the Middle Atmosphere*, Reidel, Dordrecht (1986).
2. W.B. DeMore, S.P. Sander, D.M. Golden, R.F. Hampson, M.J. Kurylo, C.J. Howard, A.R. Ravishankara and C.E. Kolb, *Chemical Kinetics and Photochemical Data for Use in Stratospheric Modelling*, Evaluation Number 11, NASA-JPL Publication 94-26, (1994).
3. C.A. Cantrell, R.E. Shetterr, A.H. McDaniel and J.G. Calvert, in ACS Advances in Chemistry Series, No. 232, American Chemical Society, Washington DC (1993), p. 291.
4. D. Mihelcic, D.H. Ehhalt, J. Klomfass, U. Schmidt and M. Trainer, *Ber. Bunsenges. Phys. Chem.* **82**, 16 (1978).
5. D. Mihelcic, P. Müsgen and D.H. Ehhalt, *J. Atmos. Res.* **3**, 341 (1985).
6. D. Mihelcic, A. Volt-Thomas, H.W. Pätz, D. Kley and M. Mihelcic, *J. Atmos. Chem.* **11**, 271 (1990).
7. D. Mihelcic, D. Klemp, P. Müsgen, H.W. Pätz and A. Volz-Thomas, *J. Atmos. Chem.* **16**, 313 (1993).
8. C.A. Cantrell and D.H. Stedman, *Geophys. Res. Lett.* **9**, 846 (1982).
9. C.A. Cantrell, D.H. Stedman and G.J. Wendel, *Anal. Chem.* **56**, 1496 (1984).
10. C.A. Cantrell, J.A. Lind, R.E. Shetter, J.G. Calvert, P.D. Goldan, W. Kuster, F.C. Fehsenfeld, S.A. Montzka, D.D. Parrish, E.J. Williams, M.P. Buhr, H.H. Westberg, G. Allwine and R. Martin, *J. Geophys. Res.* **97**, 20671 (1992).
11. C.A. Cantrell, R.E. Shetter, J.A. Lind, A.H. McDaniel, G.J. Calvert, D.D. Parrish, F.C. Fehsenfeld, M.P. Buhr and M. Trainer, *J. Geophys. Res.* **98**, 2897 (1993).
12. C.A. Cantrell, R.E. Shetter, J.G. Calvert, F.L. Eisele, E. Williams, K. Baumann, W.H. Brune, P.S. Stevens and J.H. Mather, *J. Geophys. Res.* (1996), in press.
13. D.R. Hastie, M. Weissenmayer, J.P. Burrows, and G.W. Harris, *Anal. Chem.* **63**, 2048 (1991).
14. C. Anastasi, R.V. Gladstone and M.G. Sanderson, *Environ. Sci. Technol.* **27**, 474 (1993).
15. Y.K. Maeda, K. Aoki and M. Munemori *Anal. Chem.* **52**, 307 (1980).
16. G.L. Wendel, D.H. Stedman, C.A. Cantrell and L. Damrauer, *Anal. Chem.* **55**, 937 (1983).
17. B.J. Finlayson-Pitts and J.N. Pitts, *Atmospheric Chemistry: Fundamentals and Experimental Techniques*, John Wiley & Sons, New York (1986).
18. C.A. Cantrell, R.E. Shetter, A.J. McDaniel, J.A. Davidson, J.G. Calvert, D.D. Parrish, M.B. Buhr, F.C. Fehsenfeld and M. Trainer, *EOS Trans. Am. Geophys. Union* **69**, 1056 (1988).
19. C.A. Cantrell, R.E. Shetter, T.M. Gilpin, and J.C. Calvert, *J. Geophys. Res.* (1996), in press.
20. P. Leighton, *Photochemistry of Air Pollution*, ● (1961).
21. D.D. Parrish, M. Trainer, E.J. Williams, D.W. Fahey, G. Hübler, C.S. Eubank, S.C. Liu, P.C. Murphy, D.L. Albritton and F.C. Fehsenfeld, *J. Geophys. Res.* **91**, 5361 (1986).
22. D. Mihelcic, D.H. Ehhalt, G.F. Kulessa, J. Klomfass, M. Trainer, U. Schmidt and H. Rohrs, *Pure Appl. Geophhys.* **116**, 530 (1978).
23. M. Helten, W. Platz, M. Trainer, H. Fark, E. Klein and D.H. Ehhalt, *J. Atmos. Chem.* **2**, 191 (1984).
24. J.G. Anderson, H.J. Grassl, R.E. Shetter and J.J. Margitan, *Geophys. Res. Lett.* **8**, 289 (1981).

25. R.M. Stimpfle, P.O. Wenneberg, L.B. Lapson and J.G. Anderson, *Geophys. Res. Lett.* **17**, 1905 (1990).
26. W.A. Traub, D.G. Johnson and K.V. Chance, *Science* **247**, 446 (1990).
27. R.L. de Zafra, A. Parrish, P.M. Solomon and J.W. Barrett, *J. Geophys. Res.* **89**, 1321 (1984).
28. C.A. Cantrell, R.E. Shetter, J.G. Calvert, F.L. Eisele, E. Williams, K. Baumann, W.H. Brune, P.S. Stevens and J.H. Mather, *J. Geophys. Res.* (1996), in press.
29. C.A. Cantrell, R.E. Shetter, T.M. Gilpin, J.G. Calvert, F.L. Eisele and D.J. Tanner, *J. Geophys. Res.* (1996), in press.
30. M. Weissenmayer, J.P. Burrows and M. Schupp in *Physico-Chemical Behavior of Atmospheric Pollutants*, edited by G. Angeletti and G. Restelli, (1993), p. 575.
31. T. Behmann, M. Weissenmayer and J.P. Burrows, *Atmos. Environ.*, submitted.
32. S. Madronich and J.G. Calvert, *J. Geophys. Res.* **95**, 5697 (1990).
33. R.P. Wayne *Chemistry of Atmospheres*, 2nd Ed, Oxford University Press, Oxford (1991).
34. S. Chapman, *Mem. R. Meteorol. Soc.* **3**, 103 (1930).
35. D.R. Bates and M. Nicolet, *J. Geophys. Res.* **55**, 301 (1950).
36. P.J. Crutzen, in *Composition, Chemistry, and Climate of the Atmosphere*, edited by H.B. Singh, Van Nostrand Reinhold, New York (1995).
37. R. Atkinson, S.M. Aschmann, W.P.L. Carter, A.M. Winer and J.N. Pitts, Jr., *J. Phys. Chem.* **86**, 4563 (1982).
38. C.T. Pate, B.J. Finlayson and J.N. Pits, *J. Am. Chem. Soc.* **96**, 6554 (1974).
39. K.H. Becker, H. Geiger and P. Wiesen, *Chem. Phys. Lett.* **184**, 256 (1991).
40. P.D. Lightfoot, R.A. Cox, J.N. Crowley, M. Destriau, G.D. Hayman, K.E. Jenkin, G.K. Moortgat and F. Zabel, *Atmos. Environ.* **26A**, 1805 (1995).
41. P.J. Crutzen, *Annu. Rev. Earth Planet Sci.* **7**, 443 (1979).
42. H.B. Singh and P.L. Hanst, *J. Geophys. Res.* **8**, 941 (1981).
43. J.M. Roberts in *Composition, Chemistry, and Climate of the Atmosphere*, edited by H.B. Singh, Van Nosrand Reinhold, New York (1995).
44. A.P. Altshuller, *J. Air Waste Manag. Assoc.* **43**, 1221 (1993).
45. T.J. Wallington and S.M. Japar, *Chem. Phys. Lett.* **167**, 513 (1990).
46. T.J. Wallington and S.M. Japar, *Chem. Phys. Lett.* **166**, 495 (1990).
47. T.J. Wallington, M.D. Hurley, W.F. Schneider, J. Sehested and O.J. Nielsen, *Chem. Phys. Lett.* **218**, 34 (1994).
48. T.J. Wallington, M.D. Hurley, J.C. Ball and M.E. Jenkin, *Chem. Phys. Lett.* **211**, 41 (1993).
49. S.C. Liu, M. Trainer, M.A. Carroll, G. Hubler, D.D. Montzka, R.B. Norton, B.A. Ridley, J.G. Walega, E.L. Atlas, B.G. Heikes, B.J. Huebert and W. Warren, *J. Geophys. Res.* **97**, 10463 (1992).
50. M.E. Jenkin, *PhD Thesis*, University of East Anglia (1991).
51. H.E. Jeffries, *Composition, Chemistry and Climate of the Atmosphere*, edited by H.B. Singh, Van Nostrand Reinhold, New York (1995).
52. T.Y. Chang, D.P. Chock, R.H. Hammerle, S.M. Japar and I.T. Salmeen, *Critical Rev. Environ. Control* **22**, 27 (1992).
53. W.P.L. Carter and R. Atkinson, *Environ. Sci. Technol.* **23**, 864 (1989).
54. R.G. Derwent and M.E. Jenkin, *Atmos. Environ.* **25A**, 1661 (1991).
55. D. Simpson, *J. Atmos. Chem.* **20**, 163 (1995).
56. R.G. Derwent, M.E. Jenkin and S.M. Saunders, *Atmos. Environ.* **30**, 181 (1996).
57. C.A. Cantrell, W.R. Stockwell, L.G. Anderson, K.L. Busarow, D. Perner, A. Schmeltekopf, J.G. Calvert, H.S. Johnston, *J. Phys. Chem.* **89**, 139 (1985).
58. U. Platt, G. LeBras, J.P. Burrows and G. Moortgat, *Nature (London)* **348**, 147 (1990).

59. R.P. Wayne, I. Barnes, P. Biggs, J.P. Burrows, C.E. Canosa-Mas, J. Hjorth, G. LeBras, G.K. Moortgat, D. Perner, G. Poulet, G. Restelli and H. Sidebottom, *Atmos. Environ.* **25A**, 1 (1991).
60. E. Becker, M.M. Rahman, R.N. Schindler, *Ber. Bunsenges. Phys. Chem.* **96**, 776 (1992).
61. R. Atkinson, *J. Phys. Chem. Ref. Data, Monograph 2* (1994).
62. R. Atkinson, *Atmos. Environ.* **24A**, 1 (1990).
63. A. Mellouki, G. LeBras and G. Poulet, *J. Phys. Chem.* **92**, 2229 (1988).
64. I.W. Hall, R.P. Wayne, R.A. Cox, M.E. Jenkin and G.D. Hayman, *J. Phys. Chem.* **92**, 5049 (1988).
65. A. Mellouki, R.K. Talukdar, A.M.R.P. Bopegedera and C.J. Howard, *Int. J. Chem. Kinet.* **25**, 25 (1993).
66. J. Crowley, J.P. Burrows, G.K. Moortgat, G. Poulet and G. LeBras, *Int. J. Chem. Kinet.* **22**, 673 (1990).
67. J. Hjorth, C. Lohse, C.J. Nielsen, H. Skov and G. Restelli, *J. Phys. Chem.* **94**, 7494 (1990).
68. P. Biggs, C.E. Canosa-Mas, J.M. Fracheboud, D. Shallcross and R.P. Wayne, *J. Chem. Soc. Faraday Trans.* **90**, 1205 (1994).
69. F. Daële, G. Laverdet, G. Le Bras and G. Poulet, *J. Phys. Chem.* **99**, 1470 (1995).
70. S.A. Penkett, B.M.R. Jones, K.A. Brice and A.E.J. Eggleton, *Atmos. Environ.* **13**, 123 (1979).
71. M.W. Gallagher, R.M. Downer, T.W. Choularton, M.J. Gay, I. Stromberg, C.S. Mill, M. Radojevic, B.J. Tyler, B.J. Bandy, S.A. Penkett, T.J. Davies, G.J. Dollard and B.M.R. Jones, *J. Geophys. Res.* **95**, 18517 (1990).
72. S.A. Penkett, in *The Chemistry of the Atmosphere and its Impact on Global Change*, edited by J.G. Calvert, Blackwell, Oxford, (1994).
73. R.J. Charlson, J. Langner, H. Rodhe, C.B. Leovy and S.G. Warren, *Tellus* **43B**, 152 (1991).

16 Peroxyl Radicals in the Treatment of Waste Solutions

NIKOLA GETOFF
University of Vienna, Austria

1 INTRODUCTION

The sequel to the strong development of various industries and the growth in the world population is a global overloading of the environment, particularly water resources. The major contribution in this respect is the increased disposal of large quantities of chemical and other waste in rivers, seas and oceans, which is leading to destruction of the rather sensitive marine life. This will now need a long period of time to recover. In addition, the world-wide use of fertilizers, pesticides, etc. in modern agriculture has led to pollution of ground water in various regions. It should also be mentioned that the chlorination of drinking water (containing humic substances) for disinfection purposes is leading to the formation of various toxicological organic compounds [1–4]. Furthermore, it has been found that the metabolism of such pollutants in human organisms is implicated in several diseases [5].

It has been demonstrated in recent years that complete degradation of biologically resistant pollutants in water can be achieved by irradiation treatment (e.g. Refs 6–22, and references therein). In this context, peroxyl radicals have been found to be strongly involved in the radiation-induced decomposition processes of pollutants in water in the presence of air. An understanding of their reaction mechanisms is therefore of vital importance for the practical applications of this new technology. It might be of interest here to discuss briefly some basic requirements for the reactivity of oxygen.

Molecular oxygen, under normal conditions, is in the triplet state ($^3O_2$) and hence its direct reaction with most halogenated organic compounds is spin-forbidden. Therefore, in order to mediate a reaction, oxygen has to be activated. This can be carried out by conversion into its reactive single state ($^1O_2$), which can be achieved either by irradiation with vacuum-ultraviolet (UV) light ($\lambda < 200\,nm$), electrical discharge in the gas phase, photosensitized chemical processes, reactions of ozone with organic peroxides, reaction of H_2O_2 with hypochlorite, microwave irradiation, the formation of vibrationally excited water, or various enzymatic reactions (In Ref. 23 and references therein).

Peroxyl Radicals. Edited by Z.B. Alfassi

A further possibility for oxygen activation is its conversion into peroxyl radicals by the reaction with hydrogen atoms, 'solvated electrons' (e^-_{aq}) or organic free radicals (R), which can be formed, for example, by ionizing radiation in aqueous solution:

$$O_2 + H \longrightarrow HO_2^{\cdot} \quad (k_1 = 2.1 \times 10^{10}\, dm^3\, mol^{-1}\, s^{-1}\, [24]) \tag{1}$$

$$O_2 + e^-_{aq} \longrightarrow O_2^{-\cdot} \quad (k_2 = 2.0 \times 10^{10}\, dm^3\, mol^{-1}\, s^{-1}\, [24]) \tag{2}$$

$$HO^{\cdot}_2 \rightleftharpoons H^+ + O_2^{-\cdot} \quad (pK_a = 4.8;\ \text{Ref. 25, and references therein}) \tag{3}$$

$$O_2 + R^{\cdot} \longrightarrow RO_2^{\cdot} \quad (k_4 = 10^7\text{–}10^9\, dm^3\, mol^{-1}\, s^{-1}) \tag{4}$$

In some cases, in addition to hydrogen atoms, e^-_{aq} can be formed from electronically excited inorganic ions or organic compounds by UV irradiation in aqueous solution [26,27]. In the presence of air, reactions (1)–(4) can also take place. It might also be mentioned that in the context of formate radiolysis in the presence of air, OH radicals can be converted, almost quantitatively, into $O_2^{-\cdot}$ radicals (equations (5) and (6)), which are in equilibrium with $HO^{\cdot}_2$ species:

$$HCOO^- + OH \longrightarrow \dot{C}O_2^- + H_2O \quad (k_5 = 3.2 \times 10^9\, dm^3\, mol^{-1}\, s^{-1}\, [24]) \tag{5}$$

$$\dot{C}O_2^- + O_2 \longrightarrow CO_2 + O_2^{-\cdot} \quad (k_6 = 6.5 \times 10^9\, dm^3\, mol^{-1}\, s^{-1}\, [28]) \tag{6}$$

A further pathway for producing peroxyl radicals is the well known Fenton reaction [29,30]. In this, Fe^{2+} or Ti^{3+} ions react with H_2O_2 with the formation of OH radicals, which are then scavenged by organic compounds (RH) present in the solution, resulting in $R^{\cdot}$ species (equations (7) and (8)). The latter are then scavenged by oxygen (equation (4)):

$$Fe^{2+} + H_2O_2 \longrightarrow Fe^{3+} + OH^- + OH \tag{7}$$

$$RH + OH \longrightarrow R^{\cdot} + H_2O \tag{8}$$

OH radicals can also be produced by the Haber–Weiss reaction [31–33], where $HO_2^{\cdot}$ or $O_2^{-\cdot}$ (depending on the pH of the solution, see equation (3)) react with H_2O_2:

$$HO_2^{\cdot} + H_2O_2 \longrightarrow OH + O_2 + H_2O \tag{9}$$

$$O_2^{-\cdot} + H_2O_2 \longrightarrow OH + O_2 + OH^- \tag{10}$$

In this process, a number of competition reactions are also involved (see Ref.

32 and references therein), which are not discussed further here.

The photolysis of H_2O_2 (equation (11)) leads likewise to the formation of OH radicals, which are then subsequently consumed according to reaction (8), and finally the resulting $R^{\bullet}$ species are converted into peroxyl radicals by reaction (4):

$$H_2O_2 \longrightarrow 2\,OH \tag{11}$$

It should be mentioned here that the formation of peroxyl radicals and elucidation of their reactions in aqueous solutions, in the context of radiation-induced processes, has been well surveyed in recent reviews [34,35]. In the following, the production and reactivity of typical peroxyl radicals originating from the aliphatic, olefinic and aromatic substances, which are present as pollutants in waste water, will be discussed.

2 PEROXYL RADICALS FROM CHLORINATED ALIPHATIC POLLUTANTS

In most cases, halogenated aliphatic hydrocarbons are present as pollutants in various waste waters. In addition, they are found in drinking water, formed as a result of chlorination for disinfection purposes, and also in ground water as the metabolic products from chemicals used in modern agriculture [1–17]. The simplest peroxyl radicals investigated in this group are those produced from chlorinated methane, ethane, n-propane and n-butane in water. Their kinetic and spectroscopic characteristics are shown in Table 1. For comparison, data for the $HO_2^{\bullet}$ and $O_2^{-\bullet}$ species are also given.

Table 1 Kinetic and spectroscopic data for peroxyl radicals of methane, ethane, n-propane and n-butane, plus the $HO_2^{\bullet}/O_2^{-\bullet}$ species, in aqueous solutions

Radical	Formation $k(R + O_2)$ ($dm^3\,mol^{-1}\,s^{-1}$)	Decay $2k$ ($dm^3\,mol^{-1}\,s^{-1}$)	λ_{max} (nm)	ϵ_{max} ($dm^3\,mol^{-1}\,cm^{-1}$)	Ref.
$HO_2^{\bullet}$	2.1×10^{10} [a]	3.7×10^6	230	1300	25
$O_2^{-\bullet}$	1.9×10^{10} [b,c]	10 [d]	245	2150	24,25
$CH_3O_2^{\bullet}$	3.2×10^9 [e]	Long-lived	245 [f]	1170 [f]	36,37
		Long-lived	245	1300	9
$C_2H_5O_2^{\bullet}$	2.9×10^9	Long-lived	248	1350	37
		Long-lived	250	1380	9
n-$C_3H_7O_2^{\bullet}$	1.9×10^9	Long-lived	255	1250	11
n-$C_4H_9O_2^{\bullet}$	1.3×10^9	Long-lived	258	1480	11

[a] $R^{\bullet}$ = H.
[b] $R^{\bullet} = e_{aq}^-$.
[c] Ref. 24.
[d] Ref. 25.
[e] Ref. 36.
[f] Ref. 37.

The absorption spectra of $HO_2^{\bullet}$, $O_2^{-\bullet}$, plus the spectra of n-$\dot{C}_4H_9$ and n-$C_4H_9O_2^{\bullet}$ are illustrated in Figure 1. As can be seen from the data presented in Table 1 and Figure 1, the spectroscopic characteristics of these species are very similar (see Table 1).

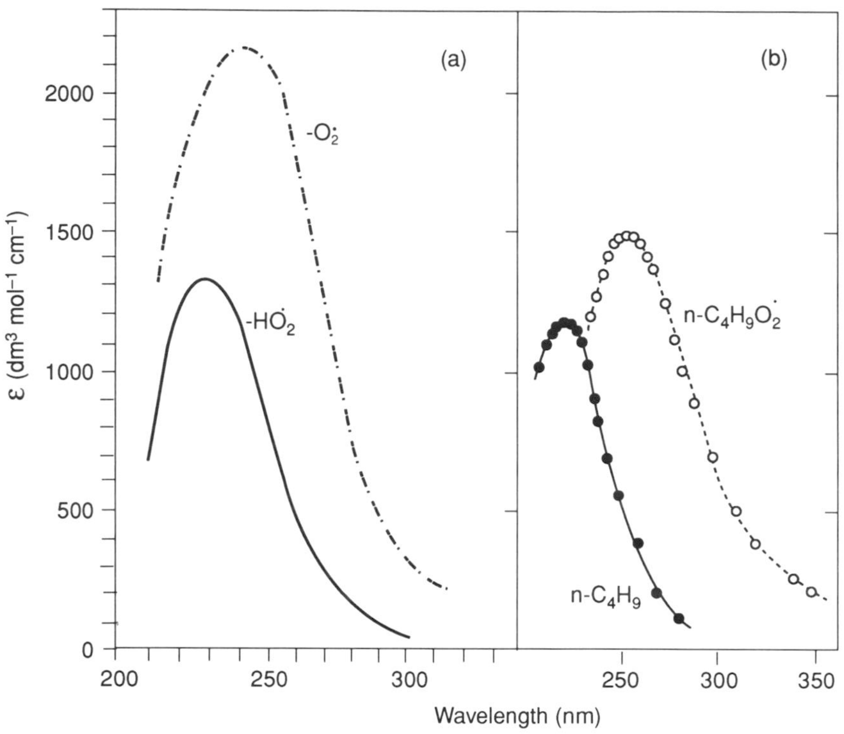

Figure 1 Absorption spectra of (A) HO_2 (pH 2.6) and $O_2^{-\bullet}$ (pH 7.2) species [25] and (B) n-C_4H_9 and n-$C_4H_9O_2$ (pH ≈ 8.0) radicals [11] in aqueous solution (reproduced from Ref. 11 by permission of Elsevier Science Ltd).

In waste water with higher pollutant concentrations containing low amounts of oxygen, the reducing species (H atoms and e^-_{aq}) can be partly consumed by the substrate, while the remainder react with oxygen to form $HO_2^{\bullet}/O_2^{-\bullet}$ radicals (see equations (1) and (2)). Hence, pollutant degradation is initiated by the attack of OH radicals, in addition to H atoms, e^-_{aq}, $HO_2^{\bullet}$, and/or $O_2^{-\bullet}$. The transients produced are then scavenged by oxygen, leading to the formation of the corresponding peroxyl radicals, and finally resulting in the various decomposition products. The yields of the latter depend strongly on the experimental conditions, such as substrate concentration, availability of oxygen, dose, dose rate, pH of the waste solution and temperature. The decomposition of a number of halogenated aliphatic hydrocarbons have been studied under such conditions [3,8,10–18,38–41].

As an example in this respect, the formation and decay of peroxyl radicals originating from *chloroform* will now be briefly discussed [3,4,9]:

$$CHCl_3 + e^-_{aq} \longrightarrow Cl^- + \dot{C}HCl_2 \quad (12)$$
$$(k_{12} = 3 \times 10^{10}\, dm^3\, mol^{-1}\, s^{-1}\, [24])$$

$$CHCl_3 + H \longrightarrow H^+ + Cl^- + \dot{C}HCl_2 \quad (13a)$$
$$ \longrightarrow H_2 + \dot{C}Cl_3 \quad (13b)$$
$$(k_{13} = 1 \times 10^6\, dm^3\, mol^{-1}\, s^{-1}\, [24])$$

It is obvious that even with higher chloroform concentrations in the solution, reaction (13) cannot compete successfully with reaction (1) in the presence of air:

$$CHCl_3 + OH \longrightarrow \dot{C}Cl_3 + H_2O \quad (14)$$
$$(k_{14} = 5 \times 10^6\, dm^3\, mol^{-1}\, s^{-1}\, [24])$$

The above mentioned $HO_2^{\bullet}$ and/or $O_2^{-\bullet}$ species (depending on the pH) are also involved in the degradation process:

$$CHCl_3 + O_2^{-\bullet} \longrightarrow \dot{C}Cl^3 + HO_2^- \quad (15)$$

The formation of substrate peroxyl radicals by the addition of O_2 to the above radicals, $\dot{C}HCl_2$ and $\dot{C}Cl_3$, is of crucial importance in substrate degradation:

$$\dot{C}HCl_2 + O_2 \longrightarrow \dot{O}_2CHCl_2 \longrightarrow Cl\dot{O} + OCHCl \quad (16)$$

$$OCHCl + H_2O \longrightarrow H^+ + Cl^- + HCOOH \quad (17)$$

$$Cl\dot{O} + CHCl_3 \longrightarrow ClOH + \dot{C}Cl_3 \quad (18)$$

$$\dot{C}Cl_3 + O_2 \longrightarrow \dot{O}_2CCl_3 \quad (19)$$

$$\dot{O}_2CCl_3 + H_2O \longrightarrow Cl_2CO\ \text{(phosgene)} + ClOH\ \text{(hypochloric acid)} + OH \quad (20)$$

$$Cl_2CO + H_2O \longrightarrow CO_2 + 2H^+ + 2Cl^- \quad (21)$$

Hydrolysis reactions (equations (17), (20) and (21)) play an important role in the decomposition process of chloroform.

Another pathway is the reaction of the $\dot{C}Cl_3$ radical with OH, since these transients are not completely scavenged by reaction (19):

$$\dot{C}Cl_3 + OH \longrightarrow CCl_3OH\ \text{(trichloromethanol)} \quad (22)$$

$$CCl_3OH \longrightarrow Cl_2CO + H^+ + Cl^- \quad (23)$$

The resulting phosgene is subsequently hydrolyzed by reaction (21). A similar reaction mechanism is also proposed for the degradation of aqueous carbon tetrachloride [38].

Studies of peroxyl radicals originating from multi-halogenated substrates, e.g. CF_3CHClO_2 [38], as well as those formed in the presence of transition metal ions [39], have also been reported.

3 PEROXYL RADICALS FROM CHLORINATED OLEFINS

It is well known that the —C=C double bond is rather sensitive to attack by practically any type of free radical, resulting in short-lived adducts. In the presence of air, the latter are scavenged by O_2 to form peroxyl radicals.

As a representative example of this class of pollutants, the formation and reactivity of the peroxyl radicals resulting from *trichloroethylene* (TCE, $Cl_2C{=}CHCl$) will be briefly discussed. This subject has been investigated by several researchers (Refs 9,13, and 41–43, and references therein). In order to illustrate the radiation-induced degradation of TCE in the presence of air the formation of Cl-ions and TCE decomposition is shown in Figure 2 as a

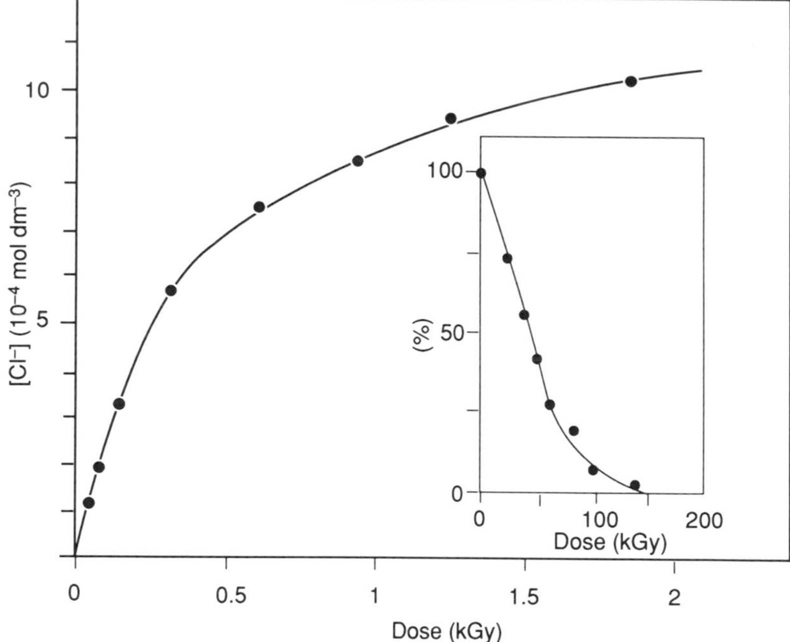

Figure 2 Formation of Cl^- ions from an aqueous solution of 1×10^{-3} mol dm^{-3} $Cl_2CH{=}CHCl$ (pH 6.4) in the presence of air as a function of radiation dose. Inset shows corresponding radiation-induced decomposition of an aqueous solution of 2.74×10^{-6} mol cm^{-3} $Cl_2CH{—}CHCl$ (pH 6.5) (reproduced from Ref. 9 by permission of Elsevier Science Ltd).

function of radiation dose. In drinking water, some of the free radicals (OH, $O_2^{-\bullet}$) are consumed naturally by bicarbonate and other components present in water. Therefore, the required radiation dose in this case is higher when compared to that used in laboratory experiments.

In aerated aqueous media TCE reacts with OH and $H\dot{O}_2/O_2^{-\bullet}$ (e^-_{aq} and H species are scavenged by O_2, see equations (1) and (2)) and the transients produced are then converted into the corresponding peroxyl radicals:

$$Cl_2C{=}CHCl + OH \longrightarrow Cl_2(OH)C{-}\dot{C}HCl \quad (24a)$$
$$\longrightarrow Cl_2\dot{C}{-}CHCl(OH) \quad (24b)$$
$$\longrightarrow Cl_2C{=}\dot{C}Cl + H_2O \quad (24c)$$
$$(k_{24} = 3.3 \times 10^9\ dm^3\,mol^{-1}\,s^{-1}\ [44])$$

These reaction steps are followed by O_2 addition to the above radicals, e.g.

$$Cl_2(OH)C{-}\dot{C}HCl + O_2 \longrightarrow Cl_2(OH)C{-}CH(O_2^\bullet)Cl \quad (25)$$

The peroxyl radicals that are obtained are unstable and can decompose as follows:

$$Cl_2(OH)C{-}CH(O_2^\bullet)Cl \longrightarrow \dot{C}Cl_2OH + CO_2 + H^+ + Cl^- \quad (26a)$$
$$\longrightarrow CO_2 + 2H^+ + 2Cl^- + Cl\dot{C}O \quad (26b)$$

$$\dot{C}Cl_2OH + H_2O \longrightarrow \dot{C}OOH + CO_2 + 2Cl^- + 2H^+ \quad (27)$$

$$Cl\dot{C}O + H_2O \longrightarrow H^+ + Cl^- + \dot{C}OOH \quad (28)$$

The fate of the $\dot{C}OOH$ radical could be as follows:

$$\dot{C}OOH \rightleftharpoons H^+ + \dot{C}O_2^- (pK_a = 1.4\ [45]) \quad (29)$$

$$\dot{C}O_2^- + O_2 \longrightarrow CO_2 + O_2^{-\bullet} \quad (30)$$
$$(k_{30} = 6.5 \times 10^9\ dm^3\,mol^{-1}\,s^{-1}\ [28])$$

Reaction (30) is to some extent in competition with reaction (31), namely:

$$2\dot{C}O_2^- \longrightarrow (CO_2^-)_2\ \text{(oxalate)} \quad (31)$$
$$(k_{31} = 1.2 \times 10^9\ dm^3\,mol^{-1}\,s^{-1}\ [28])$$

$$2\dot{C}OOH \longrightarrow HCOOH + CO_2 \quad (32a)$$
$$\longrightarrow (COOH)_2\ \text{(oxalic acid)} \quad (32b)$$

The radicals, $Cl_2\dot{C}{-}CH(OH)Cl$ (equation (24b)) and $Cl_2C{=}\dot{C}Cl$ (equation (24c)), can undergo similar oxidation reactions. The fate of the peroxyl radicals originating from the $Cl_2C{=}\dot{C}Cl$ species are briefly considered as follows:

$$Cl_2C{=}\dot{C}Cl + O_2 \longrightarrow Cl_2C{=}\underset{\displaystyle O_2^{\bullet}}{\underset{|}{C}}Cl \longrightarrow Cl_2\underset{|}{C}{-}\underset{|}{\dot{C}}Cl \quad (O{-}O) \tag{33}$$

$$Cl_2\underset{|}{C}{-}\underset{|}{\dot{C}}Cl \ (O{-}O) + H_2O \longrightarrow Cl_2CO + \dot{C}OOH + H^+ + Cl^- \tag{34}$$

The phosgene (Cl_2CO) which is produced then undergoes hydrolysis, i.e. by reaction (21).

Independently of these reaction pathways, it has been reported [9,46,47] that transients such as those resulting from reactions (24a–24c) can also undergo multiple hydrolysis reactions:

$$Cl_2(OH)C{-}\dot{C}HCl + H_2O \longrightarrow HOOC{-}\dot{C}HCl + 2H^+ + 2Cl^- \tag{35}$$

$$HOOC{-}\dot{C}HCl + H_2O \longrightarrow HCHO + \dot{C}OOH + H^+ + Cl^- \tag{36}$$

The $\dot{C}OOH$ radicals disappear according to reactions (29)–(32). A further possible hydrolysis step is:

$$Cl_2\dot{C}{-}CH(OH)Cl + H_2O \longrightarrow Cl_2\ddot{C} + HCHO + H^+ + Cl^- + OH \tag{37}$$

It has been found that the $Cl_2\ddot{C}$ species has an absorption maximum at $\lambda < 260\,nm$ and decays according to a second-order reaction, with $2k = 2.6 \times 10^8\,dm^3\,mol^{-1}\,s^{-1}$ [48]:

$$2\,Cl_2\ddot{C} \longrightarrow Cl_2C{-}CCl_2 \tag{38}$$

The perchloroethylene produced is subsequently decomposed after prolonged irradiation [11]. In air-saturated solution, a value of $G(-Cl_2C{=}CCl_2)$[†] of 4.4 has been reported [15,43].

In addition to the above discussed oxidation reactions of TCE, $HO_2^{\bullet}/O_2^{-\bullet}$, originating from the oxidation of H and e^-_{aq} (equations (1–3) and additionally formed OH radicals are involved in the degradation process [9]:

$$Cl_2C{=}CHCl + O_2^{-\bullet} \longrightarrow Cl_2\dot{C}{-}CHCl(O_2)^- \tag{39a}$$

$$Cl_2C{=}CHCl + O_2^{-\bullet} \longrightarrow Cl_2C(O_2){-}\dot{C}HCl \tag{39b}$$

$$Cl_2\dot{C}{-}CHCl(O_2)^- \longrightarrow Cl_2\ddot{C} + \dot{C}O_2^- + H^+ + Cl^- \tag{40}$$

$$Cl_2C(O_2^-){-}\dot{C}HCl + H_2O \longrightarrow H\dot{C}O + 2H^+ + 3Cl^- + CO_2 \tag{41}$$

[†] G-value = number of converted molecules per 100 eV absorbed energy.

The $H\dot{C}O$ radicals can be further involved in various reaction steps [9]:

$$2H\dot{C}O \longrightarrow CO + HCHO \text{ (aldehyde)} \quad (42a)$$
$$2H\dot{C}O \longrightarrow (HCO)_2 \text{ (glyoxal)} \quad (42b)$$

$$H\dot{C}O + O_2 \longrightarrow CO_2 + OH \quad (43a)$$
$$H\dot{C}O + O_2 \longrightarrow CO + HO_2^{\bullet} \quad (43b)$$

In addition to the above, $H\dot{C}O$ transients can first become hydrated and then undergo the following reactions:

$$H\dot{C}O + H_2O \longrightarrow H\dot{C}(OH)_2 \quad (44)$$

$$H\dot{C}(OH)_2 + O_2 \longrightarrow {}^{\bullet}O_2HC(OH)_2 \quad (45)$$
$$(k_{45} = 4.5 \times 10^9\,dm^3\,mol^{-1}\,s^{-1}\ [48])$$

$$\dot{O}_2HC(OH)_2 \longrightarrow HCOOH + HO_2^{\bullet} \quad (46)$$
$$(k_{46} = 1 \times 10^6\,s^{-1}\ [48])$$

The decomposition yield of TCE strongly depends on its concentration in water. The lower the substrate concentration, then the lower is the pollutant degradation, because the primary radicals are consumed by reacting with each other [9,13]. The main TEC decomposition products in the presence of air were found to be Cl^-, CO_2, and formic acid, dichloroacetic and oxalic acids [42], as well as aldehyde [9]. It should also be mentioned that with increasing the concentration of inorganic solutes in water, e.g. HCO_3^-, NO_3^-, etc., which partly consume the oxidizing radicals (OH, $HO_2^{\bullet}/O_2^{-\bullet}$), the degradation yield of pollutants is reduced [43].

The kinetics of several vinylperoxyl and halogenated vinylperoxyl radicals have been studied by pulse radiolysis [49]. These kinds of transients show a typical non-characteristic absorption in the UV region ($\lambda \leqslant 260$ nm), as well as a second one in the range from 440 to 580 nm. Some characteristic λ_{max} and ϵ_{max} data for this second absorption band for several vinylperoxyl radicals are given in Table 2. The measured k ($R^{\bullet} + O_2$) values are $(1–4.6) \times 10^9\,dm^3\,mol^{-1}\,s^{-1}$ [49].

Table 2 Characteristic λ_{max} and ϵ_{max} data for vinylperoxyl radicals in the visible region [49]

Radical	λ_{max} (nm)	ϵ_{max} ($dm^3\,mol^{-1}\,cm^{-1}$)
$H_2C{=}CHO_2^{\bullet}$	440	1100
$ClHC{=}CHO_2^{\bullet}$	480	900
$H_2C{=}CClO_2^{\bullet}$	480	530
$Cl_2C{=}CClO_2^{\bullet}$	580	1300
$ClHC{=}CClO_2/Cl_2{=}CHO_2(=4/1)$	540	1100

The vinylperoxyl radicals can undergo electron-transfer processes, e.g. with ascorbate ions [49]. The rate constants for such reactions were found to be similar to those previously observed for chlorinated alkylperoxyl radicals [50,51].

4 PEROXYL RADICALS FROM AROMATIC POLLUTANTS

Benzene derivatives in aerated aqueous solutions under irradiation first react with hydroxyl radicals (OH) by addition to the ring, resulting in the corresponding carbon-centred hydroxycylohexadienyl transients (see equation (47)). The latter are then scavenged by the molecular oxygen ($k \approx 10^9\, dm^3\, mol^{-1}\, s^{-1}$), providing that the O_2 concentration in the solution is sufficient, leading to the formation of peroxyl radicals ($k = (10^8–10^9)\, dm^3\, mol^{-1}\, s^{-1}$; equation (48)).

$$C_6H_5X + OH \longrightarrow \dot{C}_6H_5X(OH) \tag{47}$$

$$\dot{C}_6H_5X(OH) + O_2 \longrightarrow \dot{O}_2C_6H_5X(OH) \text{ (peroxyl radical)} \tag{48}$$

The peroxyl radicals can subsequently disappear by elimination of $HO_2^{\bullet}$(in some cases, OH), in competition to dimerization or disproportionation reactions. A similar reaction to (47) can also be initiated by the much more slowly reacting $HO_2^{\bullet}/O_2^{-\bullet}$ species [52]. It should be mentioned here that carbon-centred radicals with a high spin density at a hetero-atom, e.g. phenoxyl-type radicals [53,54], triptophan transients ([55] and references therein), etc., react very slowly with oxygen, if at all.

The formation of peroxyl radicals from aromatic pollutants is first illustrated by the example of aqueous phenol in the presence of oxygen.

4.1 PEROXYL RADICALS RESULTING ROM AERATED SOLUTIONS

The radiolysis of aromatic water pollutants has been studied under steady-state conditions, as well as by pulse radiolysis [10,27,56–60]. The radiation-induced decomposition of phenol has been investigated by several laboratories under various conditions [4,6–8,61–66]. However, the reaction mechanisms for the formation of the peroxyl radicals of phenol in oxygenated solutions are rather complicated.

The irradiation of aqueous phenol leads to the formation of several products, with the yields of the major ones being given in Table 3. Irradiation of the aqueous phenol initiates degradation by the attack of OH radicals ($k = 1.4 \times 10^{10}\, dm^3\, mol^{-1}\, s^{-1}$), resulting in the formation of hydroxycyclohexadienyl radicals (OH adducts substituted at the *o*-, *m*-, *p*-, and ipso-

Table 3 Initial yields (G_i) of the main products resulting from irradiation of 10^{-4} mol dm^{-3} aqueous solutions of phenol in the presence of 1.25×10^{-3} mol dm^{-3} O_2 (pH 6.5; dose rate = 82 Gy min^{-1}) [8]

Product	G_i^a
Pyrocatechol	0.90
Hydroquinone	0.60
Hydroxyhydroquinone	0.08
Phenol degradation	3.10

[a] Initial G-values calculated before secondary reactions take place; for conversion to SI units G is multiplied by 0.10364 to obtain yields in μmol J^{-1}.

positions). These transients can be scavenged by O_2, in competition with the other processes mentioned above, leading to the formation of peroxyl radicals. The latter reactions lead to a number of final products, e.g. 81.

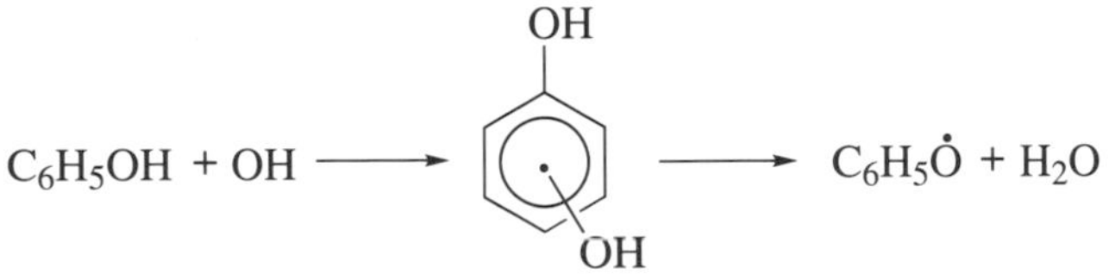

$$C_6H_5OH + OH \longrightarrow \text{[OH adduct: cyclohexadienyl radical bearing OH and OH]} \longrightarrow C_6H_5\dot{O} + H_2O \quad (49)$$

(OH adducts, at *o*-, *m*-, *p*-, and ipso-positions)
$k = 1.4 \times 10^{10}$ dm^3 mol^{-1} s^{-1})

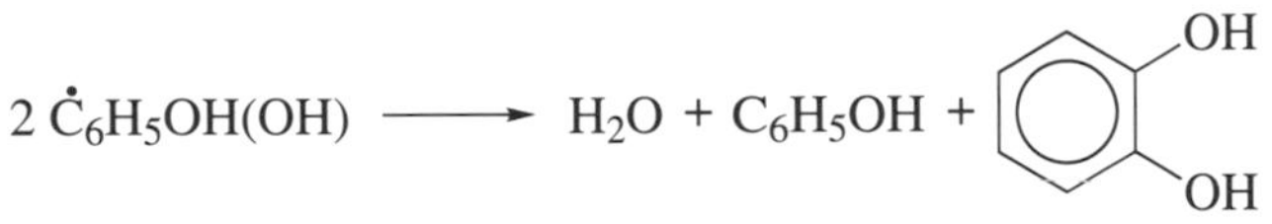

$$2\,\dot{C}_6H_5OH(OH) \longrightarrow H_2O + C_6H_5OH + \text{[1,2-dihydroxybenzene]} \quad (50)$$

$$\text{[OH adduct radical (OH, OH, H)]} + O_2 \longrightarrow \text{[peroxyl radical (OH, } O_2^{\bullet}\text{, OH, H)]} \longrightarrow \text{[1,2-dihydroxybenzene]} + HO_2^{\bullet} \quad (51a)$$

(pyrocatechol)

$$\longrightarrow \text{[hexadienedial: } \ldots C(=O)-H,\ \ldots C(=O)-H\text{]} + HO_2^{\bullet} \quad (51b)$$

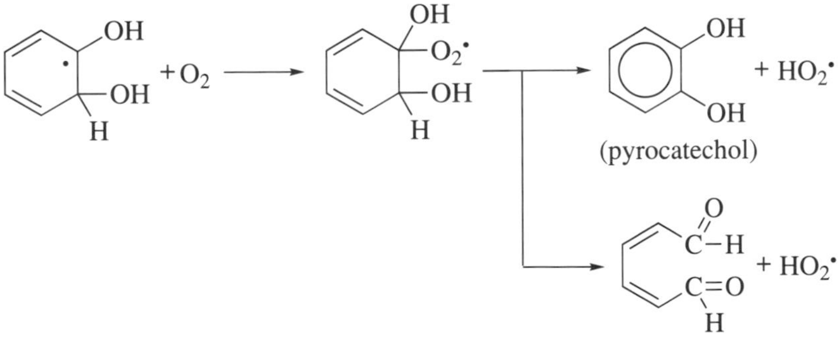

(mucondialdehyde)

OH, H, OH — cyclohexadienyl radical ·H $+ O_2 \longrightarrow$ peroxyl radical ($O_2^{\bullet}$, H; H, OH) $\longrightarrow$ (hydroquinone) $+ HO_2^{\bullet}$ (52)

(hydroquinone)

Hydroquinone (OH, OH) + OH $\longrightarrow$

OH, OH, H, H, OH radical $+ O_2 \longrightarrow$ (OH, OH, H, $O_2^{\bullet}$, H, OH) $\longrightarrow$ (hydroxyhydroquinone) $+ HO_2^{\bullet}$ (53)

(hydroxyhydroquinone)

OH, OH, H, $O_2^{\bullet}$, H, OH $\longrightarrow$

OH, OH, Ċ–H, COOH, OH $+ O_2 \longrightarrow$ (OH, $O_2^{\bullet}$, OH, C, H, COOH, OH) $\longrightarrow$ (OH, COOH, COOH, OH) $+ OH$ (54)

As shown above, the peroxyl radicals which are formed can also lead to ring opening, as well as to the elimination of $HO_2^{\bullet}$, and OH radicals, which can initiate chain reactions, such as the following:

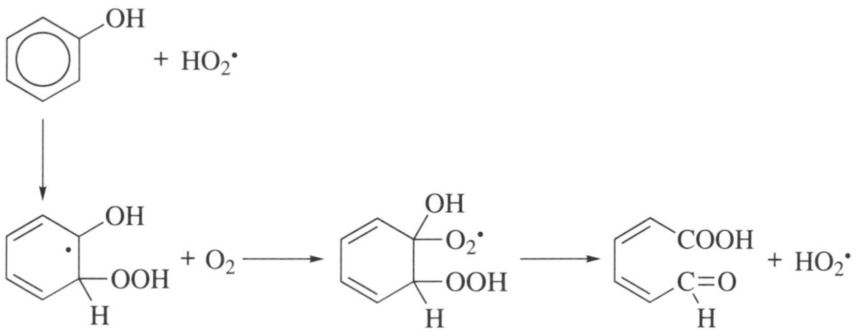

$$\text{(phenol: ring–OH)} + HO_2^{\bullet} \longrightarrow \text{(cyclohexadienyl radical: OH, –OOH, H)} + O_2 \longrightarrow \text{(OH, }{-}O_2^{\bullet}\text{, –OOH, H)} \longrightarrow \text{(COOH, C=O, H)} + HO_2^{\bullet} \quad (55)$$

Micic, Nenadovic and Makovic [60], and Pikaev and Shubin [17] reported that phenol decomposition in oxygenated solutions strongly depends on the substrate concentration and on the applied dose rate. The *G*(PhOH)-value resulting from irradiation of $2 \times 10^{-2}\,\text{mol}\,\text{dm}^{-3}$ phenol solutions increases from 2.6 to 250 by reducing the dose rate from 100 to $13\,\text{Gy}\,\text{h}^{-1}$ [60]. At low dose rates in aerated phenol solutions, *G*(PhOH)-values of 250–500 were also determined [17]. By increasing the temperature during the irradiation treatment of aqueous phenol in the presence of ozone, a 20-fold smaller radiation dose is required for its degradation [67].

As a result of prolonged irradiation of an aerated aqueous phenol solution, the formation of further products, e.g. mucon aldehyde, and muconic, maleic, oxalic and formic acids, as well as CO_2, was observed ([8] and references therein).

5 PEROXYL RADICALS INVOLVED IN THE SYNERGISTIC EFFECT OF THE RADIATION–OZONE DEGRADATION OF POLLUTANTS

5.1 GENERAL REMARKS

It is well known that ozone is a powerful oxidizing agent and therefore it is used on the industrial scale for the purification of drinking water, as well as for waste water [68,69]. Early studies [70–72] showed that ozone decomposition in water leads to the formation of hydroperoxyl ($HO_2^{\bullet}$) and hydroxy (OH) radicals, according to the chain-rection scheme proposed by

Weiss [70] (see also Refs 71 and 72):

Initiation

$$O_3 + OH^- \longrightarrow HO_2^{\bullet} + O_2^{-\bullet} \quad (56a)$$
$$\qquad\qquad \longrightarrow OH + O_3^{-\bullet} \quad (56b)$$

$$O_3 + H_2O \longrightarrow HO_3^+ + OH^- \rightleftharpoons 2HO_2^{\bullet} \quad (57)$$

$$HO_2^{\bullet} \rightleftharpoons H^+ + O_2^{-\bullet} \quad (3)$$

Propagation

$$O_3 + O_2^{-\bullet} + H^+ \longrightarrow OH + 2O_2 \quad (58)$$

$$O_3 + OH \longrightarrow HO_2^{\bullet} + O_2 \quad (59)$$

Termination

Combination of the transients, OH, $HO_2^{\bullet}$, and $O_2^{-\bullet}$:

$$OH + OH \longrightarrow H_2O_2 \quad (60)$$
$$(2k = 1.1 \times 10^{10}\,dm^3\,mol^{-1}\,s^{-1}\ [24])$$

$$OH + HO_2^{\bullet} \longrightarrow H_2O + O_2 \quad (61)$$
$$(k = 6 \times 10^9\,dm^3\,mol^{-1}\,s^{-1}\ [24]$$

$$HO_2^{\bullet} + HO_2^{\bullet} \longrightarrow H_2O_2 + O_2 \quad (62)$$
$$(2k = 3.7 \times 10^6\,dm^3\,mol^{-1}\,s^{-1}\ [25])$$

$$O_2^{-\bullet} + O_2^{-\bullet} \longrightarrow O_2^{2-} + O_2 \quad (63)$$
$$(2k < 10\,dm^3\,mol^{-1}\,s^{-1})\ [25])$$

As shown above, the kinetics of ozone decomposition in water have been clearly established (see Ref. 72).

In the combined radiation–ozone treatment of water in the presence of air, the OH transients, as well as $HO_2^{\bullet}/O_2^{-\bullet}$, are acting as carriers in the above mentioned chain reactions (55)–(58). Some radiation-induced reactions of ozone are compiled in Table 4, together with the corresponding rate constants. In addition, spectroscopic data obtained for ozone, as well as the ozone transients, are given in Table 5.

5.2 CHLORINATED OLEFINS–OZONE SYSTEMS

The formation of peroxyl radicals from olefins, with reference to the synergistic action of the combined radiation–ozone processing of polluted water, is illustrated in the following examples. The first stage of the reaction

Table 4 Some important reactions of ozone and peroxyl radicals in aqueous solutions

Reaction		k ($dm^3\,mol^{-1}\,s^{-1}$)	Ref.
$H + O_3 \longrightarrow HO_3^{\bullet}$		3.6×10^{10}	73,74
$e_{aq}^- + O_3 \longrightarrow O_3^{-\bullet}$		3.7×10^{10}	74
$HO_3 \overset{\alpha}{\underset{b}{\rightleftharpoons}} H^+ + O_3^-$ ($pK_a = 8.2$)	(a)	3.3×10^{2} [a]	75
	(b)	5.2×10^{10}	
$O_3^{-\bullet} \overset{\alpha}{\underset{b}{\rightleftharpoons}} O^{-\bullet} + O_2$	(a)	3.3×10^{3} [a]	76
	(b)	3.0×10^{9}	
$O_3^{-\bullet} + H^+ \longrightarrow OH + O_2$; $\longrightarrow HO_3$		9.0×10^{10}	77,78
$OH + O_3 \longrightarrow HO_2^{\bullet} + O_2$		1.1×10^{8}	74,79
$OH + O_3 \longrightarrow HO_4^{\bullet}$		2.0×10^{9}	80
$HO_4^{\bullet} \longrightarrow HO_2^{\bullet} + O_2$		2.8×10^{4} [a]	80
$HO_4^{\bullet} + HO_4^{\bullet} \longrightarrow 2O_3 + H_2O_2$		5.0×10^{9}	80
$HO_4^{\bullet} + HO_3^{\bullet} \longrightarrow O_3 + H_2O_2 + O_2$		5.0×10^{9}	80
$OH + O_3^{-\bullet} + OH^- \longrightarrow 2O_2^{-\bullet} + H_2O$		8.5×10^{8}	77
$HO_2^{\bullet} + O_3 \longrightarrow H_2O_2 + O_2$; $\longrightarrow OH + 2O_2$		$< 10^{4}$	79,80
$O_2^{-\bullet} + O_3 \longrightarrow O_3^{-\bullet} + O_2$		1.6×10^{9}	75
$OH + HO_2^{\bullet} \longrightarrow O_2 + H_2O$		7.9×10^{9}	74,78
$HO_2^{\bullet} + HO_2^{\bullet} \longrightarrow H_2O_2 + O_2$		7.5×10^{5}	73
		3.7×10^{6}	25
$HO_2^{\bullet} + O_2^{-\bullet} \longrightarrow HO_2^{-\bullet} + O_2$		1×10^{8}	73
$O_2^{-\bullet} + O_2^{-\bullet} \longrightarrow O_2 + O_2^{2-}$		< 10	25
$O^{-\bullet} + HO_2^- \longrightarrow O_2^{-\bullet} + OH^-$		4×10^{8}	81

[a] Units in s^{-1}.

Table 5 Spoectroscopic data for ozone and ozone transients in water

Ozone or ozone transient	λ_{max} (nm)	ϵ_{max} ($dm^3\,mol^{-1}\,cm^{-1}$)	Ref.
O_3	260	3290	82
	590	5.1	
$HO_3^{\bullet}$	350	300	75
$O_3^{-\bullet}$	430	2000	83
$O_3^{-\bullet}$	430	1900	84
HO_4	260	5100	80

of ozone with olefins can be designated as an electrophilic addition process, although a C—C cleavage reaction finally takes place with the formation of an ozonide. The reaction sequence is demonsrated by using trichloroethylene as an example:

$$\begin{array}{c} Cl_2C \\ \| \\ ClHC \end{array} + O_3 \;\longrightarrow\; \begin{array}{l} Cl_2C^{\oplus}\;\; O^{\ominus} \\ \;\;| \qquad\;\; | \\ ClHC{-}\underset{\oplus}{O}{-}O^{\ominus} \end{array} \longrightarrow \begin{array}{l} Cl_2C{-}O \\ \;\;| \qquad | \\ ClHC{-}\underset{\oplus}{O}{-}O^{\ominus} \end{array} \longrightarrow \underset{\text{(ozonide)}}{Cl_2C\overset{O}{\frown}CHCl \atop O{-}O} \tag{64}$$

The ozonide then undergoes hydrolysis, whereby two carbonyl compounds (phosgene and a halogenated aldehyde), as well as H_2O_2, are formed:

$$Cl_2C\overset{O}{\frown}CHCl \atop O{-}O \; + H_2O \longrightarrow \begin{array}{c} HO\quad OH \\ Cl_2C\;\; CHCl \\ O{-}O \end{array} \longrightarrow Cl_2CO + ClHCO + H_2O_2 \tag{65}$$

As mentioned above, phosgene hydrolyzes according to reaction (21):

$$Cl_2CO + H_2O \longrightarrow CO_2 + H^+ + Cl^- \tag{21}$$

The halogenated aldehyde can likewise undergo hydrolysis:

$$ClHCO + H_2O \longrightarrow HCOOH + H^+ + Cl^- \tag{66}$$

In addition to the above products, the formation of dichloroacetic acid has also been observed in the ozone-electron beam degradation of TCE [85]. As might be expected, the reactions (24)–(25) discussed above are also involved in this degradation process.

5.3 PHENOL- AND CHLORINATED PHENOLS–OZONE SYSTEMS

It has been previously shown ([23] and references therein) that in addition to the above mentioned reactions ((49)–(54)) the formation of ozonides takes place in aqueous phenol solutions in the presence of ozone. This is illustrated by reaction steps (67) and (68):

$$C_6H_5OH + O_3 \longrightarrow \underset{\text{(Complex)}}{[C_6H_5OH \cdot O_3]} \longrightarrow \underset{\text{(Ozonide)}}{\text{OH, H, O, O–O, H}} \tag{67}$$

$$\text{(bicyclic peroxide)} + H_2O \longrightarrow \text{(Muconic aldehyde)} + H_2O_2 \longrightarrow \text{(2-Hydroxy muconic acid)} + H_2 \quad (68)$$

(Muconic aldehyde) (2-Hydroxy muconic acid)

The formation of peroxyl radicals as a result of the synergistic effect of the combined treatment of aqueous phenol and chlorinated phenols by radiation and ozone has not yet been completely elucidated.

The peroxyl radicals originating from the two pathways, i.e. (a) oxygen addition to the hydroxycyclohexadienyl transients (49–54), and (b) from ozone with the pollutant (67 and 68), can interfere with each other. Hence, the reaction mechanisms become rather complicated. However, as a result of these processes a very efficient degradation of the pollutant is achieved. This effect is shown in Figure 3, which presents the 4-chlorophenol decomposition (curve A) and the formation of Cl^- ions (curve B) and aldehyde (curve C) as a function of the absorbed dose. Figure 3(i) shows these processes for solutions saturated with air, whereas Figure 3(ii) illustrates the same processes, but in the presence of both oxygen and small amounts of ozone [58].

It is obvious that in the second case, even at very low O_3 concentration, the required radiation dose is greatly reduced and the aldehyde is decomposed with higher efficiency. In the first case, the initial *G*-value for the substrate decomposition (G_i(4-Cl-PhOH) is 2.8, followed by $G_i(Cl^-) = 1.30$ and

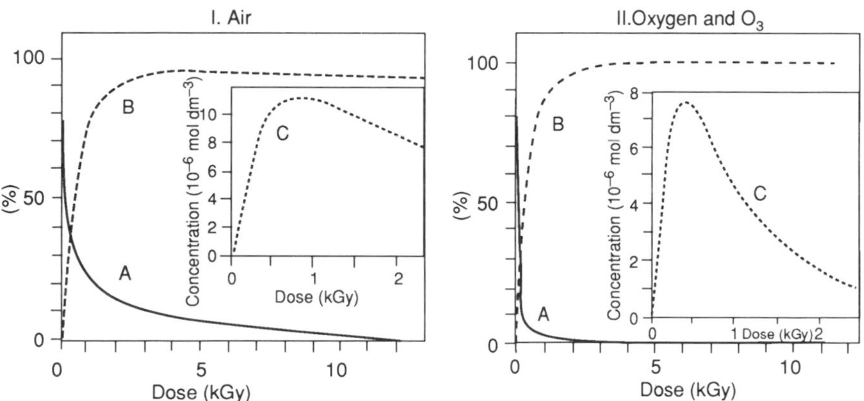

Figure 3 Radiation-induced decomposition of 10^{-4} mol dm^{-3} aqueous solutions of 4-chlorophenol (pH 6.5) as a function of dose carried out (I) in the presence of air (0.25×10^{-3} mol dm^{-3} O_2) and (II) in the presence of 1.25×10^{-3} mol dm^{-3} O_2 and 1.1×10^{-5} mol dm^{-3} O_3: (A) decomposition of 4-chlorophenol; (B) formation of Cl^- ions; (C) formation of aldehyde (reproduced from Ref. 58 by permission of Elsevier Science Ltd).

G_i(aldehyde) = 0.22, whereas in the second case the corresponding data are 3.70, 8.30 and 1.0, respectively [58].

Some major radiation-induced reactions of 4-chlorophenol are given below for illustration:

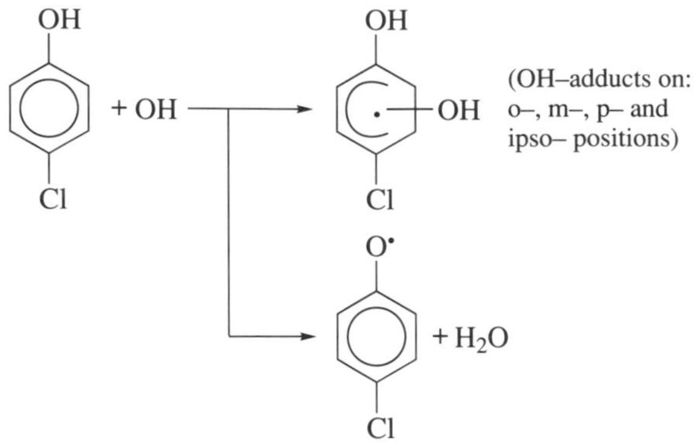

$$\text{4-Cl-PhOH} + OH \longrightarrow \text{HO(Cl)C}_6\text{H}_4^{\bullet}\text{-OH} \quad \text{(OH–adducts on: o–, m–, p– and ipso– positions)} \tag{69a}$$

$$\text{4-Cl-PhOH} + OH \longrightarrow \text{4-Cl-PhO}^{\bullet} + H_2O \tag{69b}$$

$$(k(OH + \text{4-Cl-PhOH}) = 1.5 \times 10^{10}\,dm^3\,mol^{-1}\,s^{-1})$$

In the absence of oxygen, the following reaction steps are of interest:

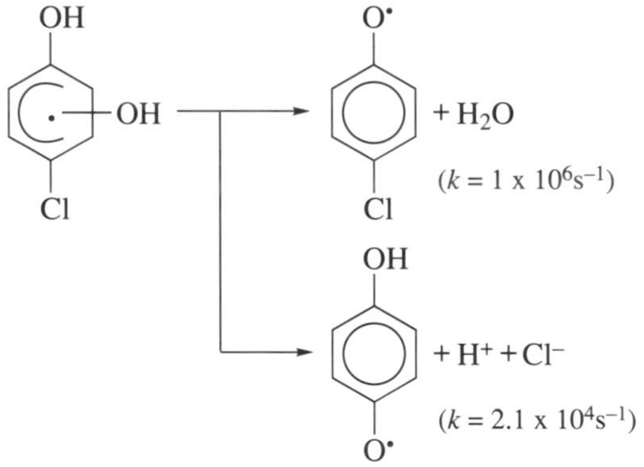

$$\text{HO(Cl)C}_6\text{H}_4^{\bullet}\text{-OH} \longrightarrow \text{4-Cl-PhO}^{\bullet} + H_2O \quad (k = 1 \times 10^6 s^{-1}) \tag{70a}$$

$$\text{HO(Cl)C}_6\text{H}_4^{\bullet}\text{-OH} \longrightarrow \text{HO-C}_6\text{H}_4\text{-O}^{\bullet} + H^+ + Cl^- \quad (k = 2.1 \times 10^4 s^{-1}) \tag{70b}$$

$$2\ \text{HO(Cl)C}_6\text{H}_4^{\bullet}\text{-OH} \longrightarrow \text{4-Cl-PhOH} + \text{4-Cl-catechol} + H_2O \tag{71}$$

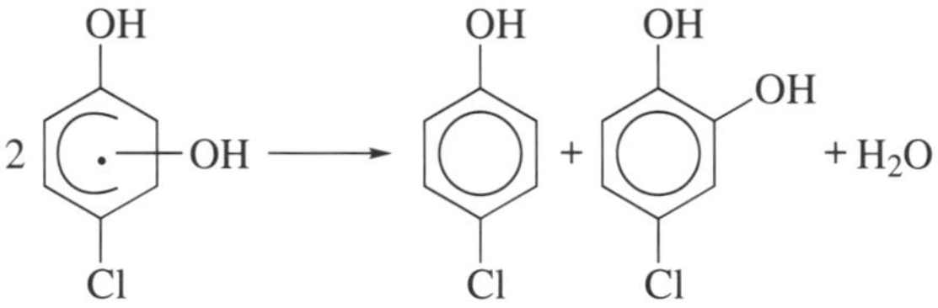

$$(2k_{total} = 3 \times 10^8\,dm^3\,mol^{-1}\,s^{-1})$$

In aerated 4-chlorophenol solutions, a number of reactions are involved in the degradation process, which are very similar to those already discussed for phenol (see reactions (49)–(54)). In this present case, however, in addition to the elimination of $HO_2^{\bullet}$ and OH radicals, strongly oxydizing acids ($HClO_2$,

$HClO_4$) can also be formed. The formation of peroxyl radicals, $HClO_2$ and $HClO_4$, and some of the final products resulting from 4-chlorophenol in the presence of oxygen are given below:

OH-adduct of 4-chlorophenol + O_2 ⟶ peroxyl radical ($\cdot O_2$) ⟶ Products (72)

$$(k(O_2 + \text{OH-adduct}) = 5.1 \times 10^8\ \text{dm}^3\,\text{mol}^{-1}\,\text{s}^{-1})$$

Examples

OH-adduct + O_2 ⟶ peroxyl radical ($O_2^{\bullet}$) ⟶ ring-opened dialdehyde (C–H, C=O) + OH (73)

ipso OH-adduct (Cl, OH) + O_2 ⟶ peroxyl radical ($O_2^{\bullet}$) ⟶ $Cl\dot{O}_2$ + hydroquinone (74)

(hydroquinone)

hydroquinone + OH ⟶ OH-adduct + O_2 ⟶ peroxyl radical ($O_2^{\bullet}$) (75)

⟶ hydroxyhydroquinone + $H\dot{O}_2$ (76)

(hydroxyhydroquinone)

$$ClO_2^{\bullet} + H_2O \longrightarrow HC1O_2 + OH \quad (77)$$

$$ClO_2^{\bullet} + HO_2^{\bullet} \longrightarrow HClO_2 + O_2 \quad (78a)$$
$$\qquad\qquad\quad \longrightarrow HClO_4 \quad (78b)$$

As a consequence of the repeated attack of OH, and $HO_2^{\bullet}/O_2^{-\bullet}$ on 4-chlorophenol, or on its resulting products, an efficient degradation of the pollutant can be achieved.

It is of interest to mention that the reactivity of peroxyl radicals is effected by the nature of the substituent. Generally, electron-withdrawing groups on a methylperoxyl radical increase its reactivity in electron-transfer, addition and H-abstraction processes. Especially high reactivities were observed in the case of polyhalogenated peroxyl radicals, e.g. CX_3OO ([86] and references therein). It has been recently found [87] that phenylperoxyl radicals have a higher reactivity than methylperoxyl species and most substituted methylperoxyls, with the exception of the halogen-substituted transients. Moreover, electron-withdrawing groups at the 4-position of the phenylperoxyl radical increase the reaction rate constant. Electron-donating groups at this position, however, decrease the rate constant for oxidation of this transient, in accord with the Hammett substituent constants [87]. It was very recently reported [88] that the oxidizing properties of monohalogenated phenylperoxyl radicals vary in the order F $<$ Cl $<$ Br, and *p*- $<$ *m*- $<$ *o*-positions. Furthermore, the reactivities of peroxyl radicals increase on going from monohalo-, dihalo-, and phenyl-, to trihalophenylperoxyl radicals. It should be finally mentioned that the spectroscopic and kinetic characteristics of several arylperoxyl radicals, as well as their reactivities, were also studied in aqueous alcohol solutions [89]. The 2-pyridylperoxyl radical (λ_{max} = 440 nm) originating from 2-chloro- or 2-bromopyridine in aqueous or organic solvent mixtures is found to be a fairly strong oxidant towards several model organic compounds [90].

5 PULSE RADIOLYSIS METHOD FOR STUDIES OF PEROXYL RADICALS

The pulse radiolysis technique [91,92] is an important method for studying the spectroscopic and kinetic characteristics of peroxyl radicals. As an example of this, the absorption spectra of various 4-chlorophenol transients are presented in Figure 4 [58]. Curve 1 in this figure represents the OH adduct of the substrate, whereas curve 2 is the superimposed spectrum of both the OH adduct and the corresponding peroxyl radicals. The kinetic traces (insets A–D) allow one to follow step-by-step some of the overlapping reactions of the peroxyl radicals which take place under the given conditions (see details given on the figure).

INDEX

Note: **bold page numbers** indicate (in this index) matter in Tables; *italic numbers* refer to Figures. Alphabetization is letter-by-letter (spaces ignored). Alternative spellings (peroxy/peroxyl) are indicated as: peroxy(l).

Index written by Paul Nash